This review and synthesis of the patterns and theories of biodiversity provide a firm scientific basis for the study and management of the Earth's biosphere. Beginning with a practical definition of biodiversity and an overview of diversity patterns on Earth, this book develops a conceptual framework that extends traditional theories of species diversity to explain both the general patterns of diversity and the apparent exceptions to these patterns. The author demonstrates the common features underlying diversity patterns at scales ranging from the polar-equatorial gradients to the margins of temporary ponds, and develops a mechanistic explanation of diversity patterns in different ecosystems. Many of the major concepts in ecology are incorporated into the framework for understanding patterns of diversity on landscapes. Case studies apply the theories and models described. This book presents ideas that will stimulate and challenge professionals and students in ecology, resource management, forestry, geography, and environmental economics.

BIOLOGICAL DIVERSITY

Biological Diversity

The coexistence of species on changing landscapes

MICHAEL A. HUSTON
Oak Ridge National Laboratory

CAMBRIDGE
UNIVERSITY PRESS

                                        ...ersity of Cambridge
                                        ...ridge CB2 1RP
                                        ...211, USA
                                        ...6, Australia

© Cambridge University Press 1994

First published 1994
Reprinted 1995

Printed in Great Britain by Athenaeum Press Ltd, Gateshead, Tyne and Wear

A catalogue record of this book is available from the British Library

Library of Congress cataloguing in publication data available

ISBN 0 521 36093 5 hardback
ISBN 0 521 36930 4 paperback

TAG

For Ann and Thomas

Open, O Lord, the eyes of all people to behold thy gracious hand in all thy works, that, rejoicing in thy whole creation, they may honor thee with their substance, and be faithful stewards of thy bounty.

The Book of Common Prayer, 1979

Contents

Preface

The endeavor of writing a book, particularly one on a topic as broad as biological diversity, quickly brings one face-to-face with one's own limitations, in terms of time, energy, memory, and ability. I realize that my ambitious goal at the outset of this effort, of creating a comprehensive synthesis of the information and ideas about biological diversity, is still far from being realized. I am left with every author's hope, that the second edition (if there is one) will be much better than the first.

A cursory review of any subfield of ecology, or any science for that matter, reveals that the literature is vast. Not only are current publications overwhelming in their number, but the past century of research and publication has left an extensive history of important and insightful work that should not be overlooked. My approach has been to look more carefully at the past literature than the current literature, undoubtedly at the cost of missing important recent work. Part of this bias stems from my impression that many of us are ignorant of the past accomplishments and ideas in the field of ecology, and often expend great effort (and publish many papers) rediscovering something that was well-known fifty years ago, or perhaps even making errors of interpretation that were corrected long ago.

While I hope that most readers will find some surprising data and exciting ideas in this book, I suspect that Frederick Clements, Arthur Tansley, S.A. Forbes, or H.C. Cowles would find little that they didn't already know or hadn't already considered (as well as various errors of fact and interpretation that they would tactfully point out). I believe that many of the 'pioneers' in the field of ecology had a much broader and clearer understanding of basic ecological processes than many of us today. It is very difficult to come up with an idea that is genuinely new.

On the premise that 'there is nothing new under the sun,' I have conscientiously tried to trace major concepts as far back into the history of ecology as possible (given my own limitations of knowledge and time), and to give credit to the originators of the ideas I discuss in this book. Undoubtedly, I have not always succeeded. It is also likely that I have missed many important recent contributions, both theortical and empirical, to the issues I attempt to address. I am hopeful that my colleagues will inform me when I have overlooked their important contributions that are relevant to issues in this book, and that they will pardon me for concentrating on the work of those who can no longer speak for themselves.

While I am interested in data that are consistent with the ideas I have developed, I am particularly anxious to try to understand data or patterns of diversity that contradict the predictions and generalizations in this book. Although I have made an honest effort to find results that do not fit the theories outlined in this book, I suspect that I will be informed of many phenomena that are inconsistent with these ideas.

Oak Ridge, Tennessee   *Michael A. Huston*

Acknowledgements

While there are many individuals throughout my academic career and personal life who have in some way contributed to this book, it would not have been written without the encouragement and support of my wife, Mary Ann. Over the much-longer-than-expected gestation of this work, with periods of intense effort and frustrating inactivity, she has kept the rest of our life functioning, encouraged me to keep going, and borne two beautiful children. It is my hope that this book will contribute to a wiser use of the world's natural resources, and a better future for many people, especially our family.

John Birks also deserves special mention, since I wouldn't have tried to write a book if he hadn't asked me to write one on biological diversity, over six years ago. I am also indebted to John, and to Robert McIntosh, for reading and carefully commenting on the entire first draft of the manuscript, an unenviable task which no one else even attempted. I did receive many valuable comments on sections of the book from my Oak Ridge colleagues Don DeAngelis, Mac Post, Antoinnette Brenkert, Jackie Griebmeier, and Kirk Winemiller, and I appreciate their help.

Since my interest in biological diversity stems from my days as graduate student at the University of Michigan, I want to acknowledge my fellow students who supported and stimulated me in the early days of my interest in this topic, and encouraged me to pursue my ideas in spite of pressure to the contrary. Those whose time and arguments were especially valuable include Doyle McKey, Tom Getty, Diane DeSteven, Lissy Coley, Hank Howe, Jim Affolter, Cathy Pringle, and George Sugihara. I was fortunate to have the opportunity of first visiting the tropics as an assistant to Dan Janzen and John Vandermeer, who did not necessarily agree with the 'insights' I gained under their tutelage. During those years the University of Michigan offered a course that brought some of the major figures in ecology to campus for two-week seminar series. I am grateful for the opportunity to have discussed some of my early ideas with these scientists, especially Henry Horn and Robert Whittaker.

As is evident from the figures and citations, this book draws heavily on past work I did as a graduate student as well as with colleagues at Oak Ridge National Laboratory. I am grateful to these friends for our many stimulating interactions and the (I believe) important papers we published together. In particular, Don DeAngelis, Tom Smith, Mac Post, John Pastor, and Phil Robertson were especially valuable in helping me to refine and express my (our) ideas, as well as educating me about a great variety of topics.

I am fortunate to have had many outstanding teachers, from elementary school through the present, who have stimulated and encouraged me. But it was my mother and father who first encouraged, then supported and tolerated my interest in natural history and later ecology. My siblings have always been very supportive and have helped me collect data in some beautiful, and sometimes uncomfortable, places. During my childhood in Newton, Iowa, Neal Deaton helped me view the world from his perspective as a naturalist and artist, from Cambrian seas to Guatemalan rain forests and African savannahs. During my graduate research, Napoleon Murillo and his wonderful family taught me about life in a tropical country and made possible my forest succession experiments in Costa Rica. Throughout my graduate career Joe Connell shared many of my interests and encouraged me to pursue career opportunities outside of traditional academia.

There are certainly many others who, as friends and teachers, have contributed to the development of my ideas and the overall course of my life. I hope they will take some satisfaction in this book, and not blame themselves for the mistakes I have made.

Portions of this book are expansions of work that I, and my colleagues, have published elsewhere, and I want to express my gratitude to the publishers of these papers for permission to use them here. In particular, the University of Chicago Press has generously allowed me to use material from my publications, with Don DeAngelis and Tom Smith, in *The American Naturalist*, specifically volume 113, 81–101, volume 129, 678–707, and volume 130, 168–198, as well as figures and tables from numerous important papers in that journal. I thank Annual Reviews, Inc. for permission to include and expand portions of my paper on coral reefs from the *Annual Reveiw of Ecology and Systematics*, 16, 149–77. I have also used major portions of a paper by Tom Smith and myself, from

Vegetatio, 83, 49–69, and I thank Kluwer Academic publishers for permission to use and expand this material. I hope that inclusion of material from these publications will help clarify the connections between these works, and increase appreciation for each of the individual contributions.

Some of the effort of writing this book has been supported as part of my activities with the Walker Branch Watershed Project at Oak Ridge National Laboratory. The Walker Branch Watershed Project is supported by the U.S. Department of Energy's Office of Health and Environmental Research as part of their Program for Ecosystem Research. While biological diversity has not traditionally been a major concern of the Department of Energy, I want to thank Dr. Clive Jorgensen of the Program for Ecosystem Research for his support, and express my appreciation to my supervisors at Oak Ridge National Laboratory who encouraged me to pursue this effort, especially Bruce Kimmel and Webb Van Winkle.

M. A. H.

1
Introduction

The extinctions resulting from human activities throughout the world have caused great concern in the scientific community and among the general public. This disappearance of species has been decried as a loss of plants and animals with potential agricultural and economic value, as a loss of medical cures not yet discovered, as a loss of the Earth's genetic diversity, as a threat to the global climate and the environment for human existence, and as a loss of species that have as much inherent right to exist as does *Homo sapiens*. The attention given this issue has led to the addition of a new word to the English language, biodiversity (a contraction of 'biological diversity', Wilson and Peter (eds.), 1988). Biological diversity is more than a scientific or economic issue. Biological diversity, in all of its manifestations, is an essential component of the quality of human existence, summarized in the ancient aphorism: 'variety is the spice of life.'

Biological diversity encompasses all levels of natural variation from the molecular and genetic levels to the species level, where we have most of our interactions with biological diversity through enjoyment of the common, strange, and beautiful forms of life or through suffering caused by the effects of pests, parasites, and diseases. Beyond the species level, biological diversity includes patterns in nature up to the landscape level. These components of biological diversity are not independent. The many flowers that form spots of color in a meadow, the songbirds that give forests a different music than fields, the various forest types that create zones of color on a mountain that we see from twenty miles away, or the variations in greenness that can be detected from satellites in space, are all ultimately the consequence of genetic diversity interacting with environmental conditions to produce differences between organisms.

There is little hope of understanding any phenomenon with as many

1

complex components and scales of spatial and temporal variability as biological diversity, *unless* it can be divided into components within which repeatable patterns and consistent behavior occur. One central premise of this book is that biological diversity can be broken down into components that have consistent and understandable behavior. The other central premise is that the various components of biological diversity are influenced by different processes, to the extent that one component may increase, while another decreases in response to the same change in conditions. If these premises are true, it is impossible to completely understand 'total biodiversity' until the regulation of each of its components is understood.

The subdivision of biological diversity into tractable components is essential for developing and testing hypotheses about its regulation. The focus of this book is on those components of biological diversity that are influenced by the number and identity of species present in a given area. My goal is to explain the regulation of species diversity and why the number of co-occurring species varies under different conditions. I will not deal with the issue of the regulation of genetic or molecular diversity within species or populations, and only note that the total genetic diversity within any area is primarily a consequence of the number and identity of species that are present. This book is about the ecological regulation of species diversity, the interaction of ecological processes with geological and evolutionary processes, and the consequences of these interacting processes for the large-scale spatial and temporal patterns on landscapes that are generally considered to be components of biological diversity.

Functional Classifications of Organisms

To apply the two premises outlined above to understanding the regulation of species diversity, the ideal components would be groups of species within which consistent patterns appear, and within which a given process will always produce the same pattern. Such groups of species could be defined on the basis of properties of the areas in which they occur, which has been a common approach in ecology and biogeography. However, with the goal of developing explanations of species diversity that are as broadly applicable as possible, my approach will be to classify species based on attributes that they have in common, such as size and physiological properties, rather than on the attributes of their environment, such as the successional stage or moisture conditions of the habitat in which they are usually found. In particular, I attempt to develop classifications that are based on 'functional' attributes, that is, attributes related to how organisms interact with each other and with their environment, rather than on phylogenetic attributes and genetic relatedness.

One very general classification scheme, which applies to many species is that of species being either 'structural' or 'interstitial'. By structural species I mean those species that create or provide the physical structure of the environment. Obvious examples include trees, reef-forming corals, giant kelp and other multicellular algae, and sessile animals such as oysters, mussels, barnacles, tubeworms, etc. These organisms create the physical structure of their environment, produce variability in physical (e.g., microclimatic) conditions, provide resources, and in general create the habitat used by many other, generally smaller, 'interstitial' organisms. Interstitial organisms would include most insects, other arthropods, birds, mammals and other vertebrates, microbes and fungi, as well as plants such as epiphytes and understory herbs.

'Structural' organisms have a major influence on the diversity of 'interstitial' organisms: in most cases the interstitial organisms would not be present in a particular area without the structural organisms. The direct influence of interstitial organisms on structural organisms and their diversity is usually minor, although the indirect effects of interstitial organisms through evolutionary and biogeochemical processes can be very important. While this size-based functional classification could be used to further subdivide interstitial organisms into smaller structural organisms and their interstitial dependents, the main point I wish to make is that the diversity of these two general types of organisms is likely to be influenced by different factors and processes.

Although the effect of structural species on the diversity of interstitial species is a major contributor to the total species diversity of virtually all communities, this explanation does not address the critical issue of the diversity of the structural species. A second and more general classification scheme provides a framework for dealing with the diversity of both structural and interstitial species, and also allows total species diversity to be broken into two components that (I will argue) operate according to different rules, and/or according to the same rules at very different temporal and spatial scales.

The total diversity of any community, or any subset of a community, can be broken into two hierarchical components: 1. the number of different *functional types* of organisms (i.e., 'guilds', Root, 1967); and 2. the number of *functionally analogous species* within each functional type (cf. Smith and Huston, 1989). Different functional types of organisms use different resources or otherwise interact with their environment in such a way that competitive interactions are minimized. Species of the same functional type use the same resources in such a way that they could *potentially* compete intensely with one another.

The critical idea behind the use of the functional type approach to species diversity is that the mechanisms that influence the number of functional types in a community are almost always different from the mechanisms that influence the number of functionally analogous species within a functional type. In addition, not all functional types respond the same way to a given change in environmental conditions. For example, competition may be important for explaining the diversity within a functional type under one set of conditions, but have no effect under other conditions. Likewise, within a community, the diversity of one functional type may be regulated by competitive interactions, while the diversity of other functional types is uninfluenced by competition, and regulated instead by immigration rates, spatial heterogeneity, evolutionary history, or other factors.

The total species diversity of a community is described by the number of functional types multiplied by the average number of species per functional type (Fig. 1.1). Clearly, the same total number of species can result from many functional types with relatively few functionally analogous species in each type or from few functional types with many functional analogues in each. In spite of the similarity in number of species, these differences in structure imply that the two communities are influenced by very different processes. These differences in functional organization have major implications for the stability and continued functioning of these communities. A large number of functionally analogous species may contribute to high stability and continuity of ecological and ecosystem function in the face of disturbances or environmental change.

This is a very flexible dichotomization, the value of which depends on having sufficient information about the organisms of interest to classify them into functional types. While a useful functional classification may require at least as much information as an accurate phylogenetic classification, the net result is likely to be of more use for understanding the regulation of species diversity. In many cases there will not be clear breaks between different functional types, and functional types will be subsets of species from a continuum or multidimensional cloud of functional combinations. The discrimination of functional types may be subjective and arbitrary, subject to as many problems and differences of

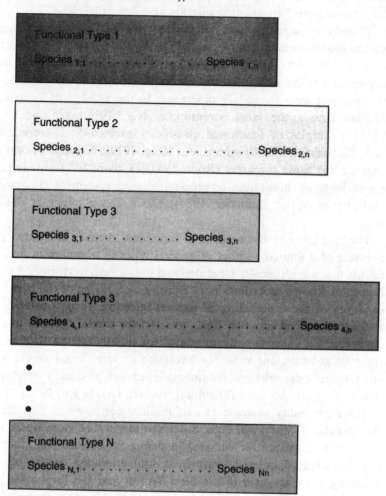

Fig. 1.1 The two components of total species diversity. The number of functional types is illustrated by the horizontal rectangles labeled 1 through N. The number of functionally analogous species within each functional type is indicated by the length of the rectangle and the species labeled $i,1$ through i,n. The total species diversity of a community is the product of the number of functional types and the average number of species per functional type. Not all functional types will have the same number of species. Because functional types are defined on the basis of resource use, competitive interactions are potentially intense among organisms of the same functional type, but very weak between organisms of different functional types.

opinion as the classification of organisms into species. The concept is useful only to the extent that it helps us to understand species diversity by resolving mechanisms that operate at different levels.

The resolution with which functional types must be classified depends on the specific component of diversity being addressed. If the object is to explain differences in the total number of animal species among several communities, the definition of functional type could be broader than if the object were to explain differences in the number of insectivorous animals among the same communities (e.g., Feinsinger and Colwell, 1978). Examples of functional classifications include 1. trophic levels (e.g., Cummins, 1973; Cummins and Klug, 1979); 2. guilds of organisms that use the same resources (Root, 1967); 3. plant life forms (e.g., tree, shrub, herb, or therophyte, cryptophyte, hemicryptophyte, chamaephyte, and phanerophyte; Raunkier, 1934); and 4. shade tolerance classes of trees.

The generality of the concept of functional types is illustrated by the presence of a limited number of general types of organism in floras and faunas that are physically separated and evolutionarily distinct. This phenomenon, known as evolutionary convergence, is one of the best known examples of the functioning of natural selection. In particular, it illustrates that natural selection tends to select for similar features in similar environments. Consequently, unrelated (or slightly related, depending on the scale at which one views the 'evolutionary tree') organisms in similar environments on different continents often are physically very similar, that is, they are the same functional type (cf. Orians and Paine, 1983).

There are many example of evolutionary convergence that illustrate the concept of functional type. Succulent stems that store water are an adaptation in plants for survival in desert conditions. This functional type has evolved independently in several unrelated groups of plants, including the Cactaceae in the New World, and the Euphorbiaceae in the Old World. To the untrained eye, a euphorb 'cactus' from Africa is indistinguishable from a true cactus from Mexico. Unrelated plants from similar environments show convergence in leaf size, shape, and structure, branching pattern, seed size and dispersal mechanism, and many other traits (Raunkier, 1934; Richards, 1952; Cody and Mooney, 1978; Smith and Young, 1987).

Among birds some examples of convergence are the toucans (Rhamphastidae) of South American and the hornbills (Bucerotidae) of Africa; the hummingbirds (Trochilidae) of South America and the sunbirds (Nectariniidae) of Africa and Asia; the meadowlarks (Icteridae) of the Americas and the yellow-throated longclaw (Motacillidae) of Africa (Cody, 1974). Many of the best known examples of convergence are from the evolutionarily isolated flora and fauna of Australia, where the marsupials have evolved into functional counterparts of the more familiar placental mammals (Keast, 1969). Striking examples of convergence are found in grassland areas around the world, which have morphologically and functionally similar, but phylogenetically distinct, subterranean mammals (Nevo, 1979).

Functional convergence may often confound taxonomic classification based on supposed genetic relatedness. Christensen and Culver (1968) discovered in their taxonomic study of the tiny insects known as Collembola, that one widespread 'species' adapted to caves was actually four separate species that had independently evolved to be virtually identical. From an evolutionary perspective, different functional types result from the effect of natural selection on adaptations to different resource conditions or adaptations to avoid competition (e.g., character displacement). Functionally analogous species may result from convergent evolution in which distantly related organisms become adapted to similar environmental conditions, or they may be related species that have not diverged functionally. From an ecological perspective, interactions such as competition or predation/cannibalism among the very similar species within a functional type are likely to operate differently than the same classes of interactions among species of different functional types.

An Approach to Understanding Biological Diversity

The approach that I will describe for explaining patterns of biological diversity is not based on a single mechanism, such as competition. Rather, it is based on understanding the conditions under which different mechanisms, such as predation, competition, dispersal, or evolutionary history, are likely to be most important in regulating the diversity of a particular group of organisms. The challenge is to identify the conditions (e.g., spatial and temporal scales, evolutionary and geological history, disturbance regime, resource availability, etc.) under which specific mechanisms are likely to have the greatest influence on the diversity of specific functionally-based subsets of organisms. For example, the explanation for short-term fluctuations in the number of bird species in a community is not likely to be the same as the explanation for long-term fluctuations. Likewise, the explanation for the number of tree species found in large areas is likely to differ in the relative importance of factors and have factors in addition to those that explain the number of tree species in a smaller subset of that area. A particular change in environmental conditions may increase the diversity of one subset of organisms within a community while decreasing the diversity of a different group of organisms. It is virtually impossible to understand variation in the total number of species in a community unless changes in the major functional groups of species can be understood.

The study of species diversity is of necessity based primarily on comparative and correlative research. The spatial area needed to define an ecological community for many organisms of interest (e.g., trees, birds, most mammals) is simply too large (and too difficult to define precisely) to be amenable to replicated experimental manipulations. Appropriately defined comparative studies can be used to identify factors or processes that are correlated with species diversity of specific groups of organisms at specific spatial and temporal scales. Once a correlation has been identified, it is critical to determine whether the correlation is based on an identifiable mechanism, or is just a coincidental, spurious relationship. The determination of whether a correlation represents a causal relationship is a difficult challenge. In most cases the evaluation must be based on strong inference (Platt, 1964) using a combination of criteria (see Weins, 1989, for a discussion of hypothesis testing in community ecology).

Criteria for the evaluation of causal relationships should include both theoretical and empirical tests. Each hypothesized causal relationship must be evaluated in comparison with all other potential causal factors and mechanisms that are correlated with it. The hypothesized causal relationship must be consistent with theoretical predictions based on specific mechanisms and assumptions that can be shown to apply to the data. The hypothesized causal relationship should be supported by data from (smaller-scale) experimental manipulations that can be shown to be relevant to the comparative data. The hypothesized causal relationship should be able to address exceptions to the expected pattern of diversity, either by clarifying why the expectation was wrong or by demonstrating why the causal mechanism should not apply. Meeting all of these criteria is a tall order, one that few hypotheses addressed in this book will completely satisfy.

My own approach to understanding species diversity has been to begin with a theoretical framework general enough to predict the relative importance of many different processes that influence local species diversity over a broad range of communities and environmental conditions, which I have called the *dynamic equilibrium model of species diversity* (Huston,

1979). My ideas were originally motivated by the well-known global patterns of species diversity, and, in particular, the high tree species diversity of tropical rain forests (Huston, 1980). However, the resulting theory made predictions about the diversity of a variety of ecological communities that contradicted the predictions of several widely accepted alternative hypotheses (e.g., Huston, 1985a). The predicted patterns for some communities were sufficiently 'counter-intuitive' that initial data in support of them were not accepted by some ecologists. In the next stage of my own pursuit of this issue, I used experimental manipulations to address specific mechanisms hypothesized to regulate small-scale, short-term changes in the species diversity of plant communities (Huston, 1982). The hypothesized mechanisms were tested for logical consistency and predictive accuracy using the mathematical formalism of computer simulation models (e.g., Huston and DeAngelis, 1987; Huston and Smith, 1987). Use of computer simulation models has allowed the predictions of mechanisms that operate at experimentally tractable scales to be extrapolated to larger spatial scales where experimental manipulation is not possible (e.g., Huston *et al.*, 1988; Smith and Huston, 1989; Huston, 1991). Understanding multiscale variation in species diversity remains a fascinating challenge, one made more stimulating by the complex feedbacks between ecosystem processes, as well as the potential responses of organisms and ecosystems to disturbances, stresses, and changing climatic conditions.

Overview

The purpose of this book is not to test the hypotheses that I propose for understanding biological diversity. Such a test would require replicated, multi-factorial experiments in many different ecosystems, and detailed comparisons of natural ecosystems over a wide range of conditions. Such data are simply not available, nor are there resources to collect the required new data. Rather, the purpose of this book is to demonstrate how these hypotheses and general framework can provide useful insights into the regulation of biological diversity, using the data and concepts developed by ecologists over the past century.

The important question throughout the book is not which explanation for species diversity is correct, since virtually every explanation that has been proposed is important under some circumstances. The critical questions are which of the many potential explanations apply to a specific diversity pattern, whether any particular mechanism is the dominant explanation for a specific pattern, and, finally, whether there are any general rules about which mechanisms are likely to be important under particular environmental conditions, among specific groups of organisms, or at particular spatial and temporal scales.

Much of the structure of this book follows the course of my own intellectual journey through the field of community ecology, and the hypotheses and models presented are those that I have found most interesting and useful. The book is divided into four sections. The first section, composed of Chapters 2 and 3, deals with the raw material for the study of biological diversity. Chapter 2 discusses the variety of diversity patterns found in ecosystems around the world at spatial scales ranging from the entire globe to soil particles. Presented along with the somewhat confusing diversity of diversity patterns is the even more confusing diversity of environmental conditions that are correlated with them. Chapter 3 briefly addresses issues related to the measurement and quantification of biological diversity.

The second section, Chapters 4 and 5, reviews the historical development of theories of species diversity, focusing on the importance of equilibrium versus non-equilibrium processes in regulating species diversity. Again, the issue is not whether an equilibrium or a non-equilibrium viewpoint is correct, but which of many potential opposing processes are likely to be in equilibrium at a particular spatial and temporal scale.

The third section, Chapters 6 through 10, addresses mechanisms of intra- and inter-specific competition, particularly among plants, which are the dominant structural organisms in most terrestrial environments and in many aquatic and marine environments. Each chapter deals with the influence of interactions among individual organisms on diversity at a different organizational level or spatiotemporal scale. Chapter 6 deals with the regulation of diversity within populations of single species, manifested primarily as variation in size. Chapter 7 extends the effect of interactions among individual organisms from the population to the community and ecosystem level. Chapters 8, 9 and 10 discuss the major spatial and temporal patterns of the distribution of organisms on landscapes, again in terms of the interactions between individual organisms. In aggregate, these chapters deal with the major components of biological diversity over the range of scales at which they are perceived by humans.

The fourth section, Chapters 11 through 14, applies the concepts developed in Parts two and three to some of the major issues in biological diversity and some of the major ecosystems in which species diversity shows the most variability. Endemism and invasions, discussed in Chapter 11, are two of the most dramatic manifestations of variation in the equilibrium between speciation and extinction. Chapter 12 discusses marine systems, ranging from the intertidal to the mid-oceanic benthos, that span the range from low to high species diversity, and illustrate the influence of virtually all the factors that are known to influence species diversity. Chapter 13 deals with a range of terrestrial plant ecosystems in which diversity is regulated by a single type of major disturbance, fire. The diversity of fire-dominated ecosystems ranges from virtually monospecific stands to ecosystems with some of the highest plant diversity found on Earth. Finally, Chapter 14 deals with the *sine qua non* of biodiversity, tropical rain forests, and comes to what may be perceived by some to be the most counterintuitive conclusion in the entire book.

I conclude with a brief discussion of some of the economic and conservation implications of the counterintuitive regulation of global diversity gradients.

Part one
Raw Material and Tools

Biological diversity is best defined by the patterns we see in the world around us. The next chapter discusses the variety of diversity patterns that have been described over a broad range of temporal and spatial scales. These patterns are the raw material for understanding biological diversity, as well as a challenge for theories of biological diversity to explain. The second chapter in this section briefly addresses some aspects of the measurement and statistical evaluation of diversity. Readers familiar with these issues may wish to move directly to Part two, 'Theories of Species Diversity'.

Part One

New Materials and Tools

2
General Patterns of Species Diversity

Diversity would not be interesting if the level of diversity were the same everywhere. Fortunately, no matter how diversity is defined, or what types of organisms are being considered, there is phenomenal variation in diversity across the entire range of living systems.

No single process or theory can explain a phenomenon as complex as biological diversity. The intellectual challenge and scientific value of the study of diversity lies in the conceptual synthesis required to understand a complex phenomenon that is influenced by many different interacting factors and processes (cf. McIntosh, 1987). Understanding diversity requires understanding many ecological, evolutionary, geological, and biogeochemical processes, and how those processes interact. Carefully framed questions about species diversity can provide insight into a wide variety of processes on spatial scales ranging from microscopic to continental, and on temporal scales of hours to epochs.

The smorgasbord of patterns in species diversity that are discussed in this chapter will not necessarily make any sense, and no attempt will be made in this chapter to explain the patterns that are described. Data that suggest some regularity in diversity patterns that might lead to an explanation will be contradicted by data that have the opposite pattern. Nearly all of these examples and counter examples will be addressed in later chapters. Our ultimate challenge is to develop an understanding of biological diversity that is both internally consistent and consistent with apparently contradictory patterns of species diversity.

In the first part of this chapter the major global patterns will be described, beginning with the latitudinal gradients of species diversity. Then, the latitudinal gradient and other patterns will be discussed in the context of the environmental conditions, or other potential causal factors, with which they are associated.

15

The Major Global Patterns of Biological Diversity

Latitudinal Gradients

The latitudinal gradient was the pattern that first attracted scientific attention to species diversity. For most groups of terrestrial plants and animals, diversity is lowest near the poles, and increases towards the tropics, reaching its peak in tropical rain forests. Understanding this diversity gradient is not simple, because there are many factors correlated with this gradient that could potentially affect species diversity, such as average temperature and precipitation, the variability in temperature and precipitation, annual net primary productivity, and geological history. Some of these factors are positively correlated with latitude, while others are negatively correlated. In addition there are a few latitudinal gradients of diversity that are the reverse of the general pattern. For example, contrary to the expectations of classical biologists who expected diversity to increase with temperature, species diversity for some groups of marine and freshwater organisms (organisms on coral reefs being an obvious exception) is lowest in tropical waters and highest near the poles (Hubendick, 1962; Paine, 1966; Sanders, 1968; Abele, 1974).

Historically, the number and variety of tropical organisms was recognized to be far greater than that found in temperate Europe as soon as Europeans began to explore the globe in pursuit of their economic interests. Linneaus was actively classifying and cataloging species, many brought back from the tropics, in the mid-eighteenth century. The contrast between the incredible diversity of tropical plants and animals and the much lower diversity of plants and animals in temperate regions, and Europe in particular, was obvious to all scientists by the mid-nineteenth century. This dramatic contrast provided much of the motivation, as well as the critical observations, for understanding why there are so many different kinds of organisms. The tropics were recognized as *the* frontier of Natural Science, and provided the initial insights and empirical support for the theory of evolution developed by Charles Darwin (1842, 1859) and Alfred Russell Wallace (Darwin and Wallace, 1858). Aspiring young naturalists, such as Wallace (1878) and Henry Walter Bates (1864), set off for the tropics to begin their careers. In 11 years in Brazil, Bates collected 14,712 species of animals, of which some 8,000 were new to science (Usinger, 1962). Given the uncertainty in current estimates of the total number of species that occur in the tropics (between 3 and 30 million), estimates of the percentage of tropical species that still remain undiscovered and undescribed by science today range from 83% to 98% (Raven, 1988).

The initial scientific question stimulated by species diversity was how the many different species arose. Darwin provided the first convincing answer to this question, which formed the basis for much of modern biology. Since the beginning of the twentieth century, the interest in species diversity from an evolutionary focus on the origin of species has been complemented by efforts to explain how so many species can coexist, and why more coexist in some areas than in others. Both evolutionary and ecological processes influence large-scale patterns of species diversity such as latitudinal gradients, and questions must be framed appropriately to allow them to be separated.

Latitudinal gradients in species diversity have been reported in a wide variety of taxa ranging from marine molluscs (Abbott, 1968; Stehli *et al.*, 1967) to mosquitoes and butterflies (Stehli, 1968), from ants (Kusenov, 1957) and stream insects (Stout and Vandermeer, 1975) to decomposers (Swift *et al.*, 1979) (Tables 2.1 and 2.2).

Plants. The most dramatic and biologically important latitudinal diversity gradient is that of plant species. Diversity of trees increases from the nearly monospecific boreal forests of the Subarctic to the overwhelming diversity of the tropical rain forests (Fig. 2.1). The same pattern of increasing diversity with decreasing latitude is found within North America (Monk, 1967; Glenn-Lewin, 1971; Currie, 1991) (Fig. 2.2) and Europe (Silvertown, 1985). Diversity increases with decreasing latitude in most groups of plants, perhaps most dramatically in the orchids (Dressler, 1981).

Vertebrates. The number of species in most vertebrate groups increases toward the tropics. This pattern has been well documented for mammals (Simpson, 1964; Wilson, 1974; Flemming, 1973; McCoy and Connor, 1980), although much of the increase in diversity results from the great tropical diversification of a few groups, such as the bats (Fig. 2.3). Similar patterns have been found for fish and reptiles (Tables 2.1 and 2.2). Within North America, recent compilations of data show latitudinal gradients that are not always monotonic for the major vertebrate groups (Figs 2.4 and 2.5).

Patterns of bird species diversity, which is much higher in the tropics than in temperate zones, raise many ecological and evolutionary issues (Cody, 1975, 1986; Diamond, 1970, 1973, 1975). The latitudinal gradient

Table 2.1. *Compilations of regional surveys showing a latitudinal gradient in species richness.*

Organism or Guild	Region	Source
Vertebrates		
Non-oceanic birds	New World	Dobzhansky, 1950
	New World	MacArthur, 1969
	New World	Cook, 1969
	Nearctic	Tramer, 1974
	Palearctic	Järvinen, 1979
Mammals	New World	Simpson, 1964
	New World	Wilson, 1974
Fish	Nearctic	Horn and Allen, 1978
Reptiles	Nearctic	Kiester, 1971
Anurans	Global	Arnold, 1972
Lizards	Global	Arnold, 1972
	Nearctic	Schall and Pianka, 1978
Snakes	New World	Dobzhansky, 1950
	Global	Arnold, 1972
Invertebrates		
Papilionid Butterflies	Global	Scriber, 1973, 1984
Sphingid moths	New World	Schrieber, 1978
Dragonflies	Global	Tillyard, 1917, cited in Williams, 1964
Wood-boring Scolytidae and Platypodidae	Global	Beaver, 1979
Planktonic foraminiferans	Nearctic	Stehli et al., 1969
Permian brachiopods	Nearctic	Stehli et al., 1969
Corals	Australian	Wells, 1955, cited in Fischer, 1960
	Global	Stehli and Wells, 1971
Tunicates	Global	Hartmeyer 1911, cited in Fischer, 1960
Calanid crustaceans	Global	Brodskij, 1959, cited in Fischer, 1960
Mollusks	Nearctic	Fischer, 1960
Plants		
Trees	Palearctic	Silvertown, 1985
Orchids	New World	Dressler, 1981

From Stevens (1989).

is complicated (and perhaps partially explained) by the fact that many tropical birds migrate to the temperate zone for a relatively brief time to breed and raise their young (MacArthur, 1959). These 'tropical' birds are a major component of temperate bird species diversity (Fig. 2.6), greatly

Table 2.2. *Compilations of point samples showing a latitudinal gradient in species richness. Compiled from studies that count the number of species at specific sampling points.*

Organism or Guild	Region	Source
Vertebrates		
Non-oceanic birds	New World	Karr, 1971
	New World	Karr and Roth, 1971
	Nearctic	Tramer, 1974
Mammals	New World	Fleming, 1973
Lizards	Nearctic	Pianka, 1967
Freshwater fish	Global	Barbour and Brown, 1974
Invertebrates		
Arthropod communities	Nearctic	Teraguchi et al., 1981
Litter mites	New World	Stanton, 1979
Stream invertebrates	Nearctic	Stout and Vandermeer, 1975
Marine invertebrates	New World	Heck, 1979
Lepidoptera	New World	Ricklefs and O'Rourke, 1975
Ants	Global	Kuzenov, 1957
	New World	MacArthur, 1972
Marine copepods	Nearctic	Turner, 1981
Polychaetes	Global	Sanders, 1968
	Old World	Ben-Eliahu and Safriel, 1982
Gastropods	New World	Spight, 1977
	Nearctic	MacDonald, 1969
Marine bivalves	Global	Sanders, 1968
Epizooplankton	Nearctic	Grice and Hart, 1962
Plants		
Trees	New World	Dobzhansky, 1950
	Nearctic	Monk, 1967
	Nearctic	Glenn-Lewin, 1977

From Stevens (1989).

increasing the temperate bird diversity during the summer breeding seasons. Oceanic birds can range over great distances, and most migrate between summer breeding grounds in the high latitudes and more tropical regions during the high latitude winters. For non-oceanic birds, there is a strong latitudinal gradient in species diversity (Dobzhansky, 1950; MacArthur, 1969; Cook, 1969; Tramer, 1974; Karr, 1971; Karr and Roth, 1971) (Fig. 2.7).

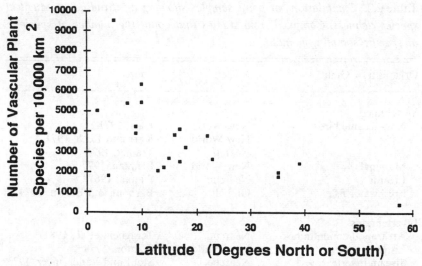

Fig. 2.1 Latitudinal gradient of plant species diversity in the New World. (Based on Reid and Miller, 1989.)

Reverse Latitudinal Gradients

Not all groups of organisms increase in diversity toward the lower latitudes (Table 2.3). Those relatively few groups that increase in diversity toward the poles should provide important insights into the mechanisms that influence species diversity. In particular, what is it about the biology of those organisms, and about the environmental conditions that influence them, that produces a pattern opposite to that of most organisms? *Sea Birds*. Sea birds that forage for fish and crustaceans along coasts and in the open ocean reach their greatest total numbers and highest diversity in the upper latitudes of both the Northern and Southern Hemisphere. In addition to wide-ranging birds such as albatrosses, which breed on sub-Antarctic islands and forage throughout the southern seas, each polar region has its own specialized group of sea birds. In the Antarctic and southern margins of the southern continents, the 17 species of flightless penguins breed along the coasts and generally spend the severest parts of the winter at sea. In the sub-Arctic, the alcids (22 species of auks, murrs, and puffins) breed along rocky coasts and move farther to sea during cold weather. These non-migratory birds are joined during winter by many species of sea birds, such as diving-ducks, oystercatchers, and plovers, that migrate south from their arctic breeding grounds to feed in the productive waters of the northern Pacific (Stehli, 1968; Scott, 1974). A

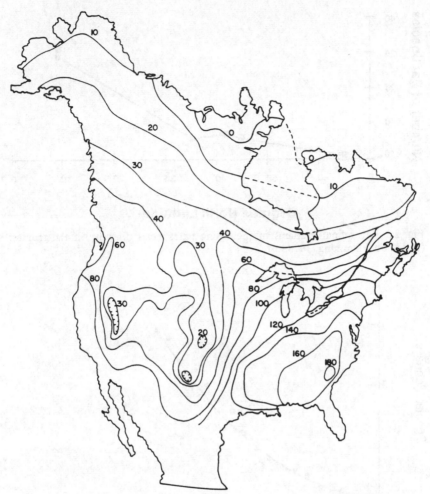

Fig. 2.2 Species richness of North American trees in relation to latitude. Data are based the number of tree species in $2\frac{1}{2} \times 2\frac{1}{2}°$ degree (5° latitude $\times 2\frac{1}{2}°$ longitude north of 50° N) cells north of the Mexican border. The number of species per cell was estimated using the range maps of 620 tree species indigenous to North America. (From Currie and Paquin, 1987.)

common feature of the high-latitude oceans that differs strikingly from tropical oceans is their high primary productivity (Steemann-Neilson, 1954; Harvey, 1955; Koblentz-Mishke *et al.*, 1970). This high primary productivity is associated with high secondary productivity of fish, which is reflected in the fact that most of the world's major fisheries and whaling grounds (with the exception of coastal upwelling areas) have been in the

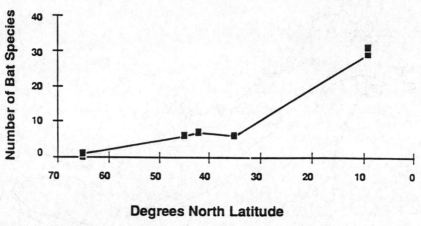

Fig. 2.3 Latitudinal gradient of increasing bat species diversity in the tropics. (Based on Flemming, 1973.)

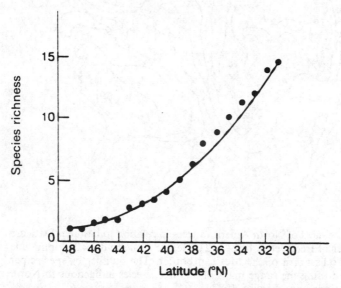

Fig. 2.4 Latitudinal gradient in lizard species diversity in North America. (From Schall and Pianka, 1978.)

higher latitudes. The high sub-polar productivity of fish, crustaceans, and molluscs may be related to the high diversity of sea birds in these regions.

Lichens. These organisms, the symbiosis of an alga and a fungus, tend to be most abundant in areas where larger vascular plants cannot survive. Regions that are dry, and/or cold generally have the greatest abundance

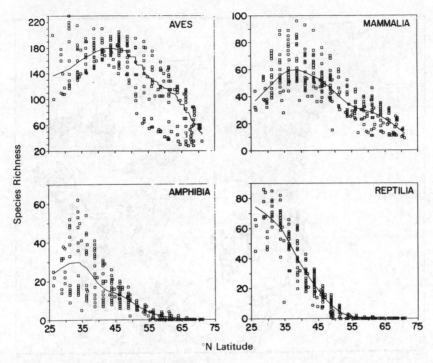

Fig. 2.5 North American latitudinal gradients in species richness of birds, mammals, amphibians, and reptiles based on $2\frac{1}{2} \times 2\frac{1}{2}°$ cells (as in Fig. 2.2). Species numbers per cell are based on range maps for individual species. (From Currie, 1991.)

and diversity of lichens (Ahti, 1977; Lindsay, 1977; Rogers, 1977). Only a few species are ever conspicuously abundant, such as the *Ramalina menziesii* of southern California and Mexico, which festoons trees in long streamers, and the species of *Cladonia*, known as reindeer moss, that cover vast areas of the Arctic tundra and serve as a major food source for the migratory caribou. Although the diversity of vascular plants is drastically reduced in the cold, dry regions of the high Arctic and Antarctic, the diversity of lichens is relatively high (Pickard and Seppelt, 1984), and may even be higher than in temperate and tropical regions. However, adequate data are not available to evaluate this hypothesis, particularly for the humid tropics (Sipman and Harris, 1989).

Marine Benthic Organisms. Among invertebrates that live within the sediment or attached to a rocky substrate, diversity is higher in the high latitudes than in the tropics. This pattern has been reported for

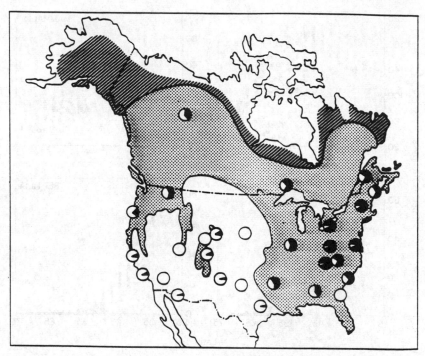

Fig. 2.6 Proportion of neotropical migrants in the breeding bird populations of North America. The proportion of the total number of breeding individuals that will migrate out of the Nearctic region in winter is indicated in black. The stippled area corresponds to the forested region. (From MacArthur, 1959.)

several groups of sediment-dwelling organisms in estuaries and shallow parts of the continental shelves (Thorson, 1957; Sanders, 1968) as well as invertebrates (Paine, 1966) and algae (Gaines and Lubchenco, 1982) from the rocky intertidal zone along coastlines and the deep-sea benthos (Sanders, 1968; Stuart and Rex, 1989).

Parasitic Wasps. Ichneumonid wasps, which are parasitic on other insect parasites, or directly parasitic on non-parasitic hosts, reach their greatest diversity at mid-latitudes (Owen and Owen, 1974; Janzen, 1981).

Soil Nematodes. The species richness of free-living soil nematodes is often higher at high latitudes that at lower latitudes, and reaches a maximum in the temperate zone (Procter, 1984). The densities, sizes, and total biomass of soil nematodes are also higher at high latitudes, where they often dominate the invertebrate faunas. According to Procter (1984), the

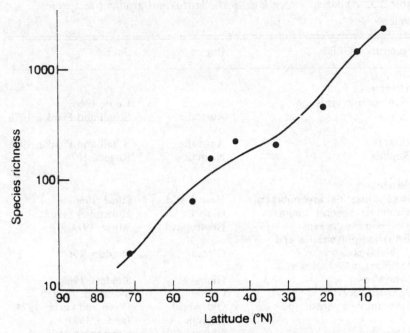

Fig. 2.7 Latitudinal gradient in species richness of breeding birds in North and Central America. (After Dobzhansky, 1950.)

high latitude nematodes feed primarily on soil microorganisms, whereas nematodes that feed on higher plants are more important in tropical faunas. Nematodes that feed on higher plants are important in most agricultural systems. Procter (1984) attributes the temperate peak in nematode diversity and abundance to 'partial ecological release due to the absence of tropical competitors'.

Altitudinal Gradients

Given the well-known correspondence between the physical conditions and natural communities that are found along a gradient of increasing latitude and those that are found along a gradient of increasing elevation (Humboldt and Bonpland, 1807), it is not surprising that species diversity generally decreases with increasing elevation. A general rule of thumb for air temperature is that an increase in elevation of 1000 meters results in a decrease in temperature (6 °C) equivalent to that associated with an increase in latitude corresponding to a linear distance of 500 to 750 kilometers (Holdridge, 1967; Terborgh, 1971; Whittaker and Niering,

Table 2.3. *Apparent exceptions to the latitudinal gradient.in species richness.*

Organism or Guild	Region	Source
Vertebrates		
Non-oceanic birds	Nearctic	Cody, 1966
	Australia	Schall and Pianka, 1978
Lizards	Australia	Schall and Pianka, 1978
Reptiles	Nearctic	Rogers, 1976
Invertebrates		
Rocky-intertidal invertebrates	New World	Paine, 1966
Basommatophoran mollusks	Global	Hubendick, 1962
Decapod crustaceans	Neotropical	Abele, 1974
Estuarine polychaetes and bivalves	Global	Sanders, 1968
Deep-sea polychaetes and bivalves	Global	Sanders, 1968
Apoidea (bees)	Global	Michener, 1979
Ichneumonid parasitoids	Old World	Owen and Owen, 1974
	Nearctic	Janzen, 1981
Collembola	Global	Rappoport, 1975

From Stevens (1989).

1965). Many physical conditions in addition to mean temperature, such as the seasonal variability of those conditions, also change along both altitudinal and latitudinal gradients.

Significant decreases in diversity over elevational gradients of 4000 meters have been reported for birds in New Guinea (Kikkawa and Williams, 1971), and for vascular plants in the Nepalese Himalayas (K. Yoda, quoted in Whittaker, 1977) (Fig. 2.8).

However, because many conditions change along elevational gradients, patterns are not always simple. Whittaker and Niering (1975) documented changes in plant diversity along an elevational gradient in the Santa Catalina mountains of Arizona. Their gradient began in the creosote bush desert at 760 m elevation and extended through grasslands to deciduous woodlands to fir forests above 2500 m. Moisture increased with increasing elevation, but the greatest number of plant species (of all growth forms together) per 0.1 ha sample was found at intermediate to low elevations (and intermediate moisture levels) in the open oak woodlands, grasslands, and Sonoran desert (Fig. 2.9).

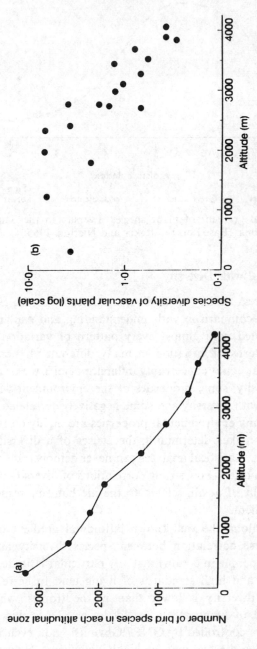

Fig. 2.8 Altitudinal gradients of plant species richness of (a) birds in New Guinea (Kikkawa and Williams, 1971) and (b) vascular plants in the Nepalese Himalayas. (K. Yoda, quoted in Whittaker, 1977.)

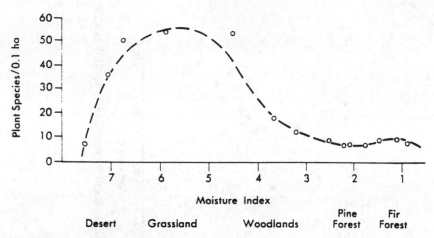

Fig. 2.9 Elevational gradient of plant species diversity in the Santa Catalina Mountains of Arizona. (Based on Whittaker and Niering, 1965.)

Factors Correlated with Diversity

Describing the great variety of biodiversity patterns on the Earth is relatively simple in comparison with understanding and explaining those patterns. Associated with almost every pattern of variation in species diversity are patterns of variation in many different physical and biological factors that could conceivably influence diversity. For almost any gradient of diversity, some properties of the environment will be positively correlated with diversity and some negatively correlated. To add to the confusion, many environmental properties are highly correlated with each other. This makes determining the causes of a diversity gradient extremely difficult. Statistical analyses can never demonstrate causal relationships, but can often show strong correlations of diversity with factors that are marginally, if at all, related to the mechanisms responsible for the diversity gradient.

For example, along the well-known latitudinal gradient of diversity there is an inverse correlation between species diversity and the per capita income of people in countries at any particular latitude. Countries with high GNPs and high standards of living tend to be in the upper latitudes, where diversity is lower than in the tropics, where GNPs and the standard of living are generally lower. Does this mean that species diversity is controlled by GNP? Obviously not, even though the correlation between the two may be highly significant. However, it may mean that both species diversity and GNP are influenced by the same factors, which is a concept that is rarely considered in either ecology or economics.

Obviously, latitude, *per se*, does not influence species diversity. Rather, the explanatory mechanisms must be related to some of the many physical and historical factors that change along the latitudinal gradient, such as sun angle, day length, seasonality, temperature means and extremes, rainfall amount and timing, winds and storms, glacial history, etc. Each of these, and other, factors can interact with organisms and their physical environment in complex ways that could potentially influence species diversity. To understand the many reasons why diversity changes along this dramatic gradient, it is necessary to examine the effect on diversity of simpler gradients, and, if possible, single factors.

The first step in understanding patterns of species diversity is to determine what factors are correlated with species diversity, independent of whether or not there is a spatial pattern such as zonation. The correlations with diversity may be either positive or negative, and may result from direct interactions that may be simple to understand, or from complex indirect interactions that require much more effort to decipher. The second step is to determine whether the correlations are causal or just coincidental, for example by experiments, comparative studies of separate locations, and studies of gradients.

Factors correlated with diversity are the raw material for identifying and potentially understanding the mechanisms that produce the diversity patterns. However, it is the theory or theories of the regulation of species diversity that will be the basis of understanding, and not simply the correlations themselves.

Productivity

Primary productivity, the solar energy that is captured by plants and converted to carbon compounds, is the basic resource that fuels life on Earth. Therefore, it is not surprising that productivity is correlated with diversity in many situations. However, there are also many cases in which productivity is negatively correlated with species diversity. Identifying the differences between situations with positive versus negative correlations of diversity and productivity is an important step in understanding how diversity is regulated.

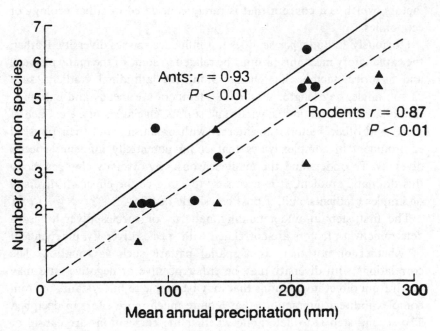

Fig. 2.10 Species richness of seed-eating ants and rodents along a precipitation gradient in southwestern North America. Triangles indicate rodents and circle indicate ants. Precipitation is positively correlated with productivity in these systems. (After Brown and Davidson, 1977.)

Positive Correlations of Diversity and Productivity

On a global scale, the primary productivity of terrestrial vegetation is positively correlated with plant species diversity (Reichle, 1970). Productive forest ecosystems generally have more species than less-productive deserts or grasslands. Among animals, there are numerous examples of diversity being positively correlated with plant productivity, or with factors that directly affect plant productivity. The species richness of seed-eating ants and rodents in North American deserts is positively correlated with precipitation (Brown and Davidson, 1977) (Fig. 2.10). Similarly, the number of lizard species is southwestern North America is positively correlated with the length of the growing season (Pianka, 1967).

Recent compilations of data on species richness and climatic conditions in North America and Europe (Currie and Paquin, 1987; Adams and Woodward, 1989; Currie, 1991) show remarkably strong correlations between the species richness of major taxonomic groups and simple

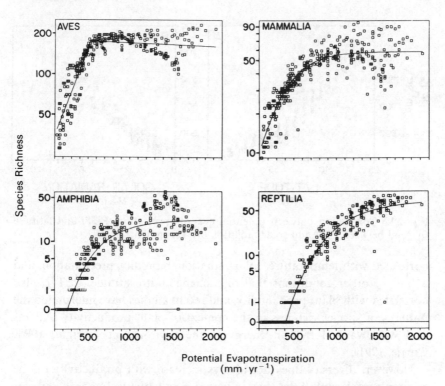

Fig. 2.11 Relationship of Potential Evapotranspiration (PET) to the species richness of birds, mammals, amphibia, and reptiles in $2\frac{1}{2} \times 2\frac{1}{2}°$ cells in North America. (From Currie, 1991.)

climatic parameters. The basic approach in these studies is to divide the region of interest into uniform cells (e.g., $2\frac{1}{2}°$ latitude $\times$ $2\frac{1}{2}°$ longitude) and estimate the number of species in each cell using published maps of the distribution of each species, and estimate the climatic conditions using climate maps and interpolating between climatic stations. These compilations demonstrate a strong correlation of the species richness of trees, birds, mammals, amphibia, and reptiles with a composite climatic variable known as potential evapotranspiration (PET) (Fig. 2.11). PET is estimated from air temperature and solar radiation, and represents the maximum amount of water that would be lost by evaporation from surfaces and transpiration of plant leaves when evapotranspiration is not limited by water availability. Not suprisingly, annual total PET is strongly correlated with latitude (Fig. 2.12).

Although the correlation between PET and species richness offers more hope for a mechanistic interpretation than latitude *per se*, PET is itself

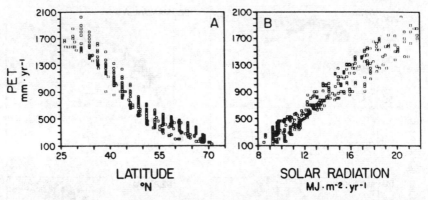

Fig. 2.12 Correlation between potential evapotranspiration (PET) and latitude (*a*), and between PET and solar radiation (*b*). (From Currie, 1991.)

correlated with temperature, solar radiation, humidity, precipitation, and a host of other conditions that also change with latitude. PET is also correlated with plant productivity, and recent studies have interpreted the patterns of species richness in the context of plant productivity (Adams and Woodward, 1989) or 'energy in the environment' (Wright, 1983; Currie, 1991).

However, diversity does not always increase with productivity or the environmental conditions that influence productivity. In the deserts of Israel, the species richness of desert rodents first increases with increasing rainfall, then decreases at higher levels of rainfall (Abramsky and Rosenzweig, 1983), with a maximum at intermediate levels (Fig. 2.13). Among vascular plants in the mountains of Arizona, Whittaker and Niering (1975) found a similar pattern of maximum diversity at elevations with intermediate levels of moisture and productivity.

Negative Correlations of Diversity and Productivity

The documentation of a large number of cases in which species diversity was negatively correlated with productivity perplexed many ecologists in the 1960s and 1970s. The unexpected phenomenon was sometimes called 'the paradox of enrichment' (Rosenzweig, 1971; Riebesell, 1974), referring to the observation that diversity often decreased when nutrients or other resources that increased productivity were added to a system. This phenomenon is particularly conspicuous in aquatic systems, where the addition of nutrients can lead to great increases in algal productivity (a process called eutrophication) and a corresponding decrease in the

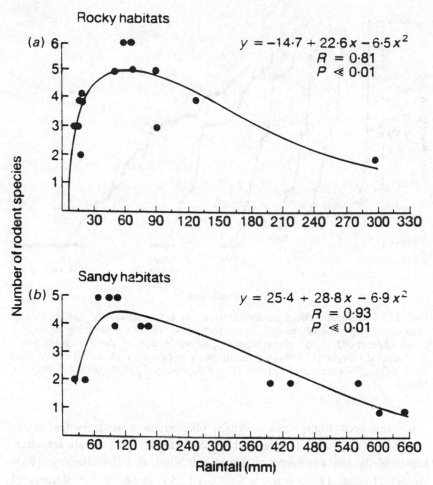

Fig. 2.13 Species richness of desert rodents along a precipitation gradient in Israel on both (*a*) rocky and (*b*) sandy habitats. (From Abramsky and Rosenzweig, 1983.)

diversity of algae and other aquatic organisms as well (Patrick, 1963). Likewise, the addition of fertilizer to herbaceous plant communities often results in a sharp decrease in species diversity (Lawes *et al.*, 1882; Thurston, 1969; Silvertown, 1980; Milton, 1940, 1947; Davies and Jones, 1930; Murphy, 1960; Willis, 1963; Grime, 1973; Aerts and Berendse, 1988; other references in Huston, 1980) (Fig. 2.14). The same pattern is often found along natural gradients of soil fertility (Mellinger and McNaughton, 1975).

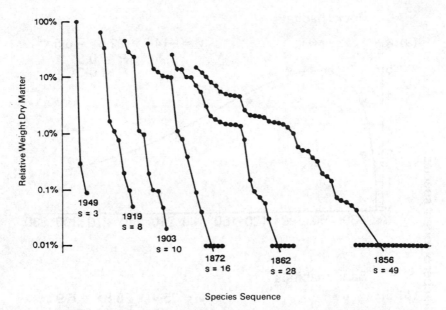

Fig. 2.14 Long-term effect of fertilization on plant species diversity. Rank/ abundance curves show changes in the patterns of relative abundance of species in an experimental plot of permanent pasture in one of the Parkgrass plots, Rothamsted, England, following continuous application of nitrogen fertilizer since 1856. s = number of species. (From Kempton, 1979, following Brenchley, 1958.)

Oceanic ecosystems show a similar phenomenon along natural gradients of productivity. The central regions of the open ocean are characterized by low productivity (Steemann-Nielson, 1954; Harvey, 1955; Russell-Hunter, 1970; Bunt, 1975; Crisp, 1975) and a low density of individual planktonic species (Hentschel and Wattenberg, 1930), but nonetheless the species diversity is high (Russell-Hunter, 1970; see Chapter 12). However, in high productivity areas, such as nutrient upwellings, lagoons, and shallow coastal areas, there is a high density of organisms but a low species diversity (Russell, 1934; Dakin and Colefax, 1940; Russell-Hunter, 1970; Sanders, 1969; Brodskji, 1959).

Productivity can be either negatively or positively correlated with species diversity. In fact, in many systems a unimodal pattern is found, with highest species diversity at intermediate levels of productivity, and diversity decreasing with either an increase or a decrease in productivity (Grime, 1973, 1979; Al-Mufti *et al.*, 1977). Furthermore, there are many examples in which the species diversity of one group of organisms in a community may be increasing along a productivity gradient, while the diversity of another group is decreasing along the same gradient. For example, along a gradient of decreasing precipitation (and plant productivity) from eastern to central North America, the diversity of trees decreases while the diversity of grasses and herbaceous plants increases. Continuing further west, the diversity of grasses and herbs begins to decrease and trees disappear almost entirely. In marine benthic communities, the diversity of foraminifera increases from the shallow (100-200 m) waters of the continental shelf to the great depths of the abyssal plain (4000-5000 m) (Buzas and Givson, 1969). Along this same depth gradient (productivity in these communities is inversely correlated with depth, Rex, 1981), the diversity of fish decreases monotonically (Haedrich *et al.*, 1980).

Diversity and the Size of the Sample Area

Productivity is not the only conspicuous environmental property that is usually related to species diversity. The relationship between species diversity and area has a long history in ecological studies and has generated at least as much interest and controversy as has the diversity-productivity relationship. With few exceptions, large areas have more species than small areas (Gleason, 1922, 1925; Preston, 1960, 1962, 1969; Williams, 1964). This may seem like a trivial observation, but the underlying mechanisms include most of those that are potentially important in regulating diversity. The pattern of how diversity increases with increasing sample size, generally called the species/area curve, can give insight into the total number of species in a region (indicated by where the curve levels off and ceases to increase with increasing area). The same type of pattern is also found when sample size is represented in terms of the total number of individual organisms, rather than the size of area from which they were collected (Fig. 2.15).

In addition to providing information needed to estimate the total number of species in a region, species/area curves can also be used to compare the rate of increase in species number with area between different regions. The exponential parameter z in the equation $S = kA^z$ (where S is the number of species, k is a constant, and A is the area of the sample), along with the constant k (cf. Lomolino, 1989), describe the rate at which the number of species increases with area. The rate at which the number of species increases with area, particularly as characterized by the exponential parameter z, has been related to a variety

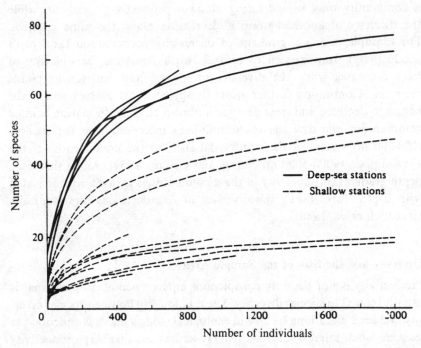

Fig. 2.15 Species-number relationships for samples of marine benthic gastropods and polychaetes from the continental shelf and slope off southern New England. (From Sanders, 1969.)

of theoretical mechanisms, including the effect of area and isolation in the equilibrium between immigration and extinction (Preston, 1960, 1962; MacArthur and Wilson, 1963, 1967), and habitat heterogeneity (Williams, 1964; Dexter, 1972; Harman, 1972; van der Werff, 1983). Some authors have suggested that the parameter z varies in a consistent manner between species area curves from different types of environments, such as mainlands versus islands (May, 1975; Gorman, 1979; Browne, 1981), but in a critical review, Connor and McCoy (1979) convincingly argued that the parameters used to describe the species/area curve had no clear theoretical significance. They demonstrated that the restriction of z values to a limited range (0.2-0.4) was a statistical artifact, and that none of the theories about the cause of the species/area relationship made any unique predictions about the shape of the curve (cf. Connor *et al.*, 1983; see Chapter 4).

Differences between species/area curves do suggest that there are differences in the ecological processes regulating the diversity of different landscapes, but those differences can only be deduced by comparative studies of the ecosystems themselves, rather than differences in the shapes of the curves. Numerous studies of species/area relationships for many communities over a wide range of scales provide little support for the interesting and plausible predictions of the equilibrium theory of island biogeography or theories on the distribution of animal abundance (Connor and McCoy, 1979; Gilbert, 1980; Connor *et al.*, 1983).

Increasing species/area curves have been described for a wide variety of organisms over a broad range of sampling scales, including: flowering plants in Britain (square miles) (Williams, 1964); birds on the Solomon Islands (km^2) (Diamond and Mayr, 1976); vascular plants on the Azores (km^2) (Eriksson *et al.*, 1974); boreal mammals living on montane islands of the Great Basin (square miles) (Brown, 1971); terrestrial invertebrates living in caves in West Virginia (km^2) (Culver *et al.*, 1973); arthropods on islands of *Spartina alterniflora* in Oyster Bay, Florida (m^2) (Rey, 1981); and mollusc species in lakes in New York State (km^2) (Browne, 1981). Relationships between several species can also show an increase in diversity with increasing area over which a host species is found. For example, the total number of phytophagous insect species that feed on bracken fern is significantly related to the area over which the plant occurs (Fig. 2.16).

There are three potential explanations for the general increase of diversity with area.

1. It is a sampling artifact. Within a homogeneous area (generally relatively small in size), samples of increasingly larger area will randomly sample an increasing proportion of the total population, and are thus likely to detect increasingly rarer species as the size of the sample increases. Diversity will increase until all species within the homogeneous area are sampled, gradually leveling off as the total number of species is approached until the species/area curve becomes horizontal.

2. It results from an equilibrium between extinction and immigration, as described by the equilibrium theory of island biogeography (MacArthur and Wilson, 1963, 1967).

3. It is a result of environmental heterogeneity. Increasing the sample area includes additional habitat types with groups of different species. Diversity will increase until all habitat types are sampled, which could be the entire Earth, if the study were not limited to a particular geographic area. In general, species/area curves of this type also tend to level off, but rarely become horizontal, because rare habitat types continue to be added to the sample.

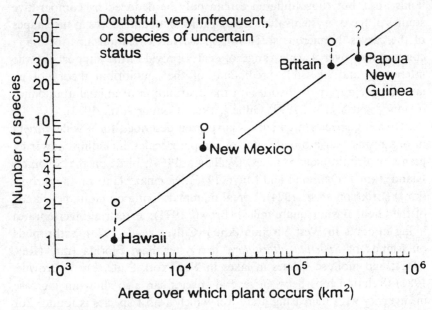

Fig. 2.16 Number of insect species that feed on Bracken in relation to the size of the area over which bracken occurs. (From Lawton, 1984.)

Because of the relationship between habitats and the number and identity of species encountered in samples, species/area curves can also reveal the spatial scale of environmental differences that influence community structure (indicated by a stairstep pattern of increase in diversity or an S-shaped curve) (Cain, 1938; Archibald, 1949; Vestal, 1949; Lassen, 1975; Shmida and Wilson, 1985) (Fig. 2.17).

Within a homogeneous area, there is evidence that the number of species reaches some sort of equilibrium over time. Even though more species could coexist over short periods, the longer-term equilibrium number is lower. This phenomenon was demonstrated by Simberloff (1976), who reduced the size of mangrove islands in Florida by removing trees and then monitored the number of insect species on the islands. Immediately following tree removal, the number of insect species on an island was the same as it was before the size was reduced. However, over several years the number of insect species declined on the islands that were reduced in size, apparently because the smaller population sizes supported by the smaller island made extinction more likely.

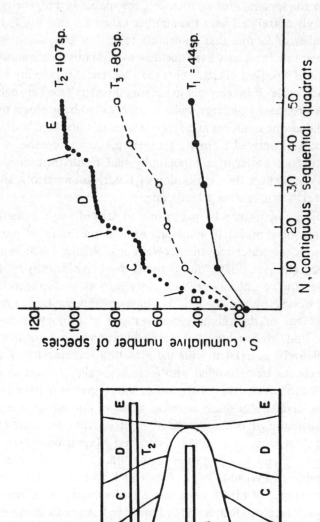

Fig. 2.17 Increase in plant species diversity with increasing sample area, in the Judean desert of Israel. Transect T_1 is in a homogeneous area. T_2 is in a heterogeneous area with four vegetation zones and demonstrates a stairstep pattern caused by complete sampling of a single homogeneous habitat followed by sampling in an different habitat. T_3 is in a homogeneous area close to transect T_2, and shows the presence of species from both the homogeneous and zoned area. (From Shmida and Wilson, 1985.)

Diversity and the Spatial Heterogeneity of the Sample Area

As suggested by the species/area relationship, the spatial heterogeneity of an area is strongly correlated with the number of species that are found there. The number of factors that contribute to spatial heterogeneity is virtually innumerable, and can even include the organisms themselves (i.e., the 'structural' species), which offers the theoretical possibility of a positive feedback cycle of indefinitely increasing diversity (see DeAngelis et al., 1986). Heterogeneity on large scales is contributed by geologic processes that influence the amounts and types of minerals found in bedrock and thus in the soil derived from it; geologic processes together with climate also influence patterns of topography that result from erosion, and topography influences the distribution of water, soil nutrients, solar energy, and other factors across a landscape.

Heterogeneity is contributed by plants in a variety of ways, including the accumulation or removal of materials around the bases of plants as a result of wind or water movement (Neite and Wittig, 1985; Wittig and Neite, 1985; Herwitz, 1986), and perhaps most significantly by the vertical structure and complexity produced by the roots, stems, branches, and leaves of woody plants, as well as the non-woody plants growing with them. One of the classic generalizations of ecology is based on MacArthur and MacArthur's (1961) observation that bird species diversity is positively correlated with the structural complexity of the vegetation, which can be quantified with a statistic called foliage height diversity (Fig. 2.18a). Since this observation, similar patterns have been reported for the diversity of many different kinds of animals in relation to vegetation structure or other aspects of environmental heterogeneity (Recher, 1969; Anderson, et al., 1983; Tonn and Magnuson, 1982), although the methods and interpretation of the MacArthurs' observation have been questioned (Wilson, 1974; James and Wamer, 1982). There have also been reports of bird species diversity decreasing with increasing foliage height diversity (Ralph, 1985; Fig. 2.18b). Animals themselves can further increase the heterogeneity created by plants by eating or killing plants, by disturbing the soil with burrows, trails, and wallows, by defecating, and by dying.

In summary, there are many causes of environmental heterogeneity, and differences in heterogeneity are almost always correlated with differences in species diversity. Environmental heterogeneity affects not only the diversity of species, but the genetic diversity within species. McArthur et al. (1988) assayed the genetic variability in ten enzymes

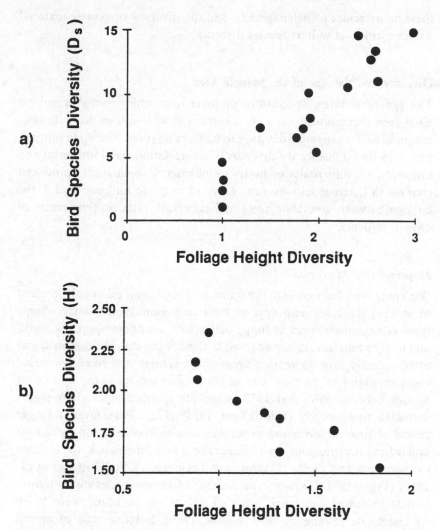

Fig. 2.18 Relationship between habitat structural complexity, measured as foliage height diversity (FHD) and the number of bird species in the habitat. (*a*) Positive correlation between FHD and bird species richness in northeastern North America (from MacArthur and MacArthur, 1961). (*b*) Negative correlation between FHD and bird species diversity (presented as the composite index, H') in northern Patagonia, Argentina. (From Ralph, 1985.)

produced by the soil bacterium *Pseudomonas cepacia* collected from five different forest types, and found that the variability within each enzyme was positively correlated with temporal variability of the chemical and physical environment. The concept of biological diversity incorporates these many scales of heterogeneity, and the diversity of genetic material within species, as well as species diversity.

Diversity and the Age of the Sample Area

The age of an area, specifically the time over which living organisms have been continuously present, is correlated with species diversity over temporal scales ranging from days to millions of years. The three primary processes that influence the diversity/age correlation are: 1. dispersal and migration on time scales of hours to millenia; 2. biologically produced changes in heterogeneity on time scales of hours to millenia; and 3. the balance between speciation rates and extinction rates on time scales of days to millenia.

Dispersal and Migration

Beginning with bare ground, the number of organisms potentially present in an area increases with time as more and more species arrive. Some types of organisms, such as fungi, plants with windborn spores or seeds, and many small insects, are adapted to rapid, long-distance dispersal and arrive quickly even in remote areas. For example, the recently-created volcanic island of Surtsey was rapidly colonized by many species of mosses between 1965 and 1973, while the number of vascular plants increased more slowly (Fridriksson, 1975) (Fig 2.19a). Over a longer period of time on the island of Krakatoa, initial colonization by mosses and lichens (Cryptogams) was followed by a rapid increase in the number of monocots and dicots (Doctors van Leeuwen, 1929; Whittaker *et al.*, 1990) (Fig. 2.19b). Likewise, the number of insect species on mangrove islands increased over time after all insects on an island were killed by insecticide (Simberloff and Wilson, 1969). Relative time of arrival of plants and insects can be predicted quite accurately based on the dispersal properties and other characteristics of the organisms.

The increase in plant species during early succession can also result from dispersal patterns (Clements, 1916; Bazzaz, 1975). Over the much longer time span following the retreat of the last continental glacier that covered parts of North America, tree species have been migrating northward at different rates. These different rates of migration, caused by various mechanisms of dispersal by wind, birds, or mammals, have resulted in diverse patterns of increase in tree species over time. Since the glaciers retreated much faster than most tree species advanced, some

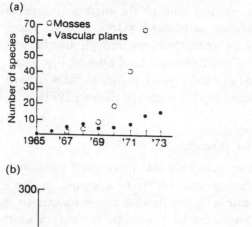

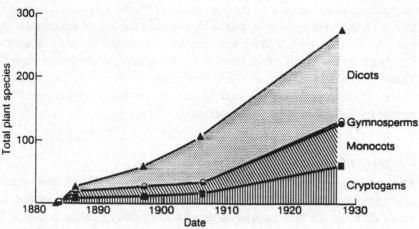

Fig. 2.19 Rates of colonization of newly formed and completely denuded islands by different types of plants. (*a*) Rapid rise in the number of moss species during the first eight years following the formation of the island of Surtsey (after Fridriksson, 1975). (*b*) Colonization of the island of Krakatoa over the first forty years following the volcanic eruption that destroyed the island. (Based on Doctors van Leeuwen, 1929.)

trees species migrating from the southern non-glaciated regions have only recently arrived in northern forests (Davis, 1976; Webb and Bernabo, 1977; Webb, 1987; Delcourt and Delcourt, 1987). The interaction between dispersal rate, time, and distance, is one of the bases of the theory of island biogeography (MacArthur and Wilson, 1967).

Successional Increase in Heterogeneity

As discussed above, environmental heterogeneity is one of the strongest correlates of species diversity. Much of the increase in species diversity that occurs during early succession is due to the increase in spatial heterogeneity caused by organisms, particularly plants. Another factor that contributes to an increase in species diversity through time in early succession is the modification of environmental conditions by organisms, such as by the addition of nitrogen and organic matter to the soil. Such processes have been called facilitation (Connell and Slatyer, 1977).

Evolutionary Increase in Species Diversity

Over long enough time periods and with the appropriate conditions, evolution can contribute to the species diversity of a region, assuming that the long-term speciation rate is higher than the long-term extinction rate. Areas that have been populated by organisms for only a short time, such as the north temperate areas from which the glaciers retreated 15,000 years ago, have had much less time for the evolution of new species than tropical regions that have been continuously occupied by organisms for much longer time periods.

Diversity and Disturbances

The age of an area is obviously related to the length of time since a disturbance eliminated some or all of the organisms and provided a 'clean slate' for the processes of dispersal and evolution. Disturbances are an important and complex topic that will be prominent in the rest of this book. As the following examples illustrate, diversity is correlated with disturbance on many different scales, with some positive correlations and some negative correlations.

Infrequent, Massive Disturbances

Disturbances such as glaciation, which totally alter large areas, killing or driving off all organisms and altering the surface of the landscape, clearly have a major influence on the species diversity of the affected area. Areas where glaciers periodically invade, particularly the upper latitudes of the northern hemisphere, differ in many ways from areas that are never affected by glaciers, such as most of the tropics. A major difference is the lower number of species in the areas that have been affected by glaciers.

While the tropics have not been subjected to extensive scouring by continental glaciers, climatic variation associated with the glacial cycles also had major effects on tropical ecosystems. During the cool, dry glacial periods, tropical deserts, grasslands, and savannahs expanded, while rain forests may have shrunk to a few isolated areas of wet mountains or coastal zones. Centers of distribution and diversity for a variety of taxa have been interpreted by some as 'refugia' where favorable conditions in the past allowed organisms to survive when most of the landscape became inhospitable (Keast, 1961; Haffer, 1969, 1982; Bigarella and Andrade-Lima, 1982; Ab'Sáber, 1982; Graham, 1982; Prance, 1982). Whether entire communities were preserved in the refugia, or whether species were distributed individually across climatic gradients, and formed their present associations in response to current climatic conditions is a subject of considerable controversy. Recent paleoecological results suggest that past climatic conditions in the supposed refugia were very different from the present conditions, so they were unlikely to have served as a refuge for the present plant communities (Bush and Colinvaux, 1990; Bush *et al.*, 1990). Notwithstanding the extinctions that undoubtedly resulted from climatic fluctuations in the tropics, these episodes clearly did not have a negative effect on species diversity comparable to continental glaciation in the temperate zone, and may have actually resulted in an increase in diversity as a result of increased speciation rates.

Frequent, Less Severe Disturbances

While massive disturbances, even if infrequent, can result in a significant reduction of species diversity, there is strong evidence that frequent, less severe, disturbances are necessary for the maintenance of diversity in some systems. Sousa (1979a,b) found that stones that were periodically rolled and tumbled by wave action in the intertidal zone had higher diversity of algae and other organisms than stones that were not disturbed at all. Likewise, mortality caused by predators has been shown to result in higher diversity of intertidal organisms than is found in the absence of predators (Paine, 1966). Disturbances or predation that are too frequent or intense can result in a reduction of species diversity. Thus, disturbance can lead to either an increase or a decrease in species diversity. In fact, many studies have found that the highest species diversity occurs at intermediate frequencies of disturbance, with low diversity at both very high and very low frequencies (Loucks, 1970; Auclair and Goff, 1971; Lubchenco, 1978; Lubchenco and Menge, 1978; Connell, 1978; Huston, 1979).

Other Spatial Patterns of Diversity

In addition to the latitudinal patterns of diversity, there are many other diversity patterns, often associated with gradients of one sort or another.

Zonation: Abrupt Changes in Species Composition and Diversity

One of the most dramatic natural patterns of biodiversity on landscapes is the zonation of organisms or groups of organisms into distinct bands. Zonation occurs at all spatial scales, ranging from the global patterns called biomes (e.g., Walter, 1973) down to zonation around individual plants, and even zonation within single particles of soil. Zones are usually distinguished by a high degree of dominance by a particular species or a particular life form of plant or animal. Dominance can be defined by a variety of different criteria, including: 1. numbers of individuals; 2. total biomass of individuals or amount of space occupied; and 3. degree of control of the community and ecosystem dynamics of the area. Striking patterns of zonation, particularly those related to the large-scale distribution of life forms of plants, attracted the attention of naturalists and biogeographers long before species diversity became a topic of interest.

Although zones are described on the basis of one or a few dominant or conspicuous organisms and by definition have sufficiently distinct boundaries to be considered zones, the distribution of species richness does not necessarily follow the zones. The distribution of individual species, and of the species richness of different groups of organisms, may vary smoothly across the same gradients along which the dominant organisms show strong zonation. If the zones are defined by the dominance of a few species, then the evenness component of diversity will inevitably be quite low.

Marine Intertidal and Subtidal Zonation. Some of the most dramatic and best-studied examples of zonation are found along the extreme physical gradient of the marine intertidal zone. This region is alternately inundated by salt water and often pounded by waves when the tide is in, and then exposed to dry air and sun when the tide is out. The areas lowest in elevation are exposed for the shortest time, while the highest areas may be exposed for long periods. Patterns of zonation are often clearest along rocky coastlines, where many of the marine plants and animals are permanently attached to the rocky substrate (Figs 2.20 and 2.21). In general, only mobile animals, such as snails, are found in the highest,

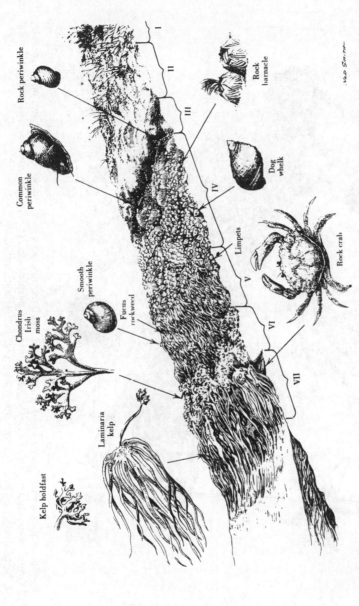

Fig. 2.20 Zonation on a rocky shore along the North Atlantic. (I) Land: lichens, herbs, grasses, etc.: (II) bare rock; (III) zone of black algae and rock periwinkles; (IV) barnacle zone: barnacles, dog whelks, common periwinkles, mussels; (V) fucoid zone: rockweed and smooth periwinkles; (VI) *Chondrus* zone: Irish moss; (VII) *Laminaria* zone: kelp. (From Smith, 1966.)

Rock periwinkle

Common periwinkle

Chondrus Irish moss

Smooth periwinkle

Fucus rockweed

Laminaria kelp

Kelp holdfast

Rock barnacle

Dog whelk

Limpets

Rock crab

I

II

III

IV

V

VI

VII

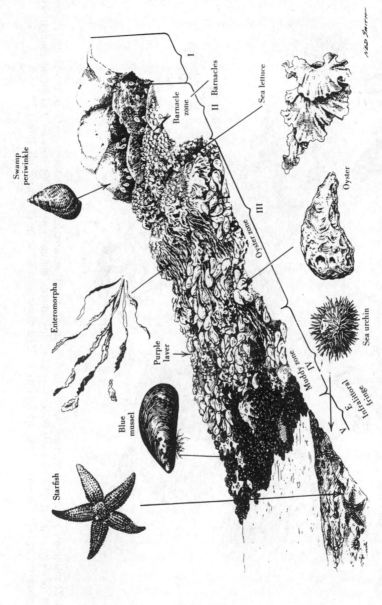

Fig. 2.21 Zonation along a rocky shore - mid-Atlantic zone. (I) Bare rock with some black algae and swamp periwinkle; (II) Barnacle zone; (III) Oyster zone: Oysters, *Enteromorpha*, sea lettuce, and purple laver; (IV) muddy zone: mussel beds; (V) infralittoral fringe: starfish, etc. Note absence of kelp in comparison with Fig. 2.20. (From Smith, 1966.)

most exposed zones. Among attached organisms, those that can close very tightly and avoid dessication, such as barnacles, are able to survive in the more exposed upper and middle zones. In lower zones, various types of algae are found, along with bivalves such as oysters and mussels (Stephenson and Stephenson, 1949, 1950, 1952, 1954a,b, 1961; Lewis, 1976). Zonation also occurs among organisms that live in the sediments at depths below the intertidal (Hughes and Thomas, 1971). In many cases the distribution of most of the individual species is not related to the visually conspicuous zones, and it is not always possible to identify a single dominant mechanism that causes the patterns of zonation and species diversity (Underwood, 1978a,b; Underwood and Denley, 1984).

Mangrove Zonation. Mangroves, with their dense tangles of prop roots and viviparous seeds with a long thick protruding root, are found along coastlines through the tropics and subtropics. Relatively few species of trees are adapted to the intertidal environment. There are only ten species of mangroves in the Neotropics and 36 species in the Indo-West-Pacific Region, from a total of only eight families. In addition to the unique adaptations of the various mangroves species, a conspicuous feature of most mangrove systems is their arrangement in nearly monospecific zones. Monospecific mangrove zones appear most clearly where there is a relatively steep shoreline gradient. However, even in very flat areas such as much of south Florida, they tend to grow in mixtures of species that can be characterized as distinct community types. The distribution of mangrove zones and mixed communities appears to be controlled by the interaction of salinity and nutrients, as affected by local patterns of tides and freshwater drainage from the land (Semeniuk, 1983; see review in Lugo and Snedecor, 1974), although patterns of dispersal of the large viviparous seeds may play some role (Rabinowitz, 1978a,b).

Terrestrial Vegetation. At the largest scale, terrestrial vegetation (along with associated animals) can be considered to be composed of types or zones known as *biomes* (Carpenter, 1939; Clements and Shelford, 1939). These broad zones are usually defined on the basis of life forms of organisms, e.g., forest versus grassland, or evergreen versus deciduous, rather than on the presence or absence of particular species. Biomes tend to be arrayed both along latitudinal gradients (related to latitudinal patterns of temperature and moisture) and also along longitudinal gradients (related to maritime/continental moisture and temperature gradients), particularly along the east and west margins of continents (Fig. 2.22).

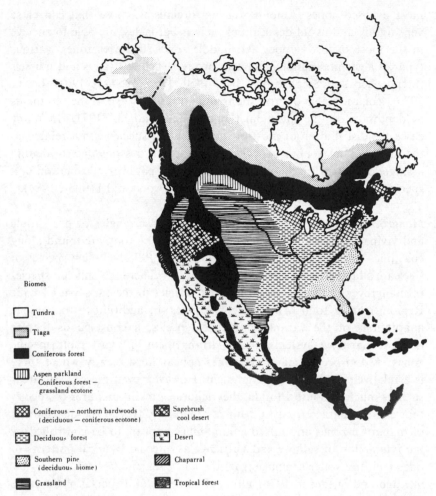

Biomes

☐ Tundra

▨ Tiaga

■ Coniferous forest

▥ Aspen parkland
     Coniferous forest —
     grassland ecotone

▨ Coniferous — northern hardwoods   ▨ Sagebrush
     (deciduous — coniferous ecotone)   cool desert

▨ Deciduous forest   ▨ Desert

▨ Southern pine   ▨ Chaparral
     (deciduous biome)

▬ Grassland   ▨ Tropical forest

Fig. 2.22 Vegetation zones in North America illustrating longitudinal patterns along the east and west continental margins and latitudinal patterns in the boreal and arctic region. (From Smith, 1966.)

Other classifications of plant communities, such as the life zone system (Holdridge, 1947), are based explicitly on temperature and precipitation. Similar zonation occurs on elevational gradients as well (Cox *et al.*, 1976).

At smaller scales, zones of differing vegetation are often associated with variation in soil properties. Zonation of vegetation along the sand dunes at the southern end of Lake Michigan led to some of the earliest insights into the process of plant succession as well as into the processes of soil formation. (Cowles, 1899; Shelford, 1913; Olson, 1958). Zones of vegetation may even occur around individual plants for a variety of reasons, including the release of phyto-toxic chemicals (Muller and del Moral, 1966; McPherson and Muller, 1969) and changes in soil chemistry induced by the plants (Wittig and Neite, 1985).

Aquatic Vegetation. The spatial patterns of vegetation along lake shores, such as from the shallow water (littoral) zone onto the dry land, or from the open-water edge of a floating bog to shrub or forest, are well-known examples of zonation (Dansereau, 1957; Dansereau and Segadas-Vianna, 1952; Spence, 1982) (Fig. 2.23).

Radial Gradients

Patterns in which diversity increases or decreases in all directions from a central location have been described, ranging in scale from gradients across continents (Brown, 1982; Hamilton, 1989; Rogers *et al.*, 1982) to gradients in a cow pat or ball of dung (Denholm-Young, 1978). Diversity of groups that are associated with geographical features, such as mountains, tend to decrease away from those features. Such gradients extend radially from a center of diversity, with diversity decreasing with distance from the center (Fig. 2.24).

Tree species diversity in eastern North America is highest in the Great Smoky Mountains of Tennessee and North Carolina, and decreases away from this area (Braun, 1950) (Fig. 2.2). In the tropics, similar centers of diversity (considered by some to represent 'refuges' during unfavorable climatic conditions in the past) have been described for plants (Prance, 1982; de Granville, 1982), birds (Haffer, 1969, 1982; Terborgh and Winter, 1982), butterflies (Benson, 1982; Brown, 1982; Turner, 1982), frogs (Duellman, 1982), primates (Kinzey, 1982), African forest mammals (Grubb, 1982; Rogers *et al.*, 1982), and African savannah mammals (Sinclair, 1983).

Deep-sea Benthic Gradients. Although it was once believed that the deep oceans were devoid of life or only populated by a small number of archaic species (Forbes, 1844; Bruun, 1957; Zenkevitch and Birstein, 1960), improved sampling methods have revealed a remarkably diverse and highly adapted fauna (Rex, 1981). In spite of the frigid temperatures, high pressures, total darkness, and nutrient-poor sediments of the deep oceans, there is a complex fauna composed of echinoderms, fish, cnidarians, polychaetes, and arthropods such as pycnogonids, decapods, and giant

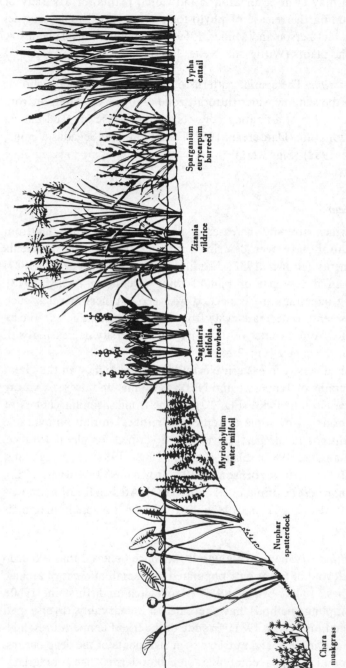

Fig. 2.23 Zonation of terrestrial and aquatic vegetation around a body of water. (Based on Dansereau, 1959, from Smith, 1966.)

Typha
cattail

Sparganium
eurycarpum
burreed

Zizania
wildrice

Sagittaria
latifolia
arrowhead

Myriophyllum
water milfoil

Nuphar
spatterdock

Chara
muskgrass

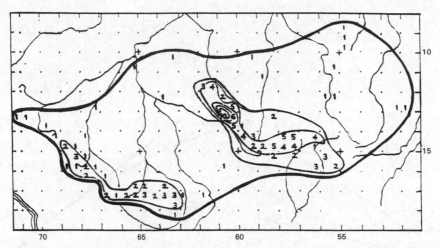

Fig. 2.24 Radial pattern of diversity decreasing away from a central location, in this case an area with a high species richness in a certain group of butterflies in the Amazon Basin. (From Brown, 1982.)

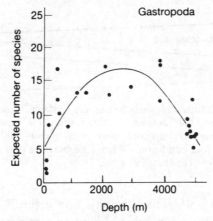

Fig. 2.25 Depth gradient in the diversity of benthic gastropod molluscs from the continental shelf to the deep ocean. (From Rex, 1981.)

amphipods, as well as small organisms such as nematodes, gastrotrichs, and kinorhynchs. The general pattern for deep-sea organisms is an increase in species diversity from the continental shelf (depths less than 200 m) to a maximum at depths of 2000 to 4000 m (the continental rise), and a decrease below 4000 m (the abyssal plain). This unimodal pattern of species diversity with depth is found in many different groups of organisms (Sanders, 1968; Rex, 1981) (Fig. 2.25). Associated with the

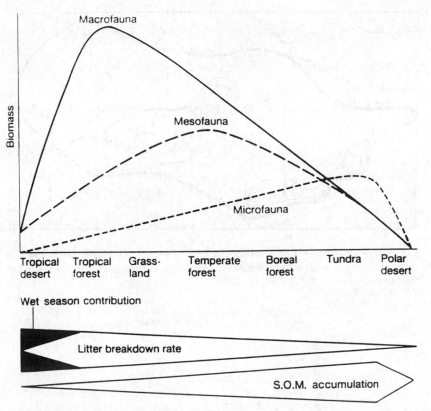

Fig. 2.26 Patterns of latitudinal variation in the relative biomass of microbes (microfauna) and larger soil organisms (mesofauna and macrofauna). Litter breakdown by organisms is impaired (and consequently soil organic matter (SOM) accumulates) under conditions of low temperatures and either insufficient or excessive soil moisture. (From Swift *et al.*, 1979.)

unimodal distribution of species diversity with depth is an exponential decrease in the density and biomass of benthic organisms (Thiel, 1979). This decrease in standing crop (and presumably productivity) is probably a result of decreasing availability of food that sinks from shallower depths, as well as the physical factors mentioned above.

Soil Gradients. Spatial variation in soil conditions is a feature of virtually all terrestrial communities. Depending on the particular soil properties of interest, and the scale at which they are measured, the spatial pattern may be almost completely random (Robertson *et al.*, 1988) or may be a clear gradient (sometimes called a 'catena'). Soil gradients occur on the

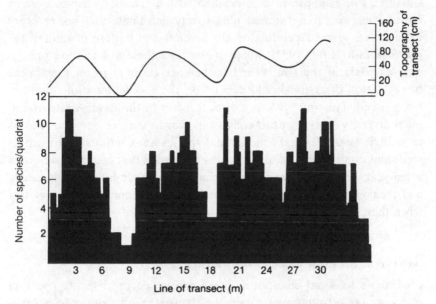

Fig. 2.27 Repeating pattern of herbaceous plant species richness in relation to topography in a ridge-and-furrow grassland in England. (From Begon *et al.*, 1986.)

scales of continents, mountain slopes, hillsides, around individual plants or animal droppings, or within single soil particles. For example, the relative importance of microbes in relation to larger soil organisms is known to affect the decomposition of plant litter and the accumulation of soil organic matter along a latitudinal gradient (Fig. 2.26).

Spatial variation in soil water, nutrients, or texture is almost always reflected by patterns in the communities of plants and other organisms growing on the soil. Numerous studies have documented changes in species composition and species diversity along many different types of soil gradients (Grime and Lloyd, 1973; Werner and Platt, 1976; Kendrick and Burges, 1962) (Fig. 2.27).

Temporal Variation in Species Diversity: Snapshots versus Movies

One of the difficulties associated with studying diversity is that diversity is constantly changing at many different scales. In some groups of organisms, diversity measured at one time of year is very different from diversity measured at another time of year. This emphasizes the extreme importance of carefully defining how diversity is evaluated in a specific situation. For example, in a desert area such as Death Valley in western North America, many annual plants only germinate and flower every three to ten years, depending on the amount and pattern of rainfall the previous fall (Hunt, 1975). Most of the time these annuals are present only as seeds in the soil. Whether they are counted in an assessment of vegetation diversity should depend on the specific questions that are being asked. The time scale is a critical factor in this case. Aboveground plant diversity can vary drastically from year to year, or between seasons in a single year. Certainly the annual species are continually present as seeds and contribute to a high number of species that could be seen over a ten-year period. However, their biomass as seeds is extremely small, and their effect on both the plant and animal communities is very low when they are buried and dormant.

Short-term Seasonal Changes in Diversity

Predictable seasonal changes in measureable species diversity occur in many groups of organisms. Often the changes result from species being in a stage of their life cycle that is not easily detected or measured (such as belowground insect larvae) or from species migrating into or out of an area.

Insect seasonal diversity. The diversity of insects that can be captured or observed has strong seasonal variation in many aquatic and terrestrial systems (Holloway, 1977; Wolda, 1988). Many insects, such as cicadas and many species of beetles, spend most of their lives as larvae feeding underground on plant roots. Since these species only appear aboveground for a relatively brief time, they contribute to seasonal variation in the aboveground insect community. Although they are difficult to census during most of the year, their most important effect on the ecosystem may actually be when they are not visible.

Throughout much of the tropics, many species of insects are most apparent at the beginning of the rainy season, when inconspicuous larvae metamorphose into conspicuous adults that disperse, mate, lay eggs and are often found at lights and in the insect traps of entomologists (Holloway, 1979; Wolda, 1983; Janzen, 1983). This high diversity of insects is obviously present throughout the year in some life history stage, but is extremely difficult to sample. Methods such as fogging the forest canopy with insecticides (Erwin, 1983) are often the only way to sample adequately the insect community for the majority of species that never have conspicuous stages. Intensity of the effort made in sampling inconspicuous organisms will strongly influence the level of diversity that is detected.

Stream insect diversity also undergoes seasonal fluctuations. Many insects that are part of the benthic community of streams are present throughout most of the year as aquatic larval forms (Cummins, 1974). These larvae are easily sampled using a variety of standard techniques. However, at some time during the year these aquatic larvae metamorphose into flying adults that leave the stream to mate and disperse. Until the adults lay eggs in the stream, the eggs hatch into larvae, and the larvae grow to sufficient size to be captured and observed, these species cannot be detected in surveys of the stream insect community. Sampling over the course of an entire year and including both flying and aquatic life cycle stages is necessary to characterize the diversity of a stream insect community.

Migratory bird diversity. The diversity of bird communities undergoes great seasonal fluctuations as a result of migrations between summer breeding areas and winter feeding areas. The increase in temperate bird species diversity that results from the influx of birds from the tropics to breed during the temperate summer is proportionately much greater than the changes in tropical bird diversity that result from the comings and goings of these same species. Evaluation and interpretation of bird species diversity obviously require careful definition of which component of the bird community is to be measured and when (Weins, 1989a).

Successional Time

The species composition and diversity of many communities change gradually over time spans of several hours to several centuries, depending on the lifespans of the dominant organisms. This pattern of gradual temporal change in species composition is generally called succession. Succession results from a variety of processes, including migration or dispersal, plant or animal growth and competition, and environmental changes caused by the organisms themselves or by forces beyond the control of the organisms (Chapter 9).

Plant species diversity generally increases during the early stages of succession and decreases during the later stages, with a maximum in mid-succession. Under different environmental conditions, and over different time spans, successional changes can occur within a single growth form

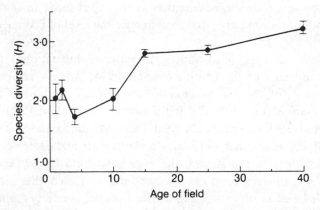

Fig. 2.28 Increase in species diversity (calculated as the composite index, H') over time during oldfield succession. (From Bazzaz, 1975.)

of plants, such as herbs, or span a range of growth forms, such as the classic grass, herb, shrub, and tree successional series.

A study of herbaceous succession in an old field is likely to show an increase in species diversity as long as the field is dominated by herbaceous species rather than woody species (e.g., Bazzaz, 1975) (Fig. 2.28). If the succession proceeds to the woody stage and a typical forest succession occurs, diversity is likely to decrease over time (Fig. 2.29).

Animal Species Diversity. The number of animal species in an area almost always changes in association with successional changes in the plant community. Bird species richness was found to increase from two to over 20 species across a series of fields representing a sequence of two to 150 years of oldfield succession (Johnston and Odum, 1956) (Fig. 2.30). Insect species richness also increases during oldfield succession, with most of the increase within the first ten years (Brown and Southwood, 1983).

Long-term Evolutionary Changes in Diversity

Over the past 600 million years the number of different types of organisms has been increasing (Signor, 1990). Beginning with bacteria, algae, and soft-bodied marine organisms that left little trace in the fossil record, diversity has increased at all but the highest levels of the phylogenetic hierarchy. Some strange (viewed from our own evolutionary perspective) phylum-level taxa disappeared prior to the Cambrian (Morris, 1979; Gould, 1989), and the increase in the number of different phyla has

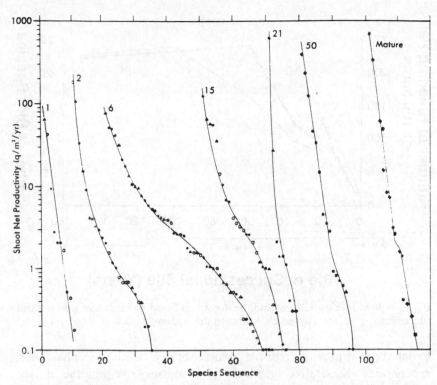

Fig. 2.29 Maximum species richness and evenness occurs early in forest succession. These 'importance value' (or 'dominance/diversity') curves for plant communities of seven different ages are based on shoot net productivity rather than biomass or abundance, which is more often presented. As species diversity increases during the herbaceous stage of succession (1- 6 years) the curve shifts from a steep curve with few species (approaching a geometric distribution) at one year to a sigmoid (lognormal) curve with many more species at six years. Growth of shrubs that overtop and suppress the herbs results in a steep, geometric curve with high dominance by the most productive species (21 years). Later in forest succession more species are added and the curve becomes less steep. The data for a mature forest are from an oak-hickory forest at Oak Ridge, Tennessee. Circles indicate herbs, triangles shrubs, and squares trees. Open symbols denote exotic species while closed ones denote native species. (B. Holt and G.M. Woodwell, unpublished, from Whittaker, 1975.)

slowed or stopped since the end of the Paleozoic (225 million years BP). All major contemporary phyla have been present since this time, although there are some phyla for which there is no fossil record, presumably because their soft bodies did not fossilize readily. Within many phyla, new classes appeared up through the Mesozoic era (ending 65 million years BP), while other classes became extinct (e.g. trilobites) or continued

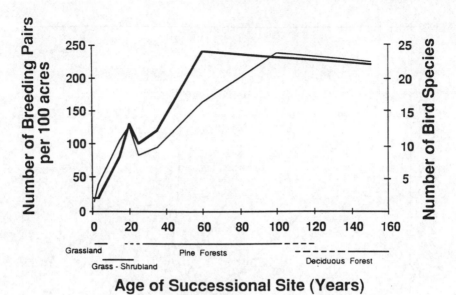

Age of Successional Site (Years)

Fig. 2.30 Bird species richness and the density of breeding birds along an oldfield successional gradient. (Based on Johnston and Odum, 1956.)

to survive in drastically reduced abundance (e.g., crinoids, cephalopods, sphenopsids (horsetails), lycopsids (club mosses)). Within the classes, at the level of orders, the composition of life on Earth has been still more dynamic. Many orders have become extinct, such as the orders of dinosaurs and marine reptiles that disappeared at the end of the Mesozoic era, while the many orders of mammals and birds have evolved in the past 65 million years (Cenozoic era) (Fig. 2.31).

The taxonomic composition of life has been most dynamic at the levels of family, genus, and species, with extinctions and speciation occurring frequently on a geologic time scale. The fossil record indicates that the rate of extinction has varied greatly, with large numbers of extinctions occurring during apparently brief periods of time (Chamberlain, 1898; Valentine et al., 1978; Raup, 1979; Raup and Sepkoski, 1982). The cause of these periodic episodes of mass extinction that have recurred throughout geologic history is one of the most fascinating and controversial topics in Earth history (Alvarez *et al.*, 1980, 1982).

Among most groups of organisms for which data have been compiled, the total number of species has increased almost continuously since the group first appeared. For insects, the number of taxa at the level of order and major suborder has increased since their appearance in the

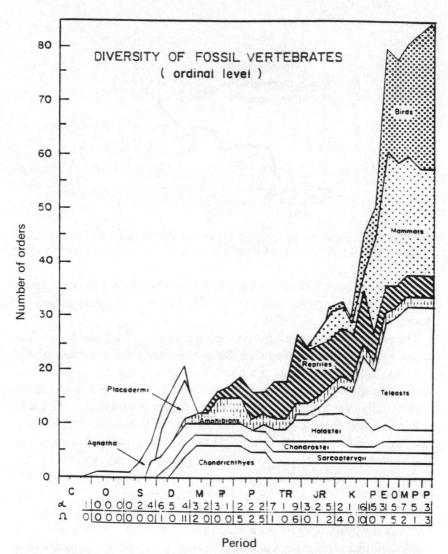

Fig. 2.31 Diversity of fossil vertebrates (in numbers of orders) from the Cambrian (600 my BP) through the present. (Based mainly on data from Romer, 1966; from Padian and Clemens, 1985.)

Devonian (350 to 413 million years BP) (Strong *et al.*, 1984). Likewise, the family-level diversity of shallow-water marine invertebrates has increased over the past 600 million years, with a slight decrease about 250 million years ago, and a recovery to the highest number at present (Valentine, 1970; Valentine *et al.*, 1975.). Evidence from the fossil record suggests

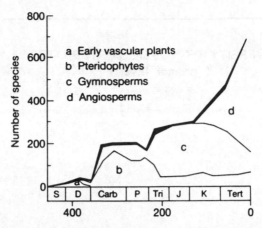

Fig. 2.32 Succession-like patterns of expansion and reduction of major plant groups over 400 million years of plant evolution. (From Niklas *et al.*, 1983.)

that there has been a latitudinal gradient of evolutionary rates over much of geological history for some groups of marine organisms, as well as mammals (Stehli *et al.*, 1969).

The evolution of species diversity among vascular land plants presents a particularly interesting pattern. Since the appearance of these plants over 400 million years ago, the number of species has increased almost monotonically, but the groups that dominate the land flora have shifted dramatically over time. The early vascular plants, the rootless and leafless Psilopsids, had become extinct by the end of the Devonian (350 my BP), and were replaced by the pteridophytes, which flourished during the Carboniferous period and decreased in abundance and variety by the early Triassic (200 m.y. B.P.). The decline of the Pteridophytes coincided with the diversification of the gymnosperms (gingko, cycads, conifers), which declined in abundance and diversity over the past 100 million years as the Angiosperms (flowering plants) diversified (Fig. 2.32). This succession-like replacement of one dominant group of plants by another over the course of evolutionary time may reflect the displacement of older taxa by newer groups that are better competitors (Niklas *et al.*, 1983), a process similar to ecological succession (cf., Clements, 1916).

Conclusion

Such regular patterns as zonation, latitudinal gradients of species diversity, and the temporal patterns of diversity found in succession, beg for an explanation. It is often obvious that changes in physical properties of the environment, such as temperature, precipitation, or soil, are correlated with the changes in diversity across the landscape or through time. There is a strong temptation to try to explain such changes in diversity on the basis of the physical factors that are changing along the same gradient. In fact, most of the explanations that have been proposed to explain species diversity are based on physical conditions that change along gradients of species diversity. However, it is important to realize that natural gradients are almost always very complex, with many different physical factors changing along the gradient, some increasing while others decrease. Likewise, the diversity of different components of the natural communities along a gradient may change in very different ways. One of the pitfalls of a complex subject such as the regulation of biological diversity is that many of the most dramatic and interesting gradients are also the most complex and difficult to explain. At this point, the most rational conclusion about patterns of species diversity might well be that they are hopelessly complex and confusing. Any hints of predictable regularity with regard to such factors as disturbance or productivity are confounded by the existence of well-documented correlations that may be either positive or negative. For those patterns that do seem to make sense, there are often several correlated factors available to provide alternative explanations.

As confusing as the details presented in this chapter may seem, the true test of the ideas to be put forward in this book is if they make sense of most of these apparently contradictory patterns. Most of the patterns described in this chapter will be discussed in more detail in later chapters, with (hopefully) convincing explanations. After completing this book, one should be able to return to this chapter, and reread it with the feeling that one is in familiar territory, where everything begins to fit together and makes some sense.

3

The Assessment of Species Diversity

The examples in the previous chapter illustrate many different methods of collecting, analyzing, and presenting data related to biological diversity. Depending on the question the data were collected to answer, any one of many different sampling designs and statistical analyses can be appropriate. However, it is also possible to collect data on species diversity that are impossible to interpret or understand. Comparison of the diversity of different groups of organisms or different communities is useful only to the extent that it contributes to an understanding of the processes that structure those communities. The goal of this book is identify the processes that regulate variation in the diversity and species composition of communities under different conditions. The data that are most useful for developing and testing explanations of species diversity are those collected in such a way that the detection of significant variation in diversity is maximized and the number of potential processes that could influence that diversity is minimized.

Far too much attention has been paid to comparison and criticism of **statistical** methods for quantifying diversity. This issue may be of interest to statisticians and mathematicians, but it has contributed virtually nothing to the ecological understanding of species diversity.

More significant than how the diversity of an ecological community is assessed statistically, is what biological properties of the community are actually measured. If several components of diversity or spatial scales are sampled simultaneously, it is likely that a variety of processes will have influenced the diversity of the sample. Since some processes may actually have opposing effects on diversity at different scales or under different environmental conditions, understanding the diversity of a sample that inappropriately combines different scales may be virtually impossible.

The Conceptual and Statistical Definitions of Diversity

The concept of diversity has two primary components, and two unavoidable value judgments. The primary components are statistical properties that are common to any mixture of different objects, whether the objects are balls of different colors, segments of DNA that code for different proteins, species or higher taxonomic levels, or soil types or habitat patches on a landscape. Each of these groups of items has two fundamental properties: 1. the number of different types of objects (e.g., species, soil types) in the mixture or sample; and 2. the relative number or amount of each different type of object. The value judgments are: 1. whether the selected classes are different enough to be considered separate types of objects; and 2. whether the objects in a particular class are similar enough to be considered the same type. On these distinctions hangs the quantification of biological diversity.

In the ecological literature, the number of species in a sample is generally called species richness (or sometimes species density when the sample size is expressed in terms of area). The relative distribution of individual objects among each of the different types is usually referred to as 'evenness,' since the more equal the numbers of objects of each type, the greater is the diversity as perceived intuitively or as measured by certain statistics. The number of species and the evenness of relative abundance (or conversely the degree of dominance by a particular species), are the two statistical properties used to quantify species diversity. These same components of diversity can be evaluated for any component of biological diversity, from the number of alleles for a particular genetic locus to the number of different landscape types on a satellite image.

Although diversity has two distinct statistical components, diversity is often described using statistical formulas that combine both components. Perhaps the best known of these composite statistics is the Shannon index (Shannon and Weaver, 1949):

$$H' = -\sum (p_i \log p_i)$$

where p_i is the proportion of the total sample (i.e., of the total number of individuals or total biomass) composed of species i. The other most commonly used statistic is Simpson's index (Simpson, 1949):

$$\lambda = \sum p_i^2$$

Because Simpson's Index is actually an index of dominance, and thus tends to be inversely related to evenness and richness, it is often expressed as a diversity index, Simpson's D, of the form

$$D = 1 - \lambda$$

These and most other diversity statistics apply to specific mathematical conditions, such as samples of infinite size or samples for which the number and indentity of all individuals is known. Since neither of these conditions is usually true for ecological samples, rigorous statistical analysis requires that the statistical formula be modified for samples that are of finite size or imperfectly known. These details are thoroughly discussed by Pielou (1975) and Peet (1974). Fortunately, in most situations the imprecision of ecological sampling probably makes the statistical nuances of finite versus infinite samples irrelevant. Diversity statistics are not likely to contain any useful information beyond the second decimal place.

The statistical measurement of species diversity became a controversial topic in the late 1970s. Mathematical ecologists discovered that some of the most popular statistics, such as the Shannon index (Shannon and Weaver, 1949), gave virtually the same value for very different patterns of species abundance. Likewise, it was possible to come up with hypothetical patterns of species abundance which different diversity indices would rank in opposite order (Hurlbert, 1971; Peet, 1974). This led some ecologists to severely criticize species diversity as measured by some of the then-standard (and still widely used) diversity indices, calling it a 'nonconcept' (Hurlbert, 1971).

While it is true that situations can be envisioned in which commonly-used diversity statistics will give contradictory results, for most sets of samples from natural communities the values for all diversity statistics are highly correlated. Some diversity indices have more desireable statistical properties than other indices (Hurlbert, 1971; Peet, 1974; Pielou, 1975, 1977; Magurran, 1988) and depending on the question being evaluated, one index may be more appropriate than another.

This controversy surrounding the statistical quantification of diversity overshadowed many of the important issues associated with species diversity, such as how diversity is generated, why and where diversity persists, and other topics that are the central issues of this book. The bruhaha was largely a result of the unreasonable expectation that a single statistic should contain all the information about the assembly of objects that it represents. This is clearly an unrealistic expectation.

Any statistic, such as an average, summarizes a certain property of the many different objects that comprise the sample group. A vast amount of information is lost in this summarization, both information about the property of interest, and information about other properties as well. The average size of balls in a box gives some expectation of what to expect when one picks a ball out of the box, but provides no information on the ranges of sizes, the proportion of large to small balls, or the color or composition of the balls. It is equally absurd to expect a single diversity statistic to express all the information we would need to characterize something as complex as the diversity of an ecological community. Biological diversity has many components in addition to the simple number of species present. Which species are present and which species are most abundant are important aspects of biological diversity that cannot be simply summarized.

Diversity statistics differ primarily in the degree to which they emphasize species richness versus species evenness. The various composite statistical indices can be classified on the basis of their emphasis on richness or evenness. Hill (1973) and Patil and Taillie (1982) have elegantly demonstrated that all diversity indices are members of a family of statistics (i.e., all can be expressed as a variant of the same general formula) that spans a continuum from pure richness to pure evenness.

In some diversity comparisons one index may not reflect any change in diversity, while another might show a statistically significant difference. For example, in some temporal or spatial comparisons the number of species does not change, but their relative abundance does, as may occur during the process of competitive displacement prior to competitive exclusion and extinction. In these situations, where changes in relative abundance clearly indicate the effect of an important process, it is appropriate to use an index that emphasizes evenness more than richness. Such indices include Simpson's D, mentioned earlier, and the inverse of Simpson's index (sometimes called Hill's N_2, Hill, 1973). The index that emphasizes the richness component of diversity most strongly, with no effect of evenness, is simply the number of species in the sample.

Some indices have been proposed to reflect 'pure' evenness. These indices are usually ratios of a composite index to the theoretical value of that index if all species in the sample were equally abundant. For the indices that have been described above, these ratios are: $H'/\log s$ (usually called 'Pielou's J', Pielou, 1966); $D/(1 - 1/s)$; and N_2/s, where s is the number of species in the sample. Because the number of species detected in a sample usually changes much more in relation to sample size or sampling intensity than does the distribution of relative abundances, these ratios are extremely sensitive to sample size and thus very difficult to interpret.

A problem with the interpretation of any diversity statistic is that, even though richness and evenness are theoretically independent properties of a collection of objects, they tend to be highly correlated in samples taken from such collections. Under ideal conditons, which rarely obtain in ecological research, a sample is a random subset of the entire collection (e.g., of an ecological community). Because the most abundant species are most likely to be detected in a sample, the number and relative abundances of the most common species can usually be estimated fairly well from a relatively small sample. However, rare species are unlikely to be detected in a small sample, and may only be detected with intensive sampling, which can shift the 'veil line' of Preston's (1948) lognormal distribution of species abundances toward those species with very low abundance. It is inevitable that rare species will always tend to be undersampled. One consequence of the failure to detect rare species is that the evenness of species abundances will almost always be overestimated. Another consequence of this relationship is that diversity and evenness will tend to be highly correlated among most communities. Thus, most diversity indices are likely to produce the same ranking of diversity, except for a few extreme situations.

Since no single statistic can ever be an adequate description of the diversity of a collection, several statistics should always be provided to represent the collection more completely. Just as we now expect the standard deviation to be provided along with the mean, several diversity indices and their statistical variance should be presented when the diversity of a community is described. At a minimum, the number of species in a sample of a specific size or area and some measure of the evenness of the distribution of species should be provided. When innovative or non-standard statistics are used to describe the diversity of a community, it is important to also calculate the simple, standard measures of species number and evenness. One simple graphical approach to visualizing the differences in richness and evenness between different samples is through the use of 'dominance-diversity' curves (Whittaker, 1965, 1975; Fig. 2.18).

Regardless of the statistics that are chosen to describe diversity, it is critical that the sample be collected using a statistical design that will allow a reliable estimate of the properties of the community that are relevant to the diversity issue being studied (Magurran, 1988). It is primarily through comparative studies of the species diversity of different communities that an understanding of the mechanisms that regulate diversity can be gained, and such comparisons will be meaningless if diversity is not sampled appropriately.

What Diversity to Measure and Where to Measure it

The statistical assessment of diversity is a relatively simple problem compared to determining what to sample and how to sample it. How this problem is addressed determines what ecological insights can be gained from the study, or conversely, what important patterns or inferred processes will be obscured (Colwell, 1979). The solution to this problem can have a much larger effect on the numerical quantification of diversity than what particular statistics are used.

The more different types of organisms, from bacteria or mites to trees or mammals, that are included in a sample, the more mechanisms are likely to have influenced the diversity of that sample. The most extreme example of an uninterpretable sample is the total diversity of a community, summing all plants (or all animals) in a single number. Besides being logistically impossible to obtain for most communities, a sample that lumps all types of organisms into a single group ignores the critical biological differences between the organisms and therefore also ignores the different ecological processes that influence each type of organism. This type of diversity sample involves not only the comparison of apples and oranges, but also of bananas, grapes, plums, and passion fruit.

The organisms to be included in a diversity sample must be chosen carefully, if an understanding of the factors that influence the diversity of those organisms is to be found. The distinction made in Chapter 1 between functional types and functional analogues within a functional type provides some guidance about how diversity should be sampled. A sample of the number of species within a particular functional type is likely to reveal different patterns than a sample that focuses on the number of functional types. Interpretation of these differences is critical for developing and testing theories about the regulation of species diversity.

Subdivision of an ecological community into functional classes allows the partitioning of diversity into components that may be influenced by totally different processes. The division of certain groups of organisms into functional groups is a familiar approach in ecological studies (Root 1967, 1973; Terborgh and Robinson 1986). For example, studies of bird communities often categorize species as frugivores, granivores, carnivores, insectivores, etc., and sometimes even subdivide these groups, e.g., hovering insectivores, hawking insectivores, gleaning insectivores, etc. The same approach is often taken for mammals, insects (e.g., Cummins and Klug, 1979), fish (e.g., Lowe-McConnell, 1987), and even nematodes (Proctor, 1984).

Plants, all of which use the same basic resources in a very similar way, have been the subject of much of the research on species diversity. However, even among plants the differences between functional types or life forms are sufficient that the mechanisms that influence diversity may vary greatly among life forms. Furthermore, the different life forms (or functional types) are likely to influence each other in ways that would obscure important mechanisms if they were lumped together. Plant life forms, such as woody and herbaceous, trees and shrubs, geophytes and epiphytes, etc., should be considered separately in diversity studies, if there is to be any hope of understanding the diversity patterns.

In the discussion so far, the basic unit of diversity has been the individual species, hence the term species diversity. However, important insights into mechanisms that influence community organization can also be gained by looking not at species, but at groups of similar species lumped into functional categories such as those mentioned above. Trophic diversity, guild diversity, life form diversity, etc., are legitimate topics of research at larger scales and higher organizational levels than the individual species. The mechanisms that influence diversity at these larger scales are often different from the mechanisms that influence diversity within groups of organisms at smaller scales.

Selection of appropriate groups of organisms so that samples of diversity can be interpreted is a major issue to which we will return repeatedly. Once an appropriate subset of a community has been selected, the problem still remains of the spatial and temporal scales over which the samples should be collected. This problem is especially critical because of the comparative nature of most studies of species diversity.

Sampling Diversity in Relation to Potential Mechanisms

The spatial and temporal scales at which organisms in a community are sampled is important not only for an appropriate sample of the diversity of the organisms themselves, but for an appropriate sample of the physical and biological factors that influence species diversity. Such factors could include the availability of resources and the spatial and temporal variability of those resources, climatic conditions such as temperature and its variation, the spatial distances over which organisms interact and the time spans of their interactions. The regulation of species diversity among a group of organisms cannot be understood independently of the effect of environmental conditions on those organisms.

Recognition of the importance of sampling scale in the evaluation of diversity has led to the widespread use of 'species/area curves', which graphically illustrate how the number of species increases as the area sampled increases (Jacard, 1912; Williams, 1947, 1964; Hopkins, 1955). The usefulness of this method comes from the fact that not all species/area curves have the same shape: some rise rapidly, some slowly, some level off at a low number of species, some show no sign of leveling off. These different shapes must mean something, but the biological meaning of these curves is not always obvious (Connor and McCoy, 1979). Species/area curves can be used to determine the minimum sample size or area needed to characterize the species diversity of a particular community adequately. These curves can also be used to compare the rate of diversity increase with area, which may allow testing of carefully chosen hypotheses.

Within an area of some relatively small size, organisms may perceive their environment as essentially homogeneous, with no significant spatial variability that would affect their growth rate or use of the area. Of course, the size of the area of perceived homogeneity will depend on the size and perception of the organisms being considered, as well as on the inherent physical variability of the environment. Within such a homogeneous area (if any really exist) individuals of different species could be expected to be distributed randomly in space. The species/area curve should increase as a sample from a spatially random distribution, and level off at the total number of species within the homogeneous area. Unless the number of species is very high, the species/area curve should level off at a size well below the total size of the area.

However, once the area of the sample exceeds the size of the environmentally homogeneous area, the species/area curve should begin to rise again as species from other environments are sampled. For increasingly larger samples, the number of species will rise as the number of different environments (or habitats) included in the sample increases (Williams, 1964). This type of increase in the species/area curve clearly results from a different cause than the increase found within a homogeneous area.

If data for a species/area curve are obtained over a sufficiently wide range of scales, from very small areas to very large areas, more complex patterns than those described above generally appear. The number of species may level off and remain constant over a range of increasing areas, then begin to rise before leveling off once more (cf. Fig. 2.17). This change in slope suggests that there has been a transition in the mechanisms that influence diversity from one set of factors at small spatial scales to a different set of factors at larger scales. The most likely explanation for this transition has to do with the scale of environmental heterogeneity (Grieg-Smith, 1957).

Efforts to compare diversity between different regions using published data are often complicated by the fact that data were collected from areas of different sizes or from samples of different numbers of individuals. One approach to this problem is a statistical technique called rarefaction (Sanders, 1968; Hurlbert, 1971; Heck et al., 1975), in which a data set from a large area or large number of individuals is randomly subsampled many times to obtain an estimate of the number of species that would be expected in a smaller sample. This method requires that the large dataset include information on the total number of individuals and their relative abundance, and involves assumptions about the original spatial and numerical distribution of the organisms that may not apply.

Processes that Influence Diversity at Different Scales

Environmental homogeneity and heterogeneity are critical not only to appropriate sampling of the organisms themselves, but also to sampling the environment in a way that can contribute to understanding species diversity (Magurran, 1988). If organisms are influenced by the environmental conditions that distinguish different homogeneous patches, then measurement of the physical environment must be scaled to those patches. An average of the environmental conditions of many different patches is unlikely to give much insight into how environmental conditions influence diversity.

The relationship between sampling scale and the processes that influence species diversity is the basis of the distinction between 'within-habitat' diversity and 'between-habitat' diversity (Whittaker, 1972, 1975). Within-habitat diversity (called 'alpha diversity' by Whittaker, 1960, 1967; Fisher et al., 1943) reflects coexistence among organisms that are interacting with one another by competing for the same resources or otherwise using the same environment. Within-habitat (alpha) diversity is measured simply as the number of species (or other components of species diversity) within an area of given size.

Between-habitat diversity (called 'beta diversity' by Whittaker 1960,

1976) reflects the way in which organisms respond to environmental heterogeneity. Between-habitat (beta) diversity is somewhat more complex to quantify, and depends not only on the number of species in the habitat, but on a comparison of the identity of those species and where they occur. Beta diversity is usually expressed in terms of a similarity index between communities or of a species turnover rate (Whittaker, 1960; Wilson and Mohler, 1983; Cody, 1986) between different habitats in the same geographical area (e.g., along some sort of gradient). High beta diversity is the result of low similarity between the species composition of different habitats or different locations along a gradient.

A third type of diversity has been defined that applies to even larger-scale phenomena, which reflect primarily evolutionary rather than ecological processes (Whittaker, 1960, 1972; Cody, 1986). Whittaker (1960) defined geographical or 'gamma' diversity as simply the number of species within a region, analgous to alpha diversity but at a regional scale. However, Cody (1986) used gamma diversity in a different sense, as a regional-scale analog of beta diversity. Like beta (between-habitat) diversity, geographical diversity (*sensu* Cody, 1986) is based on the differences in the species composition between habitats. However, Cody's gamma diversity is based on differences in species composition between similar habitats in different geographical areas (e.g., species turnover with distance separating similar habitats, Cody, 1986), rather than between dissimilar habitats in the same geographical area. The fact that ecologically similar but taxonomically different species are performing the same role in similar communities separated by a given distance implies that evolutionary processes involved in creating and maintaining separate species are operating effectively at that scale. Thus, geographical diversity must be expressed both in terms of the distance between similar habitats and the taxonomic differences between groups of ecologically similar species.

An approach to analyzing spatial patterns of diversity on landscapes that incorporates the concepts of alpha, beta, and gamma diversity is a statistical analysis of a data matrix of sites (columns) and species (rows) which has been called affinity analysis (Scheiner and Istock, 1987; Istock and Scheiner, 1987; Scheiner, 1990). This method compares the actual patterns of similarity among samples from a landscape with an expected similarity based on computer simulations that randomly sample the original data (i.e., bootstrapping, Efron, 1981). The results of this analysis have been characterized as 'mosaic diversity' and a 'multi-gradient complement to beta diversity' (Scheiner, 1990).

Another aspect of environmental heterogeneity that must be considered in sampling both organisms and environmental conditions is temporal heterogeneity (discussed in the previous chapter). Depending on the diversity questions being addressed, both the duration and timing of sampling organisms and physical conditions must be considered carefully.

Summary

As with any scientific measurement or experimental design, the critical issue in the assessment of species diversity is the question that is being asked. In general, the sampling design for the collection of diversity data is far more important than which of the many possible diversity statistics are ultimately used to describe and analyze the results. The most important considerations are what types of organisms are being sampled and at what spatial scale. These factors together influence which of many possible mechanisms could affect the diversity of the sample. Groups of organisms within a community respond differently to environmental conditions and other factors that influence diversity, so aggregate samples of the total diversity of a community are likely to be influenced by so many processes, some of which have opposing effects, that total diversity is impossible to interpret. If the goal of measuring species diversity is to understand what processes are influencing diversity, it is critical to restrict samples to groups of organisms that are likely to be influenced by the same processes.

Part two
Theories of Species Diversity

Theories based on equilibrium play a prominent role in most fields of science, with thermodynamic equilbrium being a particularly important and well-established example. The concept of a 'balance of nature' has a long history in the scientific literature, as well as in the popular press, and most explanations for patterns of biological diversity are based on some sort of equilibrium. Equilibrium, defined as a balance between opposing forces, is a general term that can be applied to situations where the opposing forces occur at scales ranging from angstroms and picoseconds in the case of chemical equilibria, to continents and geologic epochs in the case of the opposing geological forces of uplift and erosion. Equilibria that most directly affect biological diversity occur at spatial and temporal scales intermediate between these extremes. Nonetheless, ecologically important equilibria do occur over a sufficiently wide range of scales that care must be taken to distinguish between them and not to lump all equilibria under one heading.

One of the major ecological 'insights' of the 1970s was the importance of *non-equilibrium* processes in the maintenance of species diversity (e.g., Caswell, 1978; Connell, 1978; Huston, 1979). Since then, the notion of 'equilibrium' has fallen into disfavor in ecology. This chapter and the next address the roles of equilibrium and non-equilibrium processes in the regulation of species diversity. This is the general field in which most of the theoretical battles about the regulation of species diversity have been fought. I will argue that most of the disagreement about the validity of various 'diversity hypotheses' has resulted from a failure to clearly define the terms equilibrium and non-equilibrium, and more significantly, from a failure to consider the spatial and temporal scales at which specific processes are important, regardless of whether they are in equilibrium or not.

Equilibrium and non-equilibrium dynamics are not mutually exclusive. Any system that can come to an equilibrium will also exhibit some sort of behavior when it is far from its equilibrium state. The return to equilibrium of a system that has been perturbed from its equilibrium state may be either rapid or slow. If the return to equilibrium is rapid in relation to the frequency of perturbations, the system will spend most of its time at or near equilibrium. However, if the return to equilibrium is slow in relation to the frequency of the perturbations, that system will spend most of the time in a non-equilibrium state. Non-equilibrium dynamics are simply the behavior of a system far from equilibrium. Whether equilibrium is the most common state of a system, or whether it occurs rarely, will determine both the variability and the mean state of that system.

The controversy that has led to the equilibrium/non-equilibrium debate is based on a very specific type of equilibrium that can be defined mathematically for certain systems of nonlinear differential equations, including the Lotka-Volterra competition and predation equations. In the context of competition models, a competitive equilibrium can be represented mathematically in terms of differences in competitive ability among two or more species. The process of competition has been a dominant topic throughout the history of ecology (e.g., Darwin, 1859; Clements *et al.*, 1929; articles in Grace and Tilman, 1990; Connell, 1983; Schoener, 1983), and there is little doubt that, when it occurs, competition can have a dramatic effect on the coexistence of species and thus, on species diversity. I believe that a critical key to understanding patterns of species diversity is knowing when competition is likely to have a major effect on species diversity and when it is likely to be unimportant. The general conclusion of the next two chapters is that competition has a direct effect on species diversity only under a very restricted set of conditions. Under most conditions, when the direct effects of competition are relatively weak, a host of other factors, including evolutionary, geological, and human history, environmental variability, disturbances, and random population fluctuations, may assume a dominant role in regulating species diversity.

The relative importance or unimportance of competition relate directly to the two major components of species diversity discussed in Chapter 1. Because *functional types* are defined as groups of taxa (*functional analogues*) that use the same resources and respond to their environment in the same way, competition would be expected to be much more intense between organisms of the same functional type than between organisms of different functional types (e.g., between different guilds of birds, Root, 1967). Thus, explanations of species diversity based on competitive interactions would be expected to apply to the number of functionally analogous species within a functional type, rather than to the number of functional types. Explanations for the number of functional types might include the *evolutionary* effects of competition, to the extent that character or niche displacement between species is the result of competitive interactions (e.g., Darwin, 1859; Grant, 1972, 1975; Connell, 1980; Abrams, 1986, 1987a; Weins, 1989).

Chapter 4 focuses on the conditions under which a variety of processes involved in both deterministic and stochastic equilibria may be important, and on how equilibria at large and small scales interact to regulate both the functional type and the functional analogue components of species diversity. Chapter 5 focuses primarily on the non-equilibrium dynamics of processes that occur at the spatial and temporal scales at which individuals interact, that is, the scale of competitive equilibrium and the coexistence of species of the same functional type.

4

Equilibrium Processes and the Maintenance of Landscape-scale Species Diversity

Ecological equilibrium does not imply that natural systems are constant and unchanging. For example the 'theory of island biogeography' (MacArthur and Wilson, 1963, 1967) argues that species richness on islands results from a balance (or equilibrium) between the rate of extinction on that island and the rate of immigration of species to that island. Because the extinction of one species and the immigration of the same or another species are unlikely to occur simultaneously, the actual number of species on an island will fluctuate from one year or one decade to the next. Considerable evidence exists to suggest that the number of species fluctuates about a mean that may represent a balance between the immigration and extinction rates (Simberloff and Wilson, 1969; Simberloff, 1976; Gilbert, 1980; Williamson, 1981). Is this an equilibrium process? I would argue yes, while noting that many of the interesting phenomena associated with island biogeography, such as the accumulation of species on defaunated islands or the adjustment of species number on islands that have been reduced in size (e.g., Simberloff, 1976), represent the dynamics of an island community far from its equilibrium level of species richness. However, many of the factors important in regulating species diversity on islands (or any areas) are independent of any balance between immigration and extinction (Gilbert, 1980).

The theory of island biogeography is based on a stochastic dynamic equilibrium that occurs over long time scales and large spatial scales. At even larger scales one can consider the equilibrium (or non-equilibrium) between rates of speciation and rates of extinction (e.g., Flessa and Thomas, 1985; Kitchell and Carr, 1985). Application of equilibrium theories, such as the theory of island biogeography, to large spatial and temporal scales can be justified, even if it is only an approximation. On the other hand, application of deterministic equilibrium, as in competition models, to small spatial scales and short time scales has sometimes led to misunderstandings in ecology.

The predominant equilibrium of ecological theory is competitive equilibrium, which is conceptualized as a condition in which the population sizes of two or more competing species remain relatively constant as a result of a deterministic balance between the competitive abilities of the species. For competitive equilibrium, the opposing forces are the negative effects that each species has on the other (i.e., their competitive abilities). If the competitive ability of one species is substantially greater than that of the others, that species will eliminate the others and the number of species will be reduced to one by competitive exclusion.

Competitive equilibrium must occur at the spatial and temporal scales over which individual organisms and populations interact. Depending on the size and growth rates of the organisms under consideration, these scales may vary greatly. The dominant influence of competitive equilibrium in recent ecological theory is a consequence of the fact that the concept can be expressed mathematically in a form that is readily analyzed using the standard mathematical tools of algebra, calculus, and linear algebra. This led to a veritable explosion of equilibrium analyses and stability analyses of deterministic mathematical models during the 60s and 70s, to the point that much of ecological theory was based on the assumption that competitive equilibrium was the common state of natural communities.

This chapter deals first with the issue of competitive equilibrium and its failure to provide a satisfactory explanation for patterns of species diversity. The rest of the chapter considers processes that may reach equilibrium at larger spatial and longer time scales, as well as how the biology of different groups of organisms influences the scales at which different types of equilibria may occur.

Competitive Exclusion and The Theory of Competitive Equilibrium

One of the first axioms of modern ecology was the 'principle of competitive exclusion', which stated that if two species were competing for exactly the same limiting resources they could not coexist, that is, one of them would be competitively excluded by the other and would become locally extinct (Lotka, 1925; Volterra, 1926; Gause, 1934, 1935; Lack, 1944; Hardin, 1960). This principle was based on what was then, and still is, the fundamental mathematical model of ecological theory, the Lotka-Volterra competition equations (Lotka, 1925; Volterra, 1926). However, the basic concept was recognized long before the mathematical models were developed. 'Two species of approximately the same food habits are not likely to remain long evenly balanced in numbers in the same region. One will crowd out the other'. (Grinnell, 1904; see also Lawes *et al.*, 1882).

The principle was convincingly supported by many laboratory experiments that demonstrated that if populations of two similar species were competing in the same environment, one or the other would eventually die out (reviewed by Miller, 1969). These findings were substantiated in experiments with single-celled protozoa (*Paramecium*, Gause, 1934, 1935; Vandermeer, 1969), flour beetles (*Tribolium*, Park, 1948, 1954, 1962; Park *et al.*, 1961; Mertz, 1972), fruit flies (*Drosophila*, L'Hertier and Tessier, 1935; Merrell, 1951; Moore, 1952a and b), and *Daphnia* (Frank, 1952, 1957; Frank *et al.*, 1957).

The competitive exclusion principle was corroborated by careful field observations that confirmed that when similar species were apparently coexisting on the same resources, there were always differences in the food they were actually eating (Lack, 1944, 1945). These differences in resource use were often reflected in subtle morphological differences related to resource acquisition (e.g., bee tongues, Brian, 1957; bird beaks, Darwin, 1859; Lack, 1947). Even within a single species, the variability of morphology and behavior is often less when competitors are present, and shifted away from the traits of the competitor. Such patterns have been reported for ants (Davidson, 1978), snails (Fenchel, 1975; Fenchel and Kofocd, 1976), birds (Lack, 1947; Schluter, 1988), and lizards (Pianka, 1973, 1983). This phenomenon of specialization in resource use, called character displacement, is expected to be more pronounced where many species are competing for the same resources than where only one or a few species are competing (for criticisms of this interpretation, see Strong et al., 1979; Connell, 1980; Strong and Simberloff, 1981).

The implications of the competitive exclusion principle for explaining species diversity were obvious and severely constraining. High diversity depended on the coexistence of species and, if competing species could not coexist, then the explanation of species diversity had to be found in mechanisms to avoid competition. Throughout the 1960s and 1970s, much of the effort of theoretical, laboratory, and field ecologists was devoted to understanding how species avoid competition. A closely related question was how similar species could be to each other, and still be able to coexist. That is, what is the 'limiting similarity' of species (MacArthur and Levins, 1967; Bowers and Brown, 1982; Abrams, 1983)?

A critical, and often overlooked, qualifier of the competitive exclusion principle is that the extinction of one of the competing populations does not occur immediatedly, but often very slowly. In most of the early laboratory experiments, the 'losing' population decreased gradually, and extinction occurred after 30 to 70 generations (up to five years) of competition. After one population became extinct, the other population tended to stabilize at a relatively constant number, which was considered to represent the equilibrium condition for that particular experimental system. Because competitive exclusion often occurred so slowly, it was difficult to determine if equilibrium coexistence was actually occurring, or whether the apparent coexistence was simply the very gradual extinction of one of the populations.

The experimental results that showed either extinction or apparent co-existence at equilibrium were consistent with the deterministic behavior of the Lotka-Volterra competition equations analyzed at a mathematically-defined equilibrium. The concept of equilibrium was also consistent with a philosophical inclination among ecologists to believe that the continued existence of natural communities reflected some sort of 'balance of nature', stability, or equilibrium (cf. McIntosh, 1985 pg. 71-76, 178-192; Forbes, 1880a,b, 1883, 1887).

The concept of competitive equilibrium provided a framework for theories that could be applied to a wide range of ecological phenomena. Understanding the conditions that resulted in coexistence or extinction at equilibrium, the different types of equilibria, and other equilibrium properties of mathematical equations led to the development of an entire system of ecological theory based on competitive equilibrium.

Variants of the logistic equation, modified to represent competition or predation, produced a rich variety of mathematical patterns, many of which resembled natural patterns that ecologists had long been trying to explain, such as extinction or cycles of population size. The temptation was irresistable to believe that since the models seemed to behave like real populations, they must be a realistic representation of how the populations actually function and interact. The analysis of these models at or near their mathematical equilibrium points usually ignored the behavior of the theoretical populations as they approached equilibrium, as well as the length of time that it took them to reach equilibrium. Both of these factors later proved to be important components of non-equilibrium theory.

Theories of competitive equilibrium developed in two somewhat different directions. The first was use of equilibrium models to understand how species were able to coexist, and in particular, what 'balance' of properties competing species had to have in order to avoid competitive exclusion. This led to both theoretical and experimental efforts to define concepts such as the 'niche' (Hutchinson, 1957, 1965) and 'limiting similarity' (MacArthur and Levins, 1967; Rothstein, 1973; Abrams, 1983). This work is based on the well-known mathematical criteria for stable coexistence at equilibrium (e.g., $K_i < K_j/\alpha_{i,j}$, where K_i is the carrying capacity or maximum number of individuals of species i or j that the environment can support, and $\alpha_{i,j}$ is the competitive coefficient, or the effect of an individual of species j on an individual of species i expressed in units of the effect of an individual of species i on itself) that are discussed and presented as graphs in virtually every ecology textbook. The mathematical criteria for coexistence can be summarized simply as 'the negative effect of each species on itself (i.e., intraspecific competition) must be in some way greater than the negative effect of each species on the other species (i.e., interspecific competition)'.

Although equilibrium competition theory based on variations of the Lotka-Volterra equations has been strongly criticized since its introduction and has recently fallen into disfavor (see reviews in McIntosh, 1985; DeAngelis and Waterhouse, 1987), it seems to persist because of its elegance, simplicity, and heuristic value, rather than because of its mechanistic validity (Wangersky, 1978). Equilibrium theory based on Lotka-Volterra competition and predation models has recently been revived as 'a graphical-mechanistic approach to competition and predation' (Tilman, 1980, 1982; see also Abrams, 1987a,b), which adds explicit consideration of different resources, but is based on the same types of assumptions as the Lotka-Volterra models and makes predictions about the conditions for coexistence that are virtually identical to the predictions of traditional Lotka-Volterra equilibrium theory (Tilman, 1982, pg. 190ff).

The second general direction of equilibrium theory was to investigate the dynamic behavior of both predator-prey and competition equations in the vicinity of equilibrium, since it quickly became evident that there were several different kinds of equilibrium, both stable and unstable. The behavior of the equations in the vicinity of equilibrium points became the subject of extensive mathematical investigations (Rosenzweig and MacArthur, 1963; Gilpin, 1972, 1975; see Pielou, 1977 for a thorough discussion), ultimately leading to the discovery of 'chaos' in simple, deterministic models (Yorke and Li, 1975; May and Oster, 1976; Gilpin, 1979; Schaffer, 1984; Schaffer and Kot, 1985; see review in Gleick, 1987).

Stability of the mathematical models near their equilibrium points (or 'local' stability), rather than equilibrium *per se*, became a central issue in theoretical ecology in the 1970s. In practical terms, stability meant the continued coexistence of interacting populations, whether competitors or predators and prey. Instability generally led to extinction, of either some or all of the populations. As equilibrium analysis was applied to increasingly complex systems of equations that were meant to represent ecosystems, the startling conclusion was reached that large, strongly interconnected systems of interacting species were *less* stable than smaller, more weakly interconnected systems (May, 1972, 1973). This dramatic result directly contradicted the ecological common wisdom (based on observation and tradition) that large complex ecosystems were more stable than smaller simpler systems. This glaring inconsistency between theoretical prediction and the patterns thought to occur in nature highlighted a fundamental problem: the assumptions used in equilibrium analysis (e.g., linear responses, randomly assembled food webs) were unlikely to be true in most ecosystems (see discussion in May, 1973; Austin and Cook, 1974; DeAngelis, 1975; Goh, 1979; DeAngelis and Waterhouse, 1987).

The Failure of Equilibrium Competition Theory

The fundamental incompatibility between the large numbers of species that coexist in many natural communities and the stringent requirements for coexistence imposed by equilibrium competition theory were highlighted by G.E. Hutchinson in his famous paper 'Homage to Santa Rosalia' (1959). In another paper Hutchinson (1961) proposed what is now widely accepted as the explanation not only of 'the paradox of the [high] plankton [diversity]', but also of the high diversity found in many ecological communities. He suggested that species coexisted in high diversity situations because competitive equilibrium was prevented by environmental variations and other factors, that is, that coexistence was a non-equilibrium rather than an equilibrium phenomenon.

The observations that were most damaging to equilibrium competition theory were those cases of high species diversity in environments where neither spatial and temporal heterogeneity nor specialization and narrow niches could be invoked to explain coexistence. Examples include the remarkably diverse gammarids and other taxa of Lake Baikal in Russia (Zhadin and Gerd, 1961; Kozhov, 1963). Gammarids are small amphipod crustaceans that live primarily as filter feeders in the sediment, where sufficient niche specialization or spatial heterogeneity to allow the equilibrium coexistence of the over 250 species seems unlikely. Nonetheless, twenty or more species can be represented in a single sample from the benthic zone.

Another example is the diverse cichlid fish fauna of Lake Malawi and the other great rift lakes of tropical east Africa (Fryer, 1959; Fryer and Iles, 1972; Lowe-McConnell, 1987). Over 200 species of cichlids coexist in lake Malawi (11,430 square miles) which is long and narrow with apparently low habitat diversity (Fryer and Iles, 1972). Even the the most closely related cichlid species show little ecological divergence and are very similar in their food preferences in spite of specialized feeding morphologies (Fryer, 1959; Ribbink *et al.*, 1983). This seems to contradict the theoretical predictions for limiting similarity under equilibrium conditions, although there is evidence that some of these very similar species differ subtly in the depths at which they feed.

The most fundamental failure of competitive equilibrium theory has been its inability to explain the high species diversity of terrestrial plant communities. Much of the tropical increase in the diversity of birds, mammals, insects, and other organisms may be explained by the increased structural complexity and tree species diversity of tropical forests (Richards, 1952). However, the high diversity of tropical trees is apparently not explained by soil heterogeneity, which shows little relation to tree distribution in tropical forests (Hewetson, 1956; Kwan and Whitmore, 1970; Webb *et al.*, 1972). Nearly all plants are rooted in the soil, and require the same resources of light, water, CO_2, and mineral nutrients in similar proportions. The opportunities for specialization in resource use seem limited, which is apparently reflected in the morphological, physiological and genetic similarity of coexisting plants that are using the same resources (Richards, 1952; Gentry, 1986; Ashton, 1988).

'The extraordinary diversity of the terrestrial fauna, which is much greater than that of the marine fauna, is clearly due largely to the diversity provided by terrestrial plants... On the whole the problem still remains, but in the new form: why are there so many kinds of plants'. (Hutchinson, 1959)

Competitive Exclusion and Species Diversity

Competitive exclusion is a critical concept for both competitive equilibrium and non-equilibrium theories of species diversity. In equilibrium theory, competitive exclusion is one of several possible outcomes when competition is allowed to proceed to equilibrium. It is always a mathematically stable solution, in contrast to coexistence, which can be either stable or unstable (see discussion in Ricklefs, 1973; Pielou, 1975, Vandermeer, 1969). Non(competitive)equilibrium theories of species diversity generally assume that competitive exclusion occurs when competitive interactions are allowed to go to equilibrium. Thus, non-equilibrium theories usually address processes that prevent competitive equilibrium from occurring.

Efforts to explain species diversity in the context of competitive exclusion fall into two general classes. One is based on the idea that competitive exclusion does not occur instantaneously, and many factors can slow the approach to equilibrium or alter the relative strength of competitive interactions. This was essentially the argument that Hutchinson advanced in 1948, and was consistent with the second main result of the competition experiments that established the occurrence of competitive exclusion, that is, that competitive exclusion often occurred very slowly (Miller, 1969). Processes that slow or prevent the approach to competitive equilibrium include certain types of environmental fluctuations and the reduction of population sizes by predation or herbivory. These processes are considered under the heading of non(competitive)equilibrium mechanisms in the next chapter.

The other general explanation for high species diversity was the acknowledgment that competitive equilibrium and exclusion did occur, but there was sufficient patchiness in the environment that a species that became extinct on one patch would survive on another patch and thus allow high species diversity to be maintained in a stochastic equilibrium of extinction and re-immigration based on heterogeneity at large scales. Heterogeneity of environmental conditions, as a result of both spatial and temporal variation, proved a productive field of investigations for both theoretical and empirical studies (Huffaker, 1958; Huffaker et al., 1963; Levin, 1970; Levins and Culver, 1971; Levin and Paine, 1974, 1975; Paine and Levin, 1981; see review in Levin, 1976).

Environmental Heterogeneity and the Diversity of Competitors

Spatial heterogeneity, generally defined as either physically identical patches at different stages of the same series of temporal dynamics or as patches with different environmental conditions, introduces the possibility of an equilibrium at a scale completely different from that of competitive equilibrium. This type of large-scale equilibrium results from

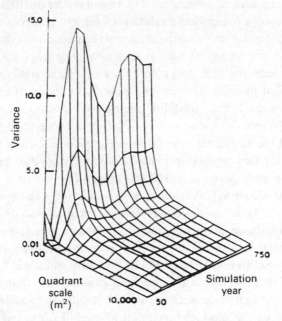

Fig. 4.1 The effect of spatial scale on temporal variability of forest structure. Results are from an individual-based forest simulation model, with spatial scale ranging from a single 10 × 10 m plot to 100 plots aggregated together. (From Smith and Urban, 1988.)

having a sufficient number of patches so that changes in one patch will be balanced by opposing changes (e.g., a population decreasing rather than increasing) in other patches, so that there is no net change over a large area. 'The forest as a whole remains the same, the changes in various parts balancing each other' (Cooper, 1913). Equilibrium at this scale is a stochastic equilibrium, analogous to the law of large numbers, and is totally different from the deterministic mathematical equilibrium of equilibrium competition theory. The processes that influence species diversity within patches (where processes such as those modeled by the Lotka-Volterra equations occur) are likely to be totally different from the processes that influence diversity at large scales (DeAngelis and Waterhouse, 1987).

For example, Smith and Urban (1988) used computer simulations to demonstrate the large-scale equilibrium of forest structure on a land-scape where the dynamics of treefall gap formation and forest succession produced extreme small-scale temporal and spatial variability (i.e., non-equilibrium conditions), as measured by tree diameter distributions. Smith and Urban used a computer simulation of the growth, competition, and death (and potential treefall gap formation) of many individual trees and 'sampled' the simulation at spatial scales ranging from 100 m^2, which was smaller than the area occupied by a single large tree, to 10,000 m^2, which included many large trees and potentially many different treefall gaps. At each scale, the variability among the replicate simulations was calculated. When the forest was sampled at a large scale (e.g., 10,000 m^2 or 1 ha) the stochastic asynchrony of gap formation eliminated the high variability that resulted from the non-equilibrium dynamics of forest succession at small scales (e.g., 100 m^2) (Fig. 4.1)

Smith and Urban's (1988) computer simulation experiment demonstrated that at large spatial scales forest structure was determined by the regional balance of gap formation, dispersal, and forest succession (the same balance also affects species diversity, although this was not addressed by Smith and Urban). Forest structure stabilized following the initiation of succession on bare plots and showed little change over time at large scales. This large-scale structural stability represented an equilibrium between the random formation of treefall gaps, and the process of forest succession, exactly as described by Cooper (1913). However, at small scales, forest structure (and also species diversity) fluctuated dramatically as a consequence of changing species composition and tree size distribution during forest succession, which are the result of the non-equilibrium dynamics of competition. These relatively slow changes resulting from succession were interrupted by rapid changes caused by treefalls, with the result that the structure and diversity of small areas (i.e., local diversity) fluctuated dramatically. Thus, community structure in these simulations was regulated by totally different processes at large versus small scales.

At landscape scales spatial heterogeneity, or patchiness, in the environment offers the possibility of regional coexistence in spite of local extinction. The issue of scale is critical to this explanation of diversity, since coexistence and diversity are being evaluated at a spatial scale (size of area) that is larger than the scale at which local extinction occurs. The applicability of asynchronous patch dynamics as an explanation of landscape scale species diversity depends on measuring diversity at a spatial scale that includes a sufficient number of patches. The area needed to incorporate enough patches (the more, the better - Crowley, 1977, 1981) will depend on the size and mobility of the organisms being considered, so samples at many different spatial scales would be needed if many different sizes and kinds of organisms were being considered simultaneously.

This mechanism is applicable to situations where extinction results from any cause, including competitive exclusion (Levins, 1969; Vandermeer, 1973; Levin, 1974; Slatkin, 1974; Abrams, 1988), predation (Hilborn, 1975; Caswell, 1978; Hogeweg and Hesper, 1981; Ziegler, 1977, 1978), or from population fluctuations caused by disturbances or stochastic variation (den Boer, 1968, 1981; Roff, 1974a,b; Reddingius and den Boer, 1970; Gurney and Nisbet, 1978; Crowley, 1977, 1981; Vance, 1984). Most mathematical models that have been developed to investigate this mechanism have also included dispersal or migration (treated as donor-dependent loss rates or diffusion), so that a species can repopulate a patch where it has become extinct by immigrating from another patch where it has survived (see review by Taylor, 1990). This phenomenon was recognized by field ecologists early in this century (e.g., Cooper, 1913; Watt, 1947), and has re-emerged more recently as a major component of both ecological theory and applied ecology (Levin, 1974, 1976; Murdoch *et al.*, 1985). The reinvasion of species that have become locally extinct has been called the 'rescue effect' (Brown and Kodric-Brown, 1977).

Temporal fluctuations are another form of environmental heterogeneity that can prevent either local or landscape-scale extinction. This mechanism can operate in a spatially homogeneous environment (Levin, 1970; Grenney *et al.*, 1973; Koch, 1974a; Armstrong and McGehee, 1976a; Sommer, 1984) or a spatially heterogeneous one (Powell and Richerson, 1985), where asynchronous temporal variation between patches can increase or even create spatial heterogeneity (Levin and Paine, 1974, 1975; Kemp and Mitsch, 1979).

The minimum differences between patches required to maintain landscape scale species diversity in spite of local extinction depends on the cause of extinction. In the case of extinction that results from predation, all patches can be considered physically identical, differing only in the presence or absence of predators and prey (e.g., Huffaker's oranges, Huffaker, 1958). Physically identical patches can also maintain diversity in cases of competitive exclusion if extinction occurs asynchronously on different patches patches and there is dispersal from occupied to unoccupied patches. An alternative mechanism occurs when the patches differ sufficiently in their physical conditions that a species that is an inferior competitor on one patch type will be a superior competitor on another patch type, where it will survive and disperse to other patches where it has become extinct (e.g., Horn and MacArthur, 1972).

Environmental heterogeneity is an important explanation for species diversity regardless of whether equilibrium or non-equilibrium mechanisms are considered to be most important. Many theoretical models of the effect of spatial heterogeneity make no assumptions about whether or not equilibrium is ever attained (e.g., Hilborn, 1975; Caswell, 1978; Hogeweg and Hesper, 1981).

Environmental Heterogeneity and the Diversity of Functional Types

In addition to facilitating the landscape scale coexistence of potentially competing species (i.e., functionally analogous species within a functional type), environmental heterogeneity has a major effect on the number of functional types that occur within a local area. Because different functional types use different resources, use the same resources in different ways, or somehow use the environment in a way that allows them to avoid competition with other functional types, greater structural heterogeneity and more types of resources in an environment will increase the number of functional types that can be present. Resource heterogeneity increases total species diversity at both the local and the landscape scales, although its primary effect is on the number of functional types, rather than the diversity within a functional type.

Spatial heterogeneity in resources may provide an alternative explanation for one of the most consistent phenomena in ecology, the lognormal distribution of species abundance in multi-species samples (Preston, 1948, 1962; Sugihara, 1980). The lognormal distribution is ubiquitous in nature and can be produced by a wide variety of processes. The lognormal distribution of species abundances is generally interpreted in terms of niche partitioning, that is, the way in which species divide up resources in an ecological (i.e., competition) or evolutionary (i.e., character displacement and speciation) context (May, 1975; Sugihara, 1980). However, if species in a sample are not actually competing, but rather represent different functional types using the environment in different ways, the lognormal distribution of species abundances may actually result from the properties of the environment itself, rather than from the properties of the species.

When the environment is viewed as a heterogeneous mosaic of resources, which may be patches of different sizes and successional ages, or plants of different types and sizes, the distribution of types and amounts of resources available to organisms is likely to be lognormal. Lognormal size distributions can be produced by random processes, such as grinding rocks or the precipitation of droplets (Randolph and Larson, 1971) and the battering of the intertidal zone by waves and logs that create open patches (Paine and Levin, 1981). Random distributions (many of which are lognormal) can also be produced by the superimposition of several predictable, deterministic processes (Cox and Smith, 1953, 1954; Getty, 1981). Finally, lognormal distributions can be produced by the interaction of deterministic processes, such as size-dependent (i.e., exponential) growth, with stochastic processes, such as food capture by fish (DeAngelis and Huston, 1987). Thus the lognormal distribution of resources in the environment, rather than the ecological subdivision of resources by organisms, may be the underlying cause of the ubiquitous lognormal distribution of species abundances.

One important type of environmental heterogeneity is that created by the organisms themselves. The structural and resource heterogeneity created by functional types that are 'structural species' increases the potential for co-occurrence of functional types that are 'interstitial species.' Physical heterogeneity (e.g., geology, soils, topography) also increases the number of functional types that can occur, as well as facilitating the landscape scale coexistence of functional analogues whose relative competitive ability is altered by environmental conditions.

Some of the best known work on the effect of habitat heterogeneity on interspecific competition was that of Robert MacArthur on the foraging behavior of warblers in a coniferous forest (MacArthur, 1958). This study, which showed different species foraging in different parts of the trees, was interpreted in the context of equilibrium competition theory as niche specialization that reduced the intensity of interspecific competition below that of intraspecific competition and allowed the species to coexist at equilibrium (MacArthur, 1958, 1964, 1965; MacArthur and MacArthur, 1961; MacArthur *et al.*, 1966). Studies of bird species diversity in different vegetation types and on different continents have consistently found a strong relationship between bird diversity and various components of the structural complexity and tree species diversity of the habitats (e.g., Johnston and Odum, 1956; Cody, 1968; Karr, 1968; Recher, 1969; Karr and Roth, 1971; Johnson, 1975), although the methods and results of MacArthur's work have been criticized (e.g., Tomoff, 1974; Willson, 1974; James and Wamer, 1982). It is particularly interesting that Willson (1974) found that the increase in bird species diversity with increasing foliage height diversity (also strongly correlated with total vegetation volume) resulted primarily from the addition of more foraging guilds (i.e., functional types) rather than from additional species within a guild.

The successional pattern that has been observed in bird communities is characterized by a temporal shift in species composition and an increase in species diversity following the abandonment of agricultural land or after some other major disturbance. These temporal changes in the bird community can be mostly explained by the increase in the structural complexity of the vegetation as the plant community undergoes succession from the simple grass and herb stage, through the shrub stage, to the high structural complexity of the forest stage of plant succession (Johnston and Odum, 1956). Thus, succession in bird communities occurs primarily in response to changes in the structure and diversity of the vegetation, and is not an analog of plant succession, in which temporal change is the consequence of interspecific interactions involving dispersal, growth, and competition (see Chapter 8).

Habitat structural complexity, provided mainly by plants, is strongly correlated with species diversity in other terrestrial vertebrates including small mammals (Rozenzweig and Winakur, 1966; Kotler and Brown, 1988) and lizards (Pianka, 1967). In benthic environments, structural complexity associated with the physical substrate or such organisms as coral, sponges, bivalve molluscs, and algae, can influence the species diversity of organisms such as fish (Gladfelter and Gladfelter, 1978; Tonn and Magnuson, 1982), snails (Kohn, 1967, 1968), and a variety of other invertebrates (Hewatt, 1935) and microbes (Pringle, 1990).

Even in purely aqueous environments, spatial heterogeneity in water temperature and chemistry that is difficult to detect can provide sufficient complexity to allow the coexistence of many different species that would not coexist in a homogeneous environment (Whiteside and Harmsworth, 1967; Richerson *et al.*, 1970; Lehman, 1980; Fenchel, 1988).

Much of the increase in species richness with increasing area sampled is a consequence of the increased number of different habitats added to the sample. On islands in the Aegean Ocean, bird species diversity was more closely correlated with the number of different habitats present on an island, than with the total area of the island (Watson, 1964). In a series of lakes in Wisconsin, fish species richness was positively correlated both with lake area and with the structural diversity of the aquatic vegetation (Tonn and Magnuson, 1982).

Patterns of insect diversity illustrate the structural role of plants in creating the environmental heterogeneity that allows the persistence of high diversity among the interstitial organisms that depend directly or indirectly on plants. In Great Britain the increase in the species richness of phytophagous insects with increasing sample area is greatest for insects

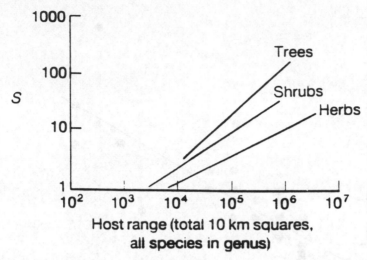

Fig. 4.2 Effect of plant structural complexity on the species/area relationship of phytophagous insects in Great Britain. Host range is based on the number of 10×10 km map grid cells in which a particular plant genus is found. (Based on Strong and Levin, 1979.)

that are found on trees, intermediate for insects found on shrubs, and lowest for insects of herbaceous plants (Strong and Levin, 1979; Fig. 4.2). This reflects both the increased structural complexity of trees in comparison to small plants at the scale of the individual insect, as well as a higher effective heterogeneity at larger spatial scales approaching the size of Great Britain. Even on plants as apparently simple as prickly pear cacti (*Opuntia*), the number of phytophagous insect species found on these plants throughout North and South America is positively correlated with the structural complexity ('architectural rating') of the plants (Moran, 1980; Fig. 4.3). Murdoch *et al.*, (1972) found that, together, plant foliage-height diversity and species diversity accounted for 79% of homopteran (plant-sucking bug) diversity in three old fields in southeastern Michigan.

The empirical examples discussed above illustrate that there are many different scales of environmental heterogeneity. The scale of the heterogeneity determines how it relates to equilibrium competition theory. In MacArthur's warbler example, the heterogeneity was interpreted as allowing sufficient niche differentiation between competing species that they could coexist stably. This could be expressed directly in terms of the Lotka-Volterra competition model as the effect on the competition parameters ($\alpha_{i,j}$) of the way in which the birds used their environment.

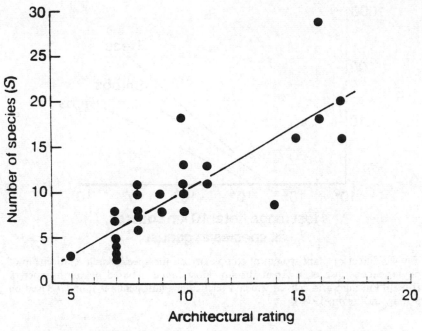

Fig. 4.3 Relationship of the number of phytophagous insect species found on *Opuntia* cactus to the 'architectural rating' of *Opuntia* species from North and South America. (Based on Moran, 1980.)

However, as the degree of heterogeneity increases, either in the spatial area being considered or the contrast between the different parts of the environment, the point is reached at which the individuals of the different species (or the same species, for that matter) no longer interact, because they are segregated into two different structural components (e.g., vegetation layers, Willson, 1974) or patches of the environment. At this scale the competitive mechanisms represented by the Lotka-Volterra model no longer apply, and other processes, such as dispersal, time lags, and stochastic effects, become important.

Relative Heterogeneity: Fundamental Differences between Animals and Plants

A mowed lawn that is a homogeneous salad to a grazing sheep is a complex, heterogenous universe to a small, flightless insect. Obviously, the issues of scale and perception are critical to understanding what kind of heterogeneity is relevant to any particular organism. The size,

Actual Physical Heterogeneity of the Environment

	LOW	HIGH
LOW	Effective Heterogeneity: LOW Competition Intensity: HIGH Species Diversity: LOW	Effective Heterogeneity: HIGH Competition Intensity: LOW Species Diversity: HIGHEST
HIGH	Effective Heterogeneity: LOW Competition Intensity: HIGH Species Diversity: LOWEST	Effective Heterogeneity: LOW Competition Intensity: HIGH Species Diversity: LOW

Mobility / Perception / Size of Organism (row axis label)

Fig. 4.4 The effective heterogeneity, and thus the species diversity, of a particular environment are a consequence of the interaction of the actual environmental heterogeneity and the ability of organisms to sense and move about the environment.

perceptive capabilities, and motility of an organism determine whether an environment is functionally homogeneous or heterogeneous. This, in turn, determines how an organism uses its environment, and how it interacts with other organisms in that environment.

Mobile organisms with well-developed perception, such as most vertebrates, tend to minimize the effects of environmental heterogeneity by concentrating their activities in favorable areas, such as those where food is available. They are able to convert a heterogeneous environment to a more homogeneous one by feeding and living in a restricted subset of the possible patch types. As a consequence, individuals of this type are likely to interact frequently with each other or with competing species under very similar conditions, leading to competitive exclusion between species and potentially strong natural selection within a species. Thus, any given degree of spatial heterogeneity is likely to be less effective in preventing competitive exclusion among these organisms than among non-motile organisms (Fig. 4.4).

In contrast, organisms that are stationary or are not effectively motile as a consequence of small size, poorly developed perceptive capabilities, or other reasons, are likely to experience the full effects of both temporal and spatial environmental heterogeneity in preventing competitive exclusion. Individuals of a sessile species are likely to experience a wider range of conditions than motile species because they are unable to continually move to favorable conditions as the environment fluctuates through time. In addition, different individuals of a sessile species may end up by chance in any one of many types of patches as a result of propagule dispersal, so they are likely to interact with a great variety of species under a great variety of conditions. This reduces both the frequency and the directionality of intraspecific and interspecific interactions that could lead to competitive exclusion or natural selection for traits that avoid competition with superior competitors (Fig. 4.4).

These differences in the effective heterogeneity of a given environment for different types of species influence the types of competitive interactions that are likely to occur. Ecologists have long distinguished different types of competitive interactions, often dichotomized into 'interference' and 'exploitation' (Miller, 1967; Park, 1954). Interference refers to the physical interference of one individual with the ability of another individual to obtain food, mates, or other resources. Such physical interference can occur through such behaviors as territorial defense, or toxic or repulsive chemicals produced by some plants and invertebrates. Exploitation (also called 'scramble' competition) refers simply to the use and depletion of resources by one individual before another individual can use them. Interference competition is likely to occur primarily among mobile organisms that congregate around concentrations of resources. Likewise, exploitation competition should also be most intense where many individuals use the same pool of resources, as is the case with mobile organisms.

Among stationary organisms exploitation of a limited pool of resources is the primary mode of competition. However, it is likely to be very different than exploitation among mobile organisms. Each stationary individual interacts only with its immediate neighbors, which will probably be a small subset of the potential competitors that will differ from one individual to the next. The overall effect of competition is likely to be nondirectional and weak when many individuals are considered. In contrast, each individual of a mobile species will regularly interact with many individuals of its own species, as well as individuals of other species that are attracted to the same resources. Competition among mobile organisms, whether by interference or exploitation, is likely to be strong and directional. Miller (1969) argued that interference competition tended to lead to competitive exclusion much more rapidly than exploitation competition.

Among mobile organisms, the high frequency of competitive interactions that could potentially lead to competitive exclusion and strong natural selection for character divergence (i.e., 'niche differentiation') is likely to have two primary consequences consistent with equilibrium competition theory: 1. Total species diversity will be relatively low; and 2. the species that do actually coexist will be sufficiently different (i.e., the differences will be equal to or greater than some 'limiting similarity') that they will avoid competing for critical resources (Hutchinson, 1959; MacArthur and Levins, 1967; Pianka, 1976; Horn and May, 1977; Abrams, 1976; Grant and Schluter, 1984). The expected pattern is many different functional types with relatively few species within a functional type. These predictions are consistent with the general patterns of vertebrate diversity. In general, total vertebrate diversity is much lower than the diversity of organisms that are less able to reduce the effective heterogeneity of their habitat, such as plants and many invertebrates. Furthermore, the classical observations of specialized adapations and niche separation that have led to the development of both evolutionary theory (e.g., Darwin, 1859; Lack, 1947, 1954, 1969, 1976; Colwell and Fuentes, 1975; Grant and Abbott, 1988) and much of equilibrium competition theory are based on patterns among vertebrates (Utida, 1957; MacArthur and MacArthur, 1961; Schoener, 1965; Root, 1967; Terborgh, 1971).

Among sessile or poorly motile organisms, the preponderance of weak and nondirectional competitive interactions that are unlikely to lead to competitive exclusion should have two primary consequences that are not consistent with equilibrium competition theory: 1. total species diversity will be relatively high; and 2. differences between species will be relatively slight, apparently insufficient to prevent competition for the same resources and competitive exclusion under equilibrium conditions. The expected pattern is relatively few functional types with many similar species within a functional type. In general, these predictions are consistent with the general patterns of plant diversity, and to some extent with the patterns in many groups of invertebrates.

These fundamental differences have consequences that lead to different patterns of evolutionary diversification. Among animals the pattern has been a steady increase in the number of functional types (at the levels of orders and classes) with relatively little evidence of newly evolved functional types replacing older functional types (Fig. 2.31). Among plants, however, the evolutionary pattern over geological time has been a succession-like replacement (at the class level) of older plant types by more recently evolved types (Fig. 2.32). This type of phytogenetic pattern has not been thoroughly investigated from a functional/evolutionary perspective.

Large-scale, Long-term Equilibria

Several hypotheses that have been proposed to explain species diversity have generated confusion and disagreement because it was never clear to what scales and what components of diversity they applied. Two of the most prominent hypotheses were based on the general concept of competitive equilibrium. Both of these hypotheses derived apparent support from the temperate-tropical diversity gradients, because important environmental properties such as stability, predictability, and productivity were thought to increase toward the tropics and make competitive equilibrium more likely. Although these hypotheses have been largely rejected as general explanations for patterns of species diversity, they do apply to some components of diversity under certain conditions.

The Productivity-stability Hypothesis

Connell and Orias (1964) proposed that diversity was highest in stable and productive environments because there was more energy available to be subdivided between competing species at equilibrium. Consequently, population sizes could be larger, which, they argued, would lead to a larger total number of mutations, along with dispersal, isolation, and ultimately speciation. The argument is based on the assumption that environments that are stable, productive, and, in general, warmer, allow more energy to be allocated to production rather than to maintenance. This assumption may be true for warm-blooded animals which must use more energy to maintain their high body temperatures in cold climates. However, for cold-blooded animals and plants the opposite may be true, since higher temperatures are inevitably associated with higher metabolic respiration rates and consequently the loss of energy that could otherwise be used for production.

As discussed in Chapter 2, diversity is postively correlated with productivity along some gradients, but there are enough clear exceptions that the original productivity hypothesis is no longer considered generally applicable. Many of the data that relate diversity to productivity actually show an inverse correlation between the two (Swingle, 1946;

Yount, 1956; Williams, 1964; Whiteside and Harmsworth, 1967; Mc-Naughton, 1968; citations in Huston, 1979, 1980). This phenomenon has been called the 'paradox of enrichment' (Rosenzweig, 1971) and has been described for both predator-prey (Rosenzweig, 1971) and competitive systems (Riebesell, 1974).

The evidence for and against the stability/productivity hypothesis illustrates the critical importance of clearly defining the scale and components of diversity in relation to the mechanisms that influence diversity. Most of the evidence used to refute this hypothesis is based upon the local diversity of competing organisms, that is, on diversity within a single functional type or a few similar functional types. Among competitors, particularly autotrophs, increasing productivity is often associated with decreasing diversity. Much of the evidence in support of this hypothesis is based upon the diversity of entire communities, including all trophic levels and thus many different functional types. Given the energetic losses imposed by the transfer of energy from lower to higher trophic levels (i.e., digestion, respiration, assimilation), it seems plausible that the number of trophic levels, and hence the functional type component of the total diversity should be limited by the total amount of energy flowing through the system (Lindeman, 1942; Rosenzweig, 1977). The larger population sizes and higher population growth rates that are possible under productive conditions should reduce the rate of extinction from disturbances or random fluctuations in population size. However, communities with high population growth rates and many trophic levels usually have a large biomass and/or high productivity of the primary producers, among which diversity may be quite low.

Thus, diversity can be either positively or negatively correlated with productivity, but the relationship is quite predictable, based on the component of diversity being considered. Lumping all organisms together into a measure of total diversity produces patterns that are uninterpretable because they result from multiple mechanisms that may have opposite effects on diversity.

The Stability-time Hypothesis

Sanders (1968, 1969; Sanders and Hessler, 1969) postulated that long periods of time with a stable, predictable environment allowed high diversity because populations would be stable and extinction rates due to population fluctuations would be low. They postulated that stable environments should also allow evolutionary specialization and equilibrium

coexistence due to resource partitioning. Although the data that originally led to the hypothesis were from the gradient of increasing diversity
of marine benthic organisms from the unstable continental shelf to the
constant conditions of the abyssal plains, the hypothesis also seemed to
apply to the temperate-tropical gradient of terrestrial diversity and to certain ancient lakes such as Lake Baikal in Russia and the great rift valley
lakes of Africa (Malawi, Victoria, and Tanganyika, see Lowe-McConnell,
1987).

However, there are numerous examples of low diversity communities
in stable environments, such as the coastal redwoods (Whittaker, 1966),
tropical forests dominated by cativo (*Prioria copaifera*), *Mora*, *Gilbertiodendron*, and other species (Holdridge *et al.*, 1971; Hart *et al.*, 1989), and
freshwater marshes. These environments have high productivity, and, at
least in the case of the forests, exist for long periods of time with little
or no disturbance.

On the other hand, communities with high diversity are often found in
environments that seem to be severe, unpredictable, or unstable, such as
the Sonoran desert (Whittaker and Niering, 1965), some benthic marine
communities (Paine, 1966; Sousa, 1979a and b), and diatoms in rivers
(Patrick, 1963). The fire-maintained sclerophyll shrub communities of
the mediterranean climate regions of southern Africa and Australia may
have higher species diversity than some rain forests (Richards, 1969).
All of these environments have low productivities and are unstable or
frequently disturbed. Recently, there has been an increasing realization
that even the tropics are not stable and predictable environments, either
on the short-term or the long-term (see Chapter 14). Similarly, other
explanations have been proposed for the high diversity of the abyssal
benthos (see Chapter 12).

The stability/time hypothesis illustrates the importance of considering
all interacting processes in explaining patterns of species diversity. This
hypothesis is based on the assumption that the rate of species accumulation over time is sufficiently similar in all areas that the primary
difference in the number of species present results from the length of
time that has been available. However, the rate of species accumulation
actually results from the difference between two rates. At evolutionary
time scales and over large areas the rate of species accumulation is the
difference between the rate of speciation and the rate of extinction (cf.,
MacArthur, 1969; Rosenzweig, 1975). There is strong evidence that neither the rate of speciation nor the rate of extinction are the same in
all groups of organisms (Mayr, 1969; Greenwood, 1964, 1974; Sepkoski

and Hulver, 1985), nor are they constant through time in a single region, or similar between different regions (Stehli, *et al.*, 1969; Eldredge and Gould, 1972; Walker and Valentine, 1984; Sepkoski and Hulver, 1985; Valentine, 1985). Differences between extinction rates, at both local and regional scales, will be a major component of the explanations of species diversity that will be developed in this book (see discussion of endemism in Chapter 10 and tropical forests in Chapter 14).

Time can also have a major effect on the accumulation of species over short time periods as a result of immigration. Immigration (the result of dispersal) often causes a rapid increase in species diversity during early succession in terrestrial communities. On isolated islands, where the immigration rate is very low, the slow increase in species diversity due to immigration following a major disturbance or the creation of the island can continue over very long time periods. However, among groups of competing organisms there is usually an inverse relationship between diversity and stability (when defined as the time between disturbances) that results from competitive exclusion during the periods between disturbances. Thus, the stability/time hypothesis is unlikely to be an important explanation of diversity patterns on the time scale of ecological interactions. However, it is consistent with known processes and with some diversity patterns at very short (dispersal) time scales and may also be associated with the accumulation of species diversity over evolutionary time scales in situations where the extinction rate is low.

Intermediate-scale Equilibria

The Theory of Island Biogeography

One of the best known and most widely studied ecological equilibria is the hypothesized dynamic equilibrium between immigration and extinction that is the basis of the theory of island biogeography (MacArthur and Wilson, 1963, 1967). Even this intuitively compelling hypothesis has come under criticism because of numerous situations (including both 'true' islands and 'habitat' islands) where its predictions fail or its assumed mechanisms cannot be demonstrated (e.g., Connor and McCoy, 1979; Gilbert, 1980). However, given the assumptions of the theory, namely an extinction rate that results soley from random population fluctuations and an immigration rate that is proportional to island size and distance from the source of propagules, it should not be surprising that the model has been consistent with actual patterns in only a few cases, and has failed to predict the diversity patterns in most situations where it has been rigorously applied.

Because of the inverse relationship of population size with the probability of extinction, and the effect of island cross-section on encounter by potential immigrants, the theory predicts a positive relationship between island size and the number of species on an island. However, this prediction is not unique to the theory of island biogeography, and can be made on the basis of other mechanisms, particularly the effect of habitat heterogeneity (Williams, 1964). Variation in the number and types of habitats on different islands could easily mask the effects of a stochastic equilibrium between immigration and extinction. In some cases, habitat heterogeneity is a better predictor of species number than is area (Power, 1972; Johnson and Simberloff, 1974; Johnson, 1975; van der Werff, 1983) and in many cases there is no significant relationship between area and species number (see Gilbert, 1980).

The assumption that immigration rate is a function of island cross-sectional area and distance to the source of immigrants (i.e., the mainland) is likely to be true only under a limited set of circumstances. Dispersal ability, and thus potential immigration rate, varies greatly among taxa, being so high for some groups, such as fungi, and so low for other groups, such as terrestrial mammals or large-seeded terrestrial plants, that immigration rates are likely to be essentially constant with regard to island properties. Among birds, the dispersal ability and habitat preferences of different species vary sufficiently that unequivocal support for the theory has been difficult to find (see Gilbert, 1980). Only among specific taxa under a specific range of conditions (e.g., distance from mainland) is there likely to be sufficient variation in immigration rate as a function of island properties that the predictions of the theory could be observed.

The strongest evidence in support of the theory of island biogeography comes from a study of the reestablishment of insects on small mangrove islands that were defaunated by fumigation (Simberloff and Wilson, 1969, 1970; Simberloff, 1976). Immigration rate was not an issue in these experiments, although immigration was clearly high enough to quickly refaunate the islands. The key to the success of these experiments in validating the theory of island biogeography is probably the structural similarity of the islands. Although structurally complex, all the mangrove islands were relatively small and had essentially the same homogeneous mixture of insect habitats. Thus, the stochastic interaction of immigration and extinction was not masked by differences in habitat heterogeneity. The relatively short life span of most insects and the small population sizes supported by the small islands probably led to high local extinction rates that contributed to the dynamic equilibrium that was apparently reestablished on these islands.

Causes of extinction other than random population fluctuations can also obscure the patterns predicted by the theory of island biogeography. Extinctions that result from competitive exclusion can result in higher extinction rates on larger islands, contrary to the assumptions of the model (Schoener and Schoener, 1978). Extinctions that result from introduced predators or major disturbances can overwhelm the patterns predicted by the model. Extinctions can also result from the disappearance of habitats as a result of plant succession (Willis, 1974; Karr, 1982). Where both immigration and extinction rates are low, speciation can determine the number of species on an island and produce patterns inconsistent with the theory's predictions (Slud, 1976).

The rate of extinction that results from failure of populations to recover from low levels caused by disturbances would be expected to reflect the interaction between the ability of populations to recover from disturbance and the frequency and intensity of the disturbances. These are conditions predicted to influence the number of co-occurring functional types. Most studies of island biogeography tend to focus on groups of organisms selected on the basis of major taxonomic divisions (e.g., passerine birds, terrestrial mammals, woody plants, etc.). Thus the values of species richness used to test the theory of island biogeography tend to reflect the total number of functional types more than the number of competing species within a functional type (see Diamond, 1975 for a discussion of island patterns among and within functional types of birds). Since the theory of island biogeography does not address extinction that results from variation in productivity and disturbance rates (two factors that may vary greatly from island to island), it is not surprising that the theory's predictions about patterns of species diversity are wrong in most cases.

In conclusion, the specific predictions of the theory of island biogeography are likely to apply only under the very limited set of circumstances (taxon-specific spatial and temporal scales and patterns of spatial heterogeneity) where the theory's assumptions are valid. As with any theory about species diversity, the theory of island biogeography only explains diversity patterns when the mechanisms on which it is based are the most important regulators of diversity.

Extinction and Diversity

Extinction, either local or at the landscape or regional scale, is the factor reducing species diversity in all of the equilibria discussed in this chapter (the emigration of all individuals of a species from a locality where they previously lived and reproduced is probably a very rare event). Extinction can result from many different processes, of which competitive exclusion and predation are not necessarily the most important. Extremes of weather and climate, the invasion or evolution of virulent pathogens, various human activities, volcanism, and the impacts of comets or asteroids have all been implicated in some extinction events.

Variation in extinction rates, regardless of the mechanisms that cause the extinction, are important in understanding species diversity at all spatial and temporal scales. The positive processes that counterbalance the extinction rate (and achieve some sort of equilibrium if the rates are sufficiently constant) are much more scale-dependent than are extinction processes (Fig. 4.5). Speciation contributes to species diversity over long time periods in comparison to immigration. Likewise, speciation will probably contribute much more to diversity at large-scales than at small scales, where immigration will be more important. The reimmigration of a species into small areas where it was temporarily absent inevitably occurs much more frequently than the immigration of a species into a large region where it was absent.

Local extinction can obviously have a strong negative effect on species diversity. Among organisms that do not compete strongly, that is, among organisms of different functional types, competitive exclusion is not likely to be an important source of mortality. The primary sources of mortality are likely to be abiotic disturbances and biotic processes such as predation, herbivory, parasites, and disease. The number of functional types present in a local patch can be understood in the context of a dynamic equilibrium between mortality-causing disturbances and population growth rates (i.e., the rate of recovery from disturbance). The probability that a population will survive random fluctuations in size, fluctuations in environmental conditions, and mortality caused by disturbances, is a function of the total population size and the rate at which it can increase in size when it is small. This general concept has a long history in ecology, and is an important component of the equilibrium theory of island biogeography (MacArthur and Wilson, 1963, 1967), as well as theories related to conservation biology (Eisenberg, 1980; Gilpin and Soule', 1986; Diamond, 1986). Assuming that the growth rates of

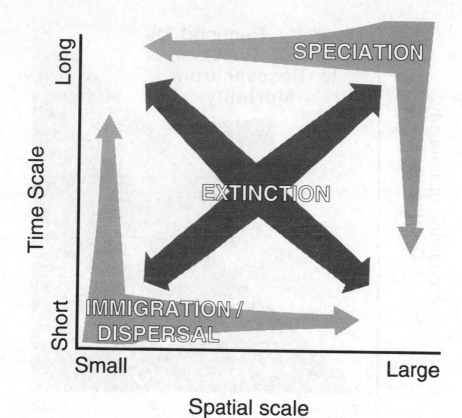

Fig. 4.5 Variation in the relative importance of extinction, speciation, and immigration on species diversity over a range of spatial and temporal scales.

populations of most functional types in an environment will be affected similarly by changes in primary productivity, or other environmental conditions that influence growth rates, it is possible to envision a dynamic equilibrium between disturbance regimes and growth conditions.

According to this hypothesis, the number of functional types, particularly at higher trophic levels, should be highest under the most productive conditions and lowest disturbance frequencies and intensities. Functional type diversity should decrease with increasing disturbance frequency or intensity and with decreasing productivity or growth rates (Fig. 4.6). Functional type diversity, particularly of 'interstitial' species, is also influenced by the physical heterogeneity and resource diversity of the environment. The effect of heterogeneity may compensate for low productivity in some situations, but in most ecosystems there should be

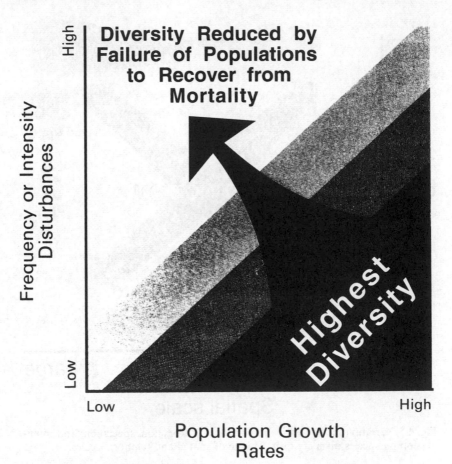

Fig. 4.6 Predicted response of the number of co-occurring functional types to variation in population growth rates (i.e., primary and secondary productivity) and the frequency and intensity of mortality-causing disturbances. Competitive exclusion does not affect the number of different functional types. Rather, the ability of populations to recover from low levels determines how many noncompeting populations can cooccur. Consequently, functional type richness decreases with decreasing growth rates and/or increasing frequency and intensity of disturbance. (Based on Huston, 1979.)

a positive correlation between primary productivity and the physical and resource heterogeneity created by the structural organisms (discussed in Chapter 5), which will enhance the pattern predicted in Fig. 4.6. The number of functional types is the component of diversity which tends to have a positive correlation with productivity (cf. Connell and Orias, 1964).

Summary

The emphasis on equilibrium in ecology has been attributed to the influential ideas of Frederick Clements (cf. McIntosh, 1985), who emphasized the vegetational 'climax,' which represented an equilibrium state of plant community development in balance with local climatic conditions (Clements, 1905, 1916, 1935). However, it must be noted that Clements also emphasized the dynamic, successional processes of plant communities (Clements, 1916). 'In nature both formations and subordinate groups are in stable equilibrium only rarely and usually for a comparatively short time.' (Pound and Clements, 1898, cited in McIntosh, 1985). The concept of equilibrium, between specific opposing processes at appropriate spatial and temporal scales, is essential to understanding patterns of community structure and species diversity.

DeAngelis and Waterhouse (1987) discussed some of the problems involved in defining and applying the concept of equilibrium. The general validity of the equilibrium assumption for competition and predation is based on the consideration of a sufficiently large number of organisms (in a correspondingly large spatial area) that random population fluctuations are relatively unimportant. However, the predatory and competitive interactions on which most competition models are based actually occur among relatively small numbers of organisms in their local neighborhoods. At these smaller scales, random fluctuations are likely to be much more important, and can be expected to eventually cause the extinction of the smaller-sized populations that occur in small areas. DeAngelis and Waterhouse pointed out that these deterministic equilibrium models cannot be extrapolated down to the small spatial scales where stochastic dynamics are likely to become much more important. However, it is at these small spatial scales where population interactions actually occur and at which most ecological observations are made.

The various large-scale equilibrium explanations of diversity discussed in this chapter are consistent with the observation of DeAngelis and Waterhouse (1987) that a robust theory of complex ecological systems should be based on the interaction of many smaller units, such as the different habitats or patches that compose landscapes. They suggest that 'appropriate extrapolation to larger areas can reproduce some of the results of traditional equilibrium theory,' but caution that these equilibrium properties emerge 'as an asymptotic limit of increasing size, not as an intrinsic property of the system.' Even though ecological properties at the landscape scale may exhibit equilibrium behavior, the behavior does not result from competitive equilibrium, but rather from the averaged interaction of many small localities over large areas and/or long times, which is essentially the ecological equivalent of the law of large numbers. The belief that equilibrium models are simply a representation of the larger scale or longer term consequences of non-equilibrium processes (e.g., Tilman, 1987b), is likely to be true only for those processes in which stochastic variation plays a major role at large scales (DeAngelis and Waterhouse, 1987). Thus, this belief is unlikely to be true for competitive equilibrium, which is based on small-scale (i.e., inter-individual) deterministic processes. Nor are large-scale stochastic processes (e.g., the theory of island biogeography) likely to be the dominant regulators of species diversity under conditions where small-scale deterministic interactions predominate (Schoener and Schoener, 1978; Smith, 1979).

The premise on which this and the following chapter are based is that when and if competitive equilibrium occurs, it usually results in competitive exclusion and the local extinction of all but the dominant competitor. Although the coexistence of many species at competitive equilibrium is theoretically possible (cf. Tilman, 1982), most evidence suggests that the coexistence of real, as opposed to theoretical, competitors under conditions of competitive equilibrium is very rare. This does not mean that the predictions of equilibrium competition theory are totally irrelevant. Some groups of organisms are likely to approach conditions of competitive equilibrium more closely than other types of organisms. In particular, mobile perceptive organisms such as birds, mammals, fish, and many insects, are likely to reflect the qualitative predictions of equilibrium competition theory to a greater degree than sessile organisms such as plants. The principal evolutionary prediction of equilibrium competition theory is character displacement, or ecological specialization, which reduces competition and facilitates coexistence (Colwell and Fuentes, 1975).

Given that the usual consequence of competitive equilibrium is local extinction, a variety of mechanisms based on connections between the many different localities (or patches) that compose a landscape can be invoked to explain the maintenance of species diversity. At both the local and the landscape scale, maintenance of diversity over long time periods requires only that extinction occurs asynchronously in different patches and that species are able to reimmigrate into localities where they have become extinct. These explanations of the maintenance of diversity on landscapes composed of patches with asynchronous extinction or other differences apply equally whether the extinctions result from competitive exclusion, predation, random population fluctuations, or other causes.

Although 'classical' equilibrium theory applies specifically to organisms that compete with one another, most of the explanations of diversity presented in this chapter also apply equally well to the regulation of the diversity of organisms that do not compete, that is, to organisms of different functional types. The focus of this chapter has been on the effect of heterogeneity at the scale of landscapes on the number of different functional types that co-occur, as well as on the number of species within a functional type. I argued that local heterogeneity in structure and resources contributes primarily to the number of different functional types in a community, and less to the number of species within the functional types. However, processes such as extinction, immigration, and speciation influence both the number of functional types and the number of organisms within a functional type.

To the extent that equilibrium was invoked as an explanation of diversity patterns in this chapter it was a stochastic equilibrium operating at large spatial scales over long time periods, rather than a local equilibrium involving deterministic processes. The next chapter returns to the issue of local diversity at the scale of organismal interactions and examines some of the consequences of the non-equilibrium dynamics of deterministic competitive processes.

5

Non-equilibrium Processes and the Maintenance of Local Species Diversity

Just as the concept of equilibrium does not imply rigid constancy of ecological properties, the argument that ecological systems are 'nonequilibrium' does not imply that nature exists in a state of randomness or disorder with no pattern or organizing principles. The previous chapter discussed a number of potential landscape-scale equilibria that could help maintain species diversity at both local and regional scales. There are clearly many predictable processes that result in similar patterns of adaptation, community structure, and species diversity around the globe. Nonetheless, the past century of ecological theory and research demonstrates that there is no single- factor explanation for all patterns of species diversity. It is evident that species diversity is influenced by a wide variety of processes that vary in importance from one community to another, from one continent to another, and from one group of organisms to another.

The relative importance of regional versus local processes in regulating species diversity has been the subject of considerable argument (see review in Ricklefs, 1987). The critical issue is not whether regional or local processes are most important, but under what conditions and for what components of diversity regional processes are most important, and under what conditions local processes are most important. Since one of the strictly local processes that has been widely considered to influence species diversity is competition, the functional classification described in Chapter 1 provides some guidance for predicting which patterns are most likely to be influenced by regional processes, and which might potentially be influenced by local processes.

By definition, competitive interactions between different functional types are weak or nonexistent, so the number of functional types in a community is not likely to be strongly influenced by the local process

110

of competition. Regional processes such as geologic and evolutionary history, speciation and extinction rates, and patterns of dispersal and immigration, are more likely to have a strong influence on the number of functional types found in a community.

The number of functionally analogous species that are able to co-exist, however, may be directly influenced by competitive interactions. Because competitive interactions between species of the same functional type (e.g., the same feeding guild or plant growth form) are *potentially* intense, local competitive interactions can have a strong influence on species diversity *under environmental conditions that allow intense competition.* Thus, among species that potentially compete, local diversity may be regulated by competition or other local interactions under some conditions, and by a variety of regional processes under other conditions. There are also strong taxonomic influences on the regulation of species diversity in different groups of organisms. Differences in mating systems, genetic structure, trophic status, and various aspects of life history, can influence whether species in a particular taxonomic (and functional) group are likely to compete intensely or not at all.

The 'non-equilibrium' theory described in this chapter can be used to predict the relative contribution of regional versus local processes to local species diversity under different environmental conditions. Rather than 'non-equilibrium', it is more appropriate to describe community structure as a dynamic equilibrium of fluctuating species diversity (and other population, community, and ecosystem properties) that results from an approximate balance among opposing local and regional processes. Important local processes include competitive exclusion and mortality, which may be caused by abiotic disturbances such as floods, windstorms, and fires and biotic interactions such as predation, herbivory, parasitism, and disease. These local processes occur at the same spatial and temporal scales as the processes invoked in models of competitive equilibrium. Regional processes include immigration and extinction, which may result from a variety of causes. *The dynamic equilibrium theory applies specifically to the local species diversity among competing organisms,* but it influences and is itself influenced by species diversity at the landscape and regional scales.

Competitive Exclusion and its Prevention: The Basis of Non-equilibrium Competition Theory

As early as 1941, Hutchinson observed that competitive exclusion could be avoided when 1. externally-imposed mortality reduces the density of competing species, or 2. environmental fluctuations alter the balance of competitive interactions and prevent the establishment of equilibrium conditions (Hutchinson, 1941, 1948). However, the significance and general applicability of his suggestion were not appreciated by most ecologists for nearly 30 years.

By the late 1970s the relevance of equilibrium competition theory to natural communities was being widely called into question by two separate lines of reasoning, one dealing with the assumptions of equilibrium competition theory and the other with the validity of its predictions about real ecosystems. First, the importance of predation, climatic fluctuations, and disturbances such as fires, floods, and treefalls, for determining the structure of communities was becoming increasingly apparent (Paine, 1966, 1969; Watts, 1973; White, 1979). Each of these factors clearly had the effect of preventing the competitive equilibrium upon which conventional ecological theory was based. Second, the failure of equilibrium competition theory to convincingly explain natural patterns of species distribution and coexistence was becoming increasingly obvious.

In spite of the abandomnent of the competitive equilibrium paradigm, competitive exclusion is still a central concept in community ecology and most theories of species diversity. Non-equilibrium theories depend just as much on the occurrence of competitive exclusion as did theories based on competitive equilibrium. Non-equilibrium theories generally assume that competitive exclusion and the consequent reduction in diversity are the almost inevitable result when competitive equilibrium occurs within homogeneous patches. Data from most natural systems and experiments are consistent with this assumption. Consequently, the key to non-equilibrium coexistence within a homogeneous patch is the prevention of competitive equilibrium, and thus the prevention of competitive exclusion. At the scale of a single patch, competitive exclusion can be prevented in three primary ways: 1. by reducing the populations of the dominant competitors before they achieve competitive exclusion; 2. by slowing the process of competitive exclusion to the point that competitive equilibrium is never reached; and 3. by changing the conditions under which competition occurs to the degree that the competitively dominant species becomes competitively subordinate before exclusion occurs. The regional-scale process of immigration may in some cases prevent competitive exclusion by 'subsidizing' the population of the inferior competitor, and even if competitive exclusion occurs, immigration may re-establish the population. In general, any factors that prevent competitive equilibrium from occurring will facilitate the coexistence of competing species and maintain species diversity.

Disturbances that Remove Dominant Species

Competitive exclusion can be simply prevented by any disturbance that reduces the population of the dominant competitors before it can drive the inferior competitors to extinction. The term 'disturbance' will be applied only to those conditions that cause mortality, either of individuals in a population or of portions of a colonial or modular organism such as corals or plants. Mortality, potentially resulting from many different biotic and abiotic causes, is the principal process that reduces the abundance of the dominant species and prevents competitive exclusion. Examples of biotic disturbances that cause mortality include predation, herbivory, parasitism, and disease. Abiotic disturbances include floods, fires, severe windstorms, and extreme weather such as droughts and severe frosts during the growing season. Other processes, such as the complete emigration of a dominant competitor, could conceivably have an equivalent effect on diversity.

Predation as a Type of Disturbance

The field observations and experiments that contributed most to the recent rejection of the equilibrium perspective were studies of the effects of predators and herbivores in the marine intertidal zone, although the positive effect of predation and herbivory on species diversity had been known for some time (e.g., Tansley and Adamson, 1925; Ridley, 1930). The sessile growth form or sedentary behavior of many intertidal organisms make them particularly amenable to study and their rapid growth rates allowed changes in community structure to occur fast enough to document during an observation period of months to years. Most of the important prey species, such as barnacles and mussels, are permanently fastened to the substrate, and most of the important predators, such as starfish, and certain snails, occur in high enough densities and move slowly enough for ecologists to observe them easily.

The predator removal experiments by Robert Paine (1966) were probably the most influential experiments in animal ecology during the second half of the twentieth century. Working at a site in the Pacific Northwest, Paine removed all individuals of the predatory starfish *Pisaster ochraceus* from sections of the intertidal zone and kept them excluded for several years. Following removal of *Pisaster* the sites were initially dominated by the barnacle *Balanus glandula*, and eventually the barnacles were crowded out by the mussels, *Mytilus californicus*, which continued to dominate the site for the rest of the experiment. Most of the other sessile organisms, including several species of algae, were also crowded out by the mussels, and the grazing species that fed on algae left the area. The number of species in the areas with starfish removed was reduced to 8, in contrast to the adjacent control areas with starfish present, which had 15 species. In the control areas, higher species diversity was maintained by the predator, which selectively removed the dominant competitors (mussels), thus preventing the occurence of competitive equilibrium and the competitive exclusion of several species. In addition to the structural species in this community, there are a large number of interstitial species that depend on the physical structure and resources provided by the mussels and other sessile organisms (Hewatt, 1935), so total diversity in the treatments either with or without starfish was undoubtedly much higher than the diversity of the structural species that were directly affected by the predator removal experiment.

Analogous results were reported by Lubchenco (1978), who found that an intermediate density of herbivorous snails resulted in the highest algal diversity in tide pools. With few or no snails present in a tide pool, a single species of algae, *Enteromorpha*, became dominant and competitively excluded most other species. With a high density of snails, all the edible species of algae were eliminated, leaving only a single unpalatable species, *Chondrus*.

Similar effects of predators or herbivores have been found in many other communities (Zeevalking and Fresco, 1977; Tansley and Adamson, 1925; Connell, 1961, 1970, 1971; Harper, 1969; Paine and Vadas, 1969; Dayton and Hessler, 1972; Menge and Sutherland, 1976). Variation in the pattern of predation or herbivory, in terms of frequency and intensity, allows these disturbances to operate as either equilibrium or non-equilibrium processes. When the frequency of predation or herbivory is so high that it is essentially continuous, it can alter the strength and direction of competitive interactions and allow the coexistence of competing species under conditions equivalent to those of competitive equilibrium (Levine, 1976; Vandermeer, 1980). Continuous predation may also prolong coexistence by slowing the approach to competitive equilibrium. However, when predation or herbivory occur at lower frequencies and act as pulses of mortality, they clearly interrupt the process of competitive exclusion and allow non-equilibrium coexistence. Variation in environmental conditions and fluctuations in the population sizes of predators and herbivores generally result in this type of mortality operating less continuously than competitive displacement, and thus usually acting as a non-equilibrium process.

The effect of a predator on species diversity depends both on the feeding behavior of the predator itself, and on the composition of the prey community. A predator that feeds on the most abundant species in a community will tend to increase the diversity of that community if the abundant species is exerting some negative effect on the populations of the other species. However, if the predator feeds on rare species in the community, the effect of predation may be a reduction in diversity.

A generalized predator is likely to prey on the most abundant species in a community (for energetic reasons, see discussions of optimal foraging in Schoener, 1969, 1974; Pyke *et al.*, 1977), so its effect is likely to be an increase in diversity independent of the species composition, as long as it continues to feed on the most abundant species. The effect of a specialized predator on diversity depends on whether its preferred prey are common or rare.

Since the role of the predator or herbivore in these examples is to reduce the population size of competitively dominant organisms, predation is simply one type of mortality-causing disturbance. Paine and Levin (1981) demonstrated that mortality resulting from such abiotic disturbances as storms waves and battering by debris had an effect analogous to predators in maintaining intertidal diversity.

Mortality that kills only the dominant competitor and does not affect the other species is the most effective type of disturbance for preventing competitive exclusion and allowing high species diversity. Such a mortality pattern could result from processes that are either 1. density-dependent, or 2. species-dependent.

General Classes of Mortality

Density-dependent Mortality. Density-dependent mortality that consistently affects the species with the highest density is extremely effective for maintaining species diversity. As soon as a particular species achieves a density that is high enough to begin to exclude other species and thereby reduce diversity, its population is reduced and diversity (evenness) increases. Even if the dominant species is actually eliminated, which reduces the number of species (richness) by one, the resulting decrease in dominance leads to higher evenness and could well allow additional species to enter the community or allow the populations of rare species to increase sufficiently that they would be detected in samples. Natural mechanisms that can produce density-dependent mortality include selective predation, particularly with predator switching (Murton et al., 1966; Rauscher, 1978; Townsend and Hildrew, 1978; Hildrew and Townsend, 1982); diseases or parasites (Janzen, 1970; Augspurger, 1983a,b; Augspurger and Kelly, 1984; Burdon et al., 1989): and physical phenomena related to crowding (such as overheating, reduced air or water circulation, etc.).

Selective grazing or predation can reduce the populations of dominant species, but probably just as often reduces the populations of a variety of palatable species and contributes to the dominance of a few unpalatable species (Paine and Vadas, 1969; Menge and Sutherland, 1976; McCormick and Stevenson, 1989; Sterner, 1989).

Size-dependent Mortality. A variation of density- dependent mortality that can be just as effective at maintaining species diversity is size-dependent mortality. In situations where the dominant competitors are the largest organisms, factors that preferentially kill the largest organisms will prevent competitive exclusion. Examples are large trees selectively killed by windstorms or lightning; large fish killed by heat stress or starvation; large corals destroyed by wave action. A wide range of size-selective feeding strategies are found among aquatic and marine planktivores, which imposes size-dependent mortality on plankton communities. Planktivorous fish that visually search for their prey often selectively remove the larger species or size classes of zooplankton (Hall, 1971; Hall et al., 1978), and different species of zooplankton selectively feed upon different sizes and species of phytoplankton (Burns, 1968; Carpenter et al., 1987; Porter, 1973).

Both density- and size-dependent mortality can also have a severe effect on small populations and small individuals, which may be more susceptible to environmental fluctuations and extreme conditions. In such cases diversity would be reduced.

Species-dependent Mortality. Species-dependent mortality can affect any species, not necessarily just the dominant competitor. However, when the population of the dominant competitor is reduced or eliminated by a species-specific disease, parasite, or predator, competitive exclusion is effectively prevented until the population of the dominant competitor recovers, or another species becomes dominant. Examples include the chestnut blight, which eliminated the dominant canopy species in much of the eastern deciduous forest of North America, or the *Cactoblastis* moth, which decimated the dense stands of the introduced prickly pear cactus in Australia (Dodd, 1940; Munro, 1967). Selective grazing of palatable species can increase diversity if the palatable species are abundant or reduce diversity if the palatable species are rare (Harper, 1969).

Species-dependent mortality reduces species richness and evenness, if the mortality occurs among rare species or species whose populations recover slowly from mortality-causing disturbances. On coral reefs, herbivores that preferentially damage slowly growing coral species may contribute to the dominance of fast-growing species and a consequent reduction in diversity (see Chapter 12).

Species-dependent mortality that affects only one of a pair of competitors can produce a result that looks like competitive exclusion. One example of this type of 'apparent competition' (Holt, 1977; Holt and Pickering, 1985) is a meningeal worm that is a parasite which infects, but does not kill, deer, and kills moose. Moose is a larger species and a dominant competitor that may exclude deer from some areas. Because the worm kills the moose it infects, a region with all moose and no deer cannot support a parasite population. However, when deer enter an area populated only by moose, they serve as a source of infections that kill the moose and produce a result opposite to that expected on the basis of competition alone (Anderson, 1972; Saunders, 1973).

Density-independent Mortality. Mortality that is independent of the density, size or identity of species can also prevent equilibrium and competitive exclusion. Such mortality generally affects a broad suite of similar organisms in a community. Examples include windstorms, landslides, and intense fires that affect trees without regard to species or size, severe freezes or droughts that affect a wide variety of plant species, or storms and ice-scouring that affect all sessile intertidal organisms (e.g., barnacles, mussels, etc.).

The effects of density-independent mortality are illustrated in Fig. 5.1, which shows the population growth through time of six hypothetical populations growing and competing according to the Lotka- Volterra competition equations. In the absence of disturbance (Fig. 5.1a) some populations (1 and 2) are nearing extinction, while other populations (5 and 6) are headed toward eventual dominance. With periodic disturbances in which all populations are reduced by half (Fig. 5.1b), neither extinction nor dominance occur as rapidly as they did in the absence

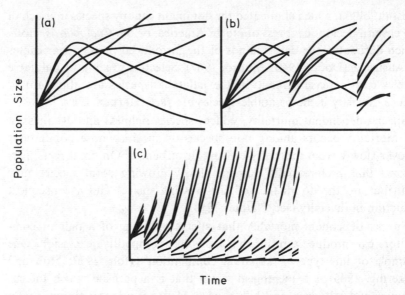

Fig. 5.1 The effect of frequency of disturbance on populations of six competing species in a Lotka-Volterra computer simulation. In each case the disturbances are population reductions of 0.5. Values for the Lotka-Volterra species parameters are the same in all three simulations. (*a*) No disturbance, diversity is reduced as the system approaches competitive equilibrium. (*b*) Periodic disturbances; diversity is maintained for a longer time than in *a*. (*c*) High frequency of disturbances (5 times greater than in *b*; diversity is reduced as slowly growing populations (low *r*) are unable to recover between disturbances. (From Huston, 1979.)

of disturbance. However, with more frequent disturbances of 50 percent mortality (Fig. 5.1*c*) some populations (3,4,5, and 6) are headed toward extinction because they cannot increase fast enough to recover from the disturbances, while other populations with high intrinsic growth rates (1 and 2) are rapidly dominating.

Density-independent mortality tends to eventually eliminate small and slowly growing populations. The lawnmower is the ultimate species- and density-independent source of mortality, and under certain conditions, some herbivores may have an equivalent effect. The competitive exclusion of one grass species by another was slowed with increased frequency of mowing (Harris and Thomas, 1972), and grazing by sheep had a similar effect (Berendse, 1985). Charles Darwin (1859) noted that frequent mowing of a patch of lawn allows the coexistence of a larger number of plant species than would be found in an unmowed patch. Milton (1940, 1947) found that mowing pastures for hay maintained the diversity of the fields, even with the addition of fertilizer, which reduced diversity in unmowed fields. However, even a mechanical mower is selective according to size, and removes proportionately more biomass from large plants than from small plants.

Perfectly density- or species-independent mortality rarely occurs, since there is almost always some selectivity that results from differences in the sizes of organisms (e.g., trees) or in physiological resistance or tolerance of stresses. Any density-*dependent* effects that reduce the most abundant competitor more than the rarer species (i.e., compensatory mortality) will be very effective in maintaining diversity.

The Timing of Mortality

The effectiveness of a given amount or frequency of mortality in preventing competitive exclusion also depends on the growth stage of the population killed. For example, when a logistically-growing population is growing slowly in the pre-exponential 'lag' phase of population growth (Fig. 5.2), periodic removal of a relatively few individuals will be enough to keep the population at a low density and maintain it in the slow growth phase. Under these conditions competitive exclusion is unlikely and diversity among competing populations can be great. A high rate of mortality applied to such a population may actually drive it to extinction.

In contrast, when a population enters the exponential phase of growth, a higher rate of mortality is necessary to reduce the population size, reduce the actual growth rate, and prevent competitive exclusion. By the time that a population's growth slows upon approaching the carrying capacity, competitive exclusion of other poorer competitors is likely to have already occurred. At this point a reduction in the population of the dominant competitor may have little effect on species diversity unless additional species are able to immigrate into the area.

The general consequences of disturbances at different stages of population growth also apply to the effect of disturbances on populations with different intrinsic growth rates. For a population with a low intrinsic growth rate, even a small amount of mortality (or a low frequency of mortality) is sufficient to prevent the population from entering the phase of exponential growth (which will be exponential at a relatively slow rate, anyway). In contrast, a population with a high intrinsic growth rate will enter a phase of rapid exponential growth quickly, and a large amount (or high frequency) of mortality will be necessary to slow population growth, and prevent the potential exclusion of competitors.

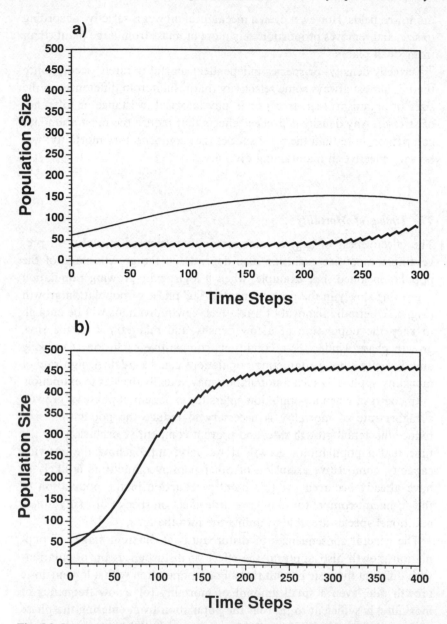

Fig. 5.2 Interaction of the timing of mortality with the growth cycle of a population. (*a*) Repeated mortality (8 individuals removed every 8 timesteps) applied to dominant competitor (heavy line) during lag phase of growth when population size was small. Poorer competitor (thin line) survives. (*b*) Mortality (8 individuals removed every 8 timesteps) of dominant competitor initiated after 100 timesteps, when population was in exponential growth phase. Poorer competitor is eliminated.

At landscape scales the timing of mortality-causing disturbances is also important because of its effect on the species pool available for re-population of disturbed areas. The process of competitive displacement leading to competitive exclusion can be viewed as a successional sequence leading to the dominance of one or a few species (see Chapters 7 and 9). For some successional sequences, particularly forest succession, diversity tends to increase to a high point in mid succession, and then decrease toward late succession (Clements, 1916; Watt, 1947; Loucks, 1970; Horn, 1974, Bormann and Likens, 1979; Whittaker, 1975; Peet and Christensen, 1980). Consequently, disturbances that occur near the point of highest diversity will not only tend to maintain the system at a high level of diversity, but will lead to increased diversity because propagules of many species will be present locally (or elsewhere in a heterogeneous land-scape) to reinitiate succession. Disturbances that occur very frequently will constantly reset the system to the early stages of succession where diversity tends to be lower, and both the local and landscape species pools for recolonization will tend to be smaller. Disturbances that con-sistently occur in late succession when diversity is lower will also tend to reduce diversity throughout the course of succession because the propag-ules of early and mid-successional species are less likely to be available during late succession, making long-distance dispersal necessary for the reintroduction of species.

The 'Intermediate' Rate of Disturbance

Most of the experimental or comparative studies of the effects of pre-dation or disturbance mentioned above found that the highest levels of species diversity were maintained at some 'intermediate' frequency or intensity of disturbance. At high rates of mortality (disturbance or predation), diversity was reduced because some species were unable to recover from the mortality. At low rates of mortality, diversity was re-duced by competitive exclusion as the dominant species eliminated poorer competitors.

Species composition of communities tends to adjust to the character-istic type and frequency of disturbance as those individuals and species best adapted to the specific disturbance type and frequency 1. have the highest survival and reproductive rates and dominate the community; 2. immigrate into the community from other areas with similar disturbance regimes; or 3. over long periods, evolve locally through natural selection imposed by the disturbance regime (e.g., Stearns, 1976, 1977; Pianka,

1970; Southwood *et al.*, 1974; Taylor, 1978; Boyce, 1984). High diversity under extremely high grazing pressure is found in some Mediterranean pastures (Whittaker, 1975) where many species seem to have evolved characteristics that allow them to survive intense grazing (McNaughton, 1984; Milchunas *et al.*, 1988). Likewise, the high fire frequency of the South African Fynbos is associated with an extremely high diversity of specially adapted shrub species (see Chapter 13).

For organisms adapted to disturbance, the distinction between chronic disturbance and stress becomes blurred. If disturbance is defined as the mortality of individuals or significant parts of modular individuals such as plants, adaptation can potentially result in a particular physical process (such as a fire) being changed from being a disturbance to being a stress or even an essential event for the completion of a life cycle. In general, if an individual is able to resist or somehow avoid the mortality caused by a physical or biological process but still suffers some reduction in growth or reproduction, that process could be considered a stress, rather than a disturbance. If however, the individual (or population) survives, but must recover lost biomass, energy or other resources, the process should be considered a disturbance. The ability to recover from, and particularly the ability to resist the damaging effects of a disturbance, is inevitably associated with adaptations that impose strong constraints on the life history or growth form of the adapted species. The consequences of such constraints on interactions between organisms and their enviroment are discussed in Chapter 7.

The phenomenon of diversity being highest at an intermediate rate of disturbance has been noted by many investigators (see Hutchinson, 1941, 1948, 1953, 1961; Watt, 1947; Andrewartha and Birch, 1954; Loucks, 1970; Dayton, 1971; Richerson *et al.*, 1970; Grime, 1973; Horn, 1975; Weins, 1977; Whittaker and Levin, 1977; Connell, 1978; Fox, 1979; Huston, 1979; Sousa, 1980; Kneidel, 1984). However, unless the 'intermediate' frequency of disturbance can be defined independently of its effect on species diversity, the argument is circular and not particularly useful.

Processes that Prevent Competitive Exclusion by Slowing its Rate

> ... many plants may grow together in comparatively peace-
> ful association because their requirements are different, or
> because the conditions are so unfavourable for the luxu-

riance of any in particular that many exist on somewhat
equal terms of limited growth.

Lawes *et al.*, 1882

As Miller (1969) documented, the time required for competitive exclusion to occur is often very long. However, the time required for
competitive exclusion to occur is not a factor in the equilibrium analysis
of competition equations; equilibrium occurs regardless of the rate at
which it is approached. The rate at which this process occurs has been
called the 'rate of competitive displacement' (Gause, 1934; DeBach and
Sundby, 1963; DeBach, 1966).

Slow rates of competitive displacement allow prolonged coexistence,
and thus the maintenance of species diversity (Huston, 1979). Repeated
sampling of a community in which competitive displacement is occurring
slowly will show little change in species diversity over time. Nonetheless,
in the absence of disturbances and other changes that affect competition,
slow rates of competitive displacement lead to competitive exclusion just
as surely as fast rates.

The effects of variation in the rate of competitive displacement are
illustrated in Fig. 5.3, which shows the changes in population size of six
hypothetical populations growing and competing according to the Lotka-
Volterra competition equations. With population growth rates doubled,
and thus the rate of competitive displacement increased, competitive
equilibrium is approached more rapidly (Fig. 5.3*c*), and evenness and
richness are reduced as a result of the dominance of the species that is
the best competitor at equilibrium. With periodic disturbances in which
all populations are reduced by half (Fig. 5.3*b*), evenness is not reduced
as rapidly as in the absence of disturbance. However, with doubled
growth rates, the increased rate of competitive displacement leads to a
reduction in both evenness and richness (one species nearly extinct, one
with extremely high biomass) even with disturbances (Fig. 5.3*d*).

Slow rates of competitive displacement also facilitate the maintenance
of high species diversity by the two other processes that prevent competitive equilibrium. First, with a longer time period prior to equilibrium,
there is a greater chance that some mortality- causing process with a finite
probability of occurrence, such as predation or a physical disturbance,
will reduce population sizes and temporarily prevent the occurence of
equilibrium and competitive exclusion. Second, with a longer time to
equilibrium, there is a greater probability that environmental conditions
will shift sufficiently to alter the balance of competitive abilities, and

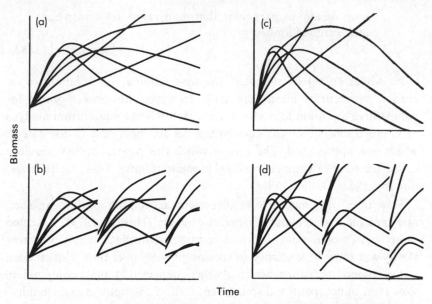

Fig. 5.3 Increased population growth rates reduce diversity both with and without disturbances. The only species parameter that differs between the simulations is the value of the intrinsic growth rate, r. (a) All species are still present at end of simulation. Same as Fig. 5.1a. (b) Intrinsic growth rates the same as in a, but periodic disturbances (0.5 mortality of each species) maintain more even distribution of abundance among species, and higher evenness than in a. (c) Intrinsic growth rate of each species is doubled from values in a. Competitive exclusion achieved more rapidly than in a, and evenness and eventually richness are reduced as a result of the high dominance of most abundant species, and the extinction of the low K species. (d) Intrinsic growth rates the same as in c. Periodic disturbances (same as in b) slow reduction of evenness, but evenness is still reduced more rapidly than in the absence of mortality.

perhaps even reverse the relative dominance of competing species. Such environmental fluctuations can potentially prevent equilibrium from ever occurring, and further slow a rate of competitive displacement that was originally slowed by other factors. Both of these processes were noted by Hutchinson over 40 years ago (1941, 1948).

The rate of competitive displacement is best visualized as a relative rate, rather than an absolute rate. A slow absolute rate of competitive displacement for phytoplankton would be an extremely rapid rate of competitive displacement for trees. The rate of competitive displacement should be considered in relation to the growth rates and generation times of the competing organisms, as well as to the rate at which environmental conditions relevant to those organisms fluctuate. While some valid generalizations may be drawn by considering all organisms together, care must be taken to make reasonable comparisons.

Slow Population Growth Rates

Competitive displacement (leading to competitive exclusion and local extinction) is the process of one population growing and using or otherwise monopolizing resources at the expense of another population which decreases in size to the point of extinction. Thus, anything that slows the population growth of competitors will tend to slow the rate of competitive displacement. Net population growth is a function of population size, the *per capita* growth rate of the population for specific environmental conditions affected by population density and resource availability, and the *per capita* mortality rate. The intrinsic growth rate of a species is considered to be an environment-specific maximum potential growth rate (represented by *r* in the Lotka-Volterra equation) that results from the interaction of the organisms' physiology and a specific set of environmental conditions. Thus, the intrinsic growth rate for a single species will vary from one environment to another, depending on resource availability and other conditions.

As conceptualized by the Lotka-Volterra competition model, the *per capita* growth rate of a population is high (near the maximum intrinsic growth rate) when population densities are low and most of the environment's resources are available, and decreases as the number of organisms increases and the available resources are taken up and made unavailable. With this type of density-dependent growth, the change in population size through time tends to follow a logistic ('s-shaped') curve. Total population size changes at a slow absolute rate when there are few individuals (even though the *per capita* rate is very high), a rapid rate at intermediate population sizes, and a slow rate as the population approaches the maximum size that can be supported by the environment and resources available for growth are reduced to near zero ('carrying capacity,' *K*, in the familiar terminology of the Lotka-Volterra competition equations).

Population growth rates can be slow for three primary reasons: 1. the absolute population size is small, which may result from many causes, including frequent or intense predation, herbivory, or mortality-causing disturbances; 2. the population is near its carrying capacity for the particular environmental conditions being considered. In this case growth is slowed because most of the environmental resources are already being used by members of the population and little is available for new growth; and 3. the intrinsic rate of population increase is low, as a result of low levels of resources caused by environmental conditions (rather than by other members of the population), as a result of environmental conditions that slow growth rates (such as low temperatures), or as a result of the inherent physiology or other adaptations of the organisms being considered. These three factors are not totally independent. For example, small population sizes can be the result of low intrinsic growth rates or low carrying capacities, as well as of high rates of mortality. Thus a low population growth rate can be self-perpetuating.

Populations with low intrinsic growth rates have relatively slow rates of increase regardless of which phase of growth the population is in. Low intrinsic growth rates are often found in situations where environmental resources are scarce. Examples include slow plant growth as a result of low water availability in deserts, slow plant growth on nutrient poor soils, and low rates of predator population growth where prey are chronically scarce.

Low intrinsic growth rates, generally due to the scarcity, but not absence, of critical resources, are often associated with high species diversity. In such cases, the intrinsic rates of increase of all competing populations in the community are usually low, since all depend on the same scarce resources. Obviously, if the intrinsic growth rates are too low, or even negative, most populations will not survive, and diversity will be reduced. The phenomenon of reduced species diversity under conditions of both low and high growth rates, with highest diversity at intermediate growth rates was emphasized by Grime (1973) in relation to soil and light conditions that influenced the biomass of herbaceous plants. The same pattern has been found along many different gradients of resources that influence growth rates, including soil nutrients (Beadle, 1966; Huston, 1980a,b, 1982; Ashton, 1977; Vermeer and Berendse, 1983), soil water (Dix and Smeins, 1967; Beard, 1983; Kutiel and Danin, 1987; Ludwig et al., 1989), light (Huston, 1985a, unpublished data) and plant biomass, which is the integrated response to all resources in the plant environment (Al-Mufti et al., 1977; Bond, 1983; Vermeer and Berendse, 1983; Kutiel and Danin, 1989) (Fig. 5.4).

In addition to being slowed by low growth rates, the rate of competitive exclusion can be slowed when competing species are very similar. When the differences in the competitive abilities, population growth rates, and other characteristics of two competing species are great, the dominant competitor usually excludes the inferior competitor very rapidly. However, when the two species are very closely matched in competitive ability,

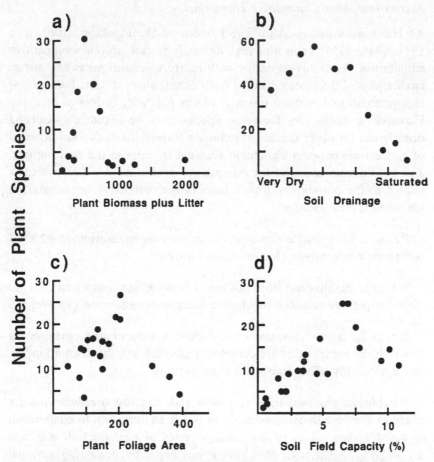

Fig. 5.4 Examples of the unimodal relationship ('hump-backed curve') of plant species richness to plant productivity and to conditions correlated with productivity, such as soil nutrients. (*a*) British herbaceous plants, productivity indexed by live and dead plant biomass (Al Mufti *et al.*, 1977). (*b*) North American prairies, productivity indexed by soil drainage conditions (Dix and Smeins (1967). (*c*) South African fynbos, productivity indexed by area under plant foliage-height profile (Bond, 1983). (*d*) Annual plants in an Israel desert, productivity indexed by soil field capacity (Kutiel and Danin, 1987).

the rate of competitive displacement by the (slightly) dominant competitor can be very slow (Huston, 1979; Caswell, 1982). This prolonged coexistence of similar species is in direct contradition to the predictions of equilibrium competition theory that there is some minimum 'limiting similarity' below which competing species cannot coexist.

Factors that Alter Competitive Hierarchies

As Hutchinson discussed in 'The Paradox of the Plankton' (1961, also 1941, 1948, 1953) environmental fluctuations can prevent competitive equilibrium. As discussed earlier with regard to disturbances and stress, environmental fluctuations differ from disturbances in that fluctuations change conditions without directly causing mortality or loss of biomass. However, a fluctuation for some species may be a mortality-causing disturbance for other species. Hutchinson framed his discussion in terms of t_e, the time between significant seasonal changes in the environment, and t_c, the time required for competitive exclusion to occur, which is related to the population growth rates of the competing organisms. He identified three conditions:

1. $t_c \ll t_e$, competitive exclusion occurs because equilibrium is achieved before the environment changes significantly.

2. $t_c \simeq t_e$, equilibrium does not occur because environmental changes alter competitive relationships before competitive exclusion can occur.

3. $t_c \gg t_e$, competitive exclusion occurs in a fluctuating environment to which the competitors are completely adapted, i.e., the fluctuations do not significantly affect competitive interactions.

Hutchinson observed that 'very slow and very fast breeders thus are likely to compete under conditions in which an approach to equilibrium is possible; organisms of intermediate rates of reproduction may not do so (1961)'. Although Hutchinson was explicitly discussing seasonal fluctuations in the environment and referring to algae in lakes, the general argument is obviously applicable to environmental fluctuations on any time scale, as well as to mortality-causing disturbances.

The types of environmental fluctuations that are relevant to understanding the competitive interactions and diversity of any particular community are clearly dependent on the organisms being considered. The critical properties of the organisms include their lifespan and growth (or reproductive) rates. Rapid environmental fluctuations that are critical to microbial populations in the soil are probably irrelevant to long-lived trees growing in the same soil.

'Predictable' Seasonal Fluctuations. Organisms are generally adapted to such predictable environmental changes as the seasonal cycles of day-

length and light intensity, temperature, nutrient, or water availability. Such seasonal changes in the availability of critical resources can alter the relative competitive abilities of organisms and prevent the occurrence of competitive exclusion among competing organisms of an appropriate range of growth rates and life spans. Among short-lived organisms this results in a seasonal 'succession' of changing species composition and dominance in certain types of communities. In lake phytoplankton the seasonal sequence of dominance in freshwater lakes is well established and highly predictable from physical conditions (Hutchinson, 1967; Wetzel, 1975; Lewis, 1978, 1987; Reynolds, 1984).

A similar seasonal pattern occurs in many plant communities. In early successional oldfields, there is a predictable seasonal sequence of growth and flowering, with rapidly growing and relatively small plants blooming in the spring, and larger, more slowly growing plants, such as thistles and goldenrods, achieving full size and blooming in the late summer and fall. Herbaceous plants, which have no permanent aboveground parts, are able to specialize for particular combinations of environmental conditions that occur at specific times in the growing season (Peterson and Bazzaz, 1978; Wieland and Bazzaz, 1975; Regehr and Bazzaz, 1976; Wesson and Wareing, 1969). The most extreme examples of specialization to temporal variation in environmental conditions are found among annual plants, many of which germinate only in response to specific cues associated with the most favorable environmental conditions. In contrast, woody plants must endure the conditions that occur throughout the year, including the favorable conditions of the growing season and the dry and/or cold conditions of the non-growing season.

Such seasonal fluctuations provide a broad range of conditions that may allow many different species to coexist over the course of the year. Seasonal sequences of abundance, which resemble succession, occur primarily among organisms with a life history (size, growth strategy, etc.) that is able to respond to the changing conditions. Seasonal successions are not found among long-lived woody plants such as trees (although there may be seasonal variation in flowering and fruiting times, see Rathcke, 1983). Between grossly different functional types, such as trees and herbs, there may be a brief period during which the fast-growing species can complete their life cycles before they are shaded by the large, long-lived trees.

One seasonal phenomenon that has a dramatic effect on species diversity and community structure is migration. The spectacular spring migrations of passerine birds, waterfowl, and predatory birds to the temperate and boreal zones with their summer burst of primary and secondary productivity produces a sudden increase in the species diversity of the bird communities of those regions (c.f. Fig. 2.6). Under favorable conditions in the northern hardwood forests, some species of warblers can produce up to three broods of four fledglings each (Holmes et al., 1992, in press) before returning to the tropics. The fact that this remarkable behavior has evolved and succeeded so effectively is an indication that conditions of food availability and predation rates are sufficiently unfavorable in the tropics that even the high mortality incurred during migration does not cancel the higher fecundity that is possible under favorable conditions in the temperate zone (Holmes et al., 1986; Sherry and Holmes, 1991).

Unpredictable Fluctuations. Environmental changes that occur at time intervals longer than the lifetime of an organism are unpredictable from the point of view of that individual organism, although the species may have evolved adaptations to the change that only are activated when the change occurs. Such adaptations are generally responses to environmental changes that occur predictably within the lifetime of a population, rather than a single individual. An example would be the formation of resting stages or special modes of reproduction in some crustaceans, in response to a pond drying out or to the onset of winter. Several generations occur during a season with no indication of the adaptation, which appears when certain environmental conditions occur. Populations of rapidly reproducing organisms, such as bacteria, may evolve in response to seasonal fluctuations (which are very slow in terms of an individual's lifespan).

Changes in physical conditions, such as temperature, can alter the outcome of competitive interactions. In his classic competition experiments with flour beetles, Park (1954) found that *Tribolium castaneum* always won under constant warm-moist conditions, while *T. confusum* always won under constant cold-dry conditions. At intermediate combinations of moisture and temperature, one or the other species always won, and the outcome was generally consistent with the results under the extreme conditions. One would predict that these species could coexist if significant changes in temperature occurred before competitive equilibrium was established.

Fluctuations in resource availability can also alter competitive interactions. Levins (1979) demonstrated theoretically that stochastic fluctuations of a single resource would allow the coexistence of two competing populations, one of which quickly suffered competitive exclusion under constant resource conditions (see also Koch, 1974a,b; Armstrong and McGehee, 1976a,b). Similar results have been obtained in computer simulation models of competition among phytoplankton (Grenney *et al.*, 1973; Kemp and Mitsch, 1979; Powell and Richerson, 1985). Controlled experiments with phytoplankton (Sommer, 1984) and studies of waterfowl populations (Tramer, 1969; Rotenberry, 1978; Nudds, 1983) have also provided empirical evidence for this phenomenon.

Resources may fluctuate as a result of environmental conditions that are completely unaffected by the presence of the organisms (i.e., allogenic or exogenous). However, resource fluctuations are often associated with the population levels of the organisms themselves (i.e., to some degree autogenic). In situations where a resource is limiting to population growth and organisms are able to compete for the resource by removing it from the environment, resource levels are generally lower when population densities are high than when population densities are low. Population fluctuations due to inherent cycles or to mortality-causing disturbances can produce variation in resource levels that allow the coexistence of species that would not coexist under constant resource conditions (Koch, 1974a,b; Armstrong and McGehee, 1976a,b). This phenomenon is particularly important in plant communities, where mortality of some plants is often associated with increased availability of nutrients, water, and light to other plants.

In summary, a variety of factors can slow the rate of competitive displacement and allow the coexistence of species that could not coexist at competitive equilibrium. Competitive exclusion can be slowed by 1. low population growth rates that result from low resource availability or other environmental conditions; 2. environmental fluctuations that alter competitive relationships and prevent any species from achieving complete dominance; and 3. sufficient similarity between competitors that competitive exclusion occurs very gradually. This explanation for coexistence among competitors is completely different from those based on a balance of competitive abilities.

A Dynamic Equilibrium of Opposing Processes: Disturbance and the Rate of Competitive Displacement

The two factors, mortality-causing disturbances and the rate of competitive displacement, that have been implicated in the maintenance of species diversity are obviously not independent. In fact, the effect on species diversity of either factor by itself produces a circular hypothesis. When the 'intermediate' rate of disturbance is defined as the rate that produces the highest species diversity, the relationship between intermediate disturbance and diversity is tautological. Similarly, the 'intermediate' rate of competitive displacement must be defined independently of species diversity if it is to be a testable hypothesis.

The circularity involved in separately considering the relationship of either of these processes to species diversity can be eliminated by considering their interaction with each other. The interaction between these two processes can be described using a three-dimensional figure, with the dependent variable, species diversity, on the vertical axis, and the two independent variables, frequency of disturbance and rate of population growth and competitive displacement, on perpendicular horizontal axes (Fig. 5.5). Diversity, as well as other community and ecosystem properties such as biomass, productivity, and resource availability, fluctuate in a range of values maintained by the interaction of growth and competitive displacement with disturbance.

This model was originally called the *dynamic equilibrium model* of species diversity, to distinguish it from competitive equilibrium models. This model applies to within-habitat (alpha) diversity among competitors, that is, among functionally analogous species within the same functional type. Consequently, the patterns predicted by the model should not be expected to appear among organisms that do not compete with one another. The model's dynamic equilibrium is based on the opposing forces of competitive interactions that lead to competitive exclusion and of mortality- causing disturbances (including predation and herbivory, as well as abiotic disturbances) that prevent competitive exclusion when the rate of competitive displacement is high and cause local extinction of slowly growing populations when population growth rates (and the rate of competitive displacement) are very low. The model does not predict a constant level of diversity for a particular combination of rates of competitive displacement and disturbance, but rather a fluctuating pattern of diversity that remains within limits set by the dynamic equilibrium. This dynamic equilibrium is analogous to, but at much smaller spatial and temporal scales than, the dynamic equilibrium model of island biogeography, where the opposing forces are immigration and extinction (MacArthur and Wilson, 1963, 1967; Simberloff, 1974).

In this dynamic equilibrium model, species diversity can be reduced by either of two processes (Fig. 5.5). On one side of the diagonal of maximum diversity, diversity is reduced by competition, while on the

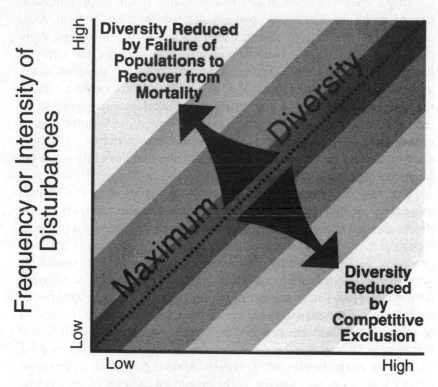

Fig. 5.5 Domains of the two primary processes that reduce species diversity in the dynamic equilibrium model. Diversity is reduced by competitive exclusion under conditions of high rates of population growth and competitive displacement and low frequencies and intensities of disturbance. Diversity is reduced by the failure of small or slowly growing populations to recover from mortality under conditions of low population growth rates and high frequencies and intensities of disturbance. Diversity is highest (and most likely to be influenced by landscape- and regional-scale processes) under conditions where neither of these processes dominate. (Based on Huston, 1979.)

other side of the diagonal, diversity is reduced by the inability of populations to recover from disturbances. The first of these processes is the competitive displacement that eventually leads to competitive exclusion and local extinction. Species diversity is reduced by this mechanism in situations where the effect of disturbance is insufficient to prevent competitive exclusion. The second process that reduces diversity is the local extinction of populations that are growing too slowly to recover from the disturbances. Just as with competitive displacement, the effect of any particular disturbance regime on the extinction rate depends on the growth rates of the populations. If most populations are growing rapidly, they will be able to survive a relatively high disturbance frequency or intensity. However, if most are growing slowly, only those with the highest growth rates will survive. This is the same mechanism that was hypothesized to underly the dynamic equilibrium of the number of co-occurring functional types in a community (Fig. 4.6).

Thus, this conceptual model is based on variation in the relative importance of different processes that influence species diversity. Local processes, such as competition, dominate and reduce diversity under conditions of high growth rates and low disturbance frequencies. Other local processes, such as mortality, reproductive failure, and local extinction, are important under conditions of low growth rates and high disturbance frequencies. However, over most of the range of conditions encompassed by the axes of the model, the effect of these local processes is weak. It is under these conditions that highest diversity occurs, and the effect of larger-scale factors that influence the size of the regional species pool, such as landscape heterogeneity, regional species pool, rates of immigration or speciation, etc., determine the maximum number of species that can coexist. This approach clearly demonstrates that the regulation of species diversity is not an issue of local processes versus regional processes, but of the expression of regional processes under some conditions and the dominance of local processes under other conditions.

Just as the relative importance of local versus regional processes is influenced by the conditions of the dynamic equilibrium, the functional relationships within a group of organisms can determine whether local or regional processes are likely to be most important. For example, in a comparison of tropical (9° S, Peruvian Amazon) and temperate (33° N, South Carolina, USA) forest bird communities, Terborgh (1980) identified 207 species in 24 guilds in the tropical forest, and 40 species in 16 guilds in the temperate forest. The nearly fivefold difference in species number is primarily a consequence of the greater average number of species in each guild in the tropical forest (8.6 versus 2.5) rather than a greater number of guilds. The potential significance of this pattern will be discussed in Chapter 14.

The dynamic equilibrium model illustrates how an increase in the

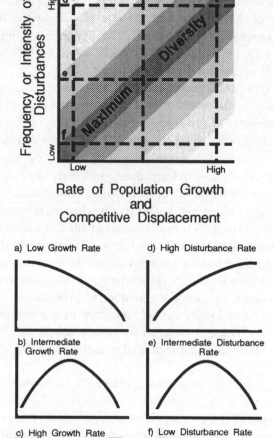

Fig. 5.6 Three-dimensional representation of the predictions of the Dynamic Equilibrium Model for species diversity in relation to the dynamic equilibria between different rates of competitive displacement (correlated with growth rate, productivity, etc.) and different frequencies of mortality-causing disturbances. Differences in diversity are represented by shading density, with the highest diversity along the diagonal ridge. The dotted lines illustrate the predicted changes in diversity when one parameter is held constant and the other varied. (Based on Huston, 1979.)

frequency or intensity of disturbance can have an effect on diversity similar to a decrease in the rate of competitive displacement. If both the frequency of disturbance and the rate of competitive displacement increase or decrease proportionately, the dynamic equilibrium would remain essentially unchanged and diversity would be little affected (i.e., the diagonal of maximum diversity in figure 5.6).

The most significant consequence of this model for understanding species diversity is the concept that the same absolute change in frequency (or intensity) of disturbance can have totally different effects on species diversity depending on the rate of competitive displacement. This is illustrated by considering the lines a, b, and c on Fig. 5.6, which represent respectively low, intermediate, and high rates of competitive displacement. As the frequency of disturbance is increased from low (level f) to high (level d) along line a, species diversity decreases from a maximum at the low level. This occurs because at the low rates of population growth that are associated with a low rate of competitive displacement, some populations are unable to recover from the mortality caused by the disturbances. As the frequency (or intensity) of mortality-causing disturbances is increased, an increasing number of populations are unable to recover and become locally extinct, resulting in a reduction in species diversity. At low rates of competitive displacement, the 'intermediate' frequency of disturbance that is needed to maintain high species diversity is very low.

In contrast, with high rates of competitive displacement (line c), species diversity *increases* as the frequency of disturbance is increased from low (level f) to high (level d). This occurs because species diversity is rapidly reduced by competitive exclusion when the rate of competitive displacement is high, and a high frequency of mortality-causing disturbance is needed to reduce the population of the competitively dominant species and allow the survival of other species, and the maintenance of species diversity. At high rates of competitive displacement, the 'intermediate' frequency of disturbance that is needed to maintain high species diversity is quite high.

The corresponding phenomenon of totally opposite effects on diversity as a result of the same increase in the rate of competitive displacement occurs depending the frequency of mortality-causing disturbance. This is illustrated by the changes in diversity along lines d, e, and f, as the rate of competitive displacement is increased from level a to level c.

The predictions of the dynamic equilibrium model can be illustrated with computer simulations, such as those of Figs 5.2 and 5.3 using

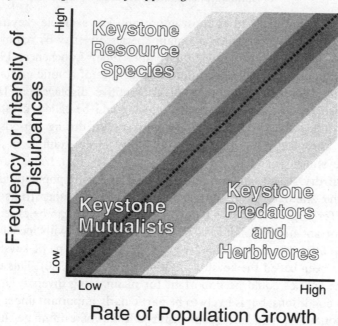

**Rate of Population Growth
and
Competitive Displacement**

Fig. 5.7 Conditions of dynamic equilibrium under which different types of 'keystone species' are expected to be important. Keystone predators will be important primarily under conditions in which the rate of competitive displacement is high and competitive exclusion is not prevented by other types of mortality, such as abiotic disturbances. Keystone resource species will be important under conditions of low growth rates that result from acute or chronically low resource availability. Keystone mutualists are likely to be most important in intermediate conditions. (Based on Huston, 1979.)

the Lotka-Volterra competition equations. The simulations were used to follow the temporal (i.e., non-equilibrium) *dynamics* of the Lotka-Volterra equations rather than to analyze their equilibrium solution, and demonstrated the totally different conclusions that could be drawn from the non-equilibrium behavior as compared to the equilibrium behavior of the equations. Many other types of analytical and simulation models can be used to demonstrate the consequences of this dynamic equilibrium (e.g., Huston and Smith, 1987; Caswell and Cohen, 1991; Chapters 7, 8, and 9).

The general framework of the dynamic equilibrium model can also be used to understand the conditions under which different types of biotic processes are likely to be most important. For example, 'keystone predators' (or herbivores) which maintain species diversity by removing or reducing the dominant competitor (Paine, 1966; Lubchenco, 1978) will only be important under conditions of low rates of abiotic mortality (e.g., Walde, 1986) and high rates of competitive displacement (Fig. 5.7). Another type of keystone species, which could be called 'keystone resources', allow the survival of dependent species during periods of acutely or chronically low resource availability. One example of this type of keystone species is the riparian fig species that fruits during the tropical dry season and supports primate and bird populations in parts of the Amazon basin during periods when no other fruits are available (Terborgh, 1986). This type of keystone species can be expected to be important under conditions of low growth rates (with increasing importance at higher disturbance frequencies). A third type of keystone species has been called the keystone mutualist (Gilbert, 1980). This type of keystone species could be important for maintaining diversity under nearly any conditions, but is likely to be particularly important under low resource conditions and infrequent disturbances, where intimate, long-term mutualistic associations can increase the efficiency of nutrient and energy acquisition by both partners (e.g., nitrogen fixation, mycorrhizal fungi; Boucher *et al.*, 1982).

Thus, the dynamic equilibrium model clarifies the conditions under which both the 'Intermediate Disturbance Hypothesis' and the 'Hump-backed Productivity-Diversity Curve' can be expected to apply. From Fig. 5.6 it is apparent that an intermediate frequency of disturbance produces a unimodal diversity response most clearly at intermediate to high rates of competitive displacement. At low rates of competitive displacement, the effect of an increasing frequency of disturbance is generally to produce a monotonically decreasing pattern of diversity. However, at high rates of competitive displacement an increasing frequency of disturbance leads to increasing diversity until the disturbances become so frequent that some species are eliminated.

It should not be surprising that virtually all the evidence in support of the 'Intermediate Disturbance Hypothesis' (Connell, 1978) comes from systems with high growth rates and high rates of competitive displacement, such as the intertidal zone (Paine, 1966; Sousa, 1979a,b), the crests of coral reefs (e.g., Connell, 1978; Connell and Keough, 1985), algae growing in shallow water (e.g., Lubchenco, 1978), or weedy plant communities. Likewise, the 'Hump-backed Productivity-Diversity Curve' (Grime, 1973, 1979) is most likely to be found in communities with intermediate to high frequencies of disturbance, such as herbaceous plant communities (Al-Mufti *et al.*, 1977). At very low frequencies of disturbance, the relationship between productivity and species diversity is usually monotonically decreasing.

Empirical Support for the Dynamic Equilibrium Model

Because the dynamic equilibrium model is essentially a synthesis of the intermediate disturbance hypothesis (Connell, 1978; Fox, 1979) and the intermediate productivity hypothesis (Grime, 1973, 1979), a test of the dynamic equilibrium model of species diversity cannot be achieved by experimentally manipulating only a *single* factor, whether it be disturbance or the rate of competitive displacement. Experimental manipulation of single factors or studies of communities along natural gradients have illustrated the limitations of hypotheses based on disturbance alone (e.g., Connell, 1978) or on productivity alone (e.g., Connell and Orias, 1964). A test of the dynamic equilibrium model in any particular system requires a *factorial* experiment with at least three levels of each factor (some mortality-causing disturbance such as mowing, fire, predation, or herbivory, and some factor that affects the rate of competitive displacement, such as resource availability or temperature). Such factorial experiments are logistically much more difficult than manipulation of a single factor, and consequently, there are, as yet, few good experimental tests of the theory. Considerable circumstantial support is provided by patterns in natural communities along interacting gradients of disturbance and rate of displacement (see Chapters 8 through 14). In addition, computer models such as mechanistic simulation models of plant competition (e.g., Botkin *et al.*, 1972) or Leslie matrix models of competition and succession (e.g., Caswell and Cohen, 1991) can be used to conduct computer 'experiments', the results of which conform closely to the predictions of the dynamic equilibrium model (see Chapters 6 and 7).

The following two examples are factorial experiments that have attempted to evaluate the dynamic equilibrium model.

Microbial Communities

Laboratory chemostats provide an ideal experimental tool to test the dynamic equilibrium model, just as simple laboratory batch cultures provided early evidence in support of the competitive exclusion principle under constant conditions. In flow-through chemostats, a nutrient-containing solution is constantly fed into an experimental vessel and excess solution, generally depleted of nutrients by the organisms, is constantly drained off. The constant outflow also carries organisms with it, acting as a chronic source of mortality that can be varied by altering the inflow (and thus the outflow) rate. Acute disturbances can be applied by simply removing some portion of the total volume of the solution in the experimental vessel (along with the organisms it contains) and replacing it with sterile solution.

Rashit and Bazin (1987) performed a factorial experiment with microbial communities using 100 ml chemostats with two levels of productivity (manipulated by the concentration of the solution fed into the chemostats) and three levels of disturbance (manipulated by the amount of solution that was removed at each disturbance). The three levels of disturbance were designated as C, which was only the amount flushed out with the continuous flow rate of 60 ml per day; $F1$, which was removal of 70% of the chemostat volume three times per week; and $F2$, which was removal of 98% of the chemostat volume twice per week. The need to maintain an equivalent nutrient supply to all three disturbance treatments necessitated an inverse relationship between the frequency of removal and the amount of material removed. However, the authors calculated that a population would have to double its numbers only twice per week to recover from the $F1$ level of disturbances, while a population would have to double its numbers five or six times per week to recover from the $F2$ level.

The microbial community used was a relatively complex mix from a local pond that was composed of bacteria, flagellate protozoa, and predatory protozoa (sarcodinians and ciliates). Both the bacteria and the flagellates obtained their nutrients directly from the medium, whereas the predators fed on the bacteria. The growth of photosynthetic organisms was prevented by conducting the experiments in the dark. There is clearly more going on in these chemostat communities than a simple competition experiment, which complicates the analysis. The authors present their results in terms of the diversity of each of the three major groups of organisms (bacteria, flagellates, and predatory protozoa), each of which was composed of many different species, and each of which achieved population densities that differed by one to two orders of magnitude.

The results of the experiments are consistent with the basic predictions of the dynamic equilibrium model. The effect on species diversity of the same increase in disturbance intensity was reversed at low versus high rates of nutrient supply. The diversity of both the bacteria and the

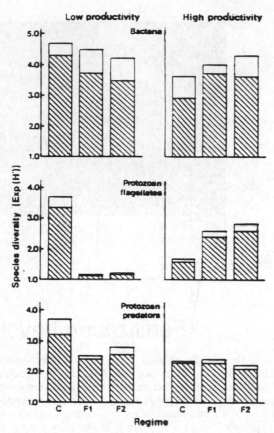

Fig. 5.8 Variation in the diversity of a microbial community in response to experimental variation of growth rates and mortality rates (chemostat washout). The treatment levels are constant (*C*), 70% removal three times per week (*F1*) and 98% removal twice per week (*F2*). The shaded areas of the bars show the average species diversities during the last four days of the experiments. The open areas show the species diversities when the data for the last four days are pooled to give larger sample sizes. In both cases, the results shown are the means from the two experiments carried out at each productivity level. (From Rashit and Bazin, 1987.)

flagellates (both of which obtained their nutrients from the chemostat solution) decreased with increasing intensity of disturbance at low levels of productivity, and increased with increasing intensity of disturbance at high levels of productivity (Fig. 5.8). Since the rate of competitive displacement should increase with the increasing population growth rates that result from higher nutrient availability, the results are consistent with the predictions of the dynamic equilibrium model. The pattern shown by

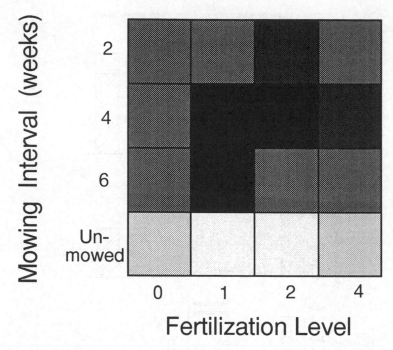

Fig. 5.9 Variation in the species richness of an oldfield plant community near Ann Arbor, Michigan, created by a two-way factorial experiment, with four levels of mowing and four levels of fertilization to manipulate growth rates and the frequency of disturbance. This pattern of diversity was measured after two seasons of treatment. With additional years of treatment, species composition continued to change, but the overall pattern of diversity remained the same. Density of shading corresponds to mean number of species per $2m^2$ sample area (three $4 \times 4m$ replicates of each treatment): from lightest to darkest, 9.0–11.3, 13.5–15.8, 20.3–22.5, 22.6–24.8, 24.9–27.2.

the more slowly growing predatory protozoans was a decrease in diversity with increasing intensity of disturbance at low levels of productivity, but no change in diversity with disturbance at high levels of productivity, a pattern consistent with the 'trophic shift' phenomenon (discussed later in this Chapter).

Experimental Fields and Grasslands

Herbaceous plant communities are easily subjected to experimental manipulations. In fact, the longest running ecological experiment ever conducted (started in 1856 and still being continued) is the Park Grass

Experiment at Rothamsted Experimental Station in England (Silvertown, 1980). These experiments investigate the effect of various types of fertilization and various systems of grazing on the yield and species composition of pastures (Lawes and Gilbert, 1880; Gilbert *et al.*, 1882; Thurston, 1969; Silvertown, 1980). The experiments have produced the results of both grazing and fertilization which would be expected based on the independent effects of disturbance and the rate of competitive displacement. Fertilization generally resulted in a decrease in species diversity (Fig. 2.14), while grazing had different effects depending on the intensity and pattern. The experiments were not designed to look at the interaction of these two factors.

Experiments to investigate the interaction of disturbance and the rate of competitive displacement were initiated in 1978 at the University of Michigan Matthaei Botanical Gardens (Huston, 1980b). The experimental manipulations were mowing at different intervals (every 2, 4, or 6 weeks, and unmowed), and different levels of fertilizer addition (0, 21, 42.5, and 85 kg Nitrogen per hectare in the form of NPK 26-3-3 fertilizer) to alter the rate of competitive displacement. Initial changes in species composition occurred within the first season of treatment, and species composition in some of the treatments continued to change during several years of treatment. The pattern of species diversity produced by the experimental treatments shows the interaction of disturbance frequency and rate of competitive displacement that is predicted by the dynamic equilibrium hypothesis (Fig. 5.9).

Synthesis

The evidence and arguments presented in this and the previous chapter suggest that competitive equilibrium, as it is mathematically defined, occurs rarely in most ecological communitites. When competitive equilibrium does occur, it is most commonly associated with competitive exclusion and a reduction in species diversity, rather than with the coexistence of many species at equilibrium. While equilibrium competition theory does permit fluctuations or limit cycles about the equilibrium levels of populations (and species diversity), these fluctuations are predicted to result from the dynamics of balanced competitive interactions, rather than from the factors such as mortality or environmental fluctuations that have been implicated in preventing competitive equilibrium from occurring. The 'non-equilibrium' community dynamics that have been emphasized in this chapter as critical to understanding local coexistence and species diversity are simply the changes in the sizes of competing populations that are not at competitive equilibrium. Thus, over most of the range of conditions considered by the dynamic equilibrium model, competition is not the dominant process structuring communities.

This concept of a dynamic equilibrium returns to ideas that were emphasized by ecologists at the beginning of the twentieth century. Cowles (1901, cited in McIntosh, 1985) emphasized the dynamic, successional natural of plant communities, stating that 'a condition of equilibrium is never reached...Succession is not a straight-line process. Its stages may be slow or rapid, direct or tortuous and often they are retrogressive'. Cooper (1913, cited in McIntosh, 1985) developed the concept of a dynamic equilibrium at the landscape scale, far in advance of the recent resurgence of interest in 'patch dynamics.' Based on his extensive studies of individually mapped and aged trees in stands on Isle Royale, he characterized the forest as a 'mosaic' of patches at different successional stages, as a result of small windfall disturbances of varying ages. 'The forest as a whole remains the same, the changes in various parts balancing each other' (Cooper, 1913).

While the processes on which the dynamic equilibrium model is based occur at the local spatial scale where competing organisms interact, this dynamic equilibrium also has important implications at the much larger spatial scales and longer temporal scales discussed in the previous chapter. If all patches have a similar frequency of disturbance and rate of competitive displacement, but are simply out of phase temporally, the species diversity of the landscape at any time should be similar to the long-term sum of diversity for any patch. Of course, the species richness of the landscape is likely to be greater than the species richness of any single patch, because the landscape represents a larger total area and its diversity is the sum of many patches that are at different points in relation to competitive equilibrium (that is, at different stages of succession). Temporal asynchrony in the timecourse of competitive exclusion in different patches allows the maintenance of diversity by dispersal of species into patches where they have become locally extinct. This process will tend to create a positive feedback that increases diversity at both the patch and landscape scales under conditions where the patch level diversity is high, and reduces landscape diversity where patch level diversity is low.

The dynamic equilibrium model predicts a spatial and temporal dynamic of patch structure and pattern on a landscape that depends on the same balance of the opposing processes as at the local scale (Fig.

5.10). The overall landscape pattern, in turn, determines which of the large-scale equilibrium processes (e.g., dispersal, immigration, speciation) can operate effectively in the maintenance or accumulation of species diversity. These landscape-scale processes can increase or decrease the level of diversity maintained by the patch-scale dynamic equilibrium. Variation in dispersal rate within different types of landscape or between different types of organisms can affect the level of diversity that can be maintained in a mosaic landscape (e.g., Caswell and Cohen, 1991).

As illustrated in Fig. 5.10, a variety of distinctly different landscape patterns can result from the dynamic equilibria of competitive displacement and disturbance represented by the four quadrants of the contour diagram of the model's predictions. In addition to the range of species diversity predicted for each of these quadrants, the model also indicates the general types of life history strategies that should be most successful under any particular dynamic equilibrium (Fig. 5.11).

For example, under conditions of high disturbance frequencies and low rates of recovery of populations from disturbance, the average plant biomass on the landscape will be very low, and variance in biomass between patches will also be low (Fig. 5.10a). Under these conditions of high disturbance frequency the very low rates of competitive displacement are unlikely to have any effect on species diversity. Instead, the local diversity of most patches on the landscape should be very low because the populations of most species will be unable to recover between disturbances. The only species present will be those that are able to complete their life cycles between the frequent disturbances and grow sufficiently rapidly for their population size to recover. These species will be a specialized subset of early successional species adapted to survive in unproductive environments with frequent disturbances (Fig. 5.11). Because the frequent disturbances prevent the survival of long-lived, late-reproducing (i.e., high K or late successional) species, these species will occur very rarely on the landscape, surviving only in patches that by chance escape disturbance for long periods, and by chance are successfully colonized by the long-lived species. Maintenance of species diversity by dispersal of species into areas where they have become extinct will be relatively unimportant because there are few refuges from which mid- and late-successional species can disperse.

Spatial variation in physical conditions on this type of mosaic landscape can have a dramatic effect on local species diversity and on the contrast between the areas with different conditions and the overall monotony of the landscape. Any conditions that increase the rate of

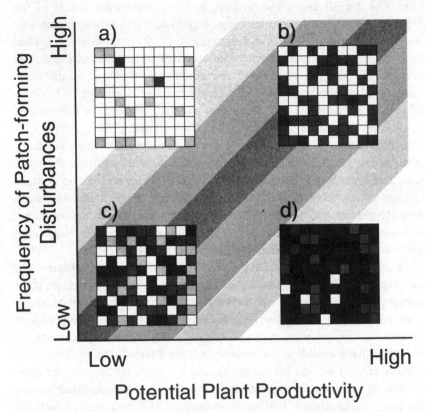

Fig. 5.10 Spatial patterns on a patch mosaic landscape predicted by the dynamic equilibrium model of species diversity (Huston, 1979). Shading density indicates the amount of plant biomass that can accumulate on a patch with a particular level of productivity between disturbances of a given mean frequency. The patterns can be produced on a landscape that is physically homogeneous, with the only structural variation resulting from temporal asynchrony of patch dynamics. While standing plant biomass provides an intuitively obvious example of the implications of the dynamic equilibrium model for this type of landscape heterogeneity, the same approach could be used to predict the spatial variation in other population and community properties, such as species diversity.

recovery from disturbance (also increasing the low rate of competitive displacement) or decrease the frequency or intensity of disturbance, should result in increased local diversity among the competitors, as well as a higher accumulation of biomass and the probable accumulation of interstitial species and non-competing functional types. For example, the effect of variation in soil moisture is particularly dramatic on arid

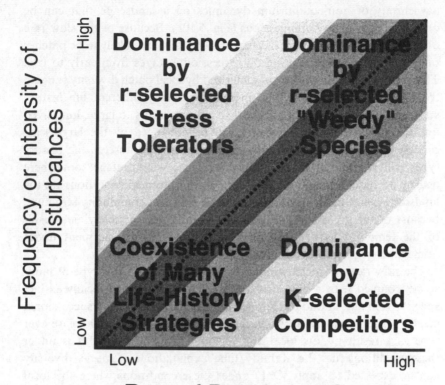

Fig. 5.11 Distribution of dominant life history strategies over the environmental conditions considered by the dynamic equilibrium model of species diversity (Huston, 1979). Under the conditions represented by the maxima along both axes, and the diagonal, different life history strategies will dominate. However, under the conditions of low disturbance frequencies and low rates of growth and competitive displacement near the origin, many different life histories will be able to coexist.

landscapes such as savannahs, where soil moisture can affect both plant growth rates and fire frequency and intensity (see Chapter 13).

Moving counterclockwise around the diagram, those conditions with a low rate of competitive displacement and also a low rate or intensity of disturbance will inevitably result in a complex spatial mosaic of standing plant biomass on the landscape that results simply from the temporal asynchrony of non-equilibrium dynamics on a landscape that can be otherwise physically homogeneous (Fig. 5.10c). Because of the slow rate of competitive displacement (i.e., succession), there are likely to be patches on the landscape representing all successional stages from early to late. This complex mosaic will be characterized by local patch diversity ranging from low to high, and by the presence of many different life history strategies, ranging from small, short- lived organisms to large, long-lived ones (Fig. 5.11). Species diversity will be enhanced by the large-scale process of immigration of locally extinct species, because all life history types will be present in an appropriate successional stage somewhere nearby on the landscape. Spatial variation in physical conditions on the landscape is not likely to have a dramatic effect on community structure, because of the great variability that is already present solely as a result of the temporal asynchrony of slow competitive displacement under homogeneous physical conditions.

The slow rates of local competitive displacement on this type of landscape, coupled with a high probability of immigration of locally extinct species from the abundant refuges on patches of different successional stages, suggests that the total species pool should be able to increase over time as a result of low local and regional extinction rates. It is under these conditions that the 'stability/time' explanations of species diversity can be expected to apply. Only under these conditions where the local and regional extinction rates are low, can species accumulate as a result of immigration or speciation over long periods.

Under conditions with high rates of growth and competitive displacement and a low frequency or intensity of disturbance, plant biomass will be quite high, with little variability between patches (Fig. 5.10d). Most of the landscape will be occupied by communities of competitors close to competitive equilibrium, with low diversity as a result of competitive exclusion. Because succession will rapidly approach competitive equilibrium, relatively few patches will be found in the higher-diversity early stages of succession. Thus, there will be few patches from which most species can disperse into areas where they have become locally extinct. Species with life histories characterized by slow growth, late reproduction, large size, efficient resource use, and other late-successional characteristics will dominate the landscape (Fig. 5.11). In spite of the low diversity among competitors, the high productivity and large amount of structural biomass that can accumulate should support many different functional types, particularly among interstitial species at higher trophic levels.

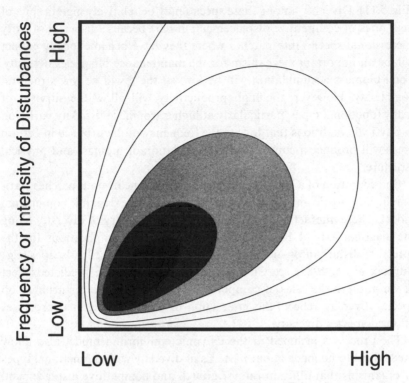

Fig. 5.12 Predicted levels of local species diversity that result from the interaction of local and landscape processes over the range of dynamic equilibria between competitive displacement and disturbance (from Huston, 1979). The pattern of local species diversity reflects the interaction between regional-scale processes on a mosaic landscape (Fig. 5.10) and local scale processes reflecting variation in the relative importance of competitive exclusion and extinction resulting from slow growth and disturbances (Fig. 5.6).

Landscapes characterized by high growth rates and high rates of competitive displacement along with a high frequency or intensity of disturbance (Fig 5.10*b*) will have a pattern complexity exceeded only by low-growth, low-disturbance conditions. Frequent disturbances will produce many patches that are denuded or have low density populations. However, recovery from disturbance will be rapid, so the contrast between recovered or recovering patches and recently disturbed patches will be great. Dominant life histories will be characterized by rapid growth, early reproductive maturity, small size, low efficiency of resource use, and other characteristics of early successional species in productive environments (Fig. 5.11). Diversity among these species may be relatively high in spite of high rates of competitive displacement, in part because dispersal of early successional species into patches where they have become locally extinct will be an important mechanism for the maintenance of species diversity. Total biomass accumulation will be low, so there will be low structural complexity. However, the high productivity will allow the survival of many functional types, particularly at higher trophic levels. Any variation in physical conditions that lowers the frequency of disturbance in certain areas will produce a conspicuous accumulation of biomass and physical structure.

Consideration of a landscape of temporally asynchronous patches clarifies the large-scale and long-term implications of the dynamic equilibrium model. These interactions contribute to the asymmetry of diversity along the diagonal axis of Fig. 5.12 and have important implications for the spread of disturbances such as fire or disease across these landscapes (e.g., Burdon, *et al.*, 1989; see Chapter 8). The shape of the predicted effect of variation in the rates of competitive displacement and disturbance on species diversity reflects the interaction of local and regional processes on local species diversity.

The processes included in the dynamic equilibrium model also allow predictions to be made about patterns of diversity within functional types of organisms that differ in rate of growth and competitive displacement. Functional types that coexist under the same disturbance regime, but have different rates of competitive displacement, should exhibit differing patterns of diversity along environmental gradients related to growth and competitive displacement. One particularly important example of this phenomenon is found by examining different trophic levels along a gradient of primary productivity. The inevitable loss of energy in transfers of resources from lower to higher trophic levels (e.g., Lindeman, 1942), means that populations at higher trophic levels will have less energy available for growth and reproduction than populations at the lower trophic levels from which they derive their energy (e.g., McNab, 1963). Thus, in comparison to lower trophic levels, populations at high trophic levels will be smaller in number and total biomass and require higher levels of primary (and secondary) productivity to maintain sufficient growth rates and population sizes to recover from disturbances and random fluctuations in population size (e.g., Eisenberg, 1980). Consequently, a given frequency and intensity of disturbance would be expected to eliminate more species at high trophic levels that at lower trophic levels. Likewise, a given frequency of disturbance will be more effective at preventing competitive exclusion and maintaining diversity among slowly-growing competitors within functional types at high trophic levels, than among the more rapidly-growing competitors at lower trophic levels.

Thus, the continuing argument about density-dependent versus density-independent regulation of community structure can be addressed in terms of the conditions under which each is most likely to be important. Density-dependent regulation by processes like competition (e.g., Lack, 1954) is likely to be most important at lower trophic levels and under conditions of high productivity and population growth rates. Density-independent regulation by disturbances and environmental fluctuations (e.g., Andrewartha and Birch, 1954) is likely to be most important at higher trophic levels and under conditions of low productivity and population growth rates.

The 'trophic shift' in maximum diversity can be illustrated using the three-dimensional figure presented earlier (Fig. 5.5). Over a fixed range of disturbance conditions, competitors at the lowest trophic levels should have the highest rates of growth and competitive displacement (Fig. 5.13, line *1*), while competitors at increasingly higher trophic levels will have decreasing rates of growth (Fig. 5.13, lines *2* and *3*). The maximum level of diversity will shift to lower disturbance frequencies at higher trophic levels, to the extent that the diversity of some trophic levels (e.g., primary producers) will increase over the same range of conditions where the diversity of other trophic levels (e.g., carnivores) decreases. At a given level of primary productivity, secondary productivity will decrease at higher trophic levels. Consequently, the effect on diversity of a given increase in primary productivity will differ among trophic levels (Fig. 5.13, lines *4,5,6*). This is demonstrated by the results of Hurd *et al.* (1971), who fertilized an abandoned field and found that the diversity of the primary producers was little affected, while the diversity of herbivores and carnivores increased by 25 to 75%. The trophic shift phenomenon provides another example of the necessity of separating biodiversity into appropriate components before it can be understood.

The same energetic constraints that underly the trophic shift in maximum diversity along disturbance and productivity gradients can also affect the interactions between trophic levels. The number of trophic levels that can be supported, as well as the potential biomass at a particular trophic level, increases with increasing primary productivity (DeAngelis

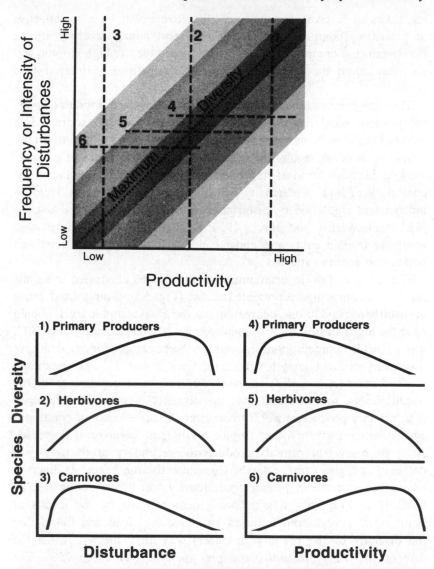

Fig. 5.13 Pattern of species diversity along a productivity gradient within functional types representing different trophic levels. The loss of energy between trophic levels (and the generally larger body sizes and lower growth rates at high trophic levels) has the inevitable consequence of reducing population growth rates with increasing trophic level in food chains based on a particular level of primary productivity (represented by the horizontal axis). Assuming a disturbance regime that affects all trophic levels proportionally, the conditions under which maximum diversity occurs should shift to higher levels of primary productivity at higher trophic levels. For a given level of primary productivity, maximum diversity of low trophic levels should occur at high disturbance frequencies, and maximum diversity of high trophic levels at low disturbance frequencies.

et al., 1978). Consequently, the negative effects that higher trophic levels can exert on lower trophic levels through predation and herbivory should be greater at higher levels of primary productivity (Oksanen *et al.*, 1981; see also Hairston *et al.*, 1960; Rosenzweig, 1973). At high levels of primary productivity, carnivores should be sufficiently abundant to control herbivore populations, which might otherwise be able to prevent competitive exclusion among primary producers (cf. Hairston *et al.*, 1960). At a somewhat lower level of primary productivity, a reduced density of carnivores may not be able to reduce herbivore numbers, with a consequently greater impact of herbivory on plant community structure. The net consequence of this type of trophic interactions (i.e., a 'trophic cascade' *sensu* Carpenter *et al.*, 1985) is to shift the maximum diversity of lower trophic levels to higher rates of primary productivity than would be associated with maximum diversity in the absence of predation or herbivory (e.g., Brooks and Dodson, 1965).

The contrasting ways in which plants and animals use resources leads to the prediction that plant diversity and animal diversity should be controlled by very different processes, and thus respond differently to the same environmental conditions. Plant diversity patterns reflect non-equilibrium competitive dynamics at the local scale of coexisting species (e.g., the dynamic equilibrium model) and larger scale (multi-patch) equilibria between local extinction and immigration (the landscape patch mosaic model, Fig. 5.10). Animal diversity is expected to reflect a situation closer to competitive equilibrium (see Chapter 4), with low diversity within functional types, but with most species specialized sufficiently to function as different types and thus avoid competitive exclusion. The functional type diversity of animals should be very strongly related to the productivity of the environment (i.e., Fig. 5.13) as well as to the spatial heterogeneity in environmental conditions and the heteogeneity in resources and structure provided by plants. For example, extremely high diversity of forest floor arthropods ($\sim$250 species/m^2) is found in the old-growth douglas fir forests of the Pacific Northwest of North America, where the highly productive forest is dominated by a single tree species (Lattin, 1990).

Finally, the dynamic equilibrium model may provide some insight into a long-standing controversy in ecology, the relationship between diversity and stability. The problematic theoretical result is that some models of complex, highly connected trophic systems are inherently less stable than models of simpler systems (Gardner and Ashby, 1970; May, 1972, 1973; Pimm and Lawton, 1977; see however, DeAngelis, 1975, 1980).

The natural phenomenon that gives credence to the theoretical result is that some diverse, complex ecosystems seem to recover slowly from disturbance, while less diverse ecosystems often recover more rapidly. A high rate of recovery is one measure of ecosystem stability (see DeAngelis, 1991). In the models and natural examples used to address this issue, the diversity of the community has been presumed to be the property that influenced the stability of the system (i.e., the rate of recovery) (May, 1973). However, the dynamic equilibrium model suggests that, at least for low levels of disturbance, high diversity communities are more likely to occur under conditions of low population growth rates, while low diversity communities are found under conditions with high population growth rates. To the extent that population growth rates are related to the rate at which the system is able to recover from a disturbance or other perturbation, the inverse relationship between diversity and some components of stability is simply a consequence of the type of environment in which high diversity communities are often found, rather than an inherent property of diversity and complexity.

Summary

The spatial scale to which the mechanisms of dynamic equilibrium model apply is the same spatial scale to which equilibrium competition models are meant to apply, that is, the scale of competing organisms. The model applies to competitive interactions within a relatively small patch on a landscape, and describes the dynamic equilibrium that occurs as a result of population fluctuations within that patch. To the extent that a patch can be characterized by a particular rate of competitive displacement (dependent on the level of resources available in the environment and other physical factors) and a particular frequency of disturbance, the model predicts that the patch should be characterized by a particular range of species diversity (and many other community and ecosystem properties). While the diversity of the patch will fluctuate over time as the community moves closer to or further from competitive equilibrium, there should be a characteristic long-term average diversity that reflects the dynamic equilibrium between competitive displacement and disturbance.

    However, it is important to keep in mind that the dynamic equilibrium within a local patch interacts with the constellation of patches that form the landscape. These landscape-scale interactions include the large-scale equilibrium processes described in the previous chapter. The maximum diversity within a local patch will be influenced by the number of species available on the landscape, i.e., the regional species pool, for dispersal into the patch. However, the effect of differences in the size of regional species pools is likely to be expressed primarily under conditions when the effects of the local processes of competitive exclusion and extreme mortality are weak. Likewise, the total number of species maintained in the entire landscape will be higher when most patches are characterized by a dynamic equilibrium with high species diversity.

Having considered the general processes that influence species diversity at different spatial and temporal scales, the next section will address the specific mechanisms that lead to competitive displacement and exclusion, as well as to population survival under various disturbance regimes and resource conditions. The focus will continue to be on the local scale of coexistence between competing species, specifically at the scale of interactions between individual organisms.

Part three

Mechanisms that Regulate Diversity at Various Spatial and Temporal Scales

Whether considered to be equilibrium or non-equilibrium, most of the explanations for patterns of species diversity discussed in the previous two chapters were based on processes that varied in their rate or relative importance in different environments. The operation of these processes, as well as the classification of components of species diversity (e.g., functional types versus functional analogues) has been addressed primarily in terms of differences between species. It is not surprising that explanations for species diversity should focus on the properties of species and interactions between species populations. However, a fruitful method in many fields of science has been to search for explanations to phenomena at one level of organization by studying mechanisms that operate at lower levels. The success of the reductionistic approach depends on avoiding excessive reductionism (i.e., the study of processes at levels too far below the phenomenon of interest), while also considering the effects of conditions imposed by processes that operate at levels above the phenomenon.

The actual mechanisms underlying the processes on which general theories of species diversity are based must be understood before those theories can be adequately tested. A mechanistic understanding of predator/prey and plant/herbivore interactions, competition, extinction, immigration and establishment, and other ecological processes is essential before specific predictions about patterns of community structure can be made and tested. The general theories of Part two make qualitative predictions about patterns of species diversity. However, quantitative predictions about species diversity must be based on the interaction of well-defined mechanisms that can be tested for internal logical consistency and for consistency with the results of experimental and comparative studies of community structure.

The five chapters of Part three describe components of a mechanistic theory of the structure of communities and their distribution through time and space. While I believe that the approach I describe can be applied to most groups of organisms, the focus of these chapters will be plants. The justification for emphasizing plants is the predominant role of plants as the structural organisms of most terrestrial (and a few aquatic and marine) environments. I believe that understanding the structure, productivity, and diversity of plants is the key to understanding the diversity of the many 'interstitial' organisms, the insects, birds, mammals, fungi, and small plants that make up the majority of the total number of species in any community. The distinction between functional types and functional analogues is critical to understanding the community structure of woody plants and plants in general, and variation in the importance of competition in relation to other processes will be a major component of these chapters. Although the goal of these chapters is to explain patterns of species diversity in communities and across landscapes, the theory is not based on species *per se*.

Species are an abstraction created for the purpose of organization and classification of knowledge about organisms. This abstraction is based on our understanding of evolutionary processes, and our observations of the patterns of variability among organisms. However, it is virtually impossible to study a species. The objects of species-level study are inevitably individual organisms that are classified as the same species because they are morphologically similar and come from a set of presumably interbreeding individuals (a population) or from a group of several such locally interbreeding populations (a metapopulation).

Neither the total extinction nor the successful establishment of a population is a process that occurs suddenly. Populations decrease in size by the death of one individual after another, until the last individual capable of reproducing dies and the local population or the species becomes extinct. Competition, predation, dispersal, establishment, extinction, and all of the fundamental ecological processes that influence species diversity are ultimately based on the actions and interactions of individual organisms. While the concept of species diversity is obviously based on the presence, absence, and relative abundance of the groups of similar organisms that we classify as species, an understanding of the causes and consequences of species diversity must extend to higher and lower levels of biological organization.

The organizational level below the species, that is, the individual organism, can be considered the basic unit of all ecological processes, and is in fact the basic unit of nearly all ecological research. With the exceptions of microorganisms, which are rarely studied as individuals; of colonial organisms, in which the object of study is usually a group of genetically identical individuals; and of clonal organisms, where the object of study is often a subunit of the individual (defined by its genome), most ecological studies are based on individual organisms. Admittedly, most ecologists and biologists try to study as many individuals as possible, particularly when the goal of the study is to understand what is happening to an entire population of individuals. However, the individual is still the basic unit of study, because it is individuals that are born, grow, die, eat, reproduce, are eaten, etc., and the aggregate fates of many individuals define such population-level phenomena as fecundity and mortality rates.

Above the species level are the ecological phenomena that are influenced by the presence of different types of organisms and that in turn influence what types of organisms are present in any area. Community structure, of which species diversity is only one component, nutrient and energy cycling, and the dynamic temporal and spatial patterns of ecosystems ranging in scale from clusters of microorganisms to the entire biosphere, are all influenced by the net effect of interacting individuals of many different species (Huston *et al.*, 1988). While there are important feedbacks on species diversity (and many other phenomena) from the aggregate effect of many species at these large scales and higher levels of organization (DeAngelis *et al.*, 1986), the feedbacks generally involve individual organisms, rather than the population as a whole.

The next five chapters describe a mechanistic approach to understanding patterns of biological diversity that is based on processes that occur at the scale of the individual organism. The basic approach dates back to the early 1970s (Botkin *et al.*, 1972), predating by more than a decade the recent enthusiasm for 'mechanistic' theories (cf. Schoener, 1986; Tilman, 1987a; see also Huston, 1992). Chapter 6 describes the interactions among individual organisms that influence the structure of monospecific populations, and concludes that in many cases the mechanisms of intraspecific interactions are the same as the mechanisms of interspecific interactions. Chapter 7 focuses on the properties of individual plants, and how these properties influence the structure of populations, communities, and ecosystems. The object of these chapters is to develop an understanding not only of the number of species that are present in a particular community, but of the properties of the species that compose a community, how they are distributed across landscapes, and how that distribution changes through time.

Chapters 8 through 10 address address spatial and temporal variation in biological diversity on both local and landscape scales. Landscape patterns of species distributions and biological diversity are constantly changing at several very different rates (see Cowles 1899, 1901; Clements, 1916; Clements *et al.*, 1929; Gleason, 1923):

1. *Rapid Change.* The fastest rate of change results from mortality-causing disturbances such as fires, volcanic eruptions, landslides, floods, severe weather such as windstorms or unusually low temperatures, and disease epidemics. Such rapid processes generally occur infrequently at a specific location, but have a dramatic effect when they do occur. These disturbances may alter the topography and the distribution of resources, as well as affect organisms directly.

2. *Moderate Rate of Change.* The next highest rate of change is the result of plant growth. The 'recovery' of a plant and animal community from mortality-causing disturbances, or simply the revegetation and refauna-tion of devegetated areas, is also known as secondary succession. Biotic change resulting from plant succession occurs continuously and is the predominant mechanism of change on all landscapes that can support vegetation.

3. *Slow Change.* Slower change results from gradual changes in climate or from other slow processes such as soil development during primary succession, and nutrient loss due to the weathering and leaching of mineral nutrients. These slow processes interact with plant community dynamics by altering the physical conditions under which primary and secondary succession occur, and thus altering the course of succession.

4. *Slowest Change.* The slowest changes can be attributed to the geologic processes of formation, alteration, and movement of land masses and the biological processes of the evolution of new species and the extinction of other species.

Chapter 8 addresses the most rapid rates of change on landscape, those due to sudden disturbances. Slower rates of change that result from successional and geological processes are considered in the Chapter 9. Spatial patterns such as zonation, and how those patterns change through time, are discussed in Chapter 10.

6

Diversity within Populations

If it is true that most fundamental ecological processes operate at the level of individual organisms, then understanding these processes at the individual level should be the key to understanding populations, communities, ecosystems, or any higher level of ecological organization that we choose. There should be strong parallels between mechanisms and phenomena at different levels. In particular, the diversity of populations and the diversity of species, which are essentially different ways of grouping individual organisms, should show similar patterns with similar explanations.

In order to understand the mechanisms of interactions between individual organisms, it makes sense to begin to study the interactions between individuals in as simple a situation as possible. By limiting consideration to populations of a single species, it is possible to greatly reduce the types of variability than can contribute to diversity among individuals within a group. The fact that all individuals in a population are similar enough to be considered the same species means that the total variability in size, morphology, behavior, genetic material, etc., among those individuals is necessarily much lower than would be found in a sample of the same number of individuals of many different species. Nonetheless, considerable variability can exist within populations, and the variability of different properties can be influenced by many different factors.

Morphological differences between individuals can result from their response to environmental conditions or from genetic variability (Schluter, 1988; Grant, 1981, 1986; Grant and Boag, 1980; Schluter and Grant, 1984; Rosatti, 1982, 1987). An extreme example of the first phenomenon is the determination of the sex of some reptile and insect species by the temperature at which the eggs are incubated (Bull and Vogt, 1979;

161

Ferguson and Joanen, 1982). Genetic variability among individuals of the
same species can impose differences in size, morphology, behavior, and
other phenotypic traits independent of environmental conditions, as well
as impose limits on the range of phenotypic variability that can occur in
response to environmental conditions (Clausen *et al.*, 1941).

Application of the concept of diversity at the species level to genetic
diversity or to the distribution of material among different individuals
within a species population, is very similar to application of the concept
at higher levels. If all individuals within a population are very similar in
size, the evenness (of the size distribution) of that population would be
high. Genetic diversity within a population (Allendorf and Leary, 1986;
Ledig, 1986) represents part of the continuum of genetic variation that is
the basic material for natural selection, as well as the basis for taxonomic
classification.

Genetic Diversity

Genetic diversity can be evaluated at many levels, including population,
community, ecosystem, and biome, as well as within individual organisms such as trees, which are composed of repeated modular units that
may differ genetically because of somatic mutations (Whitham, 1981;
Whitham and Slobodchikoff, 1981; Whitham *et al.*, 1984; Gill, 1986).
While an individual's size can be simply characterized, its genetic code is
composed of a hierarchical set of units whose variability can be characterized in several different ways. Genetic diversity can be measured using
the same statistics that have been applied to species diversity, with the
basic units being single bases, base triplets, alleles, genes, gene complexes,
chromosomes, etc. (Allendorf and Leary, 1986; Ledig, 1986).

The genetic structure of a population or metapopulation is influenced
by many factors (e.g., Loveless and Hamrick, 1984), and a detailed discussion of a property as complex as genetic diversity is beyond the scope
of this book. Nonetheless, it is important to keep in mind the potential relationships between genetic diversity and species diversity. Genetic
diversity within a species is in some cases positively correlated with environmental variability (Levins, 1968; Hedrick *et al.*, 1976; McArthur *et
al.*, 1988), which may also influence species diversity. The genetic component of adaptation of a species to different environmental conditions
(e.g., ecotypes) has long been recognized (Clausen *et al.*, 1941).

The genetic diversity of a community is inevitably correlated with the
species diversity of that community, because the genetic diversity of a

community is simply the sum of the genetic diversity of its component species. However, the genetic diversity of the individual species within the community has no fixed relationship to species diversity. One can hypothesize that the genetic diversity of a species should be *positively* correlated with the probability that the species will be able to persist in the community by adapting to changing conditions or competing effectively (Vrijenhoek, 1985). This hypothetical relationship between the genetic diversity and the long-term survival of a species is a major concern of conservation biology, but is largely speculative, since virtually no data are available to test this hypothesis.

Linhart (1974, 1976) reports differences in genetic variability of annual herbs along a spatial gradient from the center to the edge of temporary vernal pools in California. Phenotypic variability was higher in the plants from the periphery of the pool than from those from the center. When grown in a greenhouse, the plants from the periphery weighed significantly more, produced more branches, devoted a higher percentage of their biomass to vegetative parts rather than reproductive parts, and produced more but lighter seeds. There are many alternative mechanisms that might explain these patterns (Thompson, 1985).

One could also speculate that the average genetic diversity of each species in a community should be *negatively* correlated with the species diversity of the community. The rationale behind this hypothesis is that with few species, each species should be able to occupy a broader niche, which would require greater phenotypic variability than a high diversity community if competition restricted the species to a narrower niche. Presumably greater phenotypic variability would be associated with the persistance of greater genetic diversity. While this is a plausible and intuitively attractive hypothesis, it is likely to apply only to a subset of communities in which the diversity is regulated by mechanisms compatible with the assumptions of the hypothesis. The critical assumption is that diversity is related to the way in which resources are divided between organisms, that is, niche breadth.

Evidence on the niche breadth of finch communities from island and continents supports this hypothesis. Schluter (1988) found that the niche breadth (morphology, food selection, behavior) of finches in Kenya was narrower than that of Darwin's finches in the Galapagos. This contradicted their hypothesis that niche breadth would be greater in Kenya because of the greater diversity of food resources. He concluded that competition from other non-finch bird species in Kenya was responsible for the reduced niche breadth in comparison to the Galapagos, where the finches (Darwin's finches) have diversified remarkably in the absence of other bird species.

If increasing species diversity is associated with decreasing niche breadth, the hypothesis of an inverse correlation between species diversity and the genetic diversity of individual species in the community may hold. However, niche theory is based on the assumption of equilibrium, which is more likely to apply in animal communities, particularly vertebrates and other mobile organisms with well developed perception. The relationship between species diversity and niche breadth is less likely to hold in non-equilibrium situations, which are likely to occur among plants and organisms with low mobility and perception. Within some insect species, genetic diversity is positively correlated with the diversity of their host plants (Halkka *et al.*, 1975; Halkka, 1978).

Clear differences in genetic variability between species from different successional stages have been found (Zangerl *et al.*, 1977; Hamrick *et al.*, 1979; Brown and Marshall, 1981). A basic pattern seems to be that 'species with large ranges, high fecundities, an outcrossing mode of reproduction, wind-pollination, a long generation time, and from habitats representing later stages of succession have more genetic variation' (Hamrick *et al.*, 1979).

Size Diversity

An obvious, easily, and therefore, often measured feature of an individual is its size. In organisms with indeterminate growth, such as many species of trees, fish, and colonial organisms, size can vary over many orders of magnitude between different individuals, simply as a function of age or growth rate.

Individuals of species with determinate growth, such as mammals and birds, vary much less in size. However, even in populations of these organisms, there may be significant variation in size as a result of genetic differences, sex differences, differences in resources available during growth, and many other factors (Huston and DeAngelis, 1987).

Size diversity is a simpler and much more thoroughly studied type of variation within populations than is genetic diversity. Size diversity can differ greatly between populations of organisms with indeterminant growth, and it is in these organisms that most of the research on size distributions has been done. Plants in general, and trees in particular, exhibit extreme variability in size of individuals in a population. Since plants are not only the energy source for all animal life, but also provide most of the physical structure in nearly all terrestrial communities and many aquatic communities, understanding the causes of size diversity in plant populations is an important step in understanding the species diversity of communities dominated by plants.

Before discussing the mechanisms that influence size distributions, it may be useful to discuss how to evaluate size diversity. The distribution of resources among organisms has already been identified as an important aspect of both individual size and species diversity. In communities with high species diversity, biomass (representing the resources carbon, water, and nutrients) is divided relatively evenly among many species. Even distribution of biomass in a population implies that all individuals are the same size.

The characterization of size diversity is inherently different than the characterization of species diversity, because consideration of individuals confounds the relationship between richness and evenness. Since individuals are identical except for size differences (when size distribution within a population is the topic of interest), size richness would logically be defined as the number of different sizes in the population. However, the intuitive definition of size evenness would assign maximum evenness to the situation in which all individuals were identical in size, that is, the distribution of biomass among individuals in the population was perfectly even.

These definitions result in the counter-intuitive result that size evenness is inversely correlated with size richness, which is the reverse of the pattern that is generally found with regard to species diversity. The problem arises primarily because there is no obvious unit of size that corresponds to species. If one were to establish a standard size-class unit, it would be possible to characterize size-class richness by the number of different size classes represented in a single population, and size-class evenness by the relative distribution of biomass or number of individuals among the size classes. This still leaves the problem of whether evenness is to be defined on the basis of individuals, or size classes of individuals. The approach I will take is to focus on the evenness of the distribution of biomass among individuals in a population, and simply recognize that a great range in size (high 'size richness') will be associated with low evenness.

Quantification of size evenness has focused on description of the shape of the histogram or curve that represents the size distribution. Standard statistical descriptors of distributions, such as variance, coefficient of variation, skewness, and kurtosis, have been used to describe size distributions, but do not capture some of the biologically relevant properties of these distributions (see Weiner, 1985; Weiner and Solbrig, 1984).

The relative number of large versus small individuals and the relative difference in size between large and small individuals is biologically important because it often reflects the past, as well as predicts the future, division of resources between individuals of different sizes (Huston and DeAngelis, 1987). A statistic from economics that is used to describe the inequality of the distribution of wealth between the rich and poor, called the Gini coefficient, has been adopted by some population biologists to describe the relative equality of plant size distributions (Weiner, 1985; Weiner and Solbrig, 1984; Bonan, 1988) (Fig. 6.1). This statistic is zero when all individuals are the same size (i.e., wealth is evenly distributed) and increases towards 1.0 as inequality of resource distribution (size or wealth) increases.

Extreme inequality in size distributions can produce a bimodal size distribution (Fig. 6.1c) with no overlapping size classes between the 'haves' and 'have nots'. Although no good statistical methods yet exist to quantify the degree of bimodality, its presence indicates extreme size inequality.

Given this definition of the equivalence between size evenness and species evenness, an interesting question is whether, or under what circumstances, high species evenness is positively correlated with high size evenness within the populations in a community.

Factors that Influence Size Diversity

Changes in the size distribution of a population result from the aggregate effect of factors or processes that influence the growth of individuals. The most important factors are 1. the initial distribution of individual sizes; 2. the distribution of growth rates among the individuals; 3. the dependence of the growth rate of each individual on the size of that individual and the time at which growth occurs; and 4. mortality that may differentially affect individuals of different sizes. Each of these factors may be influenced by an individual's genetic makeup, as well as by abiotic and biotic environmental effects.

Huston and DeAngelis (1987) classified the biological processes that influence size distributions as either 'inherent' (i.e., genetic) or 'imposed' (i.e., by the external environment), and as either 'noninteractive' or 'interactive', depending on whether they require interactions among the individual organisms in order to be expressed (Table 6.1). Nearly all of

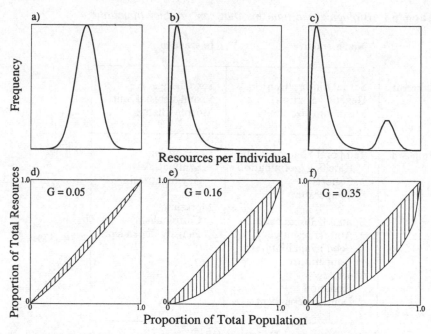

Fig. 6.1 Two methods for assessing the evenness of population size Upper row: Size frequency distributions showing (*a*) a normal unimodal distribution of sizes (or resources), which indicates a relatively even distribution of resources; (*b*) a skewed distribution of sizes, indicating that most individuals are small and have few resources, while a few are large; and (*c*) a bimodal distribution, which indicates extreme inequality of sizes, with separate subpopulations of small and large individuals. Lower row: Graphs illustrating the cumulative resource function (Lorenz curve) for the three frequency distributions above. The diagonal line indicates an equal distribution of resources (or size) among all members of the population. Increasing proportion of the total populations. The Gini coefficient (G) is the ratio of the shaded area to the total area below the diagonal line (Sen, 1973).

these processes influence the distribution of biomass between individuals and species within a community of competitors, as well as the distribution of biomass between individuals of the same population.

In most ecological interactions, the size of an organism is an extremely important determinant of its growth and survival. Large individuals are often able to capture a larger proportion of the resources and consequently grow faster and reproduce more than smaller individuals. Larger individuals are often more resistant to environmental stresses and less vulnerable to predation, although there are advantages to being small, such as being too small to be seen by predators (Hall *et al.*, 1970; Timms

Table 6.1. *Biological mechanisms that can produce bimodality*

	Noninteractive	Interactive
Inherent	Sexual dimorphism Genetic variation Maximum size limits	Sex Change Sexual/asexual Shifts Morph Change

Imposed	Temporal Heterogeneity
	   Hatching/germination
	   Cohorts
	   Temperature
	
	Spatial heterogeneity
	   Abiotic resources
	   Food availability/size
	   Temperature
	Mortality
	   Predation
	   Density-dependent
		Competition
		   Symmetrical
		   Asymmetrical
		
		Mortality
		   Cannibalism
		   Density-dependent

From Huston and DeAngelis (1987).

and Moss, 1984; Kerfoot (ed.), 1980; Kerfoot and Sih (eds.), 1987) and the shorter generation time usually associated with small size. Because of the influence of size on so many ecological interactions, both growth and mortality are often size-dependent. Consequently, any processes that influence the sizes of organisms are likely to be important in determining the structure of both populations and communities.

The processes that influence variation in size evenness are primarily those that are imposed by the external environment, including both interactive and noninteractive processes. Among imposed noninteractive processes, spatial and temporal heterogeneity of the environment can have dramatic effects on population size structure, just as they can have a significant influence on species diversity (Chapters 4 and 5).

Effects of Spatial Heterogeneity on Population Size Diversity

The effects of spatial heterogeneity on population size structure are analogous to the effects of spatial heterogeneity on species diversity. At the community level, organisms in similar, but spatially separate, environments are not likely to be competing with each other, so a high level of spatial heterogeneity can potentially allow the coexistence of many different species, and thus high species diversity.

At the population level, the physical separation of groups of otherwise identical individuals into environments with different growth conditions can result in a wide range of sizes (low evenness of size distribution) in the population as a whole. Individuals in one environment are not likely to be interacting with individuals in another environment, although mobile individuals may move between environments or all individuals may move into the same environment, either intermittently or permanently. When individuals of different size come together in the same environment, any size differences are likely to have a major influence on the outcome of competitive or cannibalistic interactions among individuals in the population (Huston and DeAngelis, 1987; Polis, 1988; Polis *et al.*, 1989).

In some cases organisms may interact with their environment and with other organisms at many different spatial scales over the course of their lifetime. This is most likely to be the case with organisms that have indeterminate growth and become very large, such as trees or some fish. In the case of trees, spatial heterogeneity in such factors as light, water, or nutrients can have a major effect on population size structure when the trees are very small (Huston, 1986; Huston and DeAngelis, 1987). However, as an individual tree increases in size, spatial heterogeneity at the small scale becomes irrelevant since the larger tree is able to draw resources from a large area that includes the many small patches of the initial small-scale heterogeneity. Large trees affect other trees that were too far away for interactions when the trees were small. Thus, the early effects of small-scale heterogeneity can lead to size differences that are critical to interactions between individuals when they reach a larger size.

Effects of Temporal Heterogeneity on Population Size Structure.

For interactions that are influenced by the relative size of organisms, such as predation and many types of competition, the timing of growth can be just as important as the effect of different growth rates on size. It is not uncommon that some individuals start growth later than others, such as when there are temporary delays in the germination of seeds or the spawning of fish that may result from periods of unfavorable weather. In these situations, the temporal variation in the initiation of growth results in an increase in the variability of the initial size distribution of the organisms. If the growth of the individuals is size-dependent for any reason, the initial variation in the size distribution can be magnified because the larger early individuals grow faster than the smaller late individuals. Such situations can result in bimodal size distributions, as have been frequently reported for fish populations (Cooper, 1936; Konikoff and Lewis, 1974; Shelton *et al.*, 1979; Timmons *et al.*, 1980; Keast and Eadie, 1985).

Timmons *et al.* (1980) studied the changes in the size distribution of large-mouth bass during their first year of growth in a reservoir in the southern United States. They also also measured the availability of prey during the same time period, which seemed to explain the pattern of bass growth. The bass were spawned in May and the initial size distribution was unimodal, with no great disparities in size. During the first two months of growth, abundant food was available for young bass of all sizes. However, in July food available to bass less than 75 mm in length was reduced by growth of most prey beyond a size that could be ingested by small bass, while food available to bass larger than 75 mm remained abundant. This size-dependent shift in food availability apparently affected the relative growth rates of the large and small bass, and by the end of the summer the size distribution of the bass was bimodal, with a group of larger, fast-growing individuals, and another group of small, slowly-growing individuals.

A computer simulation model was developed by Adams and DeAngelis (1987) to investigate the phenomenon described above. The model simulated the feeding and growth of individual large-mouth bass in relation to the availability of their primary prey species, first-year threadfin shad. The growth of each bass depended on the number and size of prey that it consumed. On each day of the simulation an individual bass encountered a random number of shad drawn from the size distribution of shad in the reservoir at that time. However, the bass were not able to eat all the shad they encountered, but only those shad that were small enough. Based on feeding experiments, it was determined that a bass could eat any shad that was less than or equal to half its length, so the growth of each individual bass was influenced by its own size in relation to the size distribution of available prey shad. Because the shad grew continuously throughout the season, the size distribution of the prey changed. Bass that, by chance, grew slowly at the beginning of the season experienced an increasing scarcity of suitable prey, as most shad grew to sizes too large to be consumed by small bass.

Many different simulations were performed which followed the feeding and growth of individual bass throughout the growing season, beginning with different initial size distributions of the prey. If the initial size

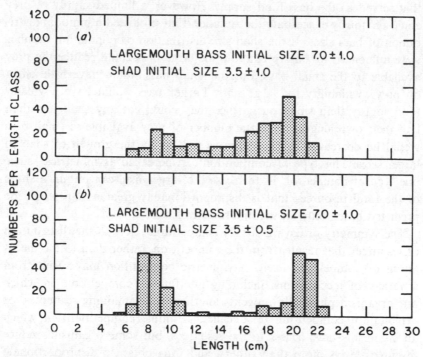

Fig. 6.2 Effect of initial prey size variability on the size distribution of a predator population. Each size distribution was produced by an individual-based simulation model of predator-prey interactions in which bass young-of-the-year randomly encountered their prey, shad, but were only able to capture and ingest shad that were less than or equal to half their own length. Both size distributions are from the end of the growing season (assumed to be November 15, in the simulated reservoir) and developed from size distibutions that were unimodel at the beginning of the simulations on July 1. Note the more extreme size bimodality in bass when the initial shad size distribution had lower variance (6.2*b*). (Based on Adams and DeAngelis, 1987.)

distribution of the shad was broad enough, that is, there were many shad in the small size classes, most of the small bass were able to encounter enough small prey to grow throughout the season. However, if there was little variation in the initial shad size distribution, the shad 'outgrew' the small bass and the small bass became a slowly-growing stunted subpopulation (Fig. 6.2).

Effects of Competition on Population Size Structure

The above simulation did not involve competitive interactions among the bass, since bass predation did not affect the much larger shad population that served as the bass food supply. However, a limited supply of prey could obviously exacerbate the situation that produced a bimodal distribution of bass sizes. If the shad size distribution or population number were influenced by bass predation, the large bass could reduce the prey available to the small bass, but the small bass would have little effect on prey availability for large bass. Larger bass would have a greater food supply than small bass, since they could eat any shad less than half their own length, while the number of prey available to small bass would be decreased by predation as well as by the growth of shad to larger sizes. This type of competition can be called asymmetrical, since one type of competitor, in this case the larger fish, can obtain a share of the total resources that is disproportionately greater than would be predicted based only on the difference in sizes.

The symmetry or asymmetry of competition refers to the distribution of resources that results from the competition, rather than to the mechanism of competition itself. Asymmetric competition may result from a variety of mechanisms, including interference competition, in which one organism physically prevents another from obtaining resources, or exploitation competition (also called scramble competition) in which all organisms have access to the resources, but some organisms reduce resources levels more than others, and thus obtain a disproportionate share. Symmetric competition is more likely to result from exploitation in which all organisms deplete the resource pool on an equal basis, and none have a disproportionate advantage.

Asymmetric competition between large and small plants provides a close analogy with the asymmetric competition between large and small bass. Among plants, competition for light is almost always asymmetric, since the supply of light comes primarily from overhead and the tallest plants can intercept most of the light, leaving very little for smaller plants underneath them, which become 'suppressed' and grow even more slowly. Asymmetric competition for light often leads to a bimodal or otherwise inequitable size distribution in populations that were initially unimodal, such as even-aged monocultures (Ford and Newbould, 1970; Ford, 1975; Mohler et al., 1978; Rabinowitz, 1979; Bradbury, 1981; West and Borough, 1983; Weiner, 1985) and also among the shade-tolerant species in all-aged stands (Lorimer, 1980; Lorimer and Krug, 1983; Glitzenstein et al., 1986).

Many different models have been developed to explain the appearance of bimodal size distributions in plants. Nearly all the models follow the interactions between individual plants, and include a mechanism that al-

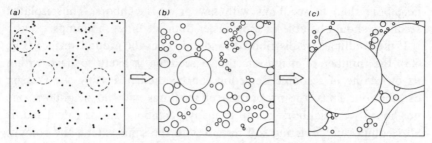

Fig. 6.3 Typical random distribution of seedlings on a plot of ground, illustrating variation in the number of neighbors of each seedling and the consequences for subsequent growth of the plants. (*a*) Some seedlings will initially grow under crowded conditions and others will grow under uncrowded conditions. (*b*) Plants with fewer neighbors are less affected by symmetric competition for belowground resources, and are able to grow more rapidly. (*c*) Plants that by chance escaped from symmetric competition for belowground resources and achieved a larger size are able to completely dominate smaller plants in asymmtric competition for light. (From Huston *et al.*, 1988, based on Huston, 1986.)

lows the larger plants to grow much more rapidly than the smaller plants (Diggle, 1976; Gates, 1978; Aikman and Watkinson, 1980; Ford and Diggle, 1981; Huston and DeAngelis, 1987; Huston, 1992). With asymmetric competition, the small stunted individuals generally experience the highest rates of mortality, which may actually reduce the bimodality of the size distribution.

In symmetrical competition there is no advantage conferred by size or other properties of the individual organism. While larger individuals may capture more resources, the increase in resource capture is proportional to the size difference between large and small individuals. Symmetrical competition among plants can also produce bimodal size distributions in plant populations (Huston, 1986; Huston and DeAngelis, 1987) by causing spatial heterogeneity in resources that leads to variation in growth rates and eventually to asymmetric competition for light.

Consider the situation of plant seedlings competing for a limited supply of nutrients or water in the soil. Each individual can reduce the resources available in some area surrounding itself, which could be called a zone of resource depletion. Depending on the size of the zones of resource depletion and the density of plant seedlings, the zones of neighboring seedlings may overlap. Some seedlings will by chance have many neighbors and some will have none. If all seedlings deplete the resources, there will be fewer resources available to seedlings with many neighbors than to seedlings with few or no neighbors. Thus isolated seedlings should be able to grow faster than seedlings in clumps.

If the seedlings are distributed randomly over the ground surface (Fig. 6.3), the number of neighbors that a seedling is likely to have within its own zone of resource depletion can be predicted using the Poisson distribution. Thus, the proportion of seedlings with no neighbors, one neighbor, two neighbors, on up to many neighbors can be calculated. If all seedlings are depleting soil resources to the same extent, the amount of resources available to each seedling should be inversely proportional to the number of neighbors it has. Since plant growth rate is closely tied to resource availability, the distribution of growth rates can then be predicted based on the number of neighbors (Fig. 6.4).

The density of the randomly distributed seedlings determines the distribution of growth rates, and thus the effect on the population size distribution. If the density is high, most seedlings will have many neighbors and grow slowly. If the density is low, most seedlings will have few neighbors and be able to grow rapidly. At intermediate densities, there can be a wide range of growth rates, since some seedlings will by chance have many neighbors, and some will have none. Under these conditions, a bimodal size distribution may result from symmetric competition through the effect of plant spatial distribution on resource heterogeneity (Huston, 1986).

The relevant spatial density of plants in terms of resource use depends on the size of the area in which resource depletion occurs, in relation to the total area available for growth. For a given total area, then, a high density can result from either many small plants, or a few large plants. Obviously, as plants increase in size (actually, as their zones of resource depletion increase in size), the effective density increases, and the distribution of growth rates will change as predicted by the Poisson distribution, as long as the spatial pattern remains random.

Plants do not withdraw resources from the entire zone of resource depletion at the same rate, but generally reduce resources from the center more rapidly than from the periphery (Wu *et al.*, 1985; Sharpe *et al.*, 1986; Walker *et al.*, 1989). For a resource depletion zone of a given radius, the pattern of resource withdrawal can be used to predict the distribution of resource availability, growth rates, and the resulting size distribution (Huston and DeAngelis, 1987) (Fig. 6.5).

An interesting example of bimodality within dense, monospecific patches is given by Thompson (1978). In patches of the weedy cow parsnip (*Pastinaca sativa*), individuals at the edge of the patch were less

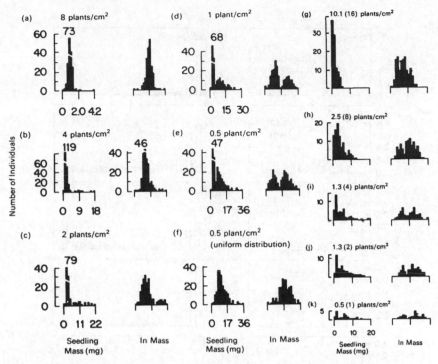

Fig. 6.4 Simulated and actual size distributions produced by symmetric competition among plant seedlings randomly distributed at different densities. (*a-f*) Simulated size histograms based on Poisson distribution of neighbors in plots of different density. (From Huston, 1986.) (*g-k*) Size histograms of seedlings of *Festuca paradoxa* sown into flats at different densities. (Based on Rabinowitz, 1979.)

crowded than those in the center. The peripheral individuals grew more rapidly, achieved a larger size, and set more seed than individuals from the center. Size variation (based on basal stem diameter) was higher at the edge of the patches than in the centers, where most individuals were uniformly small, except in the largest patches where the centers were breaking up and size variation was high. There were significant differences in phenology and life history between individuals from the center versus the edge. There may also have been genetic differences that developed from different selective pressures, as in the case of the annual plants from vernal ponds (Linhart, 1974, 1976).

The much-discussed relationship between plant density and plant size, known as the '-3/2 thinning rule' (Westoby, 1984; Weller, 1987) is not directly relevant to the issue of size evenness. The -3/2 relationship is

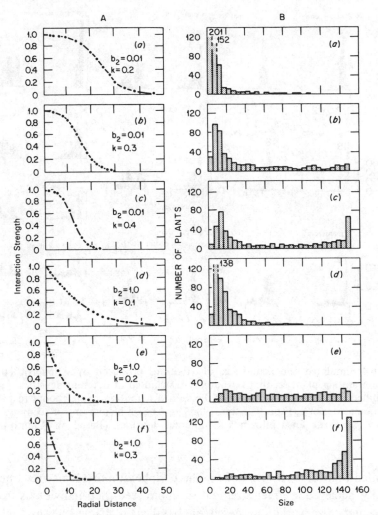

Fig. 6.5 Effect of the local resource depletion pattern on the simulated size distribution of plants that are competing symmetrically and are distributed randomly in space. The resource depletion function used in these simulations is

$$I = b_1(1 + b_2)/[1 + b_2 exp(kd)]$$

where b_1, b_2, and k are constants, and d is the radial distance from the center of the plant. In all cases, $b_1 = 1$. For $b_2 = 0.01$, there is a plateau before the interaction intensity falls off (Aa,b,c). For $b_2 = 1.0$, the interaction falls off exponentially with distance (Ad,e,f). The size distributions Ba-f were produced from the corresponding interaction functions, with the density of the plants specified by the Poisson parameter $\lambda = 0.001$. (From Huston and DeAngelis, 1987.)

based on average plant size and plant density (usually in plantations or experimental plantings of a single species), and so does not provide any information about size distributions. Nonetheless, the processes that produce thinning in plant stands, namely the suppression and mortality of small individuals, are the same processes that influence the distribution of sizes within a population.

Symmetric versus Asymmetric Competition

The spatial patterns of resource availability produced by symmetric versus asymmetric competition result in different dynamics of population size distributions. With asymmetric competition, relatively few individuals are able to capture most or all of the available resources, and most of the individuals are left with very few resources. Consequently, increasing the total density of individuals has little effect on resource availability, either for the few dominant individuals or the many suppressed individuals. As total density increases, size bimodality may appear as the subpopulation of small, resource-deprived individuals increases in number. Among plants, competition for light can produce this effect, while among animals size-dependent predation such as that described above can produce size bimodality.

In contrast to the increasingly bimodal size distribution that occurs with an increasing density of individuals competing asymmetrically, when competition is symmetric, bimodality actually decreases at high densities. Since all individuals deplete resources in proportion to their size when competition is symmetric, increasing the total density of individuals reduces the resources available to all individuals whose zones of resource depletion overlap. Thus, the growth of all individuals is potentially reduced, and if the spatial distribution is random or uniform, all individuals may have a sufficient number of neighbors that the entire population grows slowly.

With symmetric competition, the pattern of the growth and size distribution shifts from a unimodal distribution of rapidly growing large individuals at low densities, to a bimodal distribution with separate subpopulations of small slowly growing individuals and larger rapidly growing individuals at intermediate densities, back to a unimodal distribution at high densities, with many crowded, slowly-growing individuals (Fig. 6.4).

A Unifying Theory of Ecology based on Individual Organisms

All of the processes that have been described in terms of their effect on size diversity also can influence species diversity. Among plants, the small, shaded individuals generally suffer higher mortality than large individuals of the same species in full sunlight, which can affect the size distribution of a population, or the species diversity of a community. The importance of individual properties is just as great for the structure of animal populations and communities as it is for plants. Just as small individuals of a predatory species can become stunted and die when appropriate-sized food is not available, so can individuals of a small species competing with individuals of a larger species for the same food supply (Myers, 1976; Myers and Campbell, 1976; Polis and McCormick, 1986; DeAngelis and Huston, 1987).

The basic rules of competition and predation among individuals are the same regardless of whether the individuals are of the same or of different species. The outcome of competition or cannibalism within a population is determined by differences between the individuals (DeAngelis et al., 1979; DeAngelis et al., 1984; Adams and DeAngelis, 1987; DeAngelis and Huston, 1987; Polis, 1988; Polis et al., 1989). These differences may be imposed by the environment or result from intrinsic variation among the individuals. Cannibalism is essentially equivalent to predation in species with indeterminant growth (many different sizes present in the population). Predation is effectively different from cannibalism (which is very rare) in species with determinant growth (most individuals in the population are the same size).

For competition or predation among species within a community, the effect of differences imposed by the environment is important, but the influence of intrinsic (i.e., species specific) differences is much greater than it is between individuals of the same species. The greater genetic differences between individuals of different species are reflected in greater differences in maximum size, growth rate, reproductive rate, response to environmental conditions, resource requirements, and many other traits. The differences between intraspecific competition and interspecific competition result not from different mechanisms, but from the fact that the differences between the individuals are greater.

The hypothesis that interactions and processes at the level of individual organisms produce ecological phenomena at the multiple levels of populations, communities, and ecosystems, can be expressed in mathematical form as a model. Mathematical models are simply tools for developing and exploring the potential consequences of ecological theories (see Caswell, 1988). Theories about living organisms, whether or not they are expressed as mathematical models, can only be tested through experiments with the organisms themselves.

Mathematical models based on interactions among individual organisms present problems of analytical complexity much greater than traditional ecological models based on species or groups of species. The number of individuals of even a single species that must be modeled typically far exceeds the number of species modeled in a traditional ecological model (e.g., a Lotka-Volterra competition or predation model). One consequence of the complexity of many interacting units is that an analytical solution to the many equations of these models is rarely possible. Fortunately, the increasing speed and capacity of modern computers allows the dynamics of a great many equations to be followed simultaneously. These models with many interacting equations solved incrementally in time steps on a computer are known as simulation models. Simulation models are also mathematical models, but they are generally distinguished from analytical mathematical models, whose solutions and properties can be derived using tools of algebra, calculus, linear algegra, and other fields of mathematics (see refs. in DeAngelis and Gross (eds.), 1992).

Another feature of computer simulation models is that they generally focus on the dynamic behavior of the model, that is, how the solutions of the system of interacting equations changes through time, rather than on the equilibrium 'endpoint' of the equations, if there is one. Traditional ecological models of the analytical type have tended to focus on the equilibrium solution of the equations, and such properties of the equilibrium solution as its stability (see Chapter 4). Simulation models can also be used to explore the equilibrium properties of complex systems of equations (see Gardner et al., 1981; O'Neill and Giddings, 1979). Nonetheless, a major advantage of computer simulation models is that they can be used to look at the temporal dynamics of complex mathematical models.

The field of systems ecology has traditionally used computer simulation models that were not based on individual organisms, but rather on groups such as trophic levels, which include not only many individuals, but many different species as well. Ecosystem simulation models often subdivide trophic levels into units that represent parts or processes of individual living organisms, such as roots, branches, and leaves, or growth, respiration, and reproduction (Reichle and Auerbach, 1972; O'Neill et al., 1975).

Individual-based simulation models represent a significant departure from both traditional ecosystem simulation models and traditional analytical mathematical models of populations, communities, and ecosystems. The focus on individual organisms avoids two major unrealistic assumptions that are implicit in the traditional models. The first assumption is that all individuals of a species or trophic level are identical, and their properties can each be described with a single value that represents all individuals. This assumption violates the basic biological principal of individuality, that is, that each individual organism represents a unique interaction of genetic material and environmental conditions. The second assumption is that all individuals in a group interact identically with all other individuals, and that their interaction can be characterized by a single number that represents all interactions. This assumption violates the basic biological principal of locality, that is, that each organism interacts primarily with the other organisms and environmental conditions within its own local neighborhood, rather than all organisms and all environmental conditions (Huston *et al.*, 1988).

The problems inherent in these assumptions can be addressed under the general topic of aggregation error (see O'Neill and Rust, 1979; Gardner *et al.*, 1982). Aggregation errors arise when many different values are lumped into a single value, generally using some averaging techique. A classic example of an aggregation problem is the derivation of the growth rate of a population (Cohen, 1977, 1979a). A population is obviously composed of many different individuals that may reproduce at different rates, some relatively high, others low. An estimate of the intrinsic population growth rate based on the average of all individuals may significantly underestimate the actual population growth rate because the rapidly reproducing individuals contribute disproportionately to the total population growth. Likewise, the rapidly reproducing individuals will contribute disproportionately to the future genetic content of the population. Aggregation errors can be avoided by proper weighting of the different values that are used to derive the average (Cohen, 1979a, 1979b). These errors are also avoided by individual-based models, which follow the contribution of each individual to the total population, and therefore produce an estimate of total population growth that is not subject to this type of error.

All models involve assumptions, many of which represent a compromise between biological reality and the limits of current mathematical capabilities. The two unrealistic assumptions discussed above were unavoidable at the time the traditional modeling approaches were developed, because computers and mathematical techniques were not available to solve the many equations that would be needed to make the assumptions unnecessary. However, it may be that the use of these obviously unrealistic assumptions in virtually all ecological models has lead to the generally perceived failure of ecological theory to contribute to the solution of important ecological and environmental problems (Suter, 1981), and to the general skepticism of many field and experimental ecologists toward theory and models (Peters, 1980; Simberloff, 1981, 1982).

In summary, individual-based computer simulation models offer an approach to ecological modeling that has several important advantages over traditional ecological and ecosystem models:

1. Individual-based models are inherently dynamic and non-equilibrium.

2. Individual-based models can address both genetic and environmental differences between individuals, rather than assuming that all individuals of a species are identical.

3. Individual-based models permit consideration of local interactions between neighboring individuals, rather than assuming that all individuals interact with all other individuals.

4. Individual-based models use a single set of mechanisms and assumptions to model ecological processes at the individual, population, community, and ecosystem levels.

Individual-based Models Link Physiology, Population Ecology, Species Diversity, and Ecosystem Dynamics

Individual-based models are based on the premise that individuals interact according to the same basic rules regardless of whether they are of the same species or of different species. Thus the size distribution of a population or the species composition of a community can be investigated using a single model structure that implements a theory of individual interactions. Various implementations of individual-based models are being used to investigate ecological phenomena as diverse as the effect of physiological constraints on vegetation pattern (Huston and Smith, 1987; Smith and Huston, 1989), recovery of fish populations from exploitation or other causes of mortality (DeAngelis *et al.*, 1991; Madenjian and Carpenter, 1991; Madenjian *et al.*, 1991), interaction of herbivores such as mice and deer with tree seedlings on a landscape (Hyman *et al.*, 1991), and the effect of global environmental change on vegetation structure and productivity (Solomon *et al.*, 1980; Solomon, 1986; Pastor and Post, 1988).

Because each of these models is based on individual organisms, model output can be summarized on any level from the fate of a single individual to the net effect of all individuals on the resources in their environment. Thus a single model can investigate not only species diversity (one type of summary of all individuals), but many other conditions that may influence or be associated with species diversity. Such conditions could include the number and sizes of individuals, the spatial distribution of individuals or species, the resources contained in each individual or species, the levels of different resources remaining in the environment, the total productivity of the system, etc.

Individual-based models of forest succession, designed to look at long-term trends in forest structure (Botkin *et al.*, 1972; Shugart, 1984) can also be used to look at individual growth and survival, and the resulting population size structure. Figure 6.6 shows the pattern of size (diameter) distributions that were produced by a computer simulation of a tulip poplar (*Liriodendron tulipifera*) plantation at different planting densities. Tulip poplar is a shade-intolerant species, and most of the small shaded individuals die in the simulation. After 40 years of simulated growth, the higher density population shows a clearly bimodal size distribution among the large trees (Fig. 6.6e). This bimodality results from a higher intensity of asymmetric competition in comparison with the lower density population (Fig. 6.6d). In a third simulation, the physiological parameters of the trees were altered to make them very shade tolerant (more similar to sugar maple (*Acer saccharum*) than to tulip poplar), so that small shaded individuals would not die. When a high density plantation of the shade tolerant trees was simulated, a strongly bimodal size distribution appeared after only twenty years (Fig. 6.6c). The bimodality resulted from a large group of small trees that grew slowly in the understory, but did not die, i.e., the well-known ecological phenomenon of suppression. This result contrasts with the lower-density simulation of shade-intolerant trees, in which bimodality took longer to develop. Bimodality is thus a consequence of the intrinsic properties of the species (i.e., degree of shade tolerance) that determines whether they are able to survive in a suppressed state, and the intensity of competition for light, which in these examples was increased by increasing the simulated planting density (Huston and DeAngelis, 1987).

The above example involved a monospecific population, in which the size structure was influenced by an experimentally imposed factor,

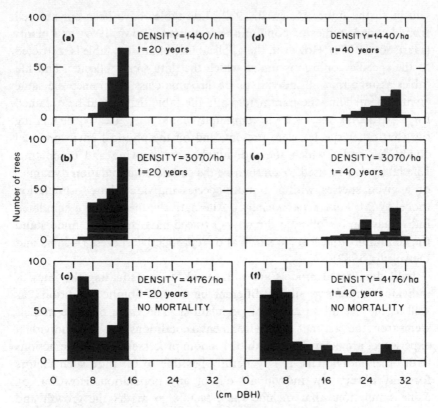

Fig. 6.6 The effect of planting density and shade tolerance on size bimodality produced by asymmetric competition for light in an individual-based computer simulation of a tulip poplar plantation. Histograms show diameter distributions after 20 and 40 years of simulated growth of monospecific stands. Simulation with no mortality illustrates size dynamics of a hypothetical species (not tulip poplar) with sufficiently high shade tolerance that there is no mortality of seedlings due to shading by larger trees. (From Huston and DeAngelis, 1987.)

namely, the planting density. Among individuals within a population that are competing with individuals of many other species in a community, the variety of biotic and abiotic conditions that each individual experiences can have a significant effect on the population structure. A standard tool of the demography of both human populations and animals and plants is the life table, a matrix of the probabilities of an individual growing from one size to another, reproducing, or dying. Life tables can be constructed from detailed information on many individuals in a population, and used to predict the probable fate of a given individual (such as an applicant for an insurance policy) or to predict what the population structure will be in the future (Caswell, 1989). Life tables have been constructed from detailed studies of populations of animals (Lowe, 1969) and plants (Hartshorn, 1972). However, the applicability of any life table is restricted to the specific conditions under which the data were collected, and life tables assume that all individuals of the same class experience the same environment. Thus the parameters of a life table that would be obtained for a particular species in a community could vary greatly, depending on what other species were present, and on the physical environmental conditions under which the individuals were growing and competing. Life tables can be used to summarize the simulated population dynamics of a given species within a multi-species individual-based simulation model (W.M. Post, *pers. commun.*). Alternatively, the fate of a simulated individual can be followed directly, or traced back in time to understand the conditions that lead to survival or reproductive success (Adams and DeAngelis, 1987).

Individual-based models are well-suited for investigating the fates of individual organisms under different biotic and abiotic conditions, as well as the overall structure of populations. The value of these models stems from the fact that the environmental conditions that each individual experiences are determined both by random processes and by interactions with other individuals. It is generally possible to measure such factors for only a very few individuals out of any population. However, by using simulations that include these factors to model the growth and interactions of many individuals, it is possible to derive generalizations about both the fates of individual organisms and the overall dynamics of the population. There is a developing consensus that differences between individuals are critical for understanding and predicting the consequences of most ecological processes (Chesson, 1978; Lomnicki, 1988; Huston *et al.*, 1988; Caswell, 1989; DeAngelis and Gross, 1992).

Summary

The size structure, or size diversity, of populations is influenced by interactions between individuals, which can be more easily studied in populations than in more complex communities. The basic mechanisms of intra-specific interactions between individuals in a population are also the principal mechanisms of inter-specific interactions. Thus, a single conceptual structure based on interactions between individuals can be applied to understanding both population and community patterns. This approach incorporates into a single conceptual framework what were previously the separate fields of population, community, and ecosystem ecology. The growth, survival, and reproduction of individual organisms, influenced by the physical environment and by interactions with other individuals, are the basis for understanding all levels of ecological phenomena, just as they have traditionally been recognized as the basis of natural selection and all evolutionary phenomena.

7

Individual Properties and the Structure of Communities and Ecosystems

The previous chapter established the importance of interactions between individual organisms for understanding patterns at the population level, and potentially at higher levels such as the community and ecosystem. The individual-based approach is based on the assumption that once the basic rules for interactions among organisms and between organisms and their environment have been specified, the outcome of the interactions depends on the differences between the individual organisms. This chapter focuses on the attributes of individuals of different species that influence growth and survival over the range of conditions considered by the dynamic equilibrium model of species diversity. The emphasis in this chapter is on differences between types of plants, and the consequences of plant physiology for patterns of species diversity and ecosystem processes.

Applying Individual-based Models to Plant Species Diversity

Individual-based models of plant communities, particularly forests, were the earliest and are now the most widely used models based on computer simulation of many individual organisms (Botkin et al., 1972; Shugart and West, 1977; Aber et al., 1979, 1982; Shugart, 1984; Huston, 1992). While these models have rarely been used to look at the issue of species diversity (exceptions include Doyle, 1981), they have been applied to a wide range of forest types, including southern Appalachian deciduous forest (Shugart and West, 1977), Arkansas upland forest (Mielke et al., 1977, 1978), Mississippi River floodplain forest (Tharp, 1978), northern hardwood forest (Botkin et al., 1972), Australian montane eucalypt forests (Shugart and Noble, 1981), Australian subtropical rain forest (Shugart et al., 1981), boreal forest (Pastor and Post, 1988), and Puerto

Rican montane rain forest (Doyle, 1981), as well as grasslands (Coffin and Lauenroth, 1990).

This type of model allows the simultaneous evaluation of ecological processes at many levels of ecosystem organization (Huston *et al.*, 1988). For example, the effect of altered climatic conditions on forest population structure, species diversity, primary productivity, and soil resources can be evaluated simultaneously with a single individual-based model (e.g., Pastor and Post, 1985, 1986, 1988; Huston and Smith, 1987). The generality of these plant-competition models, and the underlying theories, have been amply demonstrated by their power to reproduce well-known ecological patterns of many different types using the same simple model structure (Shugart, 1984; Huston and Smith, 1987; Huston and DeAngelis, 1987; Huston *et al.*, 1988; Smith and Huston, 1989; Huston, 1992).

Recently, individual-based models have been used as tools to investigate theories of plant community structure by explicitly defining the relationships between parameters used to describe species in the model, and then examining the consequences of altering these relationships (Huston and Smith, 1987). Huston and Smith demonstrated that a great variety of successional patterns could be produced by the single mechanism of competition for light, simply by varying the pattern of correlations between such plant characteristics as maximum size, maximum growth rate, shade tolerance, and fecundity. The conclusion of this simulation experiment was that the confounding variety of successional patterns found in nature did not require many different mechanisms, but could be explained by a single mechanism operating on plants with different patterns of physiological and life history characteristics. Huston and Smith (1987) demonstrated that plant succession could be simply explained as the result of competition for light among plants whose own growth altered the availability of light over time, and thus produced a shift in the relative competitive ability of different species.

Extending this theoretical approach, Smith and Huston (1989) investigated the consequences of a specific pattern of correlations among life history and physiological charactistics, using the same type of individual-based model of competition for light. The following sections describe the empirical and theoretical basis of this approach, and its implications for the structure of populations, communities, and ecosystems. This work deals primarily with the dominant plants that influence the structure and environmental conditions of the ecosystem, but it has important implications for the animals and plants that live within the environment defined by the dominant plants.

Universal Constraints Produce Universal Patterns

Much of the difficulty in interpreting patterns of biological diversity found in nature is that the patterns result from the interaction of predictable deterministic processes, such as growth and competition, with unpredictable or stochastic processes, such as climatic fluctuations, disturbances, and the vagaries of dispersal. Unless we are able to separate the predictable components of biological diversity from the unpredictable components, we will never be able to explain the patterns of biodiversity (cf. Colwell, 1979).

Fortunately, there are some strong regularities in the properties of organisms that are related to the specific environmental conditions under which they are found. There is also a consistent relationship between the different characteristics of single organisms. These regularities are the basis for the models of population growth that gave rise to the concept of r- and K-selection (MacArthur, 1962; see Chapters 4 and 5). According to the concept of r- and K-selection, properties that lead to a high intrinsic rate of population increase (r), such as small size, early reproductive maturity, and large numbers of propagules, are inversely correlated with properties that lead to a high carrying capacity (K), such as large size, later reproductive maturity, and a smaller number of propagules produced over a longer time period. This inverse relationship is presumed to be primarily the result of differences in the allocation of limited resources (MacArthur, 1962; MacArthur and Wilson, 1967).

Given that any organism has a finite amount of energy or any other resource available to it, it can use the resource in different ways that have very different consequences for population dynamics. For example, a fish of a given size can produce either a large number of small eggs, or a smaller number of larger eggs. It cannot produce a large number of large eggs, even though that combination would likely result in the greatest probability of some offspring surviving to maturity. Likewise, a tree with a given amount of solar energy, water, and mineral nutrients could either produce low-density wood and grow rapidly in height and volume (e.g., a balsa tree, *Ochroma lagopus*),or grow slowly and produce high-density wood that would allow it to achieve a larger size and greater longevity because of enhanced resistance to physical damage and attack by insects or pathogens (e.g., oak, *Quercus* spp., or lignum vitae, *Guaiacum sanctum* (Zygophyllaceae)). With a fixed amount of resources, a tree cannot both grow rapidly in size and produce high-density wood.

The constraints imposed by a finite amount of available resources result in inverse correlations between properties such as maximum growth rate and maximum size, and the total number of propagules and the size of each individual propagule. The theory of r- and K-selection predicts that different types of environments will favor organisms with specific combinations of characteristics (Pianka, 1970). In particular, in environments with frequent disturbances that kill most organisms, species that require a long time to reach reproductive maturity (i.e., high K, low r) will be eliminated, and species that reach maturity rapidly (i.e., high r, low K) will be favored. However, in environments with infrequent disturbances, species with larger size, greater longevity, etc. (i.e., high K, low r) will be favored.

This general pattern has been the basis of most previous classifications of plant strategies or vegetation types, including r, K, and adversity strategies (MacArthur and Wilson, 1967; Southwood, 1977; Greenslade, 1983); early- and late-successional types (Clements, 1916; Budowski, 1965, 1970; Whittaker, 1975; Bazzaz, 1979; Finegan, 1984); exploitative and conservative responses (Borman and Likens, 1979); ruderal, stress-tolerant, and competitive strategies (Grime, 1977, 1979); gap and non-gap species (Hartshorn, 1978; Brokaw 1985a,b); and structural characteristics (Raunkiaer, 1934; Hallé, 1974; Hallé and Oldeman, 1975; Webb *et al.*, 1970; Walker *et al.*, 1981). Most of these schemes are based on plant responses to different environmental conditions, such as resource availability and disturbance regime, rather than on inherent properties of the plants themselves.

These generalizations have been summarized as tables that list the characteristics of organisms that are found in frequently or recently disturbed (i.e., early-successional) environments versus characteristics of organisms found in infrequently disturbed or long undisturbed (i.e., late-successional environments) (Table 7.1). Understanding the energetic and physiological mechanisms underlying these general patterns is the key to understanding the distribution of individual species and of species diversity across landscapes.

Physiological Constraints on Plants

In addition to the energetic constraints that apply to all organisms, plants are subject to additional physiological constraints that impose inverse relationships on other sets of characteristics. Together these fundamental physiological and energetic constraints on the capture and use of resources by plants constrain the ability of a plant to 1. tolerate low levels

Table 7.1. *Physiological and life history characteristics of early and late successional plants.*

Characteristic	Early Succession	Late Succession
Photosynthesis		
Light saturation intensity	high	low
Light compensation point	high	low
Efficiency at low light	low	high
Photosynthetic rates	high	low
Dark Respiration rates	high	low
Water Use Efficiency		
Transpiration rates	high	low
Mesophyll resistance	low	high
Seeds		
Number	many	few
Size	small	large
Dispersal distance	large	small
Dispersal mechanism	wind, birds, bats	gravity, mammals
Viability	long	short
Induced dormancy	common	uncommon?
Resource acquisition rates	high	low?
Recovery from nutrient stress	fast	slow
Root/shoot ratio	low	high
Mature size	small	large
Structural strength	low	high
Growth rate	rapid	slow
Maximum lifespan	short	long

Based on Budowski (1965, 1970), Pianka (1970), Ricklefs (1973), and Bazzaz (1979).

of a resource while maintaining the ability to grow rapidly at high levels of the same resource; and 2. use both light and water efficiently when levels of both are low.

The tradeoff between tolerance to low resource conditions and maximum potential growth rate under high resource conditions (Fig. 7.1) is a well-documented consequence of the physiological and energetic constraints on plants (e.g., Parsons, 1968a). This type of tradeoff has been reported for different light conditions (Grime and Jeffrey, 1965; Loach, 1967; Boardman, 1977; Bazzaz, 1979), different soil water conditions (Ellenberg, 1953, 1954; Gates, 1968; Parsons, 1968b; Kozlowski, 1982; Zimmermann and Brown, 1971; Zimmermann and Milburn, 1982), and different nutrient supply conditions (Mitchell and Chandler, 1939; Grime, 1974, 1977, 1979; Chapin, 1980; Chapin et al., 1986, 1987; Bryant *et al.*, 1983). This pattern of correlations among plant characteristics is consis-

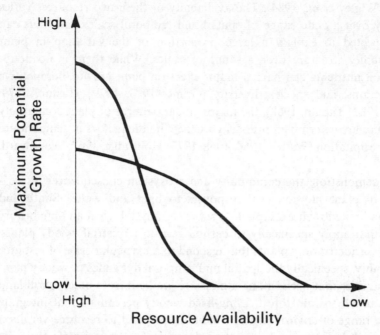

Fig. 7.1 General pattern of plant growth rate in relation to resource availability for plants with two levels of tolerance to low resource availability. Note the inverse relationship between the resource level where the growth rate is zero (x intercept of the curves) and the maximum rate of growth achieved under high resource conditions. (From Smith and Huston, 1989, based on Larcher, 1975; Orians and Solbrig, 1977; Bazzaz, 1979; Chapin *et al.*, 1986.)

tent with the general principles of cost-benefit analysis as it has been used for explanations of such characteristics as leaf size and shape (Parkhurst and Loucks, 1972; Givnish, 1978, 1979), leaf type in arid environments (Orians and Solbrig, 1977), plant height (Givnish, 1982; Chazdon, 1986), plant photosynthesis (Mooney and Gulmon, 1979; Cowan, 1986), the ability to use multiple resources (Chapin *et al.*, 1987), and herbivore defense and nutrient use (Mooney and Gulmon, 1982; Bryant *et al.*, 1983; Bloom *et al.*, 1985; Coley *et al.*, 1985).

Smith and Huston (1989) used this basic pattern of plant properties to categorize the growth responses of different species of plants, and different individuals of the same species. To illustrate the application of the 'trade-off model' to vegetation dynamics, Smith and Huston (1989) considered only traits related to the use of water and light, two resources that often limit plant growth (Clements *et al.*, 1929; Daubenmire, 1947; Donald, 1958; Walter, 1964, 1968, 1971, 1973; Gates, 1980; Kramer,

1983; Wigley *et al.*, 1984). The availability of these two resources varies greatly over a wide range of spatial and temporal scales, and thus can be expected to explain a large proportion of the variation in plant community structure over a range of scales. While there is no doubt that soil nutrients can have a major effect on plant growth, competitive interactions, and species diversity (Grime, 1973, 1979; Huston, 1979, 1980, 1982; Tilman, 1988), the major global patterns of plant community structure are determined primarily by the climatic factors of temperature and precipitation (Walter, 1964, 1968, 1973; Holdridge, 1967; Woodward, 1987).

To demonstrate the community and ecosystem consequences of variation in plant physiological responses to light and water, Smith and Huston limited their example to a subset of physiological and life history traits that apply specifically to carbon gain in terrestrial woody plants. They considered only plants that respond to a particular scale of resource variability, specifically the spatial and temporal dynamics of woody plant communities that occur along a moisture gradient from desert shrubland to savannah to rain forest. Long-lived woody perennials that integrate a wide range of environmental conditions respond to resource variation over much longer time scales than do annuals that complete their life cycle following a single rainstorm.

Perhaps the most fundamental physiological constraint upon plants results from the inherent conflict between the need to take up CO_2 through the stomata directly from the atmosphere into the leaf interior, and the need to maintain moist conditions within the leaf interior. How plants solve this dilemma is one of the most exciting questions in plant physiology, where most research has focused on enzyme systems and morphology at the leaf level (Farquhar and Sharkey, 1982; Cowan, 1982, 1986; Schulze *et al.*, 1987).

This fundamental conflict underlies another well-known tradeoff in plant growth, the allocation of energy to roots versus structures aboveground. The implications of the root:shoot ratio have been extensively discussed (Monk, 1966; Aung, 1974; Kramer and Kozlowski, 1979; Fitter and Hay, 1981; Schulze, 1982, 1986; Givnish, 1986; Hunt and Nicholls, 1986; Tilman, 1988). In general, when water and nutrients are plentiful in relation to light, plants invest relatively little energy in roots, but spend most of their energy on aboveground parts to capture light. As a result, competition for light is often intense under these conditions. In contrast, when water and nutrients are limiting to plant growth, plants must invest heavily in roots at the expense of their aboveground parts (e.g., Vitousek and Sanford, 1986). Competition for light becomes relatively unimportant under these conditions, but competition for water and nutrients may be intense. Thus root:shoot ratios vary widely in proportion to the relative availability of light in relation to moisture and nutrients. This variation in root:shoot ratios occurs both between species adapted to different resource conditions and between individuals of a single species that are grown under different conditions.

Resource allocation responses of plants to reduced light include increases in 1. leaf area/leaf weight; 2. leaf weight/whole plant weight; 3. shoot/root biomass; 4. leaf area/root surface; and 5. stem height/stem biomass (Loach, 1967; Kozlowski, 1976, 1982; Kramer, 1983; Kramer and Kozlowski, 1979; Zimmermann and Brown, 1971; Fitter and Hay, 1981; Withers, 1979). These responses tend to be greater in plants that are less tolerant of shade (i.e., they have greater morphological plasticity).

Responses to moisture or nutrient limitation are often the opposite of the responses to reduced light (Kramer, 1969; Struik and Bray, 1970; Kozlowski, 1976, 1982; Fitter and Hay, 1981; Chapin, 1980; Chapin *et al.*, 1986; Schulze, 1986; Schulze *et al.*, 1987), with corresponding decreases in the efficiency of light use. Each of these responses or mechanisms can impose constraints on the plant's ability to survive other stresses such as reduced levels of a particular resource, or loss of biomass resulting from fire or herbivory (see Oosting and Kramer, 1946; Keever, 1950; Bryant *et al.*, 1983; Chapin *et al.*, 1987; Osmond *et al.*, 1987).

Constraints on the use of resources at different levels of availability also affect growth rate and size. In general, species adapted to high levels of resource availability respond strongly to variation in resource availability and can have a wide range of growth rates, depending on patterns of energy allocation to growth versus reproduction, mechanical structure, chemical defenses, other physiological processes, the rate of respiration, etc. (Paul, 1930; Monsi, 1968; Monsi and Murata, 1970; Zahner, 1970; Mooney, 1972; Whittaker, 1975; Mooney and Gulmon, 1979; McLaughlin and Shriner, 1980; Gifford and Evans, 1981; Bazzaz *et al.*, 1987; Loehle, 1988b). However, among plants adapted to low resource conditions, the range of growth rates is narrowly restricted and the rates are much lower because of the adaptations required for growth under such conditions (Chapin, 1980).

Smith and Huston summarized the consequences of constraints on the simultaneous use of light and water by individual plants in the following three premises.

Premise 1. A plant that can photosynthesize at high rates and grow rapidly under conditions of high light is unable to survive at low light levels (i.e., it is shade-intolerant). Conversely, a plant that is able to grow in low light (shade-tolerant plant) is unable to respond to increased light and has a low maximum rate of growth and photosynthesis even under high light conditions (Bazzaz, 1979; Bazzaz and Pickett, 1980; Larcher, 1980).

Premise 2. A plant that can grow rapidly and/or reproduce abundantly under conditions of high available soil water is unable to survive under dry conditions (i.e., it is intolerant to water limitation). Conversely, a plant adapted to survive and reproduce under dry conditions (i.e., it is tolerant to water limitation) is unable to grow rapidly and/or reproduce abundantly with high soil water availability (Parsons, 1968b; Orians and Solbrig, 1977).

Premise 3. Tolerances to conditions of low light and moisture are interdependent and inversely correlated. Adaptations that allow a plant to grow at low light levels restrict its ability to survive under dry conditions. Conversely, adaptations that allow survival under conditions of water stress reduce the plant's ability to grow in low light. Thus no woody plant can simultaneously have a high tolerance for low levels of both resources.

These premises summarize the consequences at the whole-plant level of a large suite of physiological processes and life history strategies involved in the efficient use of different levels of the same resource and of different resources. The mechanisms involved in these tradeoffs include all levels of a plant's structure, from its enzyme systems and organelle structure to its branching pattern and leaf angle. Few of these specific mechanisms apply over the entire range of conditions under which plants are found, and few apply to all levels or sizes of plant structure (e.g., leaves versus whole plants, herbaceous versus woody plants, and annuals versus perennials). Yet taken together, these diverse mechanisms form a consistent pattern of tradeoffs that influence the outcome of interactions between individual plants and result in the fundamental patterns of vegetation dynamics that are found in ecosystems around the world.

Plant Growth: The Interaction of Environmental Conditions and Physiological Constraints

The consequences of the three premises for use of light and water are summarized in Fig. 7.2*a*, which illustrates the two main results of energetic and physiological cost-benefit tradeoffs in plants. First, growth rate decreases as tolerance to either low light or low water availabiltiy increases. Thus, the highest growth rate is found in the upper right-hand corner, which represents the lowest tolerance for low levels of light and water. Second, there is a limit to the combined tolerance to low light and water levels. This is illustrated by the diagonal boundary, which limits woody plant strategies to the combination of traits represented by the upper right half of the figure. 'Plant strategy' refers to a combination of plant characteristics related to the use of light and water, including the resource allocation patterns reflected in maximum growth rate, maximum size, and maximum age, along with the plant's growth response to all combinations of light and water availability.

Smith and Huston arbitrarily divided the continuum of plant strategies into 15 discrete plant types (Fig. 7.2*b*) and used these types to represent the range of woody plant strategies for light and water use. The parameters that define these plant types are used in equations to describe the growth response of each functional type for all combinations of light and water availability.

Some important consequences of the tradeoffs described earlier are evident in Fig. 7.2. Those plants that dominate in late succession because of their high tolerance to shade, the so-called 'climax' species, are generally not the species with the highest growth rates or the largest sizes. The adaptations required for shade tolerance, and also for the resistance to insects and pathogens that contributes to longevity, reduce the energy available for rapid growth and the achievement of large sizes. In the Pacific Northwest, the late successional dominants, *Tsuga* and *Abies* are smaller (but also shorter-lived) than the less shade-tolerant *Pseudotsuga*, which grows rapidly and achieves larger sizes and greater longevity, but cannot reproduce in the shade of the late-successional species. Other examples of shade-intolerant species that grow rapidly and achieve great size are *Liriodendron tulipifera* of eastern North America and *Ceiba pentrandra* of the neotropics. The growth of an individual plant of a specific functional type under different combinations of light and water availability can be represented as a three-dimensional response surface, with whole-plant carbon gain on the vertical axis and with light and

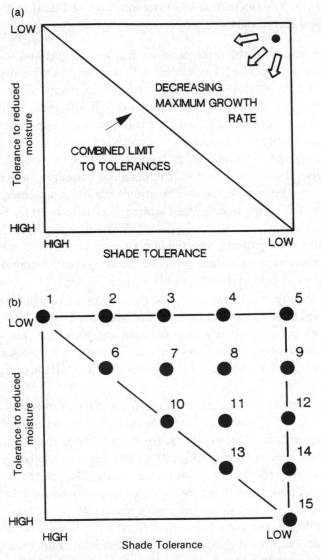

Fig. 7.2 (*a*) Potential woody plant strategies for light and water use, illustrating some of the consequences of physiological and life history tradeoffs. The highest rate of growth (carbon gain) is in the upper right corner of the figure where tolerance to drought and shade are least. Growth decreases with increasing tolerance to low levels of light and/or water. (*b*) Division of the continuum of woody plant strategies into the 15 discrete functional types used in the computer simulations. Each functional type is defined by maximum growth rate, shade tolerance, and tolerance to low moisture levels. The same labeling for these 15 functional types is used in figures in this and following chapters. (From Smith and Huston, 1989.)

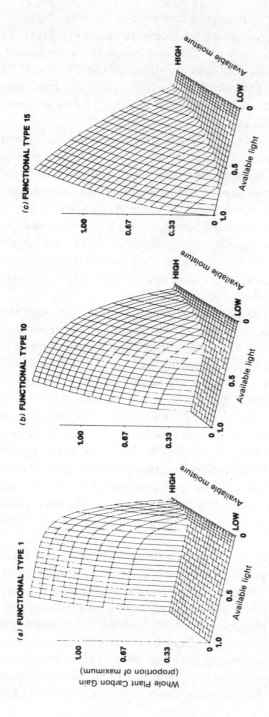

Fig. 7.3 Hypothesized growth response surfaces for three different types of plants showing carbon gain in relation to light and moisture levels. (a) Response showing high shade tolerance (photosynthetic compensation point at very low levels of available light) and low tolerance for dry conditions (zero net carbon gain under intermediate and low moisture conditions); (b) Response showing intermediate shade tolerance and intermediate tolerance to dry conditions; (c) Response showing low shade tolerance (photosynthetic compensation at approximately 20% of full sunlight) with very high tolerance to dry conditions (positive net carbon gain even under the driest conditions). The numbers of these types correspond to the numbers in Fig. 7.2 and other figures showing simulation results. (From Smith and Huston, 1989.)

water availability on the horizontal x and y axes (Fig. 7.3). Figure 7.3*b* represents the hypothesized growth- response surface for a plant with intermediate tolerance for both low light and low moisture levels, and it illustrates the interdependence of light use and water use (Premise 3). The response surface can be envisioned as a series of light-response curves (integrated to whole-plant carbon gain) that have been calculated over a range of moisture conditions.

The response surfaces of Fig. 7.3 illustrate the important consequences of the interaction between light use and water use. With low moisture availability, the level of light necessary for photosynthetic carbon gain to equal respiratory carbon loss (i.e., the whole-plant light compensation point) is higher than it is for higher moisture availability. Also, the maximum growth achieved under high light levels is lower when soil water availability is reduced. Thus, the minimum level of light required for survival is higher under dry conditions than under moist conditions, that is, an individual plant is more shade tolerant when soil water conditions are favorable than when the soil is dry.

The consequences of the tradeoffs in Premises 1 and 2 are seen most clearly in comparisons between the growth responses of different types of plants. Figure 7.3 *a* and *c* show hypothetical growth-response surfaces for two types of plants with contrasting patterns of light and water use. Figure 7.3*a* represents the growth-response surface for a plant that is tolerant of low light levels but intolerant of low moisture levels (cf. Premise 3). This plant's response to light under high moisture conditions forms a classic shade-tolerant light-response curve. This curve contrasts with the growth-response curve of the plant in Fig. 7.3*c*, which is typical of shade-intolerant plants. Note that the the shade-intolerant type requires a higher minimum light level (light compensation point) for growth under high levels of moisture availability but it is able to continue growth at much lower levels of water availability than is the shade-tolerant type.

The shapes of the response surfaces proposed by Smith and Huston actually represent hypotheses about the interaction of constraints on light and water use. The true shapes of the response surfaces for particular species could be determined with the appropriate experiments. However, the multi-factorial physiological experiments needed to quantify the precise shape of these response surfaces have not yet been performed at the whole-plant level, although leaf- and seedling-level physiological experiments show patterns similar to those predicted for whole plants (Osonubi and Davies, 1980; Brun and Cooper, 1967; Linder *et al.*, 1981) in factorial experiments involving responses to light and temperature, to light and CO_2, and to light and nutrients, respectively.

Perhaps the most impressive confirmation of the predicted response was obtained not from factorial experiments at the whole-plant level, but from careful measurement of the CO_2 influx into an entire stand of trees under a variety of weather conditions (Fig. 7.4). Jarvis (1989, citing Jarvis *et al.*, 1976) describes gas flux measurements made over a period of several weeks on a stand of Sitka spruce (*Picea sitchensis*). The conditions under which the measurements were made included variation in light intensity resulting from sun angle and cloud cover, and variation in atmospheric humidity (measured as vapor pressure deficit, VPD) resulting from dry and moist weather systems. Aggregation of a great many separate measurements into mean values representing different combinations of light and humidity conditions produced a response surface identical in overall pattern to those hypothesized by Smith and Huston (1989).

Community-level Consequences of Individual Properties

The inevitable result of adaptive tradeoffs is that no organism can be dominant over the entire range of conditions under which it can survive (cf. Darwin, 1859; Clements, 1916; Clements *et al.*, 1929). Plant types adapted to low resource conditions are at a competitive disadvantage under high resource conditions because competitors adapted to only high resource conditions have none of the constraints associated with survival at low resource levels. Likewise, functional types adapted to high resource levels are at disadvantage under conditions of low resource availability (resource levels that are near or below the minimum requirements of those types). Therefore, as resource levels change across either space or time, the distribution of plant functional types will also change.

The consequences of differences among individual plants for spatial and temporal patterns of community and ecosystem structure can be examined using the type of individual-based model of plant competition described in the previous chapter (e.g., Botkin *et al.*, 1972; Shugart, 1984; Huston and Smith, 1987; Huston, 1992). The model tracks tree growth in annual timesteps on a grid composed of plots of ground scaled to the maximum crown area of the largest type of tree (Shugart and West, 1979). Light is the only resource for which plants compete in the model, and the only way in which individuals interact is indirectly, through their effect on light availability. The birth, growth, and death of each

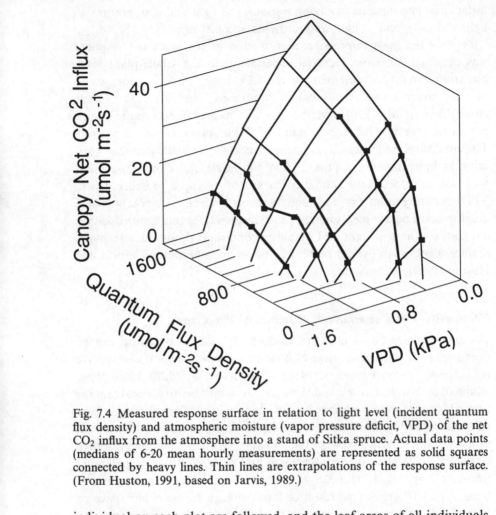

Fig. 7.4 Measured response surface in relation to light level (incident quantum flux density) and atmospheric moisture (vapor pressure deficit, VPD) of the net CO_2 influx from the atmosphere into a stand of Sitka spruce. Actual data points (medians of 6-20 mean hourly measurements) are represented as solid squares connected by heavy lines. Thin lines are extrapolations of the response surface. (From Huston, 1991, based on Jarvis, 1989.)

individual on each plot are followed, and the leaf areas of all individuals are integrated to determine the vertical light profile in each plot. The model determines the responses of each individual to available water and light according to the growth-response functions for its specific type. In a simulation for a given set of initial resource conditions, individuals of all types are allowed to interact to produce an annual record of the species composition, size distributions, resource availability, etc., that result from their particular resource-use strategies.

The output of this individual-based model can be aggregated in many additional ways to address different ecological questions. For example, it is possible to compute the spatial and temporal distribution of specific

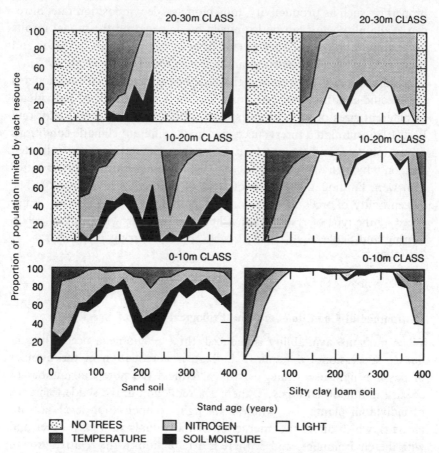

Fig. 7.5 Temporal patterns of variation in factors that limit growth of individual trees of different sizes in a forest. Growth-limiting factors were identified by calculating the environmental conditions (e.g., light, water, nutrients, temperature) experienced by each individual tree in an individual-based simulation model, and determining which factor was most limiting to the growth of each individual. Note that small trees are almost always limited by light, except following the death of large trees. Large trees tend to be limited by factors other than light. (From Pastor and Post, 1986.)

resources that are affected by the plants, such as the vertical distribution of leaf area and the available light at ground level (cf. Fig. 9.7, Chapter 9). Population dynamics under different conditions can be examined, such as the effect of different levels of resource availability or of different types of resources on size distributions (e.g., Fig. 6.5, Chapter 6). Community properties, such as height structure or species diversity, and ecosystem properties, such as productivity, total biomass, decomposition rate, nitro-gen availability, etc., can be examined by aggregating the model results for the appropriate parameters (see Aber *et al.*, 1979, 1982; Pastor and Post, 1985, 1986, 1988; Huston and Smith, 1987; Huston *et al.*, 1988).

Another type of information that can be extracted from the model is the specific conditions or resources that limit the growth of individuals of different sizes or of different species within a community. Pastor and Post (1986) simulated forest succession under different climatic conditions of precipitation and temperature. From their simulations they extracted data on which environmental factors (light, temperature, water, or nitro-gen) were limiting the growth of trees of different sizes (Fig. 7.5). The susceptibility of trees to herbivores or pathogens is also strongly influ-enced by the type of resources that limit the trees' growth (McKey, 1979; Waring and Schlesinger, 1985; Loehle, 1988a; Waring, 1987; Bazzaz *et al.*, 1987).

Environmental Conditions and the Ecological Roles of Species

When moisture availability is reduced, three phenomena occur that in-fluence successional dynamics: 1. there is a reduction in the number of possible light-use strategies, which reduces the potential number of coexisting functional types; 2. there is a reduction in the shade tolerance of individual plants (i.e., whole-plant light compensation levels are in-creased), which alters the interaction of individuals with each other and with the environment; and 3. there is a reduction in the relative growth rates of all functional types. The first two phenomena have the effect of changing the relative shade tolerances of different plant types, which allows a single functional type to have different ecological roles (e.g., early-successional versus late-successional) under different conditions.

For example, the plant functional types that dominate in late succes-sion under xeric conditions can also appear in the high-light conditions of early succession under mesic conditions, where shade tolerance is not critical. However, as light at ground level is reduced by increased leaf area during mesic succession, shade tolerance becomes more important. Therefore, a functional type that is able to dominate in late succes-sion under xeric conditions, because it is the most shade tolerant under those conditions, will be replaced by more shade-tolerant mesophytic types under mesic conditions. This trend continues as moisture avail-ability increases until eventually the xerophytic shade-intolerant types are eliminated even in early succession by the faster growing mesophytic shade-intolerant types.

This shift in the ecological role of a functional type can be visualized in Fig. 7.6, which shows the changes in abundance of five functional types over time as a function of moisture stress. Except for the type with the lowest tolerance to moisture stress and greatest shade tolerance (type 1), each functional type appears initially as an early-successional transient along the gradient of decreasing moisture availability and becomes a late-successional dominant only in communities with lower moisture availability. This shift in roles is most pronounced in the functional types with the lowest shade tolerance and highest tolerance to moisture stress (e.g. types 13 and 15).

Although these early-successional species, which are shade intolerant but tolerant to moisture stress, are usually eliminated by competition when water availability is higher, they may achieve much greater sizes and higher growth rates on an individual basis under moister conditions than they do under the drier conditions where they dominate the community. For example, *Acacia karoo* shows this pattern in southern Africa; it is a tall, fast-growing early-successional tree on the coastal sand dunes (Weisser and Marques, 1979), but it is a slower-growing tree of smaller stature in the semi-arid savannas where it is the dominant species over extensive areas (Acocks, 1975). Several pine species show this same role shift in southeastern North America, where they are early-successional transients that are replaced by hardwoods on favorable sites but persist and dominate exposed dry sites (Oosting, 1942). Likewise, as Peet and Loucks (1977) observed, communities of *Quercus macrocarpa* and *Q. velutina* persist on xeric sites, although both species are typical of early-successional stages on more productive mesic sites.

Just as the successional role of a plant functional type can change in response to conditions such as water availability, so can other plant roles that have been used to classify plants. For example, whether a particular plant is a 'gap' or a 'forest' species, where in a gap it occurs, or in what size of gap it is found, is not a constant characteristic, but rather a variable consequence of the species' particular resource-use strategy (functional type) and the environment in which it occurs. Thus, a plant can have different roles, depending on environmental conditions such as the degree of water or nutrient availability. Because traits such as successional roles, characteristic spatial position, and other aspects of a plant's interactions with its environment are variable, they cannot be the sole basis for a functional classification of plants. Inherent physiological

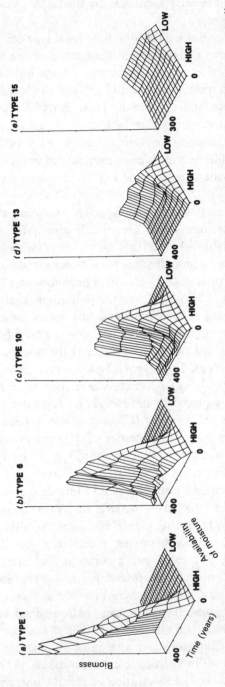

Fig. 7.6 Three-dimensional representation of the changes in the biomass of different plant types over successional time along a gradient of moisture availability. The vertical axis represents the summed biomass of all individuals of a particular type, from simulations in which all types were present. (From Smith and Huston, 1989.)

and life history characteristics, which determine how the plant responds to varying environmental conditions, are a more appropriate basis for an explanatory classification of plant strategies.

Smith and Huston's (1989) classification of plants by functional type differs from most previous classifications in that it is based on biological constraints imposed on individual organisms by processes at lower levels of system organization (e.g., physics, chemistry, physiology). These biological constraints interact with environmental constraints imposed by conditions at higher levels of system organization (e.g., climate, geology), which can be defined independently of the response of the plants themselves.

Consequences of Individual Physiology for Diversity and Ecosystem Processes

Because species differ in chemical composition and in the amounts and combinations of resources that they use, different species can have significantly different effects on the environment. For this reason, many ecosystem properties and processes can only be understood on the basis of the organisms that are present in the ecosystem. This interpretation of ecosystems differs dramatically from the traditional perspective of systems ecology, in which species were considered interchangeable components of functional element pools such as leaves, stems, and roots (Reichle and Auerbach, 1972). '... the identity of the system remains through successional changes in species ... There is no *a priori* reason to believe that the explanations of ecosystem phenomena are to be found by examining populations' (O'Neill, 1976).

Tree species can be categorized by tolerance to low nitrogen availability just as they can by shade tolerance (Mitchell and Chandler, 1939). Nitrogen responses and the interactions of the nitrogen and carbon biogeochemical cycles have been incorporated into various versions of individual-based models of forest succession (Aber *et al.*, 1979, 1982b; Pastor and Post, 1985, 1986). Soil nitrogen availability is a consequence of the relative rates of microbial immobilization of nitrogen and microbial mineralization of nitrogen through the decomposition of organic matter. The rates of mineralization and immobilization are strongly influenced by the chemical composition of the leaf litter that falls from the trees (Daubenmire and Prusso, 1963; Melillo *et al.*, 1982; Aber and Melillo, 1980; Pastor and Post, 1986). The leaf litter chemistry is controlled by the physiology of the trees, in particular by the amount of carbon allocated to lignin in the leaves in relation to the amount of nitrogen allocated to proteins and other compounds. These leaf chemical properties are correlated with a variety of other traits related to the way plants use resources, and also with resistance to insects and pathogens (which is also a consequence of resource allocation).

The effects of climatic conditions on leaf size and morphology are well known (Bailey and Sinnott, 1916; Richards, 1952; Webb, 1959; Parkhurst and Loucks, 1972; Givnish, 1979; Dolph and Dilcher, 1980; Box, 1981; Roth, 1984). Plants that are usually found under dry conditions tend to have small, thick leaves, with an entire margin and a tough and often waxy cuticle. This type of leaf is called xeromorphic. At the other extreme of wet conditions, leaves tend to be larger and thinner, with wavy or complex margins and thin cuticles. The relationship of leaf properties to climate is sufficiently strong that leaf size and shape are important tools in paleobotany for the reconstruction of past climatic patterns (Bailey and Sinnott, 1915; Wolfe and Upchurch, 1987; Upchurch and Wolfe, 1987).

These morphologic differences in leaf size, shape and structure are associated with significant differences in chemical composition, particularly in the structure of the carbon compounds. Lignin, which is more abundant in thick, xeromorphic leaves, is much more resistant to microbial decomposition than is the simpler compound, cellulose. The lignin:nitrogen ratio of plants is correlated with the drought tolerance of North American tree species, through the relationship between leaf morphology and lignin content (Post and Pastor, 1990) (Fig. 7.7). The relative amount of lignin in relation to nitrogen in plant leaves has been shown to be strongly related to the rate at which leaves decompose (Meentemeyer, 1978; Melillo *et al.*, 1982) and consequently, the rate at which nitrogen is released from decomposing leaves (Aber and Mellillo, 1980, 1982a, 1982b; McClaugherty *et al.*, 1985).

Pastor and Post (1986) incorporated the effect of leaf chemistry on nitrogen uptake and decomposition of dead leaves into an individual-based forest succession model. Their simulation experiments investigated the interactions between climate, forest species composition, nitrogen cycling, and forest productivity. In simulations of forest succession on a dry site, Pastor and Post (1986) found that the dominance of oaks and pine with slowly-decomposing leaf litter led to a reduction in soil nitrogen availability and a long-term decrease in site productivity. In contrast, on a wetter site, the dominant species had leaves that decomposed rapidly and increased nitrogen availability. Which species dominated a particular site

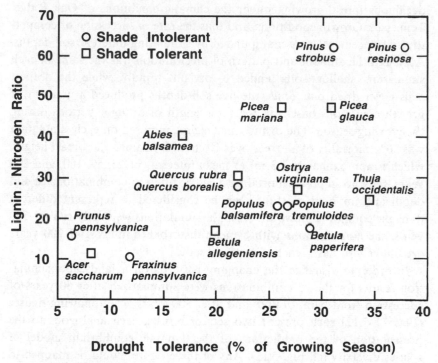

Fig. 7.7 Relationship between lignin:nitrogen ratio of fallen leaves of some North American tree species and the drought tolerance of those same tree species. (From Post and Pastor, 1990.)

was the result of competitive interactions between individuals, and the outcome of competition differed depending on the physical conditions of temperature, moisture, and nutrients. Such 'ecosystem properties' as soil nitrogen and carbon, total biomass, net primary productivity, etc., are the net result of interactions of individual organisms with each other and with the physical environment.

This type of individual-based ecosystem model can be used to investigate the landscape-scale patterns that result from different dynamic equilibria between plant succession and disturbance, not only in terms of species diversity, but also for the entire suite of population, community, and ecosystem properties that result from the interactions of individual organisms with each other and with their environment.

A two-way factorial computer simulation experiment was used to investigate the implications of the dynamic equilibrium model for landscape patterns of community structure and ecosystem processes in 'eastern deciduous forest' growing under the climatic conditions of Oak Ridge, Tennessee. Growth conditions, and thus the rate of succession or competitive displacement, were manipulated by simulating different soil depths. Although the amount and pattern of precipitation was the same in each simulation, shallow soils tended to dry out rapidly, while the deepest soils never dried out. Thus, the five soil depths produced a gradient of growth conditions based on the total length of drought periods during the growing season. The disturbance regime imposed on each simulation was 30% mortality of all trees over 10 cm dbh (diameter at breast height, which is approximately 1.3 m) at mean intervals or 25, 50, 100, and 200 years, plus no imposed mortality. The 25 different combinations of soil depth and imposed mortality can be considered to represent different dynamic equilibria. However, since the simulations were only run for 500 years, the combinations with a mean disturbance interval of 200 years are unlikely to have reached a steady state.

In order to visualize the changing patterns on a landscape, simulation results for the 25 combinations were summarized after 50 years of succession from bare ground, and after 500 years. The following figures (Figs 7.8-7.12) each present two sets of results, each analogous to the presentation of the predictions of the dynamic equilibrium model of species diversity (cf. Fig. 5.11). Any of these figures could be mapped to a landscape by matching the model predictions to the distribution of soil depth, disturbance regime, and patch age on the landscape.

The pattern of tree species richness across the interacting gradients of soil depth and disturbance frequency is shown in Fig. 7.8. The number of tree species over 10 cm dbh is highest during early succession (50 years) under favorable, but not the best, growth conditions (200 cm soil depth). Under the most favorable moisture conditions, higher growth rates have already led to a reduction in species diversity due to competitive displacement, while under less favorable conditions (25 to 50 cm soil depth) trees are growing so slowly that fewer have reached the size to be counted. After 500 years, it is reasonable to expect that a dynamic equilibrium has been established on the more frequently disturbed sites. Highest diversity on this landscape is found on the driest and most frequently disturbed sites, while diversity is lowest on the undisturbed sites. It is important to keep in mind that even sites on relatively shallow soil can support tree growth in the humid climate (140 cm mean annual precipitation) of east Tennessee. Undoubtedly, an even higher frequency of disturbances and/or shallower soils would reduce tree species richness.

Figures 7.9 and 7.10 show the distributions of total biomass and

50 Years   500 Years

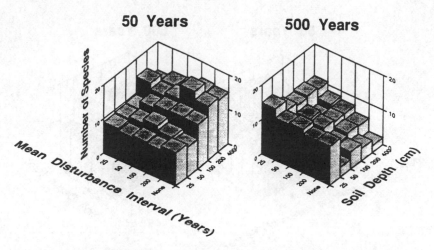

Fig. 7.8 Pattern of tree species richness (number of species with individuals over 10 cm dbh) for 25 combinations of site moisture conditions (soil depth) and mean disturbance intervals, based on simulations with an individual-based forest succession model (Pastor and Post, 1986) using an east Tennessee climate. Simulation results are shown for two different times during succession.

50 Years   500 Years

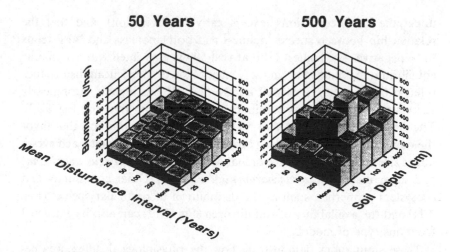

Fig. 7.9 Pattern of total biomass (summed biomass of all individual trees in tons/ha) for 25 combinations of site moisture conditions (soil depth) and mean disturbance intervals at two times during forest succession. Results are from the same simulations presented in Fig. 7.8.

net primary productivity (NPP) at two times during succession, from the same simulations from which the diversity patterns (Fig. 7.8) were taken. Note that maximum biomass and maximum NPP do not occur

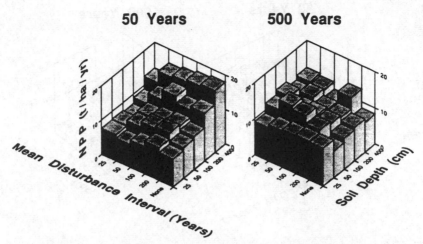

Fig. 7.10 Pattern of net primary productivity (NPP in tons/ha/year) for 25 combinations of site moisture conditions (soil depth) and mean disturbance intervals at two times during forest succession. Results are from the same simulations presented in Fig. 7.8.

under the same conditions (except early in succession), and that the relationship between species richness and both biomass and NPP tends to be negative. The highest NPP at year 500 is associated with a dynamic equilibrium maintained at an early stage of succession (mean disturbance interval of 50 years). It is well known that NPP is highest during early succession, while biomass continues to increase with time (cf. Fig. 9.13). The maximum biomass at year 500 is found under conditions that favor the dominance of tulip poplar (*Liriodendron tulipifera*), which achieves a larger size than the late-successional dominants beech (*Fagus americana*) and sugar maple (*Acer saccharum*) (cf. Fig. 7.1). Other community and ecosystem properties, such as the distribution of particular species (Fig. 7.11) and the availability of soil nitrogen (Fig. 7.12) can also be extracted from this type of model.

These simulations demonstrate how the physiological differences between plant species, manifested through the interactions of individual plants with their environment, have ramifications for the distribution of ecosystem processes and properties on landscapes. The consequences of physiological adaptations for the use of light, water, and other resources constrain the distributions of different species to particular sets of conditions on a landscape, and to particular periods of time during plant succession. The properties of the species present at a specific time and

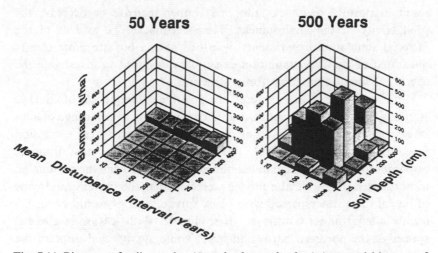

Fig. 7.11 Biomass of tulip poplar (*Liriodendron tulipifera*) (summed biomass of all individuals) under 25 combinations of site moisture conditions (soil depth) and mean disturbance intervals at two times during forest succession. Results are from the same simulations presented in Fig. 7.8.

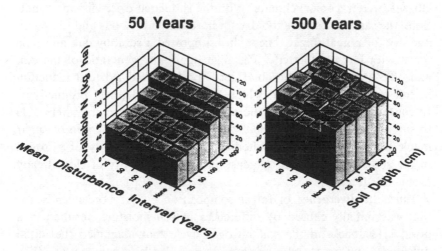

Fig. 7.12 Pattern of soil nitrogen availability (kg N / ha) for 25 combinations of site moisture conditions (soil depth) and mean disturbance intervals at two times during forest succession. Results are from the same simulations presented in Fig. 7.8.

location have a major influence on processes such as nutrient cycling and net primary productivity.

The biogeochemical interactions of plants with their environment can result in positive feedbacks that may either increase or decrease the productivity of the environment. These feedbacks are evident in the factorial simulation experiments described above, but are more clearly illustrated in another simulation experiment designed to investigate the effects of climate change on forest structure.

Pastor and Post (1988) explored the possible consequences of climatic warming on forest composition and nutrient cycling over a range of sites in eastern North America using an individual-based forest succession model (Pastor and Post, 1986). They found some of the most dramatic responses to climate change in the boreal forests. For a site they simulated in northern Minnesota, the model predicted that the current landscape of boreal forest interspersed with black spruce swamps would be significantly altered under a warmer, drier climate. With a warmer climate, species of the northern hardwood forest could survive and displace the boreal forest species. However, the fact that the climate was also drier shifted soil moisture dynamics sufficiently that sandy soils supported a very different forest type than more loamy soils, in spite of the fact that the both soil types commonly support the same forest type under the current (wetter, cooler) climate. With the simulated drier climatic conditions, the sandy sites were too dry for northern hardwoods and too warm and dry for boreal forest. These sites supported a scrubby oak and pine forest with low productivity. The high lignin:nitrogen ratio of the oak and pine litter slowed microbial decomposition and led to a reduction in soil nitrogen availability, which further reduced forest productivity. In contrast, the loamy soils held sufficient moisture that they were able to support northern hardwoods. The low lignin:nitrogen ratio of sugar maple and other species resulted in increased availability of nitrogen in the soils and led to an increase in the productivity of this already productive forest.

Thus, the divergence of forest composition on the boreal landscape that was initially caused by differences in soil moisture, resulted in a positive feedback on the soil nitrogen cycle that magnified the initial differences and produced a much more complex landscape pattern (Fig. 7.13). This demonstrates how differences in the physiological responses and chemistry of different plant species can lead to changes in ecosystem processes and landscape pattern.

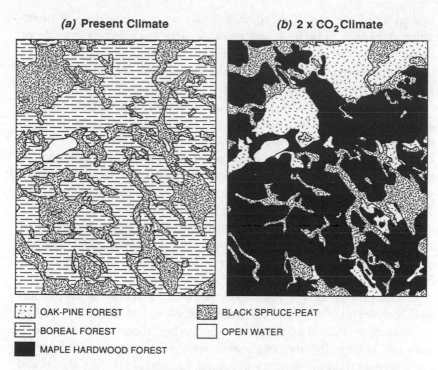

(a) **Present Climate**   *(b)* **2 x CO₂ Climate**

OAK-PINE FOREST   BLACK SPRUCE-PEAT

BOREAL FOREST   OPEN WATER

MAPLE HARDWOOD FOREST

Fig. 7.13 Projected changes in a forested landscape at the north temperate/boreal forest boundary (Carlton County, Minnesota) following a shift to a warmer, drier climate that is predicted for a doubling of atmospheric CO_2. (*a*) Under the present climate, the landscape is dominated by two forest types, the boreal forest on sandy loam soils, and black spruce on wet peat soils. (*b*) With a warmer, drier climate the model predicts that the boreal forest will be replaced by a high productivity maple-dominated deciduous forest on loamier soils, and a low productivity oak-pine forest on sandier soils. The extent of spruce forest will decrease as some of the wet peat soils dry and are invaded by maples. (From Huston *et al.*, 1989, based on Pastor and Post, 1988.)

Summary

The basic assumption of the individual-based approach to ecology is that the same mechanisms influence the structure of populations, communities, and ecosystems. The growth, survival, and reproduction of individuals in a population are determined by their inherent physiological responses to the resources available to them and by other environmental conditions, such as temperature. Resources, energy, and the physical conditions of the environment are the traditional domain of ecosystem ecology, a branch of ecology that has been largely separate from population ecology. Individual-based models demonstrate the interdependence of these two levels of ecological phenomena. Thus, an understanding of population dynamics must be based on what were traditionally considered to be ecosystem processes, and likewise, ecosystem dynamics must be understood in the context of what have been considered to be individual, population, and community processes.

The fundamental similarities (i.e., convergence) of ecological patterns around the globe are the consequence of the unavoidable constraints on all organisms that are imposed by energetics and physiology. For plants, these constraints result from the need to allocate limited resources for acquisition of different types of resources, and from the impossibility of simultaneously maximizing the uptake of CO_2 and minimizing the loss of water through the stomata. These constraints, and a suite of characteristics that are linked to them, are responsible for many of the predictable aspects of plant community structure and ecosystem dynamics, such as forest succession and the relationship between life history, physiology, and environmental conditions.

Quantifying all of the differences between all species is clearly impossible. Fortunately, such exhaustive detail is not necessary for the application of individual-based models. Certain species, and sometimes even particular individuals of a species, have a much greater influence on an ecosystem than the majority of species. In most cases, once the critical species are identified and included in an individual-based model, the outcome is determined largely by a few dominant species. The abundance of many 'interstitial' species can be predicted directly from the models or inferred based on the conditions created by the major species. For example, the abundance and species composition of understory plants can be predicted from 1. the light environment, which is directly determined by the overstory trees; 2. the soil nutrient conditions, which are strongly influenced by the overstory trees; and 3. the soil moisture conditions, which influence both overstory and understory plants, and are in turn influenced by the plants. Likewise, it should be possible to predict the abundance and distribution of many animal species from the environmental conditions and the habitat structure provided by the dominant plants, as demonstrated by the relationship between foliage height diversity and bird species diversity.

8

Landscape Patterns: Disturbance and Diversity

The dynamic equilibrium between disturbance and the rates of population growth and competitive displacement is always set in the context of a specific landscape or region. Each region has particular patterns of spatial heterogeneity and biogeographical history, as well as the temporal heterogeneity imposed by climate and its interaction with biological processes (e.g., Fleming *et al.*, 1987; Duellman and Pianka, 1990). The disturbance regimes found in any landscape are influenced by a multitude of factors, including topographic and geologic heterogeneity, geologic substrate and history, mean climatic conditions and their variability, and the activity of humans. There is no such thing as a typical disturbance. Consequently, generalizations about the effects of disturbance must be carefully qualified by the nature of the disturbance and the conditions under which it occurs.

Disturbance is a general term that must be defined rigorously if it is to contribute to understanding rather than to confusion. The definition of disturbance that is used in this book is based on the mortality of organisms (cf. Chapters 4 and 5). *Disturbance* is any process or condition external to the natural physiology of living organisms that results in the sudden mortality of biomass in a community on a time scale significantly shorter (e.g., several orders of magnitude faster) than that of the accumulation of the biomass. Thus a disturbance may kill a few, many, or all of the organisms in a community, or may simply kill a portion of a single individual, as is often the case with damage to plants. Examples of disturbances include fires, windstorms, floods, extreme cold temperatures, treefalls, ice scouring of shorelines, bulldozers, epidemics (e.g., Dutch elm disease, chestnut blight), etc. Such things as changes in average temperature, addition or removal of nutrients, or the gradual invasion of exotic species would not be considered disturbances unless they result in sudden

215

mortality. For example, mortality that results from several consecutive years of defoliation by gypsy moths would be considered a disturbance, while the gradual senescence and breakup of an old even-aged stand, perhaps exacerbated by insect pests, would not be a disturbance.

Not all mortality results from disturbance. In a forest, trees that senesce and die standing generally have an effect on resources similar to that of a small disturbance such as a single wind-throw, but without much of the associated damage or mortality to other plants. Such non-disturbance mortality is a critical process in forest succession, and marks the transition between an even-aged stand and a multi-aged shifting mosaic of forest patches. Although the transition is often accelerated by small-scale disturbances such as wind-throw, lightning strikes, and insect infestations, it would occur inevitably even in the absence of external processes (cf. Knight, 1987). Mortality of senescent individuals cannot by definition be considered disturbances, although like disturbances its effects on communities can be understood in terms of changes in resource availability.

The term 'perturbation' has been sometimes used interchangeably with disturbance. Perturbation and disturbance are *not* synonomous. Although most disturbances can be considered perturbations, a perturbation is not necessarily a disturbance. Perturbation is a more general term than disturbance and refers to the displacement of some property of a community or ecosystem, such as total biomass, reproductive rate, or nutrient influx, away from its typical value, which is generally considered to represent an equilibrium value or at least a steady state value for a specific parameter.

Sudden Change: The Effects of Disturbances on Landscape Patterns

The most rapid temporal dynamics of landscape pattern are driven by the disturbances that alter discrete portions of the landscape and create a constantly changing mosaic of recently disturbed patches, patches that are recovering from disturbances, and patches that are *apparently* undisturbed, or 'virgin' (Fig. 8.1). While an understanding of the rate and pattern of succession is critical for explaining the changing patterns on a landscape, the landscape-scale consequences of succession cannot be understood independently of the disturbances that 'reset' succession and alter the distribution of organisms and their resources on the landscape.

Disturbances that alter landscapes and their vegetation are almost never truly random. Different types of disturbance are characteristic of

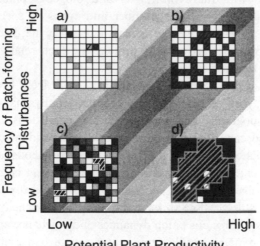

Fig. 8.1 Patterns of biomass heterogeneity that result from different dynamic equilibria between disturbance and productivity in landscapes where the primary cause of heterogeneity is asynchronous patch dynamics. Darker shading indicates higher plant biomass. Local tree species richness on patches is likely to be minimum at both very low and very high levels of biomass, i.e., early and late succession. Diagonal shading indicates the expected intensity and area of disturbances that are larger than patch-scale e.g., fires, epidemics (based on Fig. 5.10).

different regions of the globe, and the probability of particular types of disturbance, such as hurricanes, tornadoes, floods, fires, and volcanic eruptions, differs in a predictable pattern from one portion of a continent to another, and from one period of time to another (Karl, 1988; Gray, 1990; Swetnam and Betancourt, 1990). Mountaintops are rarely affected by floods, and river valleys tend to receive many fewer lightning strikes than mountaintops. The spatial and temporal pattern of disturbances on a landscape result from the interaction of (relatively) predictable patterns of weather, the topography and geomorphology of the landscape, and the structure and dynamics of the vegetation of the landscape.

Although a particular area of landscape may be characterized over the long term by a dynamic equilibrium between its characteristic disturbance regime and the rate of succession, at any particular instant the diversity and species composition of the area will be strongly influenced by the length of time since the last disturbance, as well as by the specific properties of the disturbance itself. The long-term dynamic equilibrium, in combination with the evolutionary and biogeographical history of the region, will determine the number and types of species that can potentially be present, while the actual time since the disturbance will influence which species are actually present and their relative abundance at any specific location.

The pattern of the landscape 'mosaic' can result from two fundamentally different types of heterogeneity. Heterogeneity that results from variation in topography and geologic or soil conditions can impose a pattern on landscape that will be expressed independent of the disturbance regime. Pattern on such a physically heterogeneous landscape can be envisioned as a set of patches representing different combinations of disturbance regime and rate of succession. On a physically homogeneous landscape it is still possible to have dramatic heterogeneity that is produced by the asynchronous patch dynamics of disturbance and succession (Fig. 8.1). The dynamic equilibrium between disturbance and plant productivity (i.e., rates of population growth and competitive displacement) can establish patterns of heterogeneity that strongly influence properties of the disturbance regime, as well as the recovery of ecosystems following disturbance.

Disturbances increase local species diversity primarily under conditions of rapid growth rates and high productivity, where a high rate of competitive displacement results in a rapid reduction of diversity in the absence of disturbance. The effects of disturbance on the visual pattern of a landscape are greatest where the rate of growth and productivity is low, and the vegetation takes a long time to recover from the effects of a disturbance. The effect of a disturbance on biological diversity depends not only on the properties of the disturbance itself, but also on the initial state of the ecosystem and the dynamics of population growth and competition both before and after the disturbance.

The effects of disturbance on community structure operate at several very different time scales: 1. mortality can almost instantaneously alter the diversity of a community, either decreasing or increasing diversity among the remaining organisms. 2. changes in resource availability caused by disturbance can alter species diversity more slowly through the effects on growth, reproduction, and competitive interactions; and 3. evolutionary changes and natural selection can act to increase the ability of particular species to survive the predominant type of disturbance, as well as increase the proportion of species in the community that are adapted to survive the disturbance by eliminating sensitive species and favoring resistant species. The specific properties that can be used to characterize disturbances and predict their effects are: 1. intensity; 2. frequency; 3. timing; 4. area; and 5. effect on resources.

Disturbance Intensity

Intensity of a disturbance refers to the proportion of the total biomass of a community that is killed, or in some cases, the proportion of the total biomass of a particular species that is killed. Rarely is a species made locally extinct by a single disturbance. Often an individual plant or colonial organism is not killed, although much of its biomass may be damaged or removed. As Sousa (1985) observes, 'the degree of damage caused by a disturbing force depends on a) the magnitude of the force, b) physiological and morphological characteristics of organisms in question, c) the nature of the substratum to which the organisms are attached [referring specifically to sessile marine organisms].' This useage is also consistent with the terminology of fire ecologists (Heinselman, 1973, 1981a; Gill and Groves, 1981), although fire intensity generally refers to the amount of energy released, and not simply the proportion or amount of biomass killed (Christensen, 1985).

In many situations, disturbance intensity is inversely related to disturbance frequency (Loucks, 1970; Miller, 1982; Pickett and White, 1985). Low frequency disturbances allow a larger accumulation of biomass during the intervals between disturbances, so the absolute, and often the relative, amount of biomass killed in a single disturbance event is greater with low frequency disturbances. Thus, there is usually a large autogenic component to the intensity of disturbance, that is, the intensity of the disturbance is to some extent a consequence of the organisms themselves. The autogenic component is particularly important in the case of fire, for which both the frequency and the intensity of disturbance are a consequence of accumulated dead and living biomass (Heinselman, 1981; Romme and Knight, 1981; Romme and Despain, 1989). Large trees are also more susceptible to mortality by wind-throw or lightning strike, so the probable frequency and intensity of these disturbances increases in late succession as trees mature (Clark, 1989a). Furthermore, the intensity of the disturbance can influence the proportion of the total number of species that are affected, assuming that there is some ranking of susceptiblity to disturbance or recovery from disturbance based on species-specific properties, such as individual size or population growth rates (cf. Figs 4.6 and 5.6).

The amount and pattern of plant biomass that results from a specific dynamic equilibrium of disturbance and productivity can have a significant effect on disturbance intensity, as well as on other properties of the disturbance regime. The potential intensity of fires, susceptibility to wind damage, and the proportion of biomass that is aboveground and exposed to climatic extremes, all increase with increasing biomass and height of the vegetation. The potential intensity of these and perhaps other types of disturbances should be greatest under conditions of high productivity and low disturbance frequency where large amounts of biomass can accumulate (Fig. 8.1d). To the degree that the intensity of disturbance is a function of the total area of a particular type or stage of vegetation, those dynamic equilibria that tend toward a homogeneous patch structure (e.g., Fig. 8.1 a and d) should be characterized by more intense disturbances.

The intensity of a disturbance has a major effect on the way in which resources are altered by the disturbance. In general, the amount of resources made available by the disturbance tends to increase with the intensity of disturbance up to a certain level. At high disturbance intensities, resource availability can be decreased if resources are lost from the ecosystem. Resource loss associated with intense disturbances includes volatilization of nutrients such as nitrogen and loss of soil and its minerals due to erosion.

Disturbance Frequency

Frequency is simply the number of disturbances that occur within a particular time interval, which is the inverse of the time intervals between disturbances. The time interval between disturbances determines how long the ecosystem has to recover before it is affected by another disturbance. However, the extent to which the ecosystem recovers between disturbances depends on many factors in addition to the length of time it has to recover. In particular, the intensity of the disturbance determines the starting point for the recovery, and the availability of critical resources in the environment influences the rate at which the recovery occurs. The area of the disturbance may also influence recovery if the rate of recovery depends on organisms that must invade from outside the disturbed area.

As mentioned earlier, there is often an inverse relationship between disturbance frequency and intensity. For a particular disturbance intensity and rate of recovery, the frequency of disturbances determines how far succession can proceed. To the extent that the frequency of disturbance for a particular ecosystem is relatively constant, the dynamic equilibrium between disturbance and recovery allows the prediction of many properties of the ecosystem. Such properties include the maximum biomass that can be accumulated between disturbances, the size, growth rates, reproductive 'strategies,' and other life history characteristics of the organisms, the structure of populations, and the number of species relative to an ecosystem that is otherwise similar, but has a different frequency of disturbance (cf. Chapters 5 and 7).

A critical issue is the degree to which the frequency of disturbance for a particular ecosystem is constant enough to be predictable and thus establish a dynamic equilibrium. Disturbances can be totally unpredictable at a small spatial scale, while highly predictable over a larger area. For example, it is virtually impossible to predict which individual tree will be struck by lightning during a particular year, but the total number of lightning strikes over a large area can be predicted quite accurately (Komarek, 1964). Thus, measurements to detect such patterns must be sufficiently extensive to characterize the entire area, not just a small patch.

The predictability of disturbance frequency results from three primary types of process: 1. events that occur with a constant probability; 2. predictable cycles in climate or weather; and 3. biological processes that occur at predictable rates or with predictable cycles, such as the production of biomass and the accumulation of natural fuels.

Predictable cycles of climate or weather occur at many different temporal scales (Hays *et al.*, 1976; Brubaker, 1981; Berger *et al.*, 1984; Woodward, 1987; Clark, 1989b,c). Annual patterns of storms, temperatures, and precipitation can impose highly predictable frequencies for certain types of disturbance such as fires (Clark, 1988b). Cycles such as the El Niño-Southern Oscillation increase the probability of fire in southwestern US deserts at intervals of three to seven years (Swetnam and Betancourt, 1990). Major climatic cycles have occurred at intervals of one or two hundred thousand years between glacial maxima over the past million years (Hays *et al.*, 1976; Imbrie, 1985; Berger *et al.*, 1984; Denton and Hughes, 1981). Many of the most important disturbances are directly associated with weather, including fires, floods, droughts, hurricanes, windstorms, and extreme cold. Most of these disturbances become even more predictable when geographic location and topography are considered.

Predictable biological processes include the time required by a tree to grow to a size that will create a significant gap in the canopy when it falls, and the amount of time required to produce a sufficient fuel load to support a fire (Loucks, 1970; Romme, 1982; Romme and Despain, 1989; Clark, 1989c). Of course, these biological processes interact with weather conditions to determine when a particular disturbance will occur. For example, the occurrence of a fire depends on both the availability of sufficient fuel, favorable dry and windy weather conditions, and a source of ignition. Such an interaction of predictable processes is unlikely to decrease the predictability of disturbance frequency unless the processes are moderately out of phase (occur at somewhat different frequencies, rather than the same or very different frequencies). In such cases, the disturbance interval can become chaotic and unpredictable (Cox and Smith, 1953, 1954; Getty, 1981). When the interacting processes occur at different rates or frequencies, the slowest process generally controls the frequency of disturbance.

The inherent differences between species (see Chapter 7) produce dramatic contrasts in the effect of disturbance on different species. These species-specific responses influence both the vulnerability of a species to a particular type of disturbance, and the ability of the species to recover from that disturbance. The basic pattern of these differences between species is that described by the concept of r- versus K-selection (MacArthur and Wilson, 1967) or pioneer versus climax species (Clements, 1916). It is important to keep in mind that these concepts can be applied to individual organisms, particularly those of modular construction such as plants or colonial invertebrates, as well as to entire populations. Large organisms are often more susceptible to certain types of disturbances, such as physical forces from air or water movement, than are small organisms. Likewise, small organisms may be more easily damaged by other types of disturbances, such as large temperature changes caused by fire or extreme weather. Ground fires often kill seedlings and shrubs, leaving large woody plants relatively unharmed.

Repeated disturbances impose strong natural selection for individual organisms that are either resistant to the disturbance, or recover rapidly from the disturbance. This natural selection is manifested at the species level, by the elimination of species that are sensitive to a particular disturbance type, as well as at the level of population genetics, by the elimination of those individuals or genotypes within a species that are most sensitive to the disturbance (Muir and Lotan, 1985a,b; Vrijenhoek, 1985; Rice and Jain, 1985). The evolutionary and natural history literature are replete with examples of adaptation to natural disturbance such as fires and predation (see Chapter 13). Thus, selection has the effect of creating an assemblage of species that are adapted to a particular disturbance regime, and shifting the genotype and phenotype of each species toward a configuration that is well-adapted to the disturbance regime. Of course, if selection at the genotypic level does not act sufficiently rapidly or effectively to adapt the species to the disturbance regime, that species will be eliminated.

Frequent disturbances select for species that reproduce at a young age and small size, while infrequent disturbances select species that reproduce later at a larger size and produce offspring with a high probability of survival (see review of life history selection in Stearns, 1976, 1977; also Gadgil and Solbrig, 1972; Schaffer, 1974; Sousa 1979a,b). With frequent disturbances, those species that require times longer than the inter-disturbance interval to reproduce are completely eliminated. However, even with infrequent disturbances that favor larger, longer-lived species, those species that reproduce quickly at a small size are able to reproduce shortly after the disturbance. The survival of species adapted to early-successional conditions and/or frequent disturbances depends on the ability to survive long periods between the relatively rare appearances of suitable early-successional conditions or to reach the disturbed areas quickly (Marks, 1974; Grubb, 1977). Traits such as good dispersal and the abilty to survive in a dormant stage allow early-successional species to survive in communities with low disturbance frequencies. The total species richness within a landscape with a low frequency of disturbance can potentially be very high. However, only a subset of the species is likely to be present in a particular location at a particular time, so local diversity can be low, particularly in late succession.

A particular disturbance regime can also influence the regional species pool available for recolonization. Prairie fires, which damage or kill invading woody species, can eliminate some tree species from areas in which they can potentially grow, and from which they would eventually eliminate the smaller prairie plants in the absence of fires. By eliminating the seed source for woody species, the repeated fires alter the pool of species available for recolonization, and thus alter the course of plant succession between fires.

The Timing of Disturbances

Timing refers to the coincidence of the disturbance with important cycles or events in the ecosystem affected by the disturbance. Timing can influence which species are most susceptible to damage or mortality, whether the landscape will be susceptible to additional disturbances, such as erosion, and which species are most likely to disperse into the disturbed area or germinate and become established (Hartshorn, 1978). A prairie fire that occurs in the spring has very different effects on species composition and nutrient cycling than a prairie fire that occurs in the fall (Collins and Barber, 1985; Collins and Uno, 1983). Snow or ice storms that occur in the fall when deciduous trees still have leaves do much more damage than similar storms during the leafless season.

Disturbance Area

Both the absolute and relative area of a disturbance and the shape of the disturbed area have an important effect on the recolonization of the disturbance, as a result of the general inverse relationship between the density of dispersed propagules and the distance from the propagule source (Gleason, 1925). The relative area, or percentage of the total area of a particular habitat type that is disturbed, is also an important consideration, because the undisturbed habitat may be the primary source for propagules of species able to survive in that habitat.

Disturbance area is a prime determinant of the spatial pattern of landscapes, and varies predictably between different climate/vegetation regions (biomes) that are characterized by different dynamic equilibria between productivity and disturbance (Fig. 8.1). On landscapes where the rate of recovery is high in relation to the disturbance frequency (Fig. 8.1d), most patches will have relatively high biomass and form a homogeneous matrix of mature vegetation. Under these conditions, disturbances such as fires or pathogens that spread most effectively under uniform, high density conditions are likely to affect large areas (Heinselman, 1981a; Burdon et al., 1989). In contrast, in dynamic equilibria with low rates of recovery as well as low disturbance frequencies, the complex mosaic of patches at all stages of recovery, with great variation in biomass, structure, and species composition, is likely to hinder the spread of disturbances such as fires or pathogens, and limit them to relatively small areas (Fig. 8.1c).

For example, boreal coniferous forests are characterized by fires of very large area and relatively high frequency (e.g., every 50-100 years), which tend to synchronize large areas of the landscape into even-aged stands that originate following the fires. These stands increase in fuel load, and thus fire probability, as the forests mature (Heinselman, 1973, 1981a; Rowe and Scotter, 1973; Romme, 1982; Romme and Knight, 1981;

Clark, 1989c). In contrast, broadleaf temperate forests and many tropical evergreen forests are characterized by disturbances of much smaller area, ranging from gaps caused by the windthrow of one or a few trees to fires of relatively limited extent and a high degree of patchiness (Runkle, 1981, 1985; Hartshorn, 1978; Brokaw, 1985a).

The average area of disturbance within a region obviously affects the spatial scale at which the landscape is at a dynamic equilibrium between rate of succession and frequency/intensity of disturbance. With large disturbances, much of the landscape will be synchronized in a cycle of disturbance and succession, with species composition and diversity changing synchronously with time over large areas. In contrast, on a landscape with smaller disturbances, species composition and diversity over large areas remain relatively constant over time, because the large number of patches can integrate the asynchrony in the dynamics of succession. Smaller disturbances also reduce the dispersal delays that may cause a slower accumulation of species richness and alter the patterns of species arrival and dominance during succession.

Although disturbances often create dramatic heterogeneity on landscapes, the effect of both large and small disturbances tends to be homogenization of the landscape. Large disturbances not only synchronize succession over vast areas, but they also eliminate the types of heterogeneity that are expressed primarily in late succession. As discussed in more detail in Chapter 10, species distributions along environmental gradients tend to become narrower over the course of succession. Consequently, the pattern of vegetation in response to underlying landscape heterogeneity in physical conditions becomes more prominent through successional time. Thus, disturbances that maintain large areas of vegetation in early-successional conditions can prevent the expression of heterogeneity that would appear during later succession.

Small disturbances obscure landscape patterns that result from the temporal asynchrony of succession that is initiated by larger disturbances. Landscape patterns that are the result of different portions of the landscape being at different stages of succession disappear as small disturbances reinitiate succession within late-successional patches of landscape. The occurrence of small disturbances results in the formation of a 'shifting mosaic steady state' (Cooper, 1913; Watt, 1925, 1947; Bormann and Likens, 1979; Smith and Urban, 1988), which is a dynamic equilibrium in which high spatial variability at small scales aggregates to low spatial variability (as well as temporal heterogeneity) at large spatial scales (cf. Fig. 4.1).

The spatial area over which a particular disturbance regime applies has a major influence on how strongly the disturbance regime affects community structure, which is to say, how strongly the particular dynamic equilibrium between disturbance and succession is expressed. The larger the area influenced by a particular disturbance regime, the greater will be the distance over which species adapted to other disturbance regimes must disperse, and the less likely it will be that these other species will be present at any time during succession. Whether the disturbance regime is one of a few large disturbances, or many small disturbances, species composition will be shifted toward dominance by those species best adapted to the local disturbance regime.

If the size of the area with a particular disturbance regime is small, immigration of species from other areas with different disturbance regimes can lead to a higher species richness than would be expected, as well as to the presence of species not well-adapted to that particular dynamic equilibrium. This phenomenon of immigration of species poorly adapted to local conditions has been called the 'rescue effect' (Brown and Kodric-Brown, 1977; Shmida and Wilson, 1985).

Examples of very large disturbances are the Pleistocene glaciations that caused massive local extinction of plants and the displacement of most plants and animals to the south in North America and Europe. In Europe, the barrier of the Alps prevented plants from 'escaping' to the south, and resulted in the continental-scale extinction of part of the pre-Pleistocene flora of northern Europe (Huntley and Birks, 1983; Silvertown, 1984). In eastern North America, most tree species were displaced to the southeastern United States, where the Appalachian mountains and southern coastal plain provided a large range of environmental conditions that served as a refuge during the glacial periods, as well as during the warmer, drier interglacial periods (Braun 1950; Webb, 1986).

Following the retreat of the glaciers, species reinvaded the deglaciated areas at different rates depending on their reproductive pattern and mechanism of dispersal (Davis, 1976, 1981; Huntley and Birks, 1983; Webb, 1986; Johnson *et al.*, 1981; Birks, 1989; Huntley and Webb, 1989). Some species, particularly 'early-successional' plants, which produced large numbers of small seeds that were dispersed long distances by the wind or by birds, spread rapidly and closely followed the retreating glaciers. Other species, primarily 'late-successional' plants with heavier seeds that were dispersed shorter distances, reinvaded much more slowly, and in some cases, such as that of the American beech, are still expanding their range northward (Davis, 1981; Davis *et al.*, 1986).

To the extent that species extended their ranges at different rates, the total species pool of formerly glaciated areas gradually increased in diversity (Silvertown, 1984). This increase in regional diversity did not necessarily result in an increase in local species diversity over the entire course of succession, since some of the slowly invading species may have been superior competitors that were able to reduce species diversity during later stages of succession. Slow dispersal of certain species following large-scale disturbance or climate change may be important even in the tropics, where slowly dispersing dominant species may greatly reduce the tree species diversity after they invade an area (Hart *et al.*, 1989; see Chapter 14).

The critical influence that physiological and life history properties of different species have on the outcome of interspecific interactions (Huston and Smith, 1987) emphasizes the importance of the types of species that are available to recolonize a disturbed area and participate in the subsequent succession. Under the same environmental conditions the pattern of species composition and diversity during succession can differ greatly depending on which species are present (McCune and Allen, 1985; Robinson and Dickerson, 1987; Robinson and Edgemon, 1988; Hubbell and Foster, 1986; Huston and Smith, 1987; Drake, 1991).

Disturbances and Resource Availability

Although the most dramatic effects of a disturbance are the short-term consequences of mortality, the most important effects are generally the longer-term consequences for resource availability. The recovery of the living community is generally quite rapid following a disturbance, through regrowth of surviving organisms (either from surviving vegetation or from the seed bank), reproduction, and invasion of organisms from outside the disturbed area. However, the rate at which recovery occurs, as well as the long-term structure of the community is determined primarily by the resource conditions that existed previously or were created by the disturbance. Thus, disturbances may potentially influence both axes of the dynamic equilibrium model (Chapter 5).

One of the primary mechanisms through which disturbances influence resources is altering the availability of mineral nutrient resources or energy that were contained in, or excluded by, the biomass that was killed. Disturbances influence resource availability in three primary ways: 1. conversion of nutrients that were formerly incorporated in living biomass into a form available to plants and other organisms as a result of death and subsequent decomposition; 2. eliminating organisms that previously intercepted, absorbed, or excluded large amounts of particular resources (e.g., shading by the forest canopy); and 3. altering the form of resources, sometimes leading to their loss from the system (e.g., volatilization, erosion). Since in most terrestrial communities, the physical structure and most of the biomass are contributed by plants, disturbances that directly affect plants generally have the greatest effect on resource availability.

In general, the greater the intensity of a disturbance, the greater will be the proportion of total resources that are released from the biomass. The decomposition of plant and animal matter releases mineral nutrients in the approximate proportions that are needed by organisms, and also provides a critical source of energy for many organisms, particularly those involved in decomposition (Swift et al., 1979).

Removal of biomass often increases the availability of specific resources in the disturbed area by eliminating the uptake of those resources by the biomass. For example, removal of a leafy tree canopy by a treefall allows the penetration of light to the forest floor, and has other indirect effects such as increasing soil temperature and air mixing. Disturbances that kill plants also lead to increased water availability for the remaining plants by reducing root uptake and transpiration (Oosting and Kramer, 1946; Ehleringer, 1984). Plant mortality eliminates nutrient uptake by the plants, although nutrient immobilization by microbes using the carbon from dead plant material as an energy source can reduce nutrient availability to a greater degree than plant uptake (Vitousek and Matson, 1985). In marine and aquatic systems, disturbances that kill biomass generally increase light levels to smaller organisms, but also increase the availability of nutrient- and food-bearing water. Whether space itself can legitimately be considered a resource seems doubtful. It is the resources and access to resources associated with space that make it possible for an organism to occupy an area.

However, some types of disturbances not only kill living tissue, but also cause the removal or loss of the resources in that tissue. Fires, if they are hot enough, can result in the loss of most of the carbon, and much of the nitrogen, contained in living tissues (Wells, 1971; McKee, 1982). While there may be a burst of vegetative growth following a fire as a result of the rapid release of mineral nutrients, many studies show little long-term effect of burning on forest or grassland productivity (Wells et al., 1979; Van Lear and Waldrop, 1989). The long-term effects of such a disturbance on species diversity depend on the total pre-disturbance nutrient stock, how much of the total nutrient stock was lost, and on how rapidly it is replenished by rainfall, nitrogen fixation, and mineral weathering.

When the effect of a disturbance on resource availability results primarily from the mortality of organisms and the subsequent redistribution of resources, the effects are likely to last only until the populations have recovered to the pre-disturbance level. However, disturbances that cause the physical removal or addition of resources can have long-term effects on an ecosystem. In such cases the fully recovered post-disturbance ecosystem can be different from the pre-disturbance ecosystem. Certain disturbances, such as floods or landslides, may physically remove the killed biomass, thus causing the complete loss of the resources contained in that biomass. Such disturbances may also remove all organisms and all seeds, roots, etc., thus requiring that the area be revegetated by plants invading from other areas.

Extensive erosion following major fires, the uplift or subsidence of land during earthquakes, or agricultural activities can result in such a significant change in environmental conditions that the post-disturbance community will be totally different from the pre-disturbance community (Periera, 1973; Foster, 1976; Thirgood, 1981). Depending on the pre-disturbance level of resources, and how resources are changed by the disturbance, the long-term species diversity of the ecosystem can be either increased or decreased.

Floods and landslides often affect resource availability by physically altering the landscape. Floods may result in the deposition of rocks, soil, and other materials that can increase mineral nutrient levels if the materials are high in nutrients, but can decrease nutrient availability if the deposited materials are low in mineral nutrients and cover a surface that was higher in nutrients. Likewise, floods or landslides can remove surface layers of soil and rocks, which may be either higher or lower in mineral nutrients than the underlying layers. Thus, many of the effects of disturbances on resource availability can only be predicted through a detailed knowledge of the soils and geology of the affected areas, and in some cases of more distant areas from which materials are transported.

One example of the importance of this phenomenon is the fertility of the alluvial soils in river floodplains. Such soils are periodically augmented by silts and sands deposited by the flooding river. Throughout much of the world, the most fertile agricultural soils are found in river floodplains. Examples include the Nile and Volta River valleys, where agricultural productivity has dropped dramatically since the construction of the Aswan dam prevented the annual floods that replenished the soils (Hilton and Kown-Tsei, 1972; Baxter, 1977; Abul-Atta, 1978; Petts, 1979; Walton, 1981). Other examples are the fertile river valleys and deltas of India and Southeast Asia that receive nutrient-rich alluvium from the Himalayas (Mikhailov, 1964; Tison, 1964; Milliman *et al.*, 1987). However, not all alluvial soils are fertile. In the Amazon Basin, some river floodplains receive alluvium that has been eroded from very old, highly weathered, and nutrient poor upland areas, and consequently their fertility and agricultural productivity are very low (Sioli, 1964, 1975; Fittkau *et al.*, 1975; Gibbs, 1967, 1972). The effect on mineral nutrients of disturbances that move soil can be dependent on both local and distant soils and geology, so generalizations are difficult to make.

The level of one critical resource can determine the extent to which a given type of disturbance will influence other critical resources and thus affect plant community structure. For example, disturbances that remove vegetation and increase light availability will have little impact on diversity within or between habitats in xeric environments, where only one or a few light-use strategies are possible. However, the restructuring of vertical and horizontal light availability in mesic areas can have a major impact on the diversity of functional types. The role of gap formation, in particular, is known to play a critical role in structuring both tropical and temperate forests (Hartshorn, 1978; Runkle, 1981, 1982; Orians, 1982; Runkle and Yetter, 1987; Brokaw, 1985a,b), but is apparently not a major factor in deserts.

In general, the resource effect of any particular disturbance is very predictable for some types of resource, while for other types of resource it is much less predictable, and depends on local conditions. To the extent that the effect of disturbances upon resource availability can be determined, the response of vegetation to disturbance should be predictable using a model based on plant functional responses to resources.

Summary

The effects of disturbance on landscape pattern are expressed primarily through the effect of the disturbance on the vegetation of the landscape. Disturbances can have very different effects at different points along a resource gradient (e.g., Figs 5.6 and 5.8). Some disturbances are limited to certain portions of a gradient, whereas other disturbances have a large effect only under particular resource conditions. Most disturbances, whether autogenic or allogenic, tend to occur with a characteristic frequency and intensity in different regions of a landscape. The type of disturbance that characterizes any particular landscape is determined by the interaction of climate, weather, topography, geology, and the properties of the organisms that dominate the landscape, usually plants - often with a strong human influence. This periodicity of disturbances allows the establishment of a dynamic equilibrium between the rate of vegetation change (i.e., succession) and the extent to which disturbances slow or prevent succession or reinitiate succession from some earlier stage. This dynamic equilibrium in turn strongly influences species diversity (Chapter 5) and can also influence a wide variety of other community and ecosystem properties (Chapter 7).

Along a gradient of increasing disturbance frequency or intensity, species diversity can increase, decrease, or peak in the middle of the gradient. The species present on a landscape are determined by strong natural selection exerted by the disturbance regime. Smaller landscape patches or patches closer to environments with different disturbance regimes are likely to have a higher proportion of species that do not have life histories optimal for the local disturbance regime. The effect of disturbances can be best understood in the context of the environmental conditions that influence the growth and recovery of populations, the rate of competitive displacement, and the rate of dispersal or reimmigration into disturbed areas.

9

Landscape Patterns:
Succession and Temporal Change

Disturbance and successional changes are the primary landscape processes that are observed by humans and that regularly interact with human activities. The most rapid temporal changes in landscape pattern, those caused by disturbances, were discussed in the previous chapter. The study of succession is inseparable from the study of disturbance, since it is almost always a disturbance that initiates succession and influences the conditions under which succession occurs.

Succession is the continuous change in the species composition of natural communities that results from many processes, particularly the growth and mortality of organisms under environmental conditions that are continuously changing as a result of either the actions of the organisms themselves or of externally imposed processes such as climatic cycles, or both. Succession occurs, with different rates and patterns, in all natural communities, and is the fundamental process of landscape dynamics. Because plants are the dominant organisms on most landscapes, both visually and functionally, plant succession is the dominant biological process that affects landscape patterns of biological diversity on the temporal scale of human lifetimes.

While the location and timing of any particular disturbance are difficult to predict, the course of events that are set in motion by the disturbance is much more predictable. The regrowth of the plant and animal communities following a disturbance (i.e., succession) is a consequence of the processes and interactions between organisms that have been described in previous chapters. As a result of succession, the biological diversity of a landscape is constantly changing, and the species diversity of any specific location is usually increasing or decreasing, but rarely remains constant. One fundamental process of succession that is always associated with an increase in the number (although not always the evenness) of species is immigration. The other fundamental process of succession, which is usually associated with decreasing species diversity of the dominant organisms, is the process of competitive displacement. Competitive displacement during succession generally leads toward a community in which the diversity of the dominant organisms is low; this is sometimes considered the endpoint of succession or the 'climax' community.

An inherent property of succession that is responsible for much of the confusion surrounding the interpretation of successional patterns and the development of a theory of succession is that succession creates both temporal and spatial patterns. Spatial variation in vegetation can result from a single successional sequence (sere) that occurs under similar conditions (i.e., resource availability, temperature) and follows the same pattern of species composition at different locations, but is initiated at different times in different locations. Spatial variation in vegetation can also be caused by succession that occurs under different conditions in different locations, and follows different patterns of species composition toward different endpoints. Looking at a particular landscape at a particular time, it is not always easy to determine whether the vegetation patterns result from a single sere that is occuring asynchronously on different parts of the landscape, or from different seres occurring under a variety of conditions of resource availability, either in or out of synchrony. Misinterpretation of the spatial/temporal relationship of vegetation pattern on landscapes has led to most of the disagreements about the patterns and causes of succession (e.g., Walker, 1970; Drury and Nisbet, 1973; McIntosh, 1981). Any useful theory of succession must be able to explain both spatial and temporal patterns.

This chapter focuses on the predictable temporal aspects of succession that result from the characteristics of the organisms involved in succession, and from the properties of the environment in which succession occurs. Spatial variation in successional patterns along gradients is the subject of Chapter 10.

Types and Causes of Succession

As defined above, there are two general causes of succession: 1. the relative abundances of species change as a result of changes in the environment that are caused primarily by the organisms themselves. This has traditionally been called *autogenic* succession; 2. the relative abundances of species change as a result of changes in the environment that are caused primarily by external factors that are not affected by the organisms themselves. This has traditionally been called *allogenic* succession. All successional changes on landscapes result from one or both of these causes, and while some successions can be clearly classified as one or the other, most successions are influenced by both.

Autogenic Succession

Autogenic succession results from changes caused by the organisms themselves and so occurs on a time scale that is similar to or shorter than the lifespans of the longest-lived organisms. Environmental changes caused by the organisms may be either beneficial or detrimental to the organisms that cause them, with most of the changes being simultaneously beneficial to some organisms and detrimental to others. Beneficial changes, which are sometimes called facilitation (Connell and Slatyer, 1977; McIntosh, 1981), include such things as increased soil organic matter and addition of nitrogen by plant species with nitrogen-fixing symbionts (Clements, 1916). Detrimental changes, sometimes called inhibition, include reduction of resources such as light, water, and nutrients, which are taken up by some plants and thus made less available to other plants. Detrimental changes are generally considered to be evidence of competition. Thus, autogenic succession can also be interpreted as the time-course of competitive exclusion, as described by the competition models presented in Chapters 5, 6, and 7.

Autogenic succession is most commonly thought of in terms of plants, but can also involve heterotrophic organisms that colonize, modify, and eventually cause the disappearance of dead organic matter such as feces, dead animals, and dead plant parts, a process that could be called decompositional succession. Among plants, succession has traditionally been classified based on the history of the substrate on which the succession occurs. *Primary succession* refers to succession on a substrate that has not been previously occupied by plants, such as volcanic ash or lava, bare rock, or recently exposed sands or soil. *Secondary succession* is simply succession that occurs on a site where plants have previously grown, after some or all of the plants have been removed or killed without eliminating the physical and biological changes to the substrate that resulted from the previous plant occupation.

Allogenic Successions

Because of the wide range of generation times between groups of organisms, purely abiotic environmental changes can produce successional changes over time scales ranging from days to millenia or longer. Environmental fluctuations that occur repeatedly during the lifetime of an organism are unlikely to influence the pattern of succession among species with that general lifespan. In contrast, shifts in environmental conditions that occur over periods as long or longer than the organisms' lifespans are likely to result in changes in dominance among those species, that is, succession. Whether a particular environmental change has the effect of a minor fluctuation or a major shift clearly depends on the lifespan of the organisms being considered (cf. Hutchinson, 1961; Chapter 5).

Seasonal changes in temperature, photoperiod, and light intensity produce a well-known succession of dominant phytoplankton in freshwater lakes that is repeated with little variation every year. Although independent physiological measurements demonstrate that the periods of dominance by many species are correlated with allogenic factors such as their optimal temperature, nutrient, and light conditions (Hutchinson, 1967; Lewis, 1978; Wetzel, 1983), it is also known that autogenic factors, such as competition between different phytoplankton species for nutrients (Lehman *et al.*, 1975; Titman, 1976; Tilman *et al.*, 1981) and complex interactions with higher trophic levels (Porter, 1973, 1976; Carpenter and Kitchell, 1984; Carpenter *et al.*, 1985, 1987; Kitchell and Carpenter, 1987) influence the temporal patterns of phytoplankton abundance.

Changes in environmental conditions associated with long-term climatic fluctuations in temperature or precipitation can produce allogenic successional changes in terrestrial vegetation that occur much more slowly than the changes that result from autogenic succession (Cowles, 1911; Pennington, 1986). Changes from forest vegetation to prairie vegetation in central North America during the warm, dry hypsithermal period (c. 7000-5000 yr BP) is an example of this type of allogenic succession (Webb *et al.*, 1983).

Long-term changes in community structure that result from gradually changing climate are difficult or impossible to study directly. Some comparisons of vegetation change in response to climate variation on the scale of 20 to 80 years have been conducted using photographic records. On a longer time scale, analysis of the pollen record found in the sediments of lakes and bogs provides a record of vegetation change associated with glacial and post-glacial climatic changes over the past

15,000 to 20,000 years (Davis, 1969, 1981; Webb and Bryson, 1972). The plant communities revealed in the pollen record reflect gradually changing dynamic equilibria in response to changing climate, changing disturbance (particularly fire) frequencies (Clark, 1988a, 1990), and the delayed arrival of slowly migrating late-successional species (Davis, 1981). On a still longer time scale, the fossil record of leaves and wood reveals changes in plant community composition during the Mesozoic period in response to both climate change and the evolution of new species (e.g., Wolfe and Upchurch, 1987).

Gradual soil changes that take place over 10s or 100s of thousands of years also cause a slow allogenic succession. In a series of sand dunes of increasing age along the east coast of Australia, depth of soil horizon development is highest on the oldest dunes (about 25 km inland), but the total amounts of phosphorus and calcium available in the soils decrease dramatically with increasing dune age (Walker *et al.*, 1981). Biomass is highest on the dunes of intermediate age, and vegetation decreases in height and biomass toward the oldest dunes. Both the total number of species and the number of tree species decreases from the intermediate to the oldest dunes. The mechanism behind this pattern, soil weathering and depletion of nutrients, is a largely allogenic consequence of physical processes operating over long periods. However, the autogenic influence of plants in adding organic matter to the soil, which led to an increase in soil carbon and depth of A1 horizon (surface organic soil layer) from the youngest to the intermediate dunes, and perhaps also contributed to chemical conditions conducive to mineral weathering and leaching (e.g., Johnson *et al.*, 1975; Johnson *et al.*, 1977), should not be overlooked. This type of succession has been called *retrogressive* because of the decrease in biomass, height, and diversity of the plant community. The existence of this type of succession was denied by F.C. Clements, but retrogressive successions were described by Gleason (R.P. McIntosh, personal communication).

During allogenic succession, vegetation is more likely to be in equilibrium with the slowly changing environmental conditions that affect the vegetation than during autogenic successions, which are dynamic responses to the constant change in environmental conditions caused by the organisms themselves (cf. Cowles, 1911). Autogenic succession may occur continuously during the course of allogenic succession, with species composition and diversity reflecting a series of dynamic equilibria between changing environmental conditions and disturbance regimes.

The relative importance of autogenic succession during allogenic succession depends on the frequency of disturbances that initiate autogenic succession, as well as the rate of the autogenic succession in relation to the rate of allogenic succession. With frequent disturbances, the vegetation is likely to be in a continuous state of relatively rapid change, and species composition is likely to change more rapidly in response to allogenic influences than it would in the absence of disturbance (e.g., Overpeck *et al.*, 1990). The directionality or convergence of allogenic successions depends on the consistency and directionality of the environmental changes that drive the succession.

The Long History of Succession Theory: Holism versus Reductionism

Succession, particularly plant succession, was one of the first major research topics of the field of ecology (see McIntosh, 1981, 1985) and has been a challenging problem in ecological theory up to the present (Drury and Nisbet, 1973; Connell and Slatyer, 1977; West *et al.*, 1981; Huston, 1982; Tilman, 1985; Walker *et al.*, 1986; Huston and Smith 1987; Smith and Huston 1989). The many different patterns of succession in natural communities have been thoroughly described by ecologists over the past 100 years (Cowles, 1899, 1901; Clements, 1904, 1916; Cooper, 1913, 1923; Olson, 1958). More recently, there has been increased interest in changes in species diversity over the course of succession (Billings, 1938; Bazzaz, 1975; Harcombe, 1977b; Bormann and Likens, 1979). These empirical descriptions of plant succession and the associated changes in species diversity, particularly the results of experimental studies (Harcombe, 1977b; Huston, 1982; Tilman, 1987b), clearly demonstrate that species diversity changes during the course of succession, and changes with predictably different rates and patterns depending on the biotic and abiotic conditions under which succession occurs.

Nonetheless, because some successional patterns provide exceptions to generalizations based on other patterns, a general theory of succession has been slow to develop. The many patterns of succession have generated many different and sometimes contradictory explanations (Walker, 1970; Drury and Nisbet, 1973; Connell and Slatyer, 1977; MacMahon, 1981; McIntosh, 1981; Tilman, 1985) and have led some to argue that no general explanation exists (Whittaker and Levin, 1977; Grubb, 1986).

The observation of temporal changes in plant communities has a long history, and use of the term 'succession' dates back at least to Thoreau (Thoreau, 1860). The striking features of succession noted by early observers were its directionality and predictability. Even on the basis of casual observation, it was recognized that forest succession in a particular region often followed the same pattern, and that different plots of land ended up with a similar composition of tree species, regardless of their species composition at the beginning of succession. These features of succession led a number of outstanding ecologists to attribute extraordinary properties to plant communities. The most famous and most unjustly maligned of these was Frederick Clements, who in addition to carrying out a phenomenal amount of quantitative research on plant succession, developed a descriptive theory of succession that is unfortunately remembered primarily for describing plant communities as 'superorganisms', capable of self-directed development (Clements, 1916). Although many of Clements' descriptions and ideas about succession are embedded in most 'modern theories' of succession, he is remembered primarily for the discredited concept of the superorganism, and the 80 years of debate about emergent properties and holism versus reductionism that it spawned.

The superorganism concept has its modern counterpart in systems ecology (cf., Reichle and Auerbach, 1972; O'Neill, 1976; Shugart and O'Neill, 1979; McIntosh, 1981, 1985), a central tenet of which is that an ecosystem has properties that are more than the sum of its parts. These 'emergent' properties are thought to result from the organization of the ecosystem, which must be studied, modeled and understood as a whole. Specifically, it was argued that ecosystems cannot be understood simply on the basis of their component organisms: '... ecosystem behavior cannot be predicted from the laws of organism behavior. ... We cannot expect ecosystem behavior to emerge from a set of organismic equations.' (Webster, 1979).

The antithesis of the superorganism/systems ecology holism is the 'reductionistic' view that an ecosystem *can* be understood as the sum of its parts (McIntosh, 1981, 1985). Gleason (1917) argued against the superorganism concept and stated that 'the phenomena of vegetation depend completely on the phenomena of the individual.' According to Gleason, a community was an assemblage of individual species that responded independently to environmental conditions, and succession resulted from the individual responses of different species to changing environmental conditions. More recently, Harper (1977) held out the hope that 'the complex is no more than the sum of the components and their interactions.' One factor that has kept the holism/reduction argument alive has been the difficulty of rigorously testing either philosophy. The problem of summing all the parts of a very complex system may be just as difficult as determining whether a property is truly emergent or simply appears emergent because its underlying processes are poorly understood (Salt, 1979).

The holistic theory of succession based on the superorganismic climatic climax has largely given way to the individualistic theory of succession (Gleason, 1917, 1926). Whether a reductionistic or holistic approach is taken to understanding ecosystems and the process of succession, the basic features of succession outlined by Clements early in this century are still valid, although terminology has changed (see however, MacMahon, 1980). Clements classified the essential processes of succession as nudation (disturbance), migration, ecesis (establishment and growth), competition, and reaction (effect upon the environment). Clements went astray, and spawned another controversy that has persisted nearly to the present, by concluding that all successional sequences (seres, in Clementsian terminology) lead to the 'mature' community type in any particular region, that is, a stable, self-reproducing, climatically determined *climax* formation. Along the southern shore of Lake Michigan, where Cowles (1899) conducted an influential study of succession on sand dunes, the climax formation for all types of vegetation was considered to be beech-maple forest. Plant communities that did not seem ever to reach the climax were considered to be in a state of arrested development, or *preclimax*, while those that were prevented from reaching climax by grazing or fire were called *disclimax*. The appeal of the Clementsian system of classifying vegetation types in terms of developmental stages of a single climax formation was that it allowed all plant communities within a region to be classified within a single, intuitively attractive framework. The climax was considered to be the mature state of the plant community and included all the earlier 'developmental' stages as well.

The main problem with the concept of a single climatically determined regional climax formation (*monoclimax*) was that within any region there were many vegetation types that were apparently never going to become the predicted climax formations, but rather had reached stable, self-reproducing endpoints different from that predicted by regional climate. There is no doubt that Clements recognized the incredible variety of vegetation and successional patterns found within any particular region. 'No climax area lacks frequent evidence of succession, and the greater number present it in bewildering abundance. The evidence is most obvious in active physiographic areas, dunes, strands, lakes, flood-plains, badlands, etc., and in areas disturbed by man. But the most stable association is never in complete equilibrium, nor is it free from disturbed areas in which secondary succession is evident. ... Even where the final community seems most homogeneous and its factors uniform, quantitative study by quadrat and instrument reveals a swing of population and a variation in the controlling factors.' (Clements, 1916, p. 1.)

The individual-based theory of ecology described in this book is clearly a descendant of Gleason's ideas, for which the problem of 'summing the parts' is solved by computers. Nonetheless, it is important to recognize the basic truth in most of what Clements wrote. Even the notion of a climatically determined regional climax is reasonable if one considers the long-term effects of climate on mineral weathering, erosion and deposition, and soil characteristics and structure, that can eventually level a landscape and create nearly uniform conditions that would probably be dominanted by a single vegetation type. A clarification of time and spatial scales, and a clear distinction between autogenic and allogenic processes is all that is needed to give validity to the concept of a climatic climax. The various schemes for classification of vegetation on a global scale are all essentially based on the idea of a climatically determined climax vegetation (e.g., Holdridge, 1967; Walter, 1973), and the widely used term, biome, that refers to the large-scale, climatic regions with a characteristic composition of animals and plants, was originally coined by Clements.

The interaction of climate, topography, and geology to create the environmental conditions to which individual organisms respond and with which they interact is an essential feature of the individual-based approach to ecology, as it was of Clements' theory of succession. Recent work with computer simulations of individual-based models has shown that this approach provides a basis for explaining both the wide variety of successional patterns and the common features of all successions, as well as for predicting successional patterns over a wide range of environmental conditions (Huston and Smith, 1987; Smith and Huston, 1989). Following the insights of Gleason (1917, 1926, 1939), Drury and Nisbet (1973), Pickett (1976), and others, I will discuss succession as a process that is dependent on the properties of the individual organisms interacting within a community. The peculiar properties of convergence and apparent determinism that have caused so much controversy in interpreting succession can be interpreted not as properties of ecosystems, plant communities, or superorganisms, but as the consequence of the properties of individual organisms. The remarkably similar patterns of successional sequences around the world are a consequence of the fact that the properties of all organisms are shaped by the same basic process of natural selection, acting upon a suite of characteristics that are constrained by the same basic laws of physics (cf. Chapter 7).

The rest of this chapter will describe the effect of environmental conditions and variation in the physiology and life history of plants on patterns of succession. These properties not only make succession inevitable and predictable in its overall pattern, but are also responsible for the variability of successional dynamics among different combinations of plant species. Two sets of factors must be known and understood in order to understand the pattern of succession: 1. the physiological and life history properties of the species present in the succession; and 2. the environmental conditions under which the succession occurs.

Patterns of Change in Resource Availability and Species Diversity

Succession is the quintessential non-equilibrium ecological process. Species diversity almost always changes during succession, but depending on what group of organisms is being considered, the environmental conditions under which succession occurs, and which time period of succession is being evaluated, diversity may increase or decrease. In contrast to equilibrium competition models, where absolute or relative growth rates have no effect on patterns of species abundance, the role of growth rate is critical for understanding and modeling a dynamic process such as succession. In general, high growth rates result in a higher rate of successional species replacement (i.e., competitive displacement and exclusion) than do low growth rates (cf. Chapter 5, Fig. 5.3). This has important consequences for diversity patterns both within and between seres, since the availability of different resources changes predictably during succession, and succession on a landscape occurs under a wide range of initial resource conditions.

Biomass does not necessarily increase at a constant rate during succession, rather it may increase in pulses corresponding to the release of certain species from suppression (Fig. 9.5*b,e,h*). Relatively few studies have measured biomass changes over more than a few years, but pulsed or uneven rates of biomass accumulation have been observed (Odum, 1960; Ewel, 1971; Whittaker, 1975; Bormann and Likens, 1979; Peet and Christensen, 1980; Peet, 1981).

Patterns of tree species diversity may show the same fluctuations as does biomass, although they are out of phase because evenness decreases rapidly during periods of release when biomass is increasing (Fig. 9.5*c,f,i*). Following a rapid decrease, evenness rises until another species begins to dominate. Fluctuating patterns of diversity have been found in a number of successional studies (Dyrness, 1973; Bazzaz, 1975; Opler *et al.*, 1977; Holt and Woodwell, data presented in Whittaker, 1975; Pearson and Rosenberg, 1978), and are a general property of individual-based plant competition models (Shugart *et al.*, 1981)

In the early stages of primary succession, levels of certain critical soil resources are usually low, and diversity always increases during early succession as a result of the gradual immigration of new species into an area that previously had no species. Some of the increase in diversity is also a result of the improvement in soil and other conditions caused by the early colonizers that allows additional species to survive (Clements, 1916; Dansereau, 1957; McIntosh, 1981).

One of the classic primary successional sequences occurs on the silt, sand, and rocks exposed by a retreating glacier. Assuming that the glacier has been retreating for a long time, the sequence of different plant communities that are encountered as one approaches the glacier are an excellent example of the analogy between a spatial pattern of vegetation and the temporal pattern of change, or succession. Vegetation at any location along the path of glacial retreat undergoes the same sequence of succession and soil changes, but at any particular time, locations closer to the glacier will be at an earlier successional stage (Cooper, 1923; Crocker and Major, 1955; Matthews, 1978,1979).

Changes in the plants and soil following glacial retreat in Glacier Bay, Alaska, are particularly well studied (Cooper, 1923; Crocker and Major, 1955; Lawrence, 1958). Soil organic matter accumulates gradually during early succession, until a site is invaded by nitrogen-fixing alders, after which time both soil organic matter and nitrogen accumulate rapidly (Fig 9.1). Plant species diversity is highest in the early stages of succession, when nutrient availability is very low, and mosses, herbs, and small shrubs coexist. These are replaced by taller willows, and by year 50 a dense, nearly monospecific stand of alders dominates the community. Nitrogen availability increases rapidly as a result of nitrogen-fixation by the alders, and pH drops during the same period due to nitrification and nitrate leaching. The alders are slowly invaded by spruce, which are later invaded by two species of hemlock to form the 'climax' spruce-hemlock stand (Crocker and Major, 1955).

A very different type of primary succession occurs in some aquatic and marine environments. Even if a substrate is bare and unmodified by any previous occupation by plants (usually algae), colonization and growth occur rapidly because the primary source of resources is the water

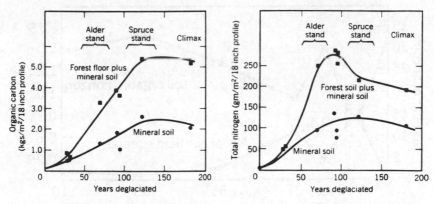

Fig. 9.1 Accumulation of soil organic matter and nitrogen during primary succession at Glacier Bay, Alaska. (From Crocker and Major, 1955.)

rather than the substrate, as is the case in most terrestrial successions (Breitburg, 1985; Steinman and McIntire, 1990; DeAngelis *et al.*, 1990). A long period of substrate improvement, or facilitation, is unnecessary, and whichever species arrive first are usually able to grow successfully and often dominate the site. In a study of colonization of bare substrate by marine algae, Sousa (1979a,b) found that diversity increased from early- to mid-succession, as an initially dominant early-successional species, *Ulva*, was invaded and eventually replaced by other species. Diversity decreased in late succession as perennial red algae came to dominate the substrate. This successional replacement of algae was clearly a result of an inverse correlation between rapid growth and the ability to resist physical disturbance or grazing by herbivores. Sousa demonstrated experimentally that the late-successional red algae could grow perfectly well on the bare substrate, but were eliminated by competition from the rapidly growing *Ulva*. It was only after low tides, grazing by crabs, and wave action had broken up the monoculture of *Ulva*, that other species could invade that were longer lived and more resistant to stress and grazing. This inverse correlation between competitive ability under different conditions is a common feature of nearly all successional sequences (see Colinvaux, 1973, 1986).

In secondary succession as well as primary succession, plant growth usually leads to change and often 'improvement' of soil properties. However, many species may be present as seeds at the beginning of succession, and almost any species that disperses into an early-successional site can potentially survive. Thus, secondary successions do not tend to have a

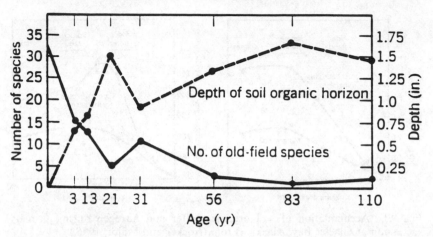

Fig. 9.2 Increase in soil depth and decrease in herb species richness over a 110-year chronosequence (a series of fields ranging from 0 to 110 years in age) of oldfield succession in the Piedmont of North Carolina. (From Billings, 1938.)

dramatic increase in diversity as a result of immigration and facilitation such as commonly occurs during primary succession. The pattern of succession beginning with nearly all species already present has been called 'initial floristics' (Egler, 1954) and requires that successional changes be explained by some process other than immigration and facilitation. The general pattern during secondary succession is a decrease in diversity among the early colonizers, rather than the increase found in primary succession. Billings (1938) studied oldfield succession in the North Carolina Piedmont and found that the depth of the soil organic horizon increased rapidly during the first 20 years of succession, while the species richness of oldfield plants decreased dramatically in the same period, and continued to decrease in sites that were up to 100 years old (Fig. 9.2).

Under certain circumstances, however, not all species that could potentially grow on a site are present following the disturbance that initiates secondary succession. In such cases, immigration of species into the site leads to an increase of species richness during early succession (e.g., Bazzaz, 1975). Which particular species happen to arrive, and the order in which they arrive, can have a major effect on species composition, diversity, ecosystem processes, and the overall pattern of succession. Successional change that results from the sequential arrival of different species, rather than from interactions among a fixed group of species, has been called 'relay floristics' (Egler, 1954). Relay floristics are likely to be important in early succession in situations where a disturbance has killed most organisms without destroying the substrate (e.g., long-term intensive agriculture). The immigration rate of species may be quite slow if the site is isolated, or if the disturbance is sufficiently large that sources of potential immigrants are a long distance away (e.g., Pavlik, 1989). Relay floristics may be particularly important in moist coniferous regions such as the Pacific Northwest of North America and boreal forests where fires may burn large areas and dominant species are not well-adapted to surviving fires (e.g., Franklin and Hemstrom, 1981; Heinselman and Wright, 1973).

Early in secondary succession, a combination of high availability of mineral nutrients, rapid growth rates, and short life spans of many invading species leads to more rapid replacement of species than occurs in later succession. This pattern is characteristic of natural successions (Shugart and Hett, 1973), and is consistent with observations that response to disturbances and other perturbations is most rapid in early succession (Hurd *et al.*, 1971; Pinder, 1975; Tomkins and Grant, 1977; Sousa, 1980).

Within a secondary sere there is usually a reduction in the maximum growth rates of dominant species through time, as larger and/or more tolerant species with slower growth rates dominate. Further reduction in growth rates of individuals of all species may result from changes in the environment, such as a reduction of resources available for growth as biomass accumulates or as nutrients are lost by leaching (Borman *et al.*, 1968; Jordan and Kline, 1972). Soil nutrients and water may be reduced in this manner (Toumey and Kienholz 1931; Korstian and Coile, 1938; Zahner, 1958; Popenoe, 1957; Bartholemew *et al.*, 1953; Marks and Bormann, 1972; Brown and Bourn, 1973; Harcombe, 1977a), and this reduction is likely to have the greatest effect where the resources involved are in short supply (e.g., water in arid regions and nutrients on infertile soils). Such a reduction in growth rates allows prolonged coexistence of competitors and increased diversity (Huston, 1979), and augments the inevitable consequences of the lower maximum growth rates of later successional species. This property of prolonged coexistence in mid succession is a widely observed feature of natural successions (Shugart and Hett, 1973; Sousa, 1980).

Although diversity among competitors is often highest during the intermediate stages of succession when the periods of coexistence become longer due to the dominance of slower-growing species, evenness and usually richness eventually decrease as the system approaches competitive equilibrium and exclusion occurs. It should be noted that under non-equilibrium conditions, the period of coexistence is longest among the most similar species (Figs 9.5, 9.7; cf. Hutchinson, 1967; Grenney *et al.*, 1973; Huston, 1979; Caswell, 1982). Without propagule input during succession, the number of competing species must inevitably decline. With propagule input, richness may increase through the intermediate stages but will generally decrease as the community approaches competitive equilibrium. While diversity among the dominant competitors in a community almost always decreases during late succession, the total diversity of the community often continues to increase, as a result of interstitial species that are able to survive because of the altered environmental conditions and the increase in environmental heterogeneity created by the dominant plants. The well-known increase in bird diversity during secondary succession (Johnson and Odum, 1956; Taylor, 1969) and the increase in the diversity of understory and epiphytic plants are examples of this phenomenon.

The increase in total diversity of the plant community as new functional types of plants enter the community is more pronounced in the humid tropics, where abundant moisture allows species with high shade tolerance to survive in the understory. In drier forests of the temperate zone, the total number of species in later succession is often lower than in the early herbaceous stages of succession. Although new herbaceous species may have entered the forest understory, they do not compensate for the large number of shade-intolerant herbaceous species that were eliminated by the growth of shrubs and trees (Holt and Woodwell, data published in Whittaker, 1975) (Fig. 2.29). Unless a community is broken down into its various functional groupings of species, the interaction of mechanisms that influence species diversity during succession is virtually impossible to decipher.

The general patterns of changing diversity among the dominant species during succession result from the population dynamics of competitors. Although the classic sequential replacement is a common successional pattern, there are many exceptions and a wide variety of successional patterns have been reported (e.g., Drury and Nisbet, 1973; Grubb, 1986). Individual-based models provide a convenient vehicle for exploring causes of variation in successional patterns. With the goal of understanding both the general patterns of population abundance and diversity during succession, as well as the exceptions to the general patterns, the next two sections explore two of the primary sources of variation in successional patterns, the properties of the species themselves, and the properties of the environment in which succession occurs.

The Effect of Species Physiology and Life History on Successional Pattern

The underlying causes and some of the consequences of inversely correlated physiological and life history traits in plants were discussed in the context of plant functional types in Chapter 7. Here the focus is on the population dynamics of competing species during forest succession and the effect of different combinations of five general traits that are important in competition among trees: maximum growth rate, maximum size, maximum longevity, maximum rate of sapling establishment, and shade tolerance. The various patterns of vegetation change, including classic successional replacement, that result from different combinations of these traits can be illustrated with a series of individual-based computer simulations of two-species competition, the simplest situation in which succession can occur. Huston and Smith (1987) created hypothetical species based on combinations of two levels of the five traits listed above.

To explore the influence of these traits on successional pattern, Huston and Smith altered the relative values of one trait at a time, while holding the other traits of the two species constant. In this way they built a four-dimensional matrix of vegetation dynamics with only one trait varying along each axis, an approach analogous to partial correlation analysis. This technique allows decoupling of the correlations among traits that are normally related in a specific pattern (Chapter 7), and determination of how different correlations among these traits affect the temporal pattern of species interactions (Fig. 9.3). The parameter values they used are representative of values for forest trees of eastern North America (see Pastor and Post, 1986; Shugart, 1984). Maximum ages were set at 175 and 350 (species 1 and 2, respectively), and the maximum size of species 2 was held constant in all runs, except where the growth rates, sizes, and ages of both species were made equal. By altering the maximum size of species 1, relative growth rates were varied from being equal, to species 1 having twice the maximum growth rate of species 2. Sapling establishment rates were either equal or a 5:1 ratio in favor of species 1, reflecting the observation that small, fast-growing species have higher seed output (and dispersibility) than large slow-growing species.

Huston and Smith (1987) classified the simulations into five groups (Fig. 9.4) based on the temporal pattern of species abundances: 1. successional replacement, 2. divergence, 3. convergence, 4. total supression, and 5. pseudo-cyclic replacement. The patterns are a consequence of

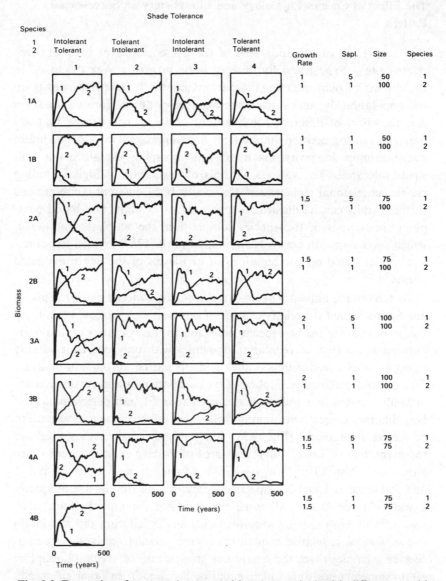

Fig. 9.3 Dynamics of two-species competition (succession) with different combinations of life history and physiological characteristics. The rows labeled 1,2,3, and 4 differ in the relationship between maximum potential growth rate and maximum size (relative values indicated in columns to right of figure). Within each pair of rows, in row A species 1 has a fivefold advantage in sapling establishment, and in row B both species have the same sapling-establishment rate (relative values in the column Sapl). The columns in each row differ in shade-tolerance relationships (see headings at tops of columns). Each simulation was run for 500 years for ten separate 0.08 ha plots. All figures are based on means for ten plots. (From Huston and Smith, 1987.)

changes in the relative competitive abilities of individuals of each species through time and therefore allow examination of the influence of each attribute on competitive ability throughout the successional sequence.

1. *Classic successional replacement* (Fig. 9.4a) results from an inverse relationship between the two species in the attributes that confer competitive advantage under the conditions of early versus late succession. Species 1 dominates during early succession, as a result of higher growth rate (Fig. 9.4a, cases 1-6), higher rate of sapling establishment (based on fecundity, not shade tolerance) (Fig. 9.4a, cases 1,2,7,8), or both (Fig. 9.4a, cases 1,2). High relative rates of establishment (resulting from high fecundity, dispersability of seeds, and tolerance to the physical conditions of early succession) also allow species to achieve early dominance by reducing the availability of sites for establishment of other species and quickly forming a dense canopy below which light is inadequate for further sapling establishment (e.g., Winsor, 1983). All of these attributes allow rapid monopolization of the limiting resource (light, in these simulations) following disturbance. As faster-growing individuals overtop individuals of the slower-growing species, the growth rates of the slow-growing individuals are reduced and their probability of mortality increased.

However, as biomass accumulates and light levels at the forest floor decrease, species 2 is able to dominate because of greater height (Fig. 9.4a, cases 1,3,5,6,7), greater sapling survival (based on shade tolerance, Fig. 9.4a, cases 1-4,7,8) or both (Fig. 9.4a, cases 1,3,7). Communities in which shade-intolerant early-successional species grow faster and become larger than shade-tolerant late-successional species (e.g., loblolly pine (*Pinus taeda*) to oak (*Quercus* spp.), white pine (*Pinus strobus*) to northern hardwoods) can be understood in this same context.

Since successful regeneration is a consequence of both sapling number and probability of survival, which is influenced by shade tolerance, successional dynamics differ when these two attributes are uncoupled. High reproductive potential (sapling establishment) *per se* does not allow a species to dominate if the probability of sapling survival is low relative to the other species (as a result of shade intolerance, Fig. 9.4a, cases 1,2,7,8). Shade tolerance is largely irrelevant in early succession, but is particularly important to regeneration during late succession when light levels at the forest floor are low. However, if both species are equally shade tolerant, high sapling establishment can allow a species to dominate in late succession even if it is shorter than the other species (Fig

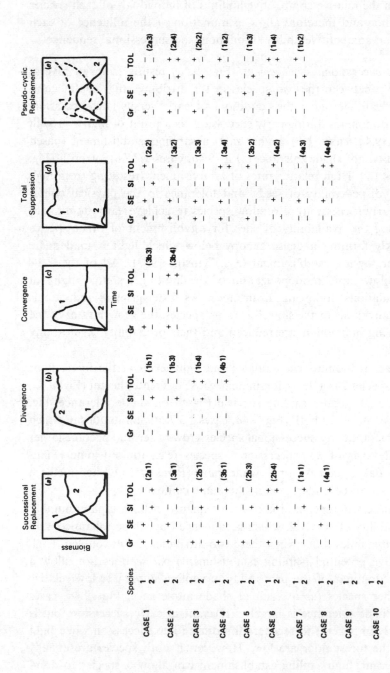

Fig. 9.4 Five possible patterns of two-species succession, summarized from Fig. 9.3. Below each pattern are the combinations of relative trait values that can produce the pattern, and, in parentheses, the location of each example in Fig. 9.3. Gr, growth rate; SE, sapling establishment; SI, size; TOL, shade tolerance. (From Huston and Smith, 1987.)

9.4*e*, cases 1,2,5,6). Together, shade tolerance and a high rate of sapling production provide a strong competitive advantage, particularly in late succession.

A large maximum height *per se* is an advantage only late in stand development and then only when regeneration rates are equal (compare Fig. 9.4*a*, cases 5 and 6 to Fig. 9.4*d*, case 1 and Fig. 9.4*e*, cases 1-7). Growth rate and longevity appear to confer no competitive advantage late in stand development, aside from their correlation with other attributes such as size, reproductive potential and shade tolerance.

The classic pattern of successional replacement (Fig. 9.4*a*) results from a specific correlation pattern among life history and physiological traits (discussed in chapter 7). When this correlation structure is altered, a variety of patterns can occur. Each of the following patterns can be found in nature, and are usually interpreted as exceptions to generalizations about succession (see Drury and Nisbet, 1973). As these simulations demonstrate, these exceptions can be interpreted simply as the result of different patterns of variation in the traits of competing species. This variation does not contradict the generalizations of chapter 7, but results from chance combinations of species with different genetic and physiological potentials or different evolutionary histories.

2. Divergence (Fig. 9.4*b*) occurs when competitive abilities are equal early in stand development (i.e., equal growth rate and reproductive potential) followed by a competitive advantage by one species or the other as a function of regenerative ability and/or size.

Examples of this pattern can be found in the simple successional sequences of subalpine conifers in the arid mountains of western North American. Over much of the intermontane west, early succession is dominated by lodgepole pine (*Pinus contorta*). The pattern of succession following the early stages is extremely sensitive to site conditions that influence soil water availability, as well as other conditions, including the proximity of seed sources for early- versus late-successional species, and the intensity of the disturbance that initiated the successional sequence. Under conditions with sufficient moisture and a seed source for both early-successional lodgepole pine, and later-successional spruce and fir, the early-successional stands can be a mixture of both types of species. As succession proceeds, the shade-tolerant spruce and fir dominate, suppressing the lodgepole pine, and causing a pattern of divergence (Romme, 1982; Romme and Knight, 1981).

3. Convergence (Fig. 9.4c) is the opposite of the process described above. One species has a competitive advantage resulting from higher growth rate early in stand development, but the competitive ability of the two species becomes equal as the stand develops. Seedling input must be equal in all cases, or it would confer an advantage in later succession due to higher actual regeneration. Convincing examples of this pattern are difficult to find, although one possible example is the spruce-fir forest of high elevations in the Great Smoky Mountains, where fast-growing Fraser fir (*Abies fraseri*) often dominate early successional areas, but later share dominance with red spruce (*Picea rubens*).

4. Total suppression (Fig. 9.4d) occurs when one species has a combination of attributes that confers a competitive advantage over the other species at all stages of stand development. In these examples, rate of sapling establishment confers an advantage in early succession (with or without an associated growth rate advantage) and in late succession as well, since species 1 always has equal or greater shade tolerance and thus higher regeneration.

Total suppression occurs frequently in nature, and appears as a very simple successional sequence, such as the 'auto-succession' found in arid deserts or polar regions. Fast-growing early-successional species that also achieve large size and great longevity, such as redwoods and tulip poplar, may dominate throughout succession under some conditions. In the intermontane west, lodgepole pine dominate all successional stages on dry sites, regenerating in the gaps formed by the mortality of individuals of the post-disturbance (fire) cohort (Romme and Knight, 1981).

5. Pseudo-cyclic replacement (Fig. 9.4e) results from a temporary period of dominance or codominance by the initial cohort of species 2 during the time between the senescence of the initial cohort of species 1 and the senescence of the species 2 cohort. Individuals of both species become established in the high light levels of early succession, and because of the greater longevity or height of species 2, the initial cohort persists longer than the initial cohort of species 1. However, because of the competitive advantage of species 1 in later succession, which is a consequence of higher regeneration, species 2 is eventually replaced by species 1. In all cases, species 1 has higher regeneration because of greater shade tolerance or greater sapling establishment with equal shade tolerance. The relative abundances of species 1 and 2 during the intermediate period are a function of initial competitive interactions and relative shade tolerances.

With the appropriate combination of site conditions and seed source availability, the subalpine succession of the intermontane west can also produce this type of pattern. Succession involving white spruce and either lodgepole pine or aspen has been reported to produce a pseudo-cyclic replacement in the mountains of British Columbia (Parminter, 1991). Under appropriate conditions, both the late-successional dominant, white spruce, and the early-successional dominant, lodgepole pine, can become established in similar proportions following a fire. The higher growth rate of the lodgepole pine allows it to overtop the spruce and dominate the stand for a period of 50 to 100 years. However, the more shade-tolerant spruce is able to survive in the light shade of the lodgepole pine, and continues to grow, eventually overtopping and eliminating the pine. A similar pattern occurs with *Pseudotsuga* and *Tsuga* in the Pacific Northwest (Franklin and Hemstrom, 1981).

Huston and Smith (1987) did not explicitly consider differences in the ability of the two species to colonize the site, but made the implicit assumption that both species become established on the site immediately following disturbance (e.g., 'initial floristics,' Egler, 1954). However, the effect of differences in colonization ability are analogous to the effects of different rates of regeneration in the simulations. Differences in ability to colonize a site can confer a major competitive advantage on the earlier colonizer, as a result of resource preemption. Early dominance based on a colonization advantage could lead to any of the five basic successional patterns mentioned above, depending on the relative values of the other attributes of the competing species.

Field studies confirm the importance of propagule input in altering the dynamics of succession. It has been repeatedly demonstrated that variation in the arrival time of species at a successional site can have a major effect on the pattern and outcome of succession (McCune and Allen, 1985; Hubbell and Foster, 1987; Huston and Smith, 1987; Robinson and Dickerson, 1987; Robinson and Edgemon, 1988; Drake, 1991; Fig. 9.5). The time of creation of the successional site is important (e.g., Salisbury, 1929; Thurston, 1951; Keever, 1950; Holt, 1972; Grubb, 1977; Harcombe, 1977a), as is the location of a site in relation to sources of propagules (Grubb, 1977; Opler *et al.*, 1977; Harcombe, 1977b). The size of the disturbance determines the relative importance of input from plants on its perimeter versus long-distance dispersal (e.g., Watt, 1934; Keever, 1950; Platt, 1975; Grubb, 1977; Brokaw, 1985a; Hartshorn, 1978), and also the availability of resources such as light and soil water within the site. The history of the site and type of disturbance can determine what, if any, seeds are already present in the soil, or whether vegetative propagules survive to have a major advantage over plants that must start from seed.

Animals can have a major effect on successional patterns through seed dispersal, seed-eating (Gill and Marks, 1991; DeSteven, 1991a,b), or herbivory (Louda, 1982a,b, 1983; Brown, 1984, 1985; Gibson et al., 1987a,b; Pastor et al., 1988; Brown et al., 1988; Louda et al., 1990). Predicting the effects of animals on successional patterns is difficult, since the effect depends on both animal abundance and plant productivity and chemistry, which are interpendent and both subject to density-independent as well as density-dependent regulation.

Although there is a strong influence of stochastic processes in the initiation of succession (as well as in later stages), the patterns of colonization tend to reinforce the pattern derived from competition among species with inversely related traits, because seed production and dispersability tend to be correlated with other traits, such as rapid growth rate, that are important in early succession. The most rapidly growing plants tend to have the most dispersible seeds, based on small size and large number (Salisbury, 1942; Harper et al., 1970; Richards, 1952; Whitmore, 1975), and usually reach a successional site before the seeds of the plants that will eventually predominate.

Just as there is an inverse correlation between traits that confer a competitive advantage in early versus late succession, there may be trade-offs within any of the broad suites of early- versus late-successional traits. While small fast-growing plants have a larger seed output per unit biomass than slow-growing larger species, some small plants may maximize seed output at the expense of growth rate, or vice versa. Likewise, there may be trade-offs, such as those between growth rate and longevity, among large plants. A fast-growing species may attain as large a size as a slowly growing species, but live for a much shorter time because it has not invested in the structural properties needed to resist storm damage, and insect or fungal attack (e.g., *Acer saccharinum, Ceiba pentandra,* and *Ochroma pyramidale*). The resistance to low-probability stresses that is necessary for long-term survival requires investment of photosynthate that makes a short-term high-growth strategy impossible.

In spite of the remarkable similarity of the basic pattern of succession in many different situations, sets of species can always be selected that demonstrate exceptions to the general pattern (see Grubb, 1986). The general pattern is most likely to break down in groups of a few species,

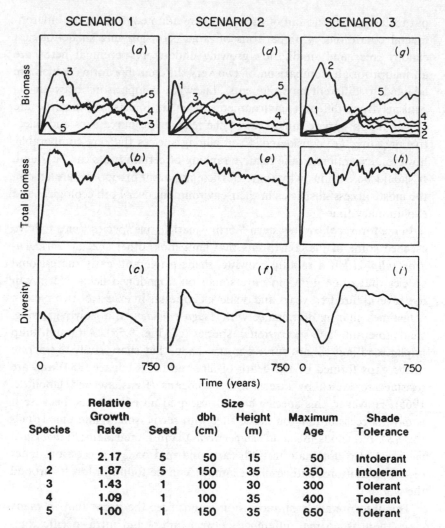

Species	Relative Growth Rate	Seed	Size dbh (cm)	Height (m)	Maximum Age	Shade Tolerance
1	2.17	5	50	15	50	Intolerant
2	1.87	5	150	35	350	Intolerant
3	1.43	1	100	30	300	Tolerant
4	1.09	1	100	35	400	Tolerant
5	1.00	1	150	35	650	Tolerant

Fig. 9.5 The effect of arrival order and species-specific properties on patterns of species abundance, total biomass, and species diversity during succession. The simulations differ only in the number and identity of species present at the beginning of succession. In Scenario 1, all three species have late-successional characteristics, but differ sufficiently in relative competitive ability to produce a 'typical' successional sequence. In Scenario 2, an early-successional species is added to the three species of Scenario 1, which results in a dramatic reduction of diversity early in succession. In Scenario 3, a 'super-species' with high growth rate and large size (but relatively short lifespan) is added to the four species in Scenario 2, which produces a pattern similar to Scenario 1, but with stronger variations in diversity. Because the number of species was predetermined, diversity was calculated as the Shannon-Weiner index ($\sum p_i \ln p_i$). (From Huston and Smith, 1987.)

particularly where not all of the species are adapted to the same environmental conditions. Comparisons of large, slow-growing shade-tolerant canopy trees and small, slow-growing understory perennial herbs are an inappropriate comparison of two very different life forms, which are adapted to different environments. Likewise, comparisons between organisms from different environments, such as forests and deserts, are clearly inappropriate, unless all of the relevant resource axes and selective pressures are considered. Even among species that are comparable, however, exceptions in the relative ranking of certain traits can be found. It should be obvious why some of these apparent exceptions are among the most successful species in their environment over both ecological and evolutionary time.

In the forests of southeastern North America, one species seems to have escaped some of the constraints that limit most other species. *Liriodendron tulipifera* is a rapidly growing, shade-intolerant, early-successional species that often grows in pure stands on abandoned fields. Yet it can live several hundred years and achieve diameters in excess of two meters, sometimes sharing dominance with *Tsuga canadensis*, a slowly-growing, shade-tolerant, 'late-successional' species (cf. Fig. 9.5g). Although tulip poplar seedlings cannot survive in the shade, they successfully regenerate in the gaps formed by the death of large adults (Shugart 1984) and are resistant to attack by insects and pathogens (Renshaw and Doolittle, 1965). However, this species has not escaped all constraints, because it is most successful only on mesic sites with fertile soil. White pine (*Pinus strobus*), redwoods (*Sequoia sempervirens*), and Douglas fir (*Pseudotsuga menzieii*) also dominate in both early and mid or late succession under certain conditions, and similar examples can be found in forests around the world.

This interpretation of succession emphasizes the point that in many cases there is no real difference between inter- and intra-specific competition. Intra-specific competition is not necessarily more intense than inter-specific competition (as required for coexistence of multiple species under conditions of competitive equilibrium). In some cases, such as competition for light, inter-specific competition may be much more intense than intra-specific competition if the individuals of the dominant species are larger, faster-growing, have denser canopies, etc. The potential intensity of competition must be evaluated for each type of resource and the sizes, life histories, and physiological properties of the species involved. Likewise, evaluation of effective population size for questions of density-dependent effects on competition and selection should not be based only on the density of conspecific individuals, but also on the density of equivalent competitors of other species.

These individual-based simulations demonstrate that a wide variety of successional patterns can be produced by a single mechanism, competition for light. The variation in successional patterns results not from different mechanisms, but from different combinations of species-specific traits that result in shifts in relative competitive ability under different environmental conditions. Thus evolutionary differences between species produce variability in successional patterns. However, evolution also produces basic similarity of all successional patterns as a result of natural selection acting upon traits that are restricted to specific relationships by energetic and physiological constraints. The fact that most successional patterns found in nature resemble the 'classic' pattern of sequential replacement suggests that the correlation structure that produces this pattern is very common in plant communities. This correlation structure of inversely related traits is the underlying basis for some durable generalizations in ecological theory, such as the r/K dichotomy of the Lotka-Volterra equations, which can produce the same types of successional patterns as the more complex individual-based models (Fig. 9.6).

The Effect of Resource Availability on Species Composition and Diversity

Because of inherent physiological differences between species, the ranking of growth rates among a group of species depends on the relative abundance of different resources. Such a growth rate advantage may or may not result in long-term dominance by a particular species, depending on its other life history characteristics and the frequency and type of disturbances in that environment. This is the non-equilibrium interpretation of the patterns of species composition that have been attributed to an equilibrium balance of inversely related competitive abilities for two or more resources (Tilman, 1980, 1982), which are hypothesized to produce succession as a consequence of sequential competitive equilibria (Tilman, 1985).

Individual-based models of succession allow predictions not only of patterns of species diversity, but also of species composition of successional sequences under different environmental conditions. In addition to light, water and soil nutrients often limit plant growth, and under some conditions plants will compete primarily for these resources rather than for light. Limitation by resources other than light is most likely to

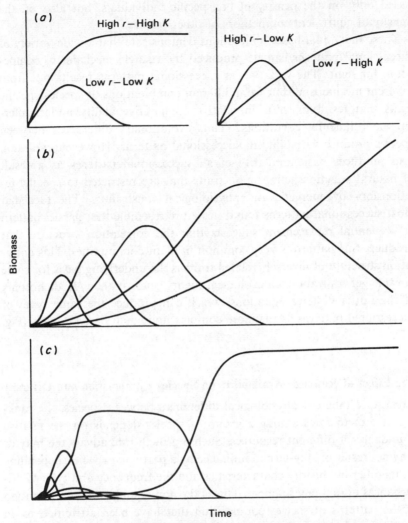

Fig. 9.6 Successional patterns produced by Lotka-Volterra competition models with two and six competing species. This approach focuses on the dynamic behavior of these equations, rather than their equilibrium solution. (a) Two dynamically different approaches to a competitive equilibrium in which only the species with the highest K value survives. The only parameters that vary between the two simulations are the r values. Parallel examples obtained using an individual-based competition model are illustrated in Figs 9.3 and 9.4. (b) Sequential species replacements gnerated by a Lotka-Volterra competition model with an inverse relationship between r and K (all $\alpha_{i,j} = 1.0$). (c) Pattern of suppression and release generated by a Lotka-Volterra competition model in which $\alpha_{i,j}$ varies depending on the relative population sizes of the competitors, and r and K are inversely related. (From Huston and Smith, 1987.)

occur among the tallest plants in a community (Pastor and Post, 1986; Fig. 7.5) and among smaller plants under conditions when canopy cover is less than 100%. Non-depletable factors such as temperature and soil pH also affect plant growth. Each species has its own requirements and optima for these factors that determine how it will perform under any set of environmental conditions. For most species, the competition with better competitors shifts their ecological range toward environmental conditions that are less favorable than the conditions under which they achieve optimal growth in the absence of competition (Ellenberg, 1953, 1954; Mueller-Dombois and Ellenberg, 1974; Austin and Smith, 1989).

Differences in resource availability between different soil types or different climates can affect growth and lead to major differences in successional patterns in different regions. A proportional increase in the growth rates of all species, (cf. Fig. 5.3a,c) results in a more rapid domination by late-successional species and, as a consequence, a more rapid reduction of species diversity.

However, there is rarely a uniform acceleration of successional dynamics under high resource conditions because fast-growing species generally show a greater response to growth stimulus, such as fertilization or increased moisture, than do slow-growing species (Mitchell and Chandler, 1939; Grime and Hunt, 1975; Mahmoud and Grime, 1976; Chapin *et al.*, 1986). A disproportionate increase in the growth rates of the fastest-growing species results in relatively greater dominance by these species during early succession. If maximum size is also increased by higher resource levels, as is often the case, this dominance may be further increased. This mechanism seems to explain the results of Harcombe (1977b) who found reduced diversity in tropical successional plots that were fertilized, as compared to plots with no nutrient addition (see Fig. 14.20). In this case, a fast- growing herbaceous species, *Phytolacca* sp., responded to fertilizer addition more strongly than woody species, and suppressed them for the course of the experiment. Similar results were obtained by Huston (1982) using only woody species.

Spatial and Temporal Patterns along Moisture Gradients

Not all resources have the same effect on population dynamics and species diversity. The previous example of the effect of soil nutrients on diversity is based on the positive effect of soil nutrient availability on the intensity of competition for light. The inverse relationship between the availability of soil resources such as nutrients and water, and the availability of light below a plant canopy is a fundamental consequence of plant growth that was a major issue of research among early plant ecologists (e.g., Clements *et al.*, 1929; Donald, 1958) and continues to be of interest (Tilman, 1982, 1987b, 1988). While soil water is essential for plant growth, and higher levels of soil water generally support higher plant biomass and leaf area, the effect of water on competition for light differs from the effect of soil nutrients such as nitrogen. Increased water availability increases shade tolerance (Chapter 7), and thus allows the survival of plants in the understory in spite of the higher leaf area supported by the canopy under moist conditions. In contrast, increased nitrogen availability does not confer any benefit on plants that must survive below the dense canopy that can develop on fertile soils. In fact, high nitrogen levels may *decrease* the shade tolerance of understory plants because of the higher respiration rates caused by higher enzyme levels (e.g., Ryan, 1991). Consequently, patterns of succession and species diversity along moisture gradients can differ significantly from patterns along nutrient gradients.

Smith and Huston (1989) used an individual-based plant competition model to investigate successional dynamics and plant community structure along a gradient of water availability. They used 15 hypothetical species whose physiological and life history traits were based on the 15 functional types described in Fig. 7.2. In each simulation for a different level of water availability all 15 types were allowed to enter the simulation repeatedly. The results of these simulations (Figs 7.6 and 9.8) illustrate that variation in environmental conditions can produce different successional patterns among a single group of species.

Under dry conditions, water stress limits the possible light-use strategies to those that are relatively shade intolerant (see Chapter 7). Light at ground level is relatively high (Fig. 9.7a) because leaf area is limited by a high light compensation point (low shade tolerance) and by the need to allocate carbon for water uptake and transport. With the reduced number of functional types capable of surviving under dry conditions, the successional sequence is simple and short (Fig. 9.8a and b); there is little change in light availability over time and little vertical stratification of the vegetation (Fig. 9.7a). Diversity among woody plants in arid regions is low in early succession and changes little over the course of succession.

In fact, arid regions are characterized by an absence of temporal shifts in species composition following disturbance (Noy Meir, 1973; Zedler, 1981; Peet and Loucks, 1977). For example, Hanes (1971) described

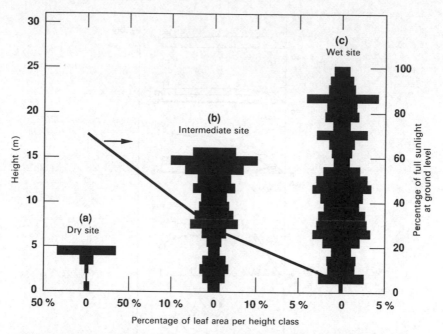

Fig. 9.7 Simulated vertical distribution of leaf area and associated light availability (solid line) at ground level for three environments along a moisture gradient. The leaf area distribution reflects the contribution of plants with different levels of shade tolerance that are present under the given moisture conditions. (a) Dry site, (b) intermediate site, (c) wet site. (From Smith and Huston, 1989.)

the patterns of vegetation dynamics in arid chaparral plant communities as 'auto-succession,' referring to the self-replacing nature of the vegetation. Diversity both within and between habitats changes relatively little through the course of succession under these dry conditions. Because of the low diversity, zonation of vegetation types in arid regions is often dramatic. Such spatial pattern results from competitive interactions along gradients (see Chapter 10) and from the results of large disturbances such as floods, droughts, and occasionally fires.

Increased water availability reduces constraints on carbon uptake imposed by water limitation and shade tolerance becomes essential if a plant is to establish and survive on the forest floor as light is reduced by the increased leaf area that plants can support under moist conditions (Fig. 9.7b). Under these moister conditions, additional species (shade-tolerant functional types that are intolerant of dry conditions) are able to enter the community, while the types that are tolerant to low moisture

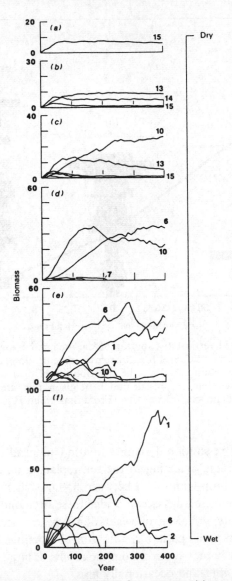

Fig. 9.8 Successional sequences resulting from competition among the same hypothetical 'functional types' of plants under different moisture conditions, representing different environments along a moisture gradient from dry to wet. The curves represent sections of the three-dimensional surfaces of Fig. 7.6, plus additional functional types, at different locations along the moisture axis. Note the similarity to the patterns in Fig. 9.10 (a) and (c), which were produced by a different individual-based competition model with parameters derived from actual species, rather than from idealized functional types. (From Smith and Huston, 1989.)

conditions are still able to survive, at least in the high light of early succession.

Only when high water availability makes shade tolerance a viable strategy can there be complex vertical stratification of forest structure. With increasing water availability, plant density and leaf area can increase and available light at ground level decreases (Fig. 9.7c). When there is sufficient water to support a closed canopy woodland, a vertical stratification of woody vegetation develops with a functionally and taxonomically distinct understory (White, 1968; Smith and Goodman, 1986, 1987). The increased leaf area results in a temporal shift in species composition (i.e., succession) because the initially dominant shade-intolerant canopy species cannot regenerate after light availability is reduced. At the grassland/forest boundary, dramatic differences in vegetation (e.g., grassland versus savannah versus closed woodland) can be produced both by subtle variations in soil conditions that influence water availability, or by disturbances such as fire or grazing that have a greater negative effect on woody vegetation than on herbs and grasses (see Chapter 13).

Tolerance to dry conditions *per se* is not a handicap under conditions of adequate moisture, as long as sufficient light is available to compensate for the low shade-tolerance of plants that are drought-tolerant. Under favorable moisture conditions, all functional types can potentially survive. However, shade-intolerant types with a low tolerance for dry conditions often dominate in early succession under mesic conditions because their growth rates are higher than those of either the more shade-tolerant types or the more xeric types (with high tolerance for dry conditions). As light is reduced at ground level in mid to late succession, the ability to regenerate and grow under shaded conditions becomes more important and the more shade-tolerant types (e.g., type 6 in Fig. 9.8d,e) are able to dominate, regardless of moisture tolerance. The most mesophytic shade-tolerant types will eventually dominate under high moisture conditions(e.g., type 1 in Fig. 9.8e,f), because their growth rate is not reduced by unnecessary (under these conditions) adaptations for low moisture availability. The highly shade-tolerant types existing in the low light environment of the forest floor are very sensitive to dry conditions and could be severely affected by extreme seasonality or occasional droughts (Nutman, 1937; Walter, 1971).

Species composition changes significantly during succession under mesic conditions, with early-successional shade-intolerant functional types being replaced by shade-tolerant functional types. Diversity in early succession can be high or low, depending on propagule input and

soil nutrient availability, and their effect on growth rates and competitive interactions. Diversity generally decreases in late succession, as the
largest and most shade-tolerant functional types dominate, reduce light
availability as a result of their high leaf area, and suppress all functional
types of lower shade tolerance.

Variation in the rate and pattern of succession under different moisture conditions is illustrated by the succession on a sandy old field in
southeastern Michigan (Evans and Dahl, 1955). Immediately following
the cessation of agriculture, the field was covered by a relatively uniform
mixture of annual weeds (Fig. 9.9a). Large-scale spatial pattern quickly
developed as a consequence of shallow swales (glacial kettleholes), which
had siltier soils and higher nutrient and moisture availability (Robertson et al., 1988). A higher biomass of perennial grasses and herbs was
associated with lower species diversity in the swales, while the drier upland areas maintained a more diverse community of prairie and old field
species (Huston and Evans, unpublished data) (Fig. 9.9c). Development
of woody vegetation on the uplands was suppressed for approximately
30 years by deer browsing on *Juniperus* spp. and *Crategus*. Shrubs and
trees grew more rapidly in the swales, were they were protected from
grazing by thickets of thorny *Rubus spp.* (Fig. 9.9e). Some woody plants
eventually reached a size large enough to escape browsing (Fig. 9.9f)
and quickly began to form a closed canopy over the field. This stage
of woody dominance will presumably lead to a loss of shade-intolerant
prairie species and the establishment of more shade-tolerant species of
forest trees and herbs.

Different strata (i.e., functional types) of vegetation can have very
different patterns of diversity over the course of succession. The reduction in diversity that is found in late succession is primarily among
those species that are actually competing, which are generally the largest
functional types in the community. Interstitial functional types, such as
understory shrubs and herbs or epiphytes, do not compete with the competitively dominant species, and do not necessarily decrease in diversity
over succession. The high shade tolerance that makes possible vertical
stratification, also allows the coexistence of shade-tolerant functional
types of different life forms. In general, smaller plants can have a lower
whole-plant light compensation level (greater shade tolerance) than large
plants because their smaller size requires less energy for maintenance respiration. The number of interstitial functional types may in fact increase
as a result of 1. creation of a reliably mesic, shaded environment that
makes shade tolerance a viable strategy; 2. continued propagule input

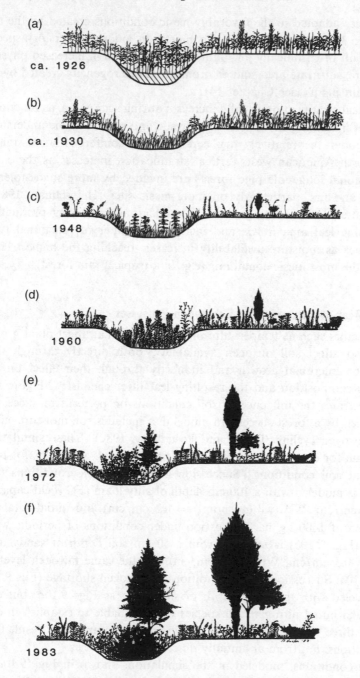

Fig. 9.9 Changes in structure of vegetation over time in moist swales and dry, sandy uplands during succession on an old field in southeastern Michigan.

of species adapted to the favorably mesic conditions created by the tree canopy; 3. reduced intensity of competition among understory species as a result of chronically low light availability; and 4. increased physical structure, substrate area, and environmental heterogeneity created by the dominant plants (see Chapter 14).

Vertical stratification usually increases during succession, as functional types of increasing shade tolerance invade and survive in the understory. Even under the relatively dry conditions of coniferous succession in the North American west, vertical stratification increases as the early-successional lodgepole pine forests are invaded by more shade-tolerant spruce and firs, particularly on more mesic sites (Heinselman, 1981a). The pattern of increasing vertical stratification, increasing production, increasing leaf area index, and increasing number of functional types continues as moisture availability increases, reaching the highest levels under the most mesic conditions (e.g., in a tropical rain forest).

The Effect of Succession on Ecosystem Processes

Any factors such as temperature or water stress that affect plant growth can also affect soil nitrogen availability, both directly through their effect on microbial growth and indirectly through their effect on tree species composition and the resulting leaf litter chemistry. Figure 9.10 demonstrates the influence of soil conditions on patterns of succession produced by a forest succession model that includes soil moisture effects and nitrogen cycling (Pastor and Post, 1985, 1986). These simulations were run for climatic conditions similar to southern Kentucky, for three different soil conditions. Succession under favorable conditions (Fig. 9.10a) is modeled with a 100 cm depth of silty loam (FC (field capacity) = 0.4 cm/cm, WP (wilting point) = 0.2 cm/cm) and initial total soil nitrogen of 1600 kg/ha. Succession under conditions of periodic water stress (Fig. 9.10c) is modeled with a 50 cm soil depth of sandy loam (FC = 0.3 cm/cm, WP = 0.2 cm/cm) and the same nitrogen levels as Fig. 9.10a. Succession under conditions of nitrogen shortage (Fig. 9.10e) is modeled with the same depth of silty loam as Fig. 9.10a, but only 10 kg/ha initial nitrogen. All species were available to establish in each of the three simulations; differences in species composition result from interactions, not from an initially different set of species.

Dry conditions, modeled in the simulations shown in Fig. 9.10c by reducing the water-holding capacity of the soil and using the same precipitation regime as in Fig. 9.10a, slows growth rates and the overall

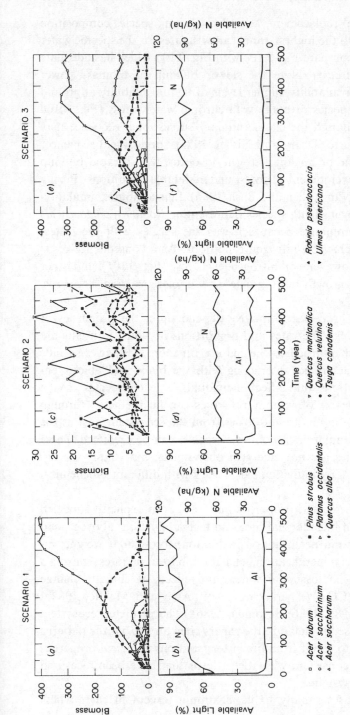

Fig. 9.10 Effect of environmental conditions on pattern of species replacement and resource levels during succession. All three successional sequences were produced by an individual-based plant competition model in which the same species were present in the same relative abundances at the beginning of each set of simulations. The only differences between the scenarios are in the moisture-holding capacity of the soil, which results in much drier conditions, and consequently much slower growth, in Scenario 2, and initial soil nitrogen, which is lower in Scenario 3 than in Scenarios 1 and 2. Scenario 1: secondary succession with no limitation by nutrients or water. Scenario 2: secondary succession under conditions of reduced water availability. Scenario 3: primary or secondary succession with initially low nitrogen availability. Differences in the levels of light (Al) and soil nitrogen (N) between the different successional series (illustrated in b, d, and f) result from the interaction of environmental conditions with the properties of the tree species that dominate in each simulation. (From Huston and Smith, 1987.)

rate of successional replacement, as well as shifting species composition to those species with the highest annual growth rates for the specific water stress conditions (compare Fig. 9.10*c* with Fig. 9.10*a* which has adequate levels of all resources). Note the slower buildup of biomass, lower levels of nitrogen availability, higher levels of light availability at ground level, and higher species diversity with chronic water stress (Fig. 9.10*d* versus 9.10*b*). Although the classic primary successional sequence may rarely occur in nature (Drury and Nisbet, 1973), successional sequences beginning with little or no soil nitrogen do occur, with a rapid buildup of nitrogen and development of forest under suitable conditions (Boring and Swank, 1984; Van Auken and Bush, 1985). In a computer simulation (Fig. 9.10*e*) a nitrogen-fixing species (representing *Robinia pseudoacacia*) achieved rapid dominance because it was the species with the fastest growth under extreme low-nitrogen conditions. As a consequence of its N-fixing activity, however, soil nitrogen levels rose (Fig. 9.10*f*), and other species with faster growth rates, greater shade tolerance, and larger size replaced it.

Biogeochemical processes ultimately depend upon the properties of individual organisms. The rates and proportions in which nutrients are taken up from and returned to the soil are functions of species-specific physiological characteristics interacting with the biotic and abiotic environmental conditions that affect individuals of each species. As a consequence, the levels of various resources can be predicted through time (incorporating inputs to and losses from the ecosystem) just as the abundances and population structures of species can be predicted. These ecosystem properties are not emergent phenomena, but simply the aggregate effects of many individuals of species with different biochemical properties. (cf. Fig. 7.7).

Ecosystem processes and properties that have been explicitly modeled using this approach include: leaf area and leaf production; aboveground biomass; aboveground net primary production; forest floor weight; nitrogen availability, mineralization, and immobilization rates; nitrogen fixation and nitrogen loss; soil water; and effects of climatic change (temperature and CO_2) (Aber *et al.*, 1979; Aber and Melillo, 1982b; Pastor and Post, 1985, 1986; Solomon, 1986). The predicted ecosystem patterns are discussed in detail in the papers cited above, but the patterns that are predicted to result from life history and physiological properties of individuals are consistent with the patterns that have been measured in a variety of ecosystems.

Leaf area increases rapidly to its maximum (except in cases where

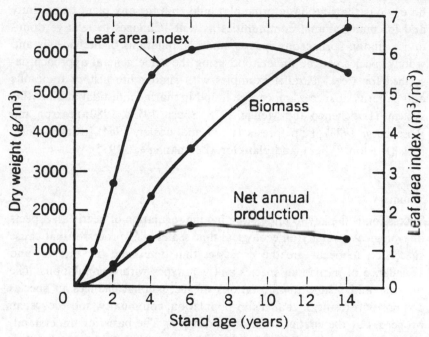

Fig. 9.11 Changes in plant biomass, primary productivity, and leaf area during forest succession. (From Kira and Shidei, 1967.)

growth is limited by nutrients or water), while total biomass increases much more slowly. This pattern has been documented in numerous ecosystem studies (Kira and Shidei, 1967; Marks, 1974; Covington and Aber, 1980; Cooper, 1981) (Fig. 9.11). An important consequence of this pattern is that light available to small plants decreases rapidly during succession (Fig. 9.10*b,d,f*).

The overall patterns of nitrogen availability produced by individual-based models of forest succession and nutrient cycling (e.g., Pastor and Post, 1985, 1986) are qualitatively similar to those that have been documented in primary and secondary succession (Robertson, 1982; Aber and Melillo, 1982b; Pastor and Post, 1986). In primary succession starting with very low nitrogen levels, nitrogen availability increases gradually through time due to N-fixation and atmospheric input to a steady-state level (Fig. 9.10*f*). In secondary succession following some sort of disturbance, nitrogen availability drops from its pre-disturbance steady-state level and gradually increases to some steady-state level (Fig. 9.10*b,d*).

Although the examples of succession in this chapter are based on forests and the physiological and life history properties of trees, it should be obvious that the basic principles hold true for any plant community and for many animal communities as well. Changes in relative dominance among herbaceous species, and combinations of herbaceous and woody species can be understood using the same general approach described for trees. Parallel examples with similar life history trade-offs and patterns of succession can be found in marine intertidal and benthic systems (Lubchenco and Menge, 1978; Sousa, 1979a, 1980; Pearson and Rosenberg, 1978), kelp forests (Reed and Foster, 1984), tropical coral reefs (Huston, 1985a), and plankton (Lehman *et al.*, 1975).

Summary

Succession is the key to understanding the regulation of nearly all aspects of biological diversity on ecological time scales. The processes that cause succession to occur are the processes that determine the survival and abundance of individual species and genotypes within populations. The physical and visible manifestations of successional changes in species composition result in changes in population, community, and ecosystem properties at the spatial scale of landscapes. The basis for understanding succession and its multilevel consequences is interactions between individual organisms and the way those interactions are shaped by the inescapable constraints of energetics and physiology that apply to all organisms. These universal constraints result in the universal similarity of the basic features of succession in plant and animal communities around the Earth. The great variation in the details of successional patterns can also be understood on the basis of these same constraints, acting on individual organisms with different combinations of properties.

10
Landscape Patterns: Gradients and Zonation

The effect of environmental conditions and disturbance regime on the rate, pattern, and synchrony of succession that occurs on different patches within a landscape creates a constantly changing kaleidoscope of color, texture, species composition and diversity. On many landscapes there is apparent regularity or order in the spatial arrangements of vegetation. This regularity is often a consequence of gradients. A gradient can be defined as a change in the value of a particular parameter, such as temperature, soil pH, or species composition, over space and is generally characterized as change along a linear distance. The term gradient implies a gradual unidirectional change, although the change may occur in discrete steps as well as in continuous small increments.

While gradients of gradual change in species diversity or species composition are common, it is the more dramatic patterns of zonation, in which the parameter of interest changes significantly over a short distance, that attract attention and invite interpretation. On most landscapes the pattern of zonation is defined by the visually dominant species, which may or may not be associated with equally dramatic changes in species composition or diversity.

In the context of individual-based ecological theory, zonation is a spatial sequence of species replacements along a spatial gradient of environmental conditions, just as succession is a temporal sequence of species replacements resulting from a temporal gradient in environmental conditions. The same trade-offs in physiological and life-history characteristics that are the explanation of successional patterns (Chapters 7 and 9) also cause spatial patterns of species composition in response to gradients in environmental conditions. In fact, the individualistic theory of ecology was first discussed in the context of spatial gradients (Gleason, 1917, 1926).

Patterns of zonation along gradients are not static, but are constantly changing as a consequence of the dynamic equilibrium between succession and disturbance. Wherever patterns of zonation are determined primarily by biotic processes along gradients, and not by underlying boundaries between substrate conditions, those patterns change through time. Patterns along certain types of gradients predictably become sharper and more definite with time, while other types of zonation blur and disappear with time. Understanding the biological mechanisms underlying different types of zonation allows prediction of how these major patterns of biological diversity on landscapes will change through time.

The focus of this chapter is on situations in which discontinuities in species dominance are not associated with any obvious discontinuity in the underlying physical conditions (e.g., the contact zone between two contrasting geological formations or soil types). Patterns of zonation can be divided into two general types: 1. those resulting from temporal asynchrony of biotic processes; and 2. those resulting from spatial variation in physical conditions.

Zonation Caused by Temporal Asynchrony of Biotic Conditions

In contrast to zonation along spatial gradients of physical conditions, zones that result from temporal asynchrony are simply different temporal stages of the same successional sequence (sere). This type of zonation does not depend on any variation of physical conditions other than that created by the organisms themselves. Over a long enough time, the vegetation in all zones will pass through essentially the same stages, in contrast to the vegetation along spatial gradients of physical conditions, which represent different dynamic equilibria and will never pass through identical stages. This zonation of successional stages is usually initiated by discrete events that may occur with a relatively constant frequency but affect a different area at each occurrence. The succession that causes this type of zonation is usually autogenic, but allogenic effects, such as changes in drainage pattern as a result of stream down-cutting, may also occur, generally in synchrony with the autogenic changes.

A classic example is the successional zonation found along the meanders of a river (Weaver, 1960; Shelford, 1963; Viereck, 1970; Nanson and Beach, 1977; Walker et al., 1986). As the river meanders and erodes soil from one location, often undercutting and destroying mature riverine forest, soil is redeposited in other locations. Since most sediment movement occurs during periods of high flow, new sites for succession

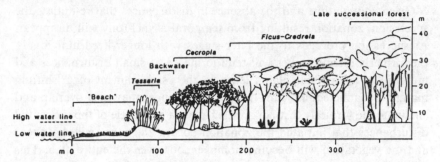

Fig. 10.1 Pattern of vegetation zonation on the floodplain of the Manu River in the Peruvian Amazon. (From Salo *et al.*, 1986.)

are created at discrete intervals. Such a pattern can create a sequence of zones of increasing age, with succession being initiated periodically at the edge along the river.

This pattern is found along the interior of meander loops of the Manu River in the Columbian Amazon. 'Large areas in the Manu basin adjacent to the river are occupied by successional vegetation. The interiors of meander loops are filled with it. Because of the great regularity with which plant species replace one another in the canopy, and because each stage reaches a greater absolute height than the previous one, the successional vegetation is strikingly zoned. As viewed from the vantage of a passing boat, one sees an orderly sequence of stands - *Tessaria, Gynerium, Cecropia,* mixed forest, and *Ficus-Cedrela* - each forming a nearly discrete band of greater or lesser width running parallel to the adjacent beach. Of these, the *Ficus-Cedrela* association occupies by far the largest area by virtue of its vastly greater longevity. Eventually, however, the *Ficus* and *Cedrela* trees begin to die off, one by one, and are replaced by an entirely different set of species which form a forest of considerably greater diversity. The succession continues, in other words, to even more advanced stages, but the details have not been studied.' (Terborgh, 1983, pp. 8-10.) (Fig. 10.1.)

Similar patterns of successional zonation are found wherever bare substrate is created or exposed sequentially. The classic example is the forest succession on sand dunes along lake Michigan (Cowles, 1899; Olson, 1958), which was extremely influential in developing early theories of succession (Clements, 1916). Other examples are the vegetation found on materials exposed by retreating glaciers (Cooper, 1923, 1931, 1939; Lawrence, 1958; Viereck, 1966) and herbaceous vegetation that grows on substrate exposed by a drying lake or pond.

Given enough time and the absence of disturbances that re-initiate the succession, zonation resulting from temporal asynchrony will disappear. As succession proceeds to the later stages with longer-lived individuals, the formerly distinct zones will tend to merge as small disturbances and mortality of individual plants leads to the establishment of a 'shifting mosaic' dynamic equilibrium. Large disturbances may also overlap and eliminate the formerly distinct zones. Unless a new cycle of the sequential disturbances that initiated the zonation occurs, the zones will disappear and the vegetation will become homogeneous over the entire area. This is in striking contrast to the temporal patterns along physical gradients, which tend to become more distinct with time.

Zonation Caused by Spatial Gradients of Physical Conditions

Probably the most common type of zonation occurs along spatial gradients that have different combinations of physical conditions at different locations along the gradient. There are virtually no gradients along which only a single factor varies, and most physical gradients have many different abiotic factors varying along their length, some with negative correlations, others with positive correlations, but very few that are totally independent. Along such gradients, vegetation within a region (e.g., many patches) can be considered to be in dynamic equilibrium with the environment. Succession may occur at different rates at different locations along the gradient, disturbances may occur at different average frequencies along the gradient, and the relative competitive ability of any particular species is likely to change along the gradient. Species composition and diversity will fluctuate within different dynamic equilibria at each location. The key characteristic of zonation along a physical gradient is that vegetation at different locations will never converge to the same composition unless environmental conditions become uniform and the physical gradient disappears.

The history of ecological interest in gradients is closely related to the history of succession theory, since the original concept of temporal change in plant communities was based on the presumed correspondence of spatial patterns of community structure along gradients with temporal changes at one location (Cowles, 1899). Efforts to quantify and understand vegetation patterns along gradients (i.e., the field of phytosociology, cf. McIntosh, 1978) led to such well-known areas of investigation in ecology as the 'continuum concept' (Curtis, 1959; Cottam and McIntosh, 1966; McIntosh, 1967; Austin, 1985) and 'gradient analysis' (Whittaker, 1967, 1973). As with most issues in community structure, whether vegetation pattern is interpreted as a continuum or as a series of distinct communities is often a matter of the scale at which the pattern is examined, and which groups of species are given most importance (cf. Daubenmire, 1966; Cottam and McIntosh, 1966).

Whittaker (1970, 1973; Whittaker *et al.*, 1973) and other ecologists (e.g., Austin, 1980, 1985; Austin and Smith, 1979) have made an effort to relate the distribution of organisms along gradients to the concept of niche (Hutchinson, 1961; MacArthur, 1968) and to theories of how organisms subdivide resources to minimize competition (Brown and Wilson, 1956). A basic issue in continuum theory is whether the abundance maxima and limits of individual species are independently distributed along gradients (Gleason, 1926), or whether 'major' species are regularly distributed along gradients (Gauch and Whittaker, 1972; Clapham, 1973; Schoener, 1974; see review in Austin, 1985). One problem with interpreting species distribution in the context of niche theory is that most species do not show the symmetrical bell-shaped curves used in the theories, but rather tend to have negatively skewed distributions, with the tail toward the most productive conditions (Austin, 1987; Austin and Smith, 1989).

Austin and Smith (1989) have clarified the issue of plant distributions along gradients by distinguishing between functionally different types of gradients, and applying the concepts of the physiological and life history trade-offs (Huston and Smith, 1987; Smith and Huston, 1989) to develop a mechanistic continuum theory. Two principal types of gradients can be defined (Austin, 1980; Austin and Smith, 1989).

(1) *Resource Gradients.* The property that varies along the gradient is a resource that must be consumed by plants in order for them to grow. An important feature of these resources is that plants are able to deplete them in the local environment. These resources include light, water, essential mineral nutrients, and potentially carbon dioxide and oxygen. The growth of plants is generally greatest at high levels of the resource, but is often limited by toxicity at very high concentrations. Growth is limited by deficiency at very low concentrations of most resources.

(2) *Regulator Gradients.* The property that varies along the gradient regulates the rate of physiological processes, but is not taken up or incorporated into the organism's structure nor is the level or amount of the factor in the environment depleted by the organisms. Air temperature and soil pH are perhaps the two best examples of regulators. The growth

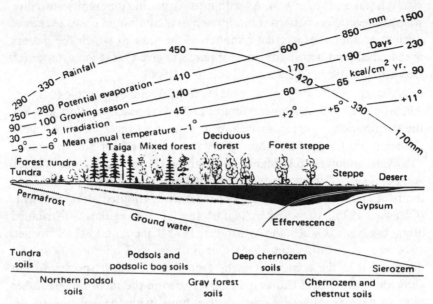

Fig. 10.2 Schematic climate, vegetation, and soil profile of the east European lowlands from northwest to southeast (black: humus horizon; diagonal shading: illuvial B-horizon). Growing season in the tundra corresponds to the number of days with a mean temperature above 0 °C, elsewhere to the number of days with a mean temperature above 10 °C. (From Walter, 1964/68, based on Schennikov.)

of plants is generally greatest at intermediate levels of the property, and plants are unable to survive at the extremes. This type of gradient was called a 'direct' gradient by Austin (1980; Austin and Smith, 1989).

A third type of gradient can be classified as *complex*, because both resources and regulators vary along the gradient. This type of gradient is equivalent to 'factor-complex' gradients (Whittaker, 1973) and 'indirect' gradients (Austin, 1980; Austin and Smith, 1989). Altitude is the best example of a complex gradient. Altitude *per se* has no direct influence on plant growth, but rather is correlated with a variety of resources and regulators that do affect plant growth, including precipitation, temperature and the intensity and spectral quality of solar radiation. Latitudinal gradients are also complex gradients that are closely analogous to altitudinal gradients (Fig. 10.2). Latitudinal gradients first focused attention on patterns of species diversity, and continue to be a major challenge for theoretical ecology. In most cases, patterns based on complex gradients cannot be generalized to other environments, unless all conditions are identical.

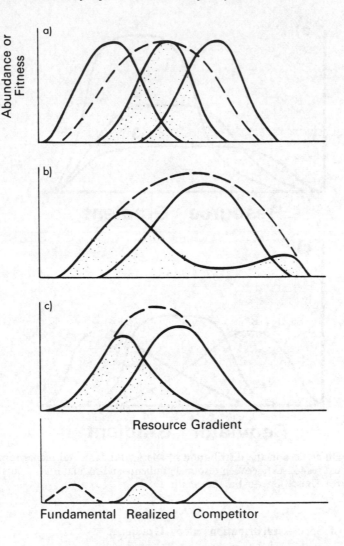

Types of Species Response
Curves

Fig. 10.3 Patterns of species response curves (realized niche) along an environmental gradient. Stippled area represents the niche of the species whose fundamental niche is indicated by the dotted curve. The solid curves with no stippling represent other competing species. (*a*) classic niche concept with Gaussian response curves, and maximum of the realized niche at the maximum of the fundamental niche. (*b*) and (*c*) Ellenberg's concept of realized niche displaced away from optimum conditions of fundamental niche by competitors. (*b*) Bimodal realized niche with maxima displaced to extremes. (*c*) Skewed realized niche with maximum displaced toward low end of resource gradient. (Based on Mueller-Dombois and Ellenberg, 1974; Austin and Smith, 1989.)

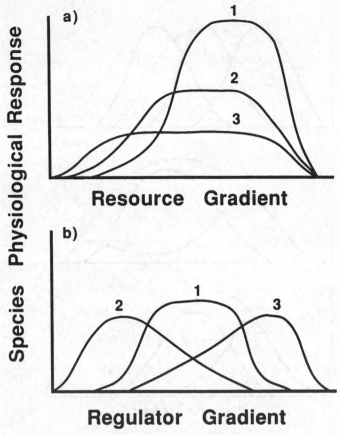

Fig. 10.4 Differences in the distribution of species fundamental niches (optimal physiological responses) between resource gradients and regulator gradients (e.g., temperature). (Based on Austin and Smith, 1989.)

Patterns of Species Distributions along Gradients

Ellenberg (1953, 1954; Mueller-Dombois and Ellenberg, 1974) makes the important distinction between the potential distribution of a species along a gradient in the absence of competition from other species, which he called the 'fundamental niche', with maximum abundance at the physiological optimum (see also Hutchinson, 1957), and the actual distribution of a species in the presence of competitors within a plant community, which can be called the 'realized niche' or 'ecological niche' (Hutchinson, 1957) with maximum abundance at the ecological optimum. Depending on which other species are present, the realized niche of a species can take a variety of forms, ranging from a symetrical bell-shaped distribution, to highly skewed, to bimodal (Fig. 10.3).

Austin and Smith (1989) note that the patterns of fundamental niches (i.e., the distribution of physiological optima) are completely different for resource gradients versus regulator gradients. Along regulator gradients, fundamental niches are distributed independently, with optima for different species occurring anywhere along the gradient, except at the highly stressful extremes (Figure 10.4). This pattern reflects natural selection on processes ranging from enzyme kinetics to size and morphology, which can produce optimal responses at any point along the gradient. For regulators, there is no absolute advantage or disadvantage at any particular level of the factor, except for a disadvantage at the extremes. A temperature of 20 °C is not intrinsically better for growth than a temperature of 25 °C, nor is pH 6 superior to pH 8. The optimal response (in terms of growth rate, biomass, fitness, etc.) is greatest near the center of gradient, and decreases toward either extreme (Fig. 10.4b)

Along regulator gradients, there is generally a close correspondence between the fundamental niche (physiological optimum) and the realized niche (ecological optimum). Species tend to be most abundant under conditions near their physiological optima, and the physiological optima of species tend to be distributed fairly evenly along regulator gradients. The ecological optimum of a species can be narrowed by competition from species at either higher or lower levels of the regulator.

In contrast to the pattern on regulator gradients, fundamental niches along resource gradients are concentrated at high resource levels. Physiological optima occur at the highest non-toxic levels of the resource, and growth decreases toward low levels. This pattern reflects the physiological response to the supply of an essential resource. High levels of an essential resource are inherently better for growth, reproduction, and survival than are low levels.

As a general rule, all plants grow best with abundant light and water (as well as mineral nutrients and CO_2). Many studies have found a great similarity in the physiological optima of different species when they are grown in monocultures along experimental gradients of nutrients (Bradshaw *et al.*, 1964; Austin and Austin, 1980) and moisture (Ellenberg, 1953, 1954; Mueller-Dombois and Sims, 1966). In spite of the high overlap of physiological optima along resource gradients, there is usually much less overlap in the resource conditions under which species are most abundant naturally (i.e., the ecological optimum) (Ellenberg, 1953, 1954; Walter, 1960, 1971; Austin, 1982). Plants are rarely most abundant in natural communities under their physiologically optimum conditions because of competition from other species (Mueller-Dombois and Sims, 1966; Austin, 1982; Austin *et al.*, 1985).

For most species, the ecological optimum is shifted toward lower resource conditions than the physiological optimum. Only species with the highest growth rate and/or maximum size (or some other relevant parameter of competitive ability or fitness) under optimal conditions have their ecological optimum close to their physiological optimum (Fig. 10.5). Species that grow more slowly or are smaller in size (i.e., low fitness) under high resource conditions tend to have their ecological optima shifted toward the lower resource levels where the species that dominate at high resource levels grow poorly or are unable to survive. Virtually all studies of plant growth and community structure along experimental or natural fertility gradients find more intense competition and lower species diversity at the high productivity end of the gradient (e.g., Reader and Best, 1989; Austin, 1979; Wilson and Keddy, 1986; Day *et al.*, 1988; Moore and Keddy, 1989; Keddy and MacLellan, 1990; Shipley *et al.*, 1991).

An individual-based simulation model of competition along a moisture gradient can produce the observed patterns of differences between the 'physiological optimum' (Fig. 10.6*a*) and the 'ecological optimum' of a species (Fig. 10.6*b*) (Salisbury, 1929; Walter, 1971; Ellenberg, 1953, 1954; Mueller-Dombois and Ellenberg, 1974; Rorison, 1968; Austin, 1982). Each functional type (with the exception of the most mesophytic) shows declining total biomass both with increased and with decreased moisture availability when they are grown together with all other functional types (Fig. 10.6*b*). The decline in relative biomass of most types as moisture availability is increased results from competition for light, since each type has its maximum *potential* growth under such favorable conditions (Fig. 10.6*a*). As moisture decreases, most functional types decline in total biomass as a consequence of both competition and physiological limitations (Fig. 10.6*b*). The fact that the computer simulation results closely match the results of the laboratory experiments suggests that the individual-based model of competition for light with physiological and life-history constraints adequately incorporates the underlying processes that produce the consistent patterns of plant distributions found in these experiments, and presumably in natural communities as well.

Even though the physiological optima of all species are clustered at the high concentration end of a resource gradient, not all species have the same response (in terms of growth rate, biomass, fitness, etc.) at

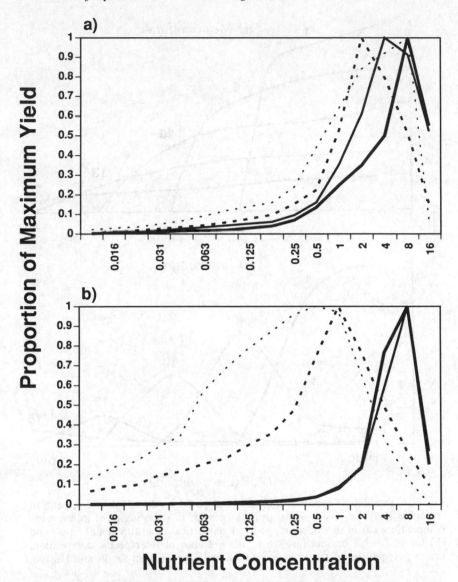

Fig. 10.5 Physiological and ecological responses of four grass species grown along an experimental nutrient gradient. (*a*) Physiological responses (potential niche) of species grown in monoculture along the nutrients gradient. (*b*) Ecological responses (realized niche) of the same species grown in a five-species mixture along the nutrient gradient. The four species are *Vulpia membranacea,* light dotted line; *Festuca ovina,* heavy dotted line; *Lolium perenne,* light solid line; and *Dactylis glomerata,* heavy solid line. (Based on Austin and Austin, 1980.) Nutrient concentrations are in relation to standard Long Ashton nutrient solution, which is composed of 172 g/l N, 41 g/l P and 156 g/l K, plus micronutrients.

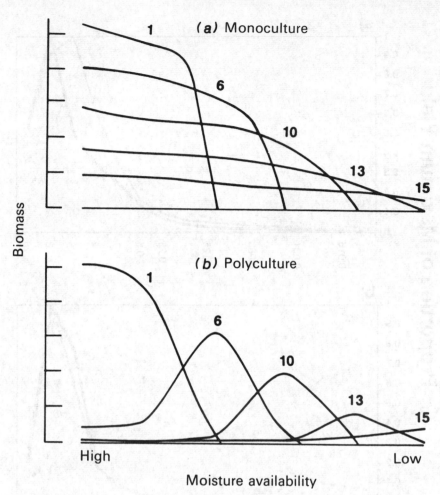

Fig. 10.6 Simulation results from individual-based model of plant competition illustrating (*a*) the physiological optima (growth in the absence of interspecific competition, i.e., in monoculture) of five hypothetical plant functional types; and (*b*) the ecological optima (growth in the presence of interspecific competition, i.e., in polyculture) of the same five functional types. (From Smith and Huston, 1989.)

their optimum. Specifically, those species with the highest responses under optimal conditions have a much poorer response under conditions of lower resource levels than do species with a lower response at the optimum (Figs 10.4*a* and 10.6*a*). This pattern reflects the trade-offs in the ability to grow at high rates under high resource conditions and the ability to grow and survive at low resource levels that are the basis of

Smith and Huston's (1989) theory of vegetation dynamics (see Chapter 7).

Bimodal distributions of species abundance are most likely to occur along regulator gradients, where certain species may be adapted to survive under a wide range of conditions, but not have a high growth rate or size under any particular conditions. Such species can be pushed to both the upper and lower extremes of the gradient by a species that has a high response near the center of the gradient, but is unable to survive at the extremes. Such a bimodal distribution is found in several tree and shrub species along the altitudinal gradient (a complex gradient that is a combination of temperature, water, and perhaps soil nutrient gradients) on Mauna Loa in Hawaii (Fig. 10.7), where the grass and tree species that dominate on the deep, rich soils of the savannah zone apparently displace some species to both higher and lower elevations (Mueller-Dombois *et al.*, 1981).

Although most of the examples in this chapter are based on plants, the same general patterns also occur among animals. As discussed in Chapters 4 and 5, the fundamentally different ways in which plants and animals utilize their environment results in different patterns of community structure. The greater mobility of animals allows both interference competition, which is rare among plants, and intense exploitation competition. This higher intensity and directionality of competitive interactions has ecological consequences in terms of competitive displacement and evolutionary consequences in the form of character displacement to minimize negative competitive interactions. Consequently, patterns of species zonation along both regulator and resource gradients often have sharper boundaries among animals than among plants (e.g., Terborgh, 1971). Miller (1964, 1967) found a strong competitive hierarchy among four species of pocket gophers of the genera *Geomys, Cratogeomys,* and *Thomomys*. Although all four species did best in deep, light soils, the dominant species, *Geomys bursarius*, displaced the other three from the optimal conditions, although it was unable to invade poorer soils. The least dominant species, *Thomomys talpoides*, was usually limited to the poorest soils, even though it could thrive on any type of soil, in the absence of more dominant species. Thus, patterns along resource gradients of the physiological and ecological optima of animals tend to be similar to those of plants, with strong overlap in physiological optima, and displacement of ecological optima (e.g., Holmes, 1961, Fig. 10.8).

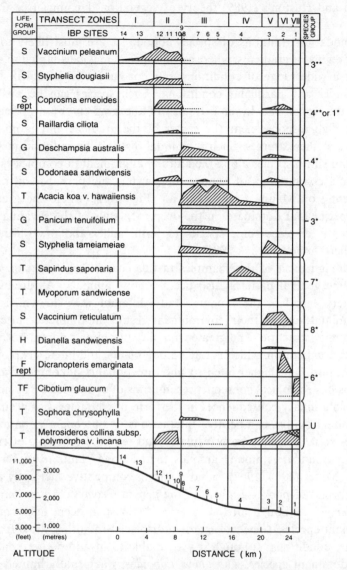

Fig. 10.7 Altitudinal distribution of plant species on Mauna Loa in Hawaii. Note the bimodal distributions of the tree *Metrosideros collina*, and the shrubs *Raillardia ciliolata* and *Dodonaea sandwicensis*, which are apparently displaced from mid-elevations by the savannah trees and grasses. Note also that diversity among shrubs and trees is higher at upper and lower elevations than in the middle of the gradient, where *Acacia koa* dominates. The number of grass and herbaceous species (not shown) is highest in the zone dominated by *Acacia koa*. Curve heights indicate relative percentage cover of different species. Life-form symbols: F - fern, G - grass, H - herbaceous plant other than grass, S - shrub, T - tree, TF - tree fern. (From Mueller-Dombois *et al.*, 1981.)

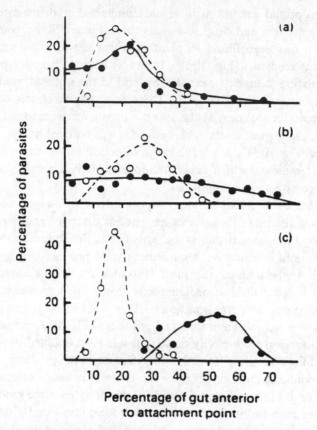

Fig. 10.8 Fundamental and realized niches of two species of intestinal parasites, a tapeworm *Hymenolepis diminuta* (solid circles) and a spiny-headed worm *Moniliformis dubius* (open circles) in the gut of a laboratory rat. Fundamental niches (from monospecific infections) at low (*a*) and high (*b*) densities. (*c*) Realized niches (from mixed-species infection). (From Colwell and Fuentes, 1975, based on data from Holmes, 1961.)

Patterns of Species Diversity along Environmental Gradients

The above arguments about the responses of individual species along environmental gradients lead to predictions about species diversity patterns that are at first glance contradictory to those discussed so far in this book. Specifically, Austin and Smith (1989) predict that species richness will show a bimodal response along both resource gradients and regulator gradients (Fig. 10.9). Implicitly invoking the relationship between productivity, competition, and species diversity that has been discussed in the previous chapters, they hypothesize that diversity will be low under optimal conditions in an environmental gradient, increase toward both extremes, and drop to zero at the extremes. Some evidence of this pattern has been found in plant mixtures grown along experimental nutrient gradients (Fig. 10.10). Evenness among 10 grass species grown in a mixture is highest near the low end of the nutrient gradient, and decreases toward the high end. There is some indication of the expected decrease in evenness at the lowest extreme of the gradient, and an increase at the highest levels, where biomass was reduced by apparent toxic effect (cf. Fig. 10.5). Such a full range of conditions is rarely found along natural gradients, which generally represent only a portion of the potential range of resource conditions.

It is clear that the bimodal pattern of species diversity along gradients actually reflects the relationship between species diversity and the rate of competitive displacement that is the basis of the dynamic equilibrium model. The rate of competitive displacement (also productivity, growth rates, etc.) is highest under the most favorable conditions along the gradient, which occur at intermediate levels along direct gradients, and at high levels along resource gradients. Thus the bimodal responses along Austin and Smith's environmental gradients (Fig. 10.9) are simply a unimodal diversity/productivity curve that has been doubled to reflect the decrease in productivity toward both extremes of the gradient.

While bimodal distributions of species diversity have been documented along regulator gradients, bimodal distributions along resource gradients are much rarer because resources rarely reach toxic levels under natural conditions. Thus the general pattern of diversity along nutrient gradients is either unimodal or monotonically decreasing toward high levels, depending on the level at the low end of the gradient. For example, in a selection of tropical rainforest sites from Costa Rica, with relatively rich volcanic soils, the highest tree species richness was found on the poorest soils, that is, there was no decrease in maximum diversity at the low end of the resource gradient (see Chapter 14) and maximum diversity decreased monotonically along the gradient (Huston, 1980a; Chapter 14). In a selection of samples from Southeast Asia, which included some very low nutrient sites, maximum diversity did decrease at the low end of the resource gradient, resulting in a unimodal diversity response (Ashton, 1977).

Walker et al. (1981) describe a series of plant communities on sand dunes of increasing age in Eastern Australia, with maximum plant diversity on dunes of intermediate age and nutrient availability. The six dune systems range in age from approximately 6000 years to around 400,000

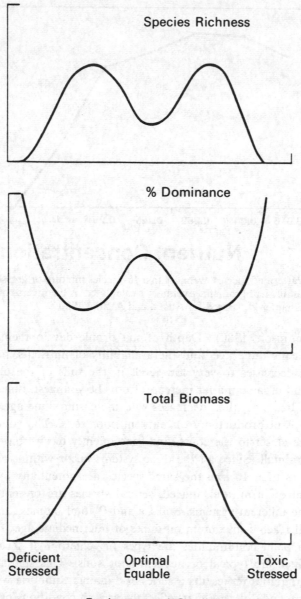

Fig. 10.9 Expected pattern of species diversity (evenness and dominance), and total biomass along an environmental gradient. The same pattern is expected on both resource gradients (along their full range from deficient to toxic) and on regulator gradients (along their full range from stress due to high levels to stress due to low levels). (From Austin and Smith, 1989.)

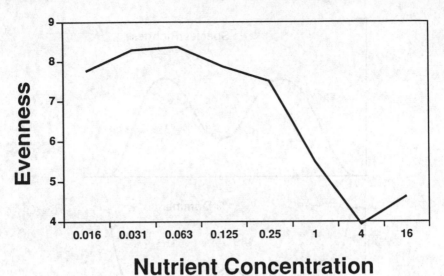

Fig. 10.10 Pattern of species evenness in a 10-species mixture of grasses along an experimental nutrient gradient. Evenness is calculated as the inverse of Simpson's Index (see Chapter 2). (Based on Austin and Austin, 1980.)

years (Thompson, 1980). Depth of soil profile development increases with increasing dune age, and the availability of nutrients, particularly Ca and P, decreases to very low levels in the soils of the oldest dune systems. Soil organic matter increases from the youngest dunes to dunes of intermediate age, then decreases with increasing dune age, reflecting the lower plant productivity on nutrient-poor soils. The younger dune systems are closer to the coast, and consequently receive more nutrient input via rainfall. Since all the dune systems occur within a linear distance of less than 10 km, the same species are potentially available to colonize all of them, and, indeed, several species are present on all or most of the different systems. Soil Ca and P, total biomass, and species richness all reach a maximum on dunes of intermediate age (Fig. 10.11). All of the plant communities are types of schlerophyll shrublands or forests, which are typical of nutrient-poor soils. This spatial series of plant communities represents a series of dynamic equilibria with species composition and diversity that reflect the growth conditions produced by long-term soil development.

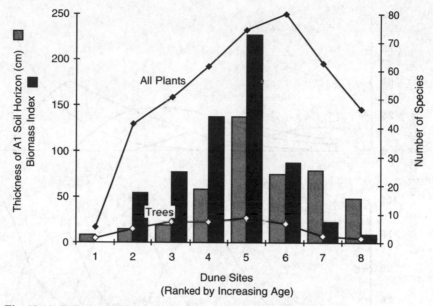

Fig. 10.11 Soil properties and vegetation structure along a 400,000 year chronosequence of soil development and degradation in Eastern Australia. (Based on Walker *et al.*, 1981.)

Shifts in Species Distributions over Time

Although zonation is a spatial phenomenon, it also has a temporal component. Succession occurs at every point along a spatial gradient, so patterns of zonation may change over time. Temporal changes in patterns of zonation reflect changes in resources, particularly light, that are caused by the plants themselves. Individual-based simulations using the plant functional types of Smith and Huston (1989) illustrate the temporal changes in plant type distribution (i.e., zonation) between early (Fig. 10.12a) and late succession (Fig. 10.12b) along a moisture gradient.

The principal changes in species distributions along a moisture gradient over time are : 1. a decrease in the range of moisture conditions over which a plant type is found, resulting primarily from competition for light at the high-resource end of the gradient; and 2. a shift in the mode of most plant distributions toward conditions of lower water availability. A significant feature of the simulation results (Smith and Huston, 1989), and of the relatively few studies that have examined this phenomenon in natural communities (e.g., Werner and Platt, 1976), is that patterns of zonation along physical gradients become more distinct through time.

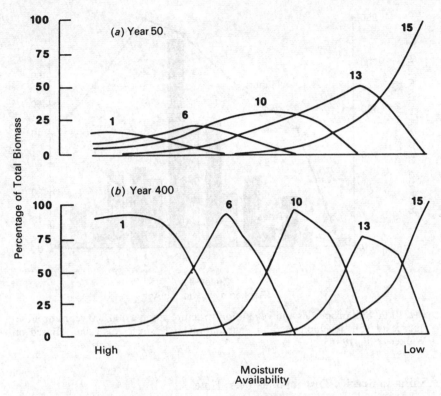

Fig. 10.12 Temporal changes in the simulated spatial distribution of plant functional types along a resource (moisture) gradient. The curves are based on the output of an individual-based computer simulation model of plant competition for light, at two different times during succession. The numbering of the functional types is based on Fig. 7.2. (From Smith and Huston, 1989.)

This is in contrast to the patterns of zonation that result from temporal asynchrony of the initiation of succession, which tend to disappear with time.

In the early stages of succession, when plants are small and leaf area is low, shade tolerance is irrelevant to competitive success, although other traits, such as growth rate and drought tolerance, which are correlated with shade tolerance, can be important. Thus a plant can become established at any point along a gradient where it is physiologically capable of growth and species distributions along gradients are broad during early succession. In later succession, as increases in plant height and leaf area reduce light levels, shade tolerance becomes more critical. Shade-intolerant xerophytes are eliminated from the wet end of the moisture gradient, where leaf area can increase and reduce light to levels lower than the xerophytes can tolerate. The distribution of these species is displaced toward the drier conditions where leaf areas are low enough to allow sufficient light penetration for their survival and reproduction.

Because all plants are potentially capable of surviving under the favorable conditions at the high resource end of the gradient, most are able to persist at low densities as a result of periodic disturbances that allow sufficient light penetration for their survival. As a result, a long tail of the distribution persists under high resource conditions where the plants can potentially survive, but are usually eliminated by competition. This same pattern of skewness has been found in many studies of plant distributions along resource gradients (Austin, 1987).

The result of this shift in plant distributions over successional time is that patterns of zonation become more distinct, as plants are restricted by competition on the high resource end of the gradient to a narrower region slightly above their physiological limit. This shift from a broad range of relatively high abundance along the gradient during early succession to a narrow zone of maximum abundance during later succession can cause zonation to develop on landscapes that appear relatively homogeneous during early succession.

These patterns generated by the simulation model for moisture gradients are similar to the pattern documented by Werner and Platt (1976) for goldenrods (Fig. 10.13). There was higher diversity at most points along the moisture gradient and greater overlap between species in an old field (considered to represent an earlier stage of herbaceous succession) than in a natural prairie. Similar patterns of decreasing 'habitat breadth' over the course of succession have been reported for other herbaceous communities (Pineda *et al.*, 1981a,b) and forests (Auclair and Goff, 1971; Christensen and Peet, 1984).

The patterns of increasing distinctness of zonation and decreasing species ranges through time along resource gradients are directly related to the phenomenon of shifting successional roles of species (or functional types) under different resource conditions (see Chapter 7). This shift from being an early-successional species under mesic conditions to a late-successional species under drier conditions is associated with a restriction of the species 'ecological optimum' to the intermediate portion of the moisture gradient.

A clear example of this phenomenon is found in the forest succession of the subalpine zone in the intermontane region of western North America. Under the cool, dry conditions of this region, tree species

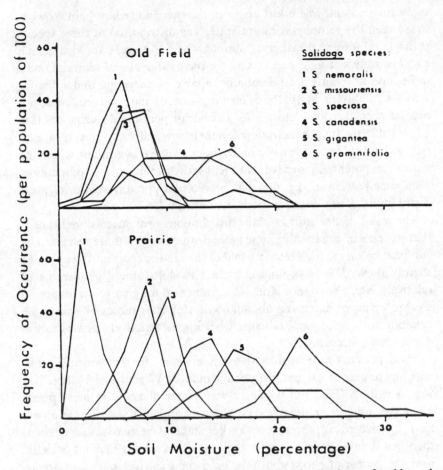

Fig. 10.13 Presumed temporal shift in the spatial distributions of goldenrod (*Solidago* spp.) along a moisture gradient. (*a*) Distributions along a moisture gradient in an old field, assumed to represent a relatively early stage in plant succession, and (*b*) distributions along a moisture gradient in a prairie, assumed to represent a later stage in plant succession. (From Werner and Platt, 1976.)

diversity is quite low, and fire is important in regulating landscape patterns (Romme and Knight, 1981; Romme and Despain, 1989). The early stages of forest succession on nearly all sites are dominated by monospecific stands of lodgepole pine (*Pinus contorta*), a rapidly growing, shade-intolerant and drought-tolerant species. On dry sites and southern exposures, lodgepole pine dominates the stands indefinitely, regenerating in the openings caused by the deaths of individuals of the cohort that established following the most recent fire. However, on more mesic sites in valley bottoms and north-facing slopes, spruce (*Picea engelmannii*) and fir (*Abies lasiocarpa*) become established in the understory of the early-successional lodgepole pine forest and eventually overtop the pines and form a self-replacing stand, until the next fire reinitiates succession with lodgepole pine. Thus, the subalpine landscape shifts through time from one dominated by a monospecific forest of lodgepole pine on sites of all types, to a patterned landscape with spruce-fir forest on mesic sites and lodgepole pine forest on dry sites (Taylor, 1969, cited in Romme and Despain, 1989; Romme and Despain, 1989).

Successional Patterns along a Resource Gradient

Just as the spatial distribution of plants along a resource gradient changes through time, the successional patterns along the gradient change from one spatial location (resource level) to another. As discussed in Chapters 7 and 9, the primary changes in successional patterns along a gradient from low to high moisture availability are: 1. an increase through time in the total number of functional types present in a sere; 2. an increase in the total biomass and leaf area that accumulates during the sere; 3. an increase in the complexity of the sere, with more shifts in species dominance and a longer time between early and late succession; and 4. the coexistence of more functional types during mid and late stages of the sere. The number of species within any specific functional type does not necessarily increase with increasing moisture availability, so the pattern of total species diversity may be difficult to predict. Only when the functional types are considered separately can the changes in species diversity along a temporal or spatial gradient be understood.

Under conditions of resource deficiency, few functional types can survive and species diversity changes little through succession. As Whittaker (1975) observed, 'Toward increasingly unfavorable [xeric] environments there is a stepping down of community structure and a reduction of stratal differentiation, with generally smaller number of growth forms arranged in fewer and lower strata.'

Under wetter conditions, diversity within habitats can be higher because more light-use strategies are possible and thus more functional types can coexist. Species composition and diversity change dramatically through the course of succession, with a decrease in diversity among the dominant life forms (cf. Fig. 9.8*a,b*). However, there may be an increase in total diversity as additional functional types, such as vines, epiphytes, and understory herbs with higher shade tolerance (but reduced size and longevity) are added to the community.

The longer and more complex successional sequence possible under high resource conditions (particularly moisture availability), allows these portions of the resource gradient to take on a series of different properties as they become dominated by plants of different successional stages. Such temporal changes in the characteristics of a particular zone occur most clearly under conditions where the entire zone (or series of adjacent zones) has been affected by a single large disturbance that synchronizes succession throughout that portion of the gradient.

Interaction of Regulator Gradients and Resource Gradients

Patterns of zonation on landscapes are almost always found along complex gradients where many different properties change. Thus most patterns of zonation reflect responses to both resource gradients and regulator gradients. Along regulator gradients, the role of resource availability is important in an evolutionary sense, because selection to maximize fitness leads to physiological specialization for the conditions where the negative effects of superior competitors are minimized (e.g., Brown and Wilson, 1956). Resources are also important on ecological time scales, since competition from species at either higher or lower levels of the regulating factor can constrain a species to a narrower range about its physiological optimum for the regulator.

By definition, resource levels are high at one end of a resource gradient, and low at the other end. However, at any point along a regulator gradient, resource levels can either be high or low. The level of resources along a regulator gradient has a significant impact on how patterns of species distributions and zonation in response to the regulator are actually expressed. Because zonation is a phenomenon of species dominance, zonation appears most clearly under conditions where dominance is high (and diversity low), which is at the low productivity extreme and the high productivity upper end of a resource gradient (Figs 10.9 and 10.10). Thus, the distinctness of zones caused by a regulating factor such as temperature should be greatest under conditions of very low or very high productivity, and least distinct under conditions of intermediate to low productivity where diversity is expected to be high.

One of the most common and dramatic type of gradient is the elevational temperature gradient. Of course, many other factors in addition to temperature change with elevation, most importantly, the availability

Fig. 10.14 Zonation of mangrove species along a tropical east African coast. HWL - high-water limit. LWL - low-water limit. (From Walter and Steiner, 1936.)

of important resources such as water and soil nutrients. Along such complex gradients it is primarily the distribution of physiological optima that determines the distribution of species and thus the location of zones. However, the distinctness of the zonation is strongly influenced by resource levels and productivity, which differ between gradients as well as along a particular gradient.

Elevational vegetation zones, particularly the transition between deciduous trees and conifers, appear clearly in the productive forests of the Great Smoky Mountains of southeastern North America and the Caucasus Mountains of southern Russia and northern Georgia (M. Huston, personal observation). Strong zonation is also found in productive marine environments, such as that among plants in saltmarshes, mangroves along tropical coastlines (Fig. 10.14), and among both plants and animals along the intertidal zone of rocky shorelines (Figs. 2.20 and 2.21). Intertidal zonation is associated both with stressful conditions, which tend to be most extreme at one end of the zone or the other (depending on whether the organisms are primarily marine or terrestrial/aquatic), and with high productivity, which is generally found near the middle of the tidal range.

Strong patterns of zonation are often present in deserts, where nearly monospecific zones of vegetation are arrayed along subtle elevational (primarily moisture) gradients, as well as in arid mountains (Peet, 1978; Margules *et al.*, 1987; Whittaker and Neiring, 1965). Vegetation dominance and zonation along an elevational gradient in the lower, arid portions of the Santa Catalina Mountains of Arizona are much stronger than in the moister, upper elevations of the Santa Catalinas or in the Siskiyou Mountains of Oregon, where moisture availability and productivity are higher (Whittaker, 1967; Whittaker and Niering, 1965; Whittaker, 1960). (Fig. 10.15).

Zonation along resource gradients should also appear most strongly under the low diversity conditions found at very low resource levels and

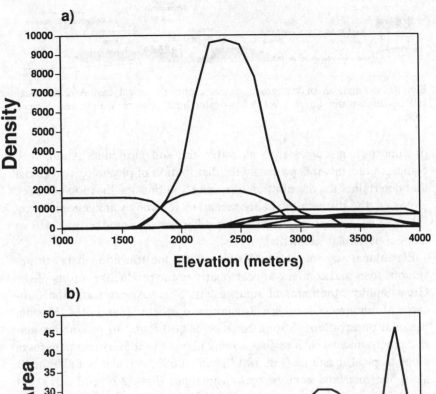

Fig. 10.15 Plant species distributions along elevational gradients. (a) Woody shrubs in low elevation deserts in the Santa Catalina Mountains of Arizona (density is number of stems / ha); (b) deciduous and coniferous trees in the Santa Catalina Mountains of Arizona (basal area is m^2/ha). (Based on Niering and Lowe, 1984.)

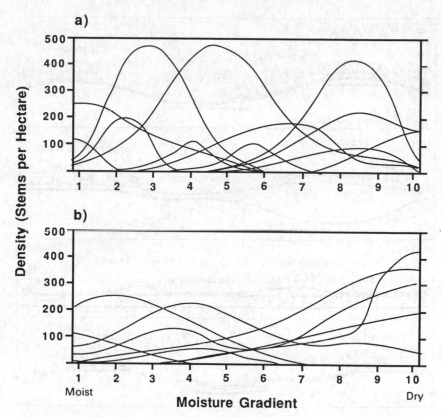

Fig. 10.16 Plant species distributions along a topographic moisture gradient from moist ravines (1) to dry southwest-facing slopes (10) in (*a*) the Santa Catalina Mountains of Arizona, and (*b*) the Siskiyou Mountains of Oregon. (Modified from Whittaker, 1967; see also Whittaker and Niering, 1965; Whittaker, 1960.)

very high (but not toxic) levels of the resource. It should be kept in mind that resource levels that are low enough to preclude the survival of a particular functional type of plant, such as trees, may be sufficient for the survival of other functional types, such as grasses. Zonation should appear most strongly in those organisms most limited by the resource. Vegetation dominance along a moisture gradient (indexed by aspect and exposure) is much stronger in the arid Santa Catalina Mountains than in the more mesic and productive Siskiyou Mountains (Fig. 10.16).

Because plants found under dry conditions have a low leaf area and small stature, light availability is relatively high at ground level (Fig. 9.7*a*). Zonation of woody plants in arid regions generally involves shade-intolerant species of increasing size along gradients of increasing moisture

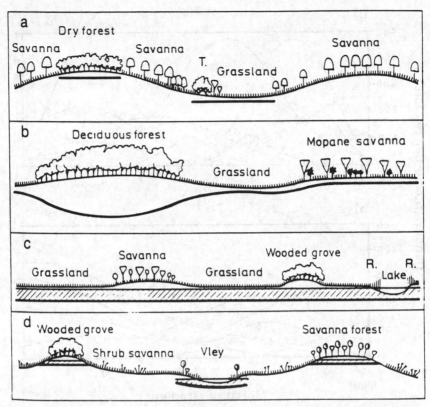

Fig. 10.17 Vegetation zones in response to soil moisture in a semi-arid region (Zonobiome II of Walter, 1971), Humido-arid tropical summer rain region with deciduous forests). Heavy dark lines indicate location of impermeable layer, diagonal hatched areas indicate water saturation during rainy season. Trees are eliminated by saturated soil conditions in (a), (b), and (c), and by insufficient water on deep sandy soils in (d). (From Tinley, 1982.)

availability. Although diversity within a zone is low, many different zones can occur because slight differences between soil types result in significant differences in soil moisture under low rainfall conditions (Fig. 10.17). Minimal overlap occurs between the zones because shade intolerance precludes coexistence through vertical stratification of light.

Summary

Zonation and the spatial distribution of species across landscapes are the result of the same types of interactions among individual organisms and the same physiological and life-history constraints that produce the temporal phenomenon of succession. Spatial patterns of species abundance are produced by relative differences between species in their ability to compete or survive over a range of environmental conditions. Plants are found in much more varied conditions than those under which they grow best without competition. Competition displaces species toward environmental conditions that they are able to tolerate, but which the species that outcompete them under optimal conditions cannot tolerate (cf. Connell, 1961). The ecological optimum of a species along a resource gradient is generally constrained on the high resource end primarily by competition, and on the low resource end primarily by physiological limitation. Thus, for many species, the ecological optimum is closer to their physiological limit than to their physiological optimum.

Patterns of zonation often change through time. Zonation that results from the responses of organisms to an underlying physical gradient tends to become more clearly defined through time during succession. Zonation that results from temporal variation in the initiation of succession tends to disappear with time. Competition for resources is potentially important along all gradients. However, there are two distinct types of gradients: *resource gradients*, along which the principle variable is a resource that can be depleted by organisms, and *regulator gradients*, along which the principle variable is a factor, such as temperature, which is unaffected by organisms.

Part four
Case Studies: Patterns and Hypotheses

These last five chapters are an effort to make sense out of some of the diversity patterns described at the beginning of the book. The framework for this effort is the general theories described in Part two and the mechanisms described in Part three. Some of the patterns that I will address are phenomena that have been known and discussed by ecologists for many years. Other patterns have been described only recently, in some cases because the ideas described in this book led me and others to look for the patterns where we hypothesized that they should occur. This selection of patterns and ecological issues is certainly neither a comprehensive nor a random survey of biological diversity. These are the patterns that interest me, and for which I believe the concepts in this book provide some insight.

These case studies are not intended to serve as tests of specific hypotheses. In few cases are sufficient data available to carry out statistically rigorous analysis, and such an effort is far beyond the scope of a single book. If these case studies support the conceptual framework that I have developed, it is by weight of anecdote rather than experimental rigor. For most ecosystems, comparison of pattern and structure under different conditions is the best we can do at present. In the few cases in which experiments have been conducted, they apply to specific systems under a narrow range of environmental conditions.

I have attempted to address a broad range of patterns and issues, rather than select only those that fit the hypotheses. However, it is also true that one tends to focus on those issues that one thinks one understands.

To the extent that the general framework and specific mechanisms that I have described provide useful insights into the examples discussed in these chapters, there is hope that these components of biological diversity can be understood. To the extent that the framework and mechanisms fail to clarify the regulation of these patterns, or other patterns not included in these chapters, there are still many challenges to be met before we understand the complex regulation of biological diversity.

11

Case Studies: Endemism and Invasions

Invasions and endemism represent the two most extreme conditions of the possible geographical distributions of species. Invaders are species that have spread far beyond their original distribution, to distant continents or sometimes around the entire globe. Endemic species are restricted to extremely small ranges, often a single island or mountaintop, or even a single rock outcrop. At first glance, there would seem to be little similarity in these phenomena, associated with the most widely distributed and 'successful' species in the world on the one hand, and with the most restricted species, which are often on the brink of extinction, on the other hand. However, there are some strong similarities in the patterns of endemism and invasion, and a comparative examination of physical conditions of these environments may provide insights into the mechanisms that regulate both patterns of endemism and the frequency of invasions.

These two issues are critical to the understanding and conservation of biological diversity. Areas with high rates of endemism also have high species diversity, although this is not always the case. Invasions of organisms into areas where they did not evolve have been the cause of the most extinctions both before and after the advent of *Homo sapiens,* at least until the current era of massive habitat destruction. In this chapter I attempt to demonstrate how the dynamic equilibrium model of species diversity can be used to predict the conditions under which endemic species are most likely to be found, as well as those conditions that are most susceptible to invasion by particular types of organisms.

Table 11.1. *Endemism on some islands.*

Island	Area (km^2)	Genera	Endemic Genera	Species	Endemic Species	Percentage Endemism
Cuba[1]	114,914	1308	62	5900[2]	2700	46
Hispaniola[1]	77,914	1281	35	5000	1800	36
Jamaica[1]	10,991	1150	4	3247	735	23
Puerto Rico[1]	8,897	885	2	2809	332	12
Galapagos[3]	7,900	250	7	701	175	25
Hawaii[4]	16,600	253	31	970	883	91
New Zealand[3]	268,000	393	39	1996	1618	81
New Caledonia[5]	17,000	787	108	3256	2474	76

[1] Liogier, 1981
[2] Includes invaders
[3] Raven and Axelrod, 1978
[4] Wagner *et al.*, 1985
[5] Morat *et al.*, 1984
From Gentry, 1986

Patterns of Endemism

An endemic species is defined as one whose geographic range is below a specific size or confined to a specific area. The minimum area used to define endemism varies greatly depending on the purpose of the definition, but may range from 'continential endemics' that are limited to a particular continent, to 'extreme local endemics' that are found on only a single mountaintop. A geographical range of less than 50,000 km^2 has been used as a criterion for 'local endemism' in tropical birds (Terborgh and Winter, 1980) and plants (Gentry, 1986).

The characteristics of centers of endemism, and of the endemic species themselves, should provide insights into conditions that allow the survival of endemic species, if not the origin of those species. The most obvious and best-known centers of endemism are isolated oceanic islands, such as Hawaii (Carlquist, 1974) and New Zealand (Table 11.1). In light of accepted theories about patterns of dispersal, genetic isolation, and speciation, it is not surprising that certain islands have high levels of endemism (Williamson, 1981). However, even within an island, the distribution of endemic species is often restricted to extremely small areas that are unusual in some way. For example, Gentry (1986) reports that of the 22 species of the plant family Bignoniaceae that are found on the island Hispaniola, 19 are known only from single small areas, or a few outcrops of serpentine.

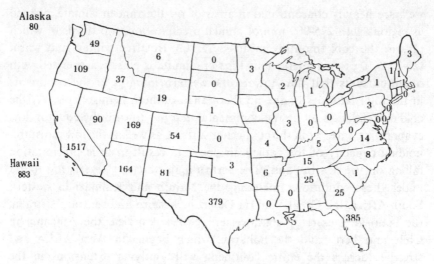

Fig. 11.1 Number of endemic plant species by state in the United States. Species with restricted distributions that span state boundaries are not included. (From Gentry, 1986.)

Although the relative number of endemic species in temperate regions is much lower than in the tropics, there are clear spatial patterns of endemism in North America that show interesting similarities to patterns of endemism in the tropics. Figure 11.1 illustrates the non-random distribution of endemic plants among the states of the United States. On the continent of North America there seems to be a clear distinction between areas of high versus low endemism. Endemism is low in the central and midwestern states, which are areas of relatively low topographic relief, fertile soils, moderate precipitation, and high agricultural productivity. Much of this area has been affected by Pleistocene glaciation, which makes the existence of long-term refuges unlikely.

The highest numbers of endemic plant species are found in Florida, Texas, and California, regions that have in common poor dry soils and/or low precipitation. Topographic relief is high in California, moderate in Texas, and very low in Florida, where the endemics are concentrated on the sandy soils of flat northern and central Florida, rather than in the subtropical southern end. Thus, while topographic relief can potentially provide barriers that prevent geneflow and allow speciation, it is apparently not essential for the survival of large numbers of endemic species.

Endemic plant species in temperate North America, and Europe as well, are heavily concentrated in areas of mediterranean climate, defined as regions with 275-900 mm of annual precipitation, two thirds of which falls in the cool (monthly mean < 15 °C), frostfree winter, and warm (20-25 °C), dry summers. This type of climate is closely associated with ocean currents that flow toward the western margins of the continents in the high latitudes of the northern and southern hemispheres. As the cold currents flow along the continental margins toward the equator, low evaporation rates from the cold water surface, as well as the low moisture-holding capacity of the cooled air above it, result in little precipitation falling on the nearby land. The earth's major coastal deserts are found under these conditions, including the Namib and Kalihari in western South Africa, the Great Victoria Desert in western and central Australia, the Sonoran Desert of southwestern North America, the Atacama in Chile and Peru, and the Saharan, which begins in West Africa and stretches across the entire continent, with outlying extensions in the mideast and central Asia. The poleward coastal margins of these desert regions, some of which receive more moisture during relatively warm, wet winters, support the unique vegetation of the mediterrean climate regions. The climatic similarities of mediterranean zones around the world has apparently led to the convergent evolution of vegetation that is similar in appearance and adaptation on each of the five continents, although only distantly related (e.g., Mooney, 1977; Mooney and Dunn, 1970; Cody and Mooney, 1978; see also Barbour and Minnich, 1990).

Mediterranean climate regions around the world are characterized by both high species diversity and high levels of endemism, reaching a maximum in the South African Cape fynbos vegetation (Goldblatt, 1978; Oliver *et al.*, 1983; Cody, 1986). Analyses of the California flora emphasize the importance of edaphic specialization, such as for serpentine soils, (Raven and Axelrod, 1978) and edaphic and climatic variability (Stebbins and Major, 1965) for promoting speciation among plants in regions of mediterranean climate.

Cody (1986) describes patterns of diversity and endemism among birds and plants of mediterranean regions at the three spatial scales originally discussed by Whittaker (1960): 1. within- habitat (or alpha) diversity; 2. between-habitat (or beta) diversity; and 3. geographical (or gamma) diversity (see Chapter 3). High geographical diversity occurs where ecologically similar, but taxonomically different, species are found in similar habitats that are geographically isolated from one another. Thus high geographical diversity is a direct consequence of high endemism. The more localized the endemism, which is to say, the smaller the geographical distance separating the ecologically similar species, the higher is the geographical diversity.

Among plants in mediterranean regions, all three components of diversity have high values. In particular, the high within-habitat diversity of fire-maintained chaparral vegetation seems to be inversely correlated with soil nutrient availability (Hopper, 1979; Bond, 1983; Milewski, 1983), as predicted by the dynamic equilibrium model. The South African chaparral (called fynbos) and southwestern Australian chaparral (called kwongan or heath) have much higher within-habitat diversity than chaparral on other continents (Kruger and Taylor, 1979). The soils in these two high diversity areas are much poorer than soils underlying this type of vegetation on other continents (Bond, 1983), while topographic heterogeneity, particularly in southwest Australia, is relatively low (Hopper, 1979).

The total diversity of these two high diversity mediterranean climate regions is a consequence of both high between-habitat diversity and high geographical diversity (Cody, 1986). Between-habitat diversity associated with soil type, slope, and slope aspect tends to increase in dry regions because the consequences of differences in soil water-holding capacity and evapotranspiration are exaggerated under conditions of low rainfall (Smith, 1950; Vincent and Thomas, 1961; Pastor and Huston, 1986; Pastor and Post, 1988; Huston *et al.*, 1988). Thus, the high between-habitat diversity in these areas may not be a consequence of species 'dividing up' the habitats more finely, but could simply reflect greater differences between soil water conditions than would be found between similar soils in a wetter climate. However, the issue of high geographical diversity, that is, the tendency for similar habitats to be occupied by different species, is of particular interest because it is also an important component of species diversity in tropical rain forests.

A large proportion of the total endemism, as well as much of the total species richness, in regions of high endemism is often contributed by a few large groups of closely related species. Such species 'swarms' in mediterranean vegetation are found in South Africa (*Protea* with 69 species and *Leucadendron* with 80 species, Rourke, 1980; Williams, 1972), southwestern Australia (*Banksia* with 58 species and *Acacia* with about 250 species, George, 1984; Maslin and Pedley, 1982) and in California (*Ceanothus* and *Arctostaphylos* with 43 species each, Munz, 1968). In the humid neotropics Gentry (1986) identifies genera such as *Anthurium*, *Piper*, *Psychotria*, and *Miconia*, as having high rates of endemism and thus contributing to high geographical diversity. 'Almost every isolated cloud forest in Panama has locally endemic species [of *Anthurium*]. There are 8 *Anthurium* species endemic to Cerro Jefe, 8 to Cerro Pirre, 3 to Cerro Sapo, 4 to Cerro Tacarcuna, and 7 endemic to the Cocle Province forests' (Croat, 1985). Thus the species that contribute to high geographical diversity are likely to be sufficiently similar that either geneflow and hybridization or competitive exclusion could potentially hinder their coexistence.

Factors that Create and Perpetuate Endemism

The occurrence of high levels of endemism is the result of three distinct, and possibly independent, processes: 1. the generation of endemics by speciation; 2. the failure of the endemic species to increase their geographic range and cease to be endemic; and 3. the survival, and possible accumulation over time, of the endemic species (i.e., low extinction rates). Thus, high endemism implies that either: 1. there is currently or has been in the past sufficient genetic separation between similar habitats that speciation can occur and genetic differences be maintained; 2. there is currently sufficient restriction of dispersal between similar habitats that species that occupy one area are unable to spread to other areas and exclude similar species; or 3. both. The significance of both the geneflow limitation hypothesis and the dispersal limitation hypothesis is indicated by the fact that species turnover between similar habitats is generally among species of the same genus, which are presumably most susceptible to hybridization and homogenization due to geneflow, and are also likely to compete strongly under equilibrium conditions.

What factors contribute to similar patterns of high endemism (and high species diversity) found in dry mediterranean chaparral communities and wet tropical rainforest communities? In particular, are there any similarities between the environmental condititions of these two dramatically different types of communities that could explain these similar patterns of endemism? Are physical barriers to gene flow sufficient to allow the persistence of related endemic species, or are there other factors that reduce geneflow and allow coexistence. The following sections address conditions relevant to the geneflow limitation and the dispersal limitation hypotheses outlined above.

Topographic Barriers

With regard to the generation of endemics, some regions of high endemism currently have, or have had in the past, sufficient topographic or environmental barriers to allow for classic allopatric speciation. Certainly, topographic heterogeneity plays a major role in isolating similar species both among the tropical sites discussed by Gentry (1986) and among the Mediterranean sites discussed by Cody (1986) and others. However, a high degree of topographic heterogeneity is also found in many areas of the world that do not have high levels of endemism or high species diversity.

Tropical rivers seem to act much more effectively as barriers to dispersal than do temperate rivers. Distribution patterns for a wide variety of rainforest taxa have boundaries or strong differentiation along rivers (Brown, 1975, 1982; Haffer, 1974; Prance, 1973). Rivers in rainforest regions carry large volumes of water and may be very broad (the Amazon is 4 km across at Manaus, ~1400 km inland). Furthermore, even for flying animals such as birds or butterflies, a river cutting through dense rain forest may pose a much greater barrier than a river cutting through broad open prairies. Although some have proposed that the rivers are actually effective barriers that allow allopatric speciation (Hershkovitz, 1978), an alternative explanation for many of the patterns is that the rivers have simply halted the expansion of populations expanding from other areas, such as the centers of endemism that have been postulated to be Pleistocene refuges (Simpson and Haffer, 1978; see Beven *et al.*, 1984).

Topographic barriers of a variety of types may contribute to the restriction of geneflow and thus allow speciation to occur. Nonetheless, the existence of many regions with high levels of endemism and no apparent topographic barriers demonstrates that such barriers are not essential to the maintenance of high levels of endemism, or perhaps even for the generation of endemism.

Low Productivity

Many areas of high endemism are also characterized by very low productivity, as a result of low levels of soil nutrients and/or soil water. Environmental properties that influence the productivity of vegetation are likely to influence both geneflow and seed dispersal, as well as the intensity of competition. Where growth is limited by resource availability, plants will have fewer resources (specifically mineral nutrients and carbon) to invest in reproduction (Bazzaz and Reekie, 1985; Reekie and Bazzaz, 1987a,b; Mooney and Bartholomew, 1974; Below *et al.*, 1981). Reduced investment in the total amount of pollen, or in the floral structures and other inducements needed to get pollen dispersed by insects or birds, could contribute to reduced geneflow between isolated populations, and thus contribute to a higher rate of speciation. Similarly, a reduction in seed production, or in the energetic inducements, such as sugary or oily arils or pulp, needed for seed dispersal by birds or mammals, could act to reduce both geneflow and the probability of seed dispersal to other suitable habitats.

In spite of the dramatic differences in climate between mediterranean and tropical rainforest areas, low productivity is a common characteristic of regions where high levels of endemism and high species diversity are found in both of these environments. Low soil fertility is associated with the highest diversity sites in the South African fynbos (Bond, 1983). Soils of the high diversity southern Cape fynbos studied by Bond were derived from quartzites and sandstone, and were acidic and low in bases (i.e., Ca, K, Mg). More fertile soils that were derived from shales and conglomerates supported low diversity vegetation types (called renosterveld and karoo). Within the fynbos, species richness of 1 m^2 plots was negatively correlated with soil fertility, as measured by the sum of exchangeable Ca, Mg, Na, and K ($r = -0.56$, $P < 0.05$, $N = 17$). Species richness was lowest on dry sites with low biomass, and in tall, high biomass communities (Fig. 11.2). Both phytovolume (an estimate of biomass) and an index of dominance were positively correlated with soil moisture, which suggests that water may influence productivity, and thus diversity, to a degree equal to or greater than the effect of soil nutrients in these communities. In southwest Australia, the highest species diversity and endemism are found in the heath communities on very poor soils (Hopper, 1979). The low fertility of soils can also influence the seed dispersal mechanisms of plants (Milewski, 1982) and the structure of animal communities (Milewski, 1981, 1983; Scholes, 1990).

Likewise, with a few exceptions such as young volcanic soils and young alluvial soils, the inherent productivity of tropical rainforest soils is quite low. The widespread failure of agricultural settlement projects in the Amazon basin is one well-known consequence of this fact (Fearnside, 1979, 1985, 1987; Goodland, 1980). The low fertility of most tropical rainforest soils, and, in particular, the correlation between high rainfall and low nutrient availability are discussed in Chapter 14. Low productiv-

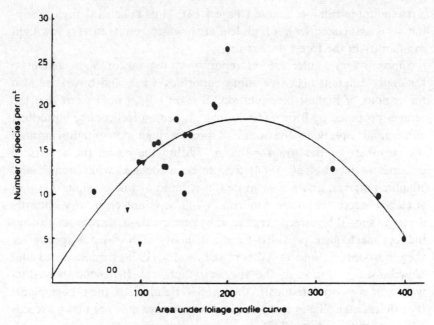

Fig. 11.2 Relationship between species richness and biomass in southern Cape fynbos. Area under the foliage profile curve is measured in arbitrary units and is correlated with total community biomass. (From Bond, 1983.)

ity resulting from low soil fertility may contribute to reduced pollen and seed dispersal, and thus be an indirect cause of high levels of endemism and high geographical diversity.

Low productivity in tropical rainforest understories may also result from reduced light availability, and many of the genera mentioned by Gentry (1982, 1986) as having 'species swarms' are primarily understory shrubs and herbs. Availability of light near the forest floor is often less than 1% of full sunlight (Richards, 1952; Schulz, 1960) which is only sufficient to maintain growth at very low levels of productivity. While individuals of some of these species are able to flower and fruit profusely when they are exposed to higher light levels in or near 'light gaps' caused by treefalls (Levey, 1988a), their reproductive output under typical understory light conditions is generally very low (Levey, 1988b).

Understory light availability is directly related to the total leaf area shading the ground surface. Leaf area index (LAI, defined as the number of units of leaf area above a unit of ground area) is largely determined by the ratio of transpirational demand to soil water availability (Woodward, 1987) and increases with increasing soil water availability, reaching a maximum in rainforests (see Chapter 14). Thus in general, high precipitation is associated with a high leaf area, which results in very low light availability in the forest understory.

Simpson and Haffer (1978) report that regions of high endemism for many different taxa (including butterflies, trees, and birds) are also the regions of highest precipitation. Benson (1982) points out that the refuges proposed by Brown (1975, 1982), based on Heliconiine butterflies, correspond closely to the areas of present high precipitation around the periphery of the Amazon basin. Within the basin, the centers of endemism (Beven *et al.*, 1984) tend to be associated with climatic and edaphic patterns, and are separated by a band of low precipitation that stretches across the lower Amazon. While it is not clear why butterfly diversity should be directly regulated by precipitation, there is an obvious indirect mechanism provided by the diversity of host-plant species for these host-specific insects. Gilbert and Smiley (1978) demonstrated that between-site variation in the species richness of *Heliconius* butterflies was strongly correlated with the species richness of their host plants (Passifloraceae). The number of *Passiflora* species at a site rarely exceeds 10-15 species (Gilbert, 1975).

The key to high levels of endemism seems to be the survival and accumulation of endemic species, which may be allowed by the low rates of competitive displacement found in low productivity environments. All of these systems can be considered non-equilibrium in the sense that competitive equilibrium is prevented by disturbances such as the fires in chaparral and fynbos, frequent droughts and fluctuating water availability on shallow or sandy soils, or simply by very slow growth and low plant biomass that reduce competition for light.

There is good reason to believe that the low productivity in these regions of high endemism restricts geneflow by pollen and seed dispersal to the degree that relatively short distances or small topographic barriers are sufficient to allow what appears to be sympatric speciation. The failure of endemics to spread widely may also be a consequence of restriction in dispersal ability under low productivity conditions. In general, weedy or early-successional species adapted to exploit high resource levels have broad ranges, in contrast to species adapted to low resource conditions (except perhaps in the circumpolar Arctic tundra).

Geographical Range, Endemism, and Species Diversity

The issue of endemism is closely related to the topic of geographical range size, since by definition, endemic species have small geographic range sizes. Correlated with the increase in endemism in the tropics is a temperate-tropical gradient of decreasing latitudinal range in the distribution of species of many different taxa (Rapoport, 1975, 1982; Stevens, 1989).

The greatly reduced annual range of temperature conditions experienced by tropical organisms, as compared with temperate organisms, has been proposed as an explanation for the relatively small geographical range of many tropical species (Stevens, 1989). This idea was developed by Janzen (1967) as an explanation of why mountains represent a greater barrier to the dispersal of organisms in the tropics than in the temperate zone. The basic argument is that for a tropical organism that experiences only a narrow range of temperatures throughout the year, the temperature conditions at other locations on an elevational temperature gradient represent conditions for which it is not adapted, and thus should pose a significant barrier. In contrast, for a temperate zone organism that experiences temperatures ranging from very warm to below freezing during the course of a year, the temperature conditions at any location on an elevational gradient are not greatly different from what the organism experiences at some time during the year.

This phenomenon is undoubtedly part of the explanation for the relatively small geographic ranges of many tropical organisms, particularly in mountanous regions. However, there are other aspects of this issue that must be considered, particularly because of the potential confusion between cause and effect with regard to species diversity and geographical range size. Whether the restricted geographical range of tropical species is a cause of high species diversity, or whether it is a consequence of the same mechanisms that produce high species diversity, is an issue that must be considered carefully.

Stevens (1989), referring to the phenomenon of reduced latitudinal ranges in the tropics as Rapoport's rule, suggests that the wider latitudinal extent of the geographic range of high-latitude species reflects an evolved broader climatic tolerance in these species. He then goes on to suggest that the higher tropical species diversity can also be explained by this phenomenon. 'If low-latitude species typically have narrower environmental tolerances than high-latitude species, then equal dispersal abilities in the two groups would place more tropical organisms out of their preferred habitat than higher-latitude species out of their preferred habitat' (Stevens, 1989). It is hypothesized that a larger number of 'accidentals' (i.e., species that are poorly suited for the habitat in which they are found) occur in tropical assemblages. The constant input of these accidentals 'artificially inflates species numbers and inhibits competitive exclusion', a phenomenon that has been called the 'mass effect' (Shmida and Wilson, 1985) or the 'rescue effect' (Brown and Kodric-Brown, 1977).

There is no doubt that the geographical ranges of tropical species tend to be smaller than those of temperate-zone species, not just in latitudinal extent, but in total area as well. However, there is no evidence that the dispersal ability of tropical organisms is equal to that of temperate organisms. Wind transport of seeds (as well as pollen) is an effective means of long distance dispersal for many plant species, particularly in the temperate zone. However, the proportion of wind-pollinated and wind-dispersed species decreases toward the tropics, for reasons that may not be as obvious as they first appear (Regal, 1982).

The majority of rainforest plant species are pollinated by insects, bats, or birds (Bawa, 1974, 1990; Frankie *et al.*, 1974; Stiles, 1975; Kubitzki, 1985; Baker *et al.*, 1983), rather than by wind. Likewise, wind dispersal of seeds is relatively unimportant in tropical forests, and decreases in importance from dry forests to wet forests (Gentry, 1982), while dispersal by birds and mammals increases along the same gradient. Among plant growth forms, the proportion of species with wind dispersal decreases from lianas to trees to shrubs, reflecting the height of the plants within the forest canopy, and perhaps the likely availability of wind as a dispersal agent. Seed dispersal by birds or mammals requires a much higher per seed investment of energy (carbon) for the dispersal inducement (e.g., pulp, aril) and for the stored reserves of the seed, as well as a higher investment of mineral nutrients for the seed, than does wind dispersal. Thus, plants dependent on bird or mammal dispersal of seeds can produce a smaller number of seeds with the same total energy investment than can a species with wind-dispersed seeds. Unless animal dispersal is much more effective than wind dispersal, there will be an inevitable reduction in the ability to invade suitable but distant habitats. The effectiveness of rivers as barriers to bird dispersal in the tropics is an indication of the limited dispersal capabilities of some tropical animals. Increased dependence on animal dispersal of pollen and seeds may decrease geneflow among isolated populations of tropical plants, and thus facilitate speciation.

These arguments, as well as the hypothesized relationship between productivity and dispersal, suggest that the dispersal ability of many

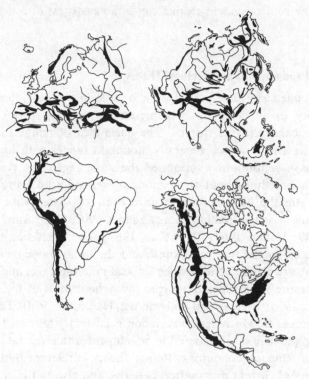

Fig. 11.3 Orientation of geographical barriers in tropical and temperate regions. Note orientation of black mountainous regions is primarily north-south in boreal and temperate zones and primarily east-west in the tropics and subtropics. (From Walter, 1985.)

tropical plants may actually be lower than that of temperate plants. Nor is there evidence that tropical habitats contain more 'accidentals' that are poorly suited to the habitat.

It should also be considered that the differences in the latitudinal ranges of tropical versus temperate species described by Rapoport (1975, 1982) may be at least partially a consequence of differences in the number and orientation of major geographic barriers. High latitudes have been planed by glaciation from north to south, reducing or eliminating east-west barriers that could restrict the latitudinal (north-south) range of species. In the boreal and temperate regions of both North America and Eurasia, the major geographical barriers (Rocky Mountains, Cascade Mountains, Appalachian Mountains, Ural Mountains, Mississippi River) tend to run in a north-south direction, with the greatest variability in an east-west direction at their southern end. In contrast, many geographical barriers in the subtropics and tropics run in an east-west direction (Fig. 11.3).

Endemism, Productivity, and Refuge Theory

Patterns of endemism have played a major role in the development of 'refuge theory' as an explanation of present patterns of species distributions, particularly in the tropics. The basic idea of refuge theory is that certain restricted areas, generally mountain ranges with high precipitation, have continuously supported the same vegetation type (and associated fauna) throughout the periods of generally unfavorable (to rain forest) climatic conditions that have been associated with glacial cycles over the past two million years (Adams, 1902, 1905; Rand, 1948; Gentilli, 1949; Haffer, 1969, 1982; Salo, 1987). The vegetation type of primary interest in the tropics is forest, and the unfavorable conditions were the cool, dry glacial periods when tropical rain forest was apparently replaced by more xeric vegetation types throughout most of the tropics (Livingstone, 1975; Livingstone and van der Hammen, 1978). The relatively small areas to which xeric vegetation is currently restricted within the humid tropics can be considered to be refuges for this vegetation type under present climatic conditions. Refuge theory is relevant both to an understanding of current distribution patterns, and also to the evolution of tropical species as a consequence of isolation and subsequent allopatric speciation (Nelson and Rosen, 1981; Nelson and Platnick, 1981).

Brown (1982) presents a detailed analysis of the distribution of neotropical butterflies (Heliconiini and Ithomiinae) and concludes that the centers of species diversity and endemism are generally separate in these groups. Species diversity is highest near the edges of endemic centers, as defined by Brown, and apparently reflects the dispersal and mixing of species from adjacent centers of endemism. According to Brown, areas of high butterfly species diversity (and low endemism) can be characterized as 'dense tropical forest in a climate characterized by heavy rainfall all year round, growing on soils which are sandy, plinthic, or coarsely alluvial, or which include stone lines [which reflect a drier climate in the past (Ab'Saber, 1982)].' In contrast, areas of high endemism and relatively low species diversity can be characterized as 'open or semideciduous forests in a strongly seasonal or low-rainfall climate, growing on rich, fine-textured forest soils'. Because 'regional endemic patterns do not correlate well with present climate or vegetation', Brown attributes high endemism to historical factors, specifically the existence of refuges with more favorable climatic conditions in the past. Benson (1982) presents an alternative explanation of the same patterns based on current climatic conditions (discussed below).

The existence of centers of endemism and/or species diversity within the tropics is well established (see Prance, 1982; Beven *et al.*, 1984). However, the validity of the refuge theory as an explanation for patterns of endemism and species diversity in the tropics has been strongly questioned (Benson, 1982; Endler, 1977, 1982a,b; Colinvaux, 1987; Wiley, 1988; Bush and Colinvaux, 1990). The arguments presented above for increased genetic isolation under low productivity conditions suggest that speciation can occur at a high rate at some sites under present climatic conditions. Thus the high levels of endemism and high species diversity (particularly geographical diversity) do not require the historical existence of refuges, but may be explained by current processes.

In summary, low productivity conditions resulting from low precipitation, low levels of mineral nutrients, or low light levels (associated with high precipitation), are associated with high levels of plant endemism (and often high species diversity) in North America as a whole, the mediterranean zones around the globe, and in tropical forests. In some, but not all, situations, high levels of topographic heterogeneity that result in barriers to dispersal are also associated with endemism and species diversity.

Evidence on the global distribution of plant species suggests high speciation rates and low extinction rates both in rain forests (Stebbins, 1974) and in marginal environments, such as arid regions (Axelrod, 1966, 1967). Evidence from the fossil record of marine organisms and mammals indicates that rates of speciation and the proportion of endemic species have been higher in tropical than in temperate and polar regions throughout much of geological history (Stehli *et al.*, 1969). Surprisingly high levels of species diversity and endemism have been reported from the deep-sea environment with very low productivity, yet few barriers to dispersal (Rex, 1981, 1983; Wilson and Hessler, 1987).

Thus, the same environmental conditions that the dynamic equilibrium hypothesis predicts should reduce the rate of competitive exclusion and allow the maintenance of high species diversity at ecological time scales may also lead to increased rates of speciation and the prolonged coexistence of these many species as a result of reduced geneflow and dispersal and lower extinction rates.

What does endemism have to do with biological invasions? Almost by definition, endemic species are not successful invaders or weeds, so the environmental conditions in which these two groups of organisms occur might be expected to be non-overlapping. However, this is not always the case. Under several different sets of conditions, the issue of endemism is closely tied to the issue of biological invasions.

Biological Invasions

Invasions are a topic of considerable concern (see Groves and Burdon, 1986; Mooney and Drake, 1986; Drake et al., 1989) because of the dramatic negative impacts that invading organisms have had on other species and even entire ecosystems. Such impacts include the extinction during the past few hundred years of thousands of species of mammals, birds, amphibians, and reptiles (most of them endemic species) on oceanic islands as a result of the invasion of rats, pigs, dogs, cats, ferrets, and other domestic and wild predators and their diseases (with transportation provided by oceangoing ships). Several plant species in North America have been nearly eliminated by invasion of pathogens from other continents, most notably the American Chestnut (*Castanea dentata*) and the American Elm (*Ulmus americana*). Even the hydrology and nutrient cycles of entire ecosystems have been changed by invading species (Vitousek, 1984).

With regard to biological diversity, invasions potentially lead to an increase in species richness, as the invading species are added to the existing species pool. However, the examples cited above demonstrate that invasions can also lead to extinctions, resulting in a decrease in species richness. Invasions usually have a negative effect on organisms of functional types other than that of the invader. Rapid extinctions result from interactions such as predator-prey or host-parasite. Invasions are less likely to result in extinction of organisms of the same functional type as the invader. The negative interactions in these cases will be primarily competition, which may allow coexistence or lead to extinction more gradually than predation.

While invasions, or the expansion of species' ranges, have occurred naturally throughout the Earth's history, invasions are currently occurring at an unprecedented rate as a direct result of human activities. The rate at which invasions are occurring has probably been increasing exponentially over the course of human history, and is now undoubtedly many orders of magnitude greater than it was before people developed the capability to alter the landscape through agriculture, forestry, and animal husbandry, and the capability and economic incentive to transport materials (and intentionally or inadvertantly, organisms) for long distances. Thus, while invasions are a natural phenomenon and operate according to the general principles that structure all ecosystems, the current rate of invasions is strictly a human phenomenom, and is perhaps the most destructive effect humans have had on natural ecosystems.

Understanding the ecology of invasions requires addressing three critical issues: 1. the rates and mechanisms of transport or movement of organisms; 2. the characteristics of organisms that allow them to be successful invaders; and 3. the properties of ecosystems that make them susceptible or resistant to invasions.

Transport of Invading Species

The so-called 'natural' invasion processes occur at very low rates that are determined by, among other factors, the well-known principles that are the basis of the 'theory of island biogeography' (MacArthur and Wilson, 1967), namely, inverse proportionality to the square of the distance between the source and the object of the invasion, and direct proportionality to the size of the source area and the cross-section of the object area perpendicular to a direct line from the source. The obvious consequences of these constraints are that small, distant islands receive many fewer potential invaders than small islands that are close to the source of invaders (generally a continent or very large island), and large islands at a given distance from the source receive more potential invaders than small islands at the same distance.

All other things being equal (cf. Gilbert, 1981; Williamson, 1989), the consequences of these processes are lower species diversity on small or distant islands than on large islands or islands that are close to a source of invading species. The level of species diversity on an island must to some extent represent a *dynamic equilibrium* between the rate of extinction on the island, which increases as population sizes (presumably related directly to island size) decrease, and the rate of invasion, which is influenced by the processes described above. At evolutionary time scales, the appearance of new species is quite high on islands, as a result of adaptation to different environmental conditions and diversification to utilize resources more efficiently. Thus, islands tend to have many endemic species, and also provide the best examples of evolutionary processes by which a single type of organism diversifies to fill many different niches, e.g., Darwin's finches on the Galapagos (Darwin, 1859;

Lack, 1947; Grant, 1975). The number of endemic species tends to be greatest on islands that are 1. sufficiently remote that mainland species adapted to the island conditions and resources are unlikely to arrive, and 2. sufficiently large that the newly evolved species can maintain large enough populations to avoid extinction due to random fluctuations in environmental conditions.

The patterns of species distributions and endemism that result from variation in these low natural rates of invasion have a major influence on the susceptibility of different ecosystems to invasion as well as on the invasive potential of different species. The basic patterns of invasions and invasibility are: 1. the high susceptibility of island communities to invasion by mainland species, with subsequent extinction of island endemics and replacement by mainland species (Elton, 1958); 2. the dominance of the earth's largest landmass, Eurasia, as the primary source of successful invaders throughout the rest of the world (Crosby, 1986); and 3. the relatively low proportion of successful invaders in Eurasia in comparison with the rest of the world (di Castri, 1989a). The mechanisms underlying these patterns, as well as many of their implications, will be addressed in the following sections of this chapter. These basic patterns are introduced here because these consequences of natural invasion processes have set the stage for the ecological cataclysm that has resulted from the recent exponential increase in invasion rates caused by human activities.

The great increase in human-caused biological invasions is indisputably centered in Eurasia, and particularly Europe, which complicates the evaluation of the relative invasiveness of Eurasian organisms and the invasibilty of Eurasian ecosystems in comparision with the rest of the world. The details and consequences of the human history of increasing populations, migrations and wars, explorations and trade, and the improving technologics of transportation and agriculture, have been thoroughly described by historians, geographers, and ecologists (e.g., Braudel, 1979a,b,c; Crosby, 1986; di Castri, 1989a,b; Sykora, 1989) (Table 11.2). It is sufficient to note that the transport of organisms, and their introduction into regions where they did not previously occur, has tremendously increased as a result of human activities, and continues to increase exponentially at present.

Properties of Organisms in Relation to Invasions

Arrival is not equivalent to invasion. While the transport and introduction of a species into an area where it has not previously occurred

Table 11.2. *Human-history driving forces in the Old World as related to biological invasions (from di Castri, 1989a)*

Before 1500 AD	After 1500 AD	From last century in a worldwide perspective
Forest clearing	Exploration, discovery and early colonization by Europeans of other territories and continents	Improvement of transportation systems (roads, railways, internal navigation canals)
Primaeval agriculture		
Sheep and cattle-raising		Large engineering works for irrigation and hydropower
Migrations and nomadism	Establishment of new market economies and crossroads places (e.g. Amsterdam, London) favouring the 'globalization' of trade exchanges	
Inshore coastal traffic		Opening of inter-oceanic canals (e.g. Suez, Panama, Volga-Don)
Settlement of islands (e.g. Corsica)		
Intensification of agriculture by ploughing	Large 'colonies' under the rule of Europeans, often entailing introduction of European-like agriculture and increasing *inter alia* intertropical exchanges	Aircraft transportation
		World wars and displacement of human populations
Offshore traffic and trade		
Coastal 'colonies' (e.g. Phoenician and Greek colonies)	'Revolution' of food customs in wealthy Europe (e.g. increased use of tea, coffee, chocolate, rice, sugar, potatoes, maize, beef and lamb)	'Decolonization'; international aid to newly independent countries following 'western' patterns
Building up of large empires (e.g. Persian, Roman, Arab, Mogul) with considerable expansion of communication and transportation systems		Emergence of multinational companies
	Increased demand in Europe of products such as cotton, tobacco, wool, etc.	Tropical deforestation and resettlement schemes
	Negro slavery; Indian and Chinese migrations	Afforestation of arid lands with exotic species
Long-ranging wars and military expansion	Missionary establishments	Environmental impacts decreasing ecosystems' resilience
	Occupation by Russians of northern and part of central Asia, up to Siberia	Increased urbanization and creation of ruderal habitats
Invasions of German and Asian people, mainly from east to west	Intentional introduction into the Old World of exotic species through activities of acclimatization societies, botanical gardens and zoos, and for agricultural, forestry, fishery or ornamental purposes	International interdependence of markets
Long-distance shipping trade		
Establishment of 'market economies' (e.g. Venice) covering the 'known world' up to the Far East		Release of genetically engineered organisms
	Large-scale emigration from the Old World due to persecution during religious conflicts, civil and 'independence' wars, and to increased demography, unemployment and famine	

is the first step of a successful invasion, it is not necessarily the most critical step. Of 100 plant species that are introduced into a new area, di Castri (1989a) estimates that only ten actually colonize the area, and of these ten only five are able to establish in existing plant communities and become naturalized, and of these five only two or three are actually able to spread beyond the site of introduction.

Why are some species successful invaders on all continents, while others are endemic species threatened with extinction? Is success as an invader merely a consequence of the 'good fortune' to be transported by humans in sufficient numbers to become established, or are there biological properties that distinguish good invaders from poor invaders? What are the relationships, if any, between the biological properties of an invading species and the properties of the ecosystem that it invades?

Although controlled, replicated experiments on plant or animal invasions are virtually impossible to perform, as well as subject to strong ethical limitations, the history of invasions as revealed through paleontology, paleoecology, biogeography, and recent ecological studies must contain clues about the answers to these questions. An important caveat in interpreting patterns of invasions is that the vast majority of examples we have are of successful invasions (Simberloff, 1986). We rarely know when a potential invasion has failed, let alone why it failed. Nonetheless, the patterns that emerge from the many invasions that have occurred may provide some insights into the biological processes underlying successful invasions.

The Relationship between Land Area and Competitive Ability

As mentioned earlier, one robust pattern in the biogeography of invasions is the dominance of Eurasian species, of many different taxa, as successful invaders around the world (Crosby, 1986; di Castri, 1989a; Simberloff, 1989). At first glance, it seems difficult to separate the biological properties of European species from the fact that they have had far greater opportunities to disperse throughout the world because of the pre-eminence of Europe as a center of exploration and trade. However, trade and transport are two-way processes, with the return of tradegoods, raw materials, and agricultural products probably containing as many potential invaders from other regions as have been dispersed out of Europe. Notwithstanding this presumed influx of potential invaders (unfortunately for research, failed invasions leave no trace), the proportion of alien species in the European and Mediterranean floras is lower than in most other regions of the world (Allan, 1937; Frenkel, 1970; Raven, 1977) and the same is true for insects and other groups of animals (Greathead, 1976; Crosby, 1986). The success of European organisms at invading other communities while resisting the invasion of their own communities has been attributed to 'the more efficient work-shops of the north' (Matthew, 1915), and the fact that continental species in general are 'steeled by competition' in comparison with island species (Carlquist, 1965).

The interpretation that Eurasian organisms tend to be better com-petitors than organisms from other continents is consistent with both prehistoric and historical patterns of extinction and invasion. The well-known phenomena of island species being poor invaders of mainland environments and also prone to extinction caused by species invading from the mainland (Wilson, 1961, 1965; Carlquist, 1965; Brown and Gibson, 1983), may be simply one extreme of a general continuum of competitive ability among organisms that extends from the smallest, most remote islands to the largest continental land mass.

The evolutionary explanations for the poor competitive ability of island species have been thoroughly expounded (Elton, 1958; Carlquist, 1965; Greathead, 1971), but rarely applied to the complete size range of land masses. The argument is based on the assumption that competitive ability is based on the efficiency with which critical resources can be obtained from the environment. Competition among the many species found on mainlands presumably selects for those that are specialized to be highly efficient at obtaining a particular resource. This is the general phenomenon of 'niche partitioning' that is clearly evidenced among animals, and less clearly among plants. Because islands have fewer (potentially competing) species than mainlands, and because islands generally have a smaller variety and total amount of resources (e.g., plant seeds, insects) than do mainlands, island species may tend to be generalized to use a broader range of resources than mainland species (e.g., Schluter, 1988) and thus are unlikely to have the high efficiency that results from specialization on particular resources.

Even when island species evolve specialized adaptations for partic-ular resources, the lack of other specialized competitors for the same resources allows them to survive with adaptations that may be inefficient in comparison with species that have evolved with strong competitors to use the same resources on mainlands. For example, the specialization of *Camarhynchus pallidus*, the Galapagos finch that obtains woodboring in-sects by probing with a thorn that it breaks from a plant (Darwin, 1857;

Lack, 1947), is undoubtedly much less efficient than a mainland wood-pecker species adapted to use the same resource by means of a strong, thin bill and a long, barbed tongue. While the adaptations evolved by island species that allow them to perform activities for which their an-cestral species were totally unsuited are remarkable and have provided critical insights into evolutionary processes, these island species are rarely able to survive competition from species that have evolved functionally similar abilities on larger land masses.

In contrast to the situation on islands, natural selection that results from competition or predation on mainland species is likely to be intense on short time scales, as a result of competition from co-occurring species, as well as on longer time scales, as a result of the invasion of potential competitors from other parts of the mainland. On large mainland areas, the number of potential competitors that can disperse into any area is greater than the number on small mainland areas. Long-term geological and climatic events such as orogenies and glacial advances provide both the opportunity for speciation, and the opportunity for dispersal and invasion. Thus, over evolutionary, as well as ecological time, mainland organisms are likely to experience more intense natural selection from interactions with other species than are island organisms, and the same is true of organisms on large continents in comparison with smaller continents.

The low 'ecological resistance' to biological invasion of islands in com-parison to mainlands (Elton, 1958) is reflected in the relative proportion of insect species that have invaded successfully on islands in relation to mainlands. In Hawaii, 29% of the total insect fauna is composed of introduced species, and on the tiny island of Tristan da Cunha the proportion is 28% (Simberloff, 1986; Sailer, 1978; Holdgate, 1960). In contrast, for North America the proportion of introduced insects is only 1.7% (Sailer, 1978). The apparently lower resistance of island faunas to invasion has been attibuted to both weaker competitive ability and lower resistance to predators and pathogens (cf. Carlquist, 1965; Elton, 1958; Simberloff, 1986).

The history of mammalian evolution, dispersal, and extinction is filled with examples of extinctions caused by mammals that evolved on large continental regions invading and causing mass extinction in smaller continental areas. The massive area of the Old World is subdivided only by natural barriers, such as deserts and mountain ranges, which change dramatically in their penetrability as climatic conditions fluctuate. North America is smaller and more isolated, but has been periodically connected to Eurasia across the Bering Land Bridge, with dramatic effects on the fauna (Simpson, 1947; Cracraft, 1974), perhaps the greatest of which was the invasion of humans.

In contrast, South America has been totally isolated from the rest of the world for most of geological history. The relatively recent connection of South to North America, with the invasion of placental mammals from the north causing the extinction of most of the marsupial fauna of South America, is one of the most dramatic mass extinctions that has resulted from purely ecological processes (Simpson, 1950, 1965; Keast *et al.*, 1972; Cracraft, 1974; Lillegraven, 1974; Webb, 1977, 1978; Marshall *et al.*, 1982). While Australia has not been physically connected to any other land mass since the radiation of the placental mammals, the diverse marsupial fauna of Australia is undergoing a wave of extinctions as a result of the human introduction of placental mammals, as well as the activities of humans themselves (MacDonald *et al.*, 1989; Myers, 1986). Thus, on a global scale, the more intense natural selection resulting from biotic interactions on large land masses in comparison with smaller land masses, has consistently resulted in the displacement or extinction of faunas that are invaded by animals from larger land masses.

The variation in competitive ability and predator avoidance between organisms on continents of different sizes is not correlated with local species diversity, since the local diversity of Eurasian communities is not particularly high in relation to communities elsewhere in the world. Likewise, the high diversity environments around the world, such as tropical rain forests and mediterranean shrublands, are not major sources of successful weeds or other invading species. Nor is the northern temperate climate the key to evolving successful invaders, since North American species have not been as successful as Eurasian species from similar climate regions. Finally, mere proximity to Eurasia is not a guarantee of resistance to invasion, since the mammalian fauna of Corsica has undergone a wave of extinctions similar to that experienced on other islands around the world (Vigne, 1983).

In spite of the general resistance to invasion of Eurasian plant and animal communities, there have been some notably successful invasions, including prickly pear cactus (*Opuntia*), the nitrogen-fixing legume black locust (*Robinia pseudoacacia*) and the fall web worm (*Hyphantria cunea*) from North America (Marcuzzi, 1989). Within Eurasia, there have been many notable invasions of Europe from the Far East and Middle East, including the black rat (*Rattus rattus*) and the house mouse (*Mus musculus*) (Cheylan *et al.*, 1989). With regard to both plants and animals, the communities that are successfully invaded in Europe are not a random selection, but represent specific environmental conditions, which will be discussed in detail in the next section on the invasibility of ecosystems.

Just as the types of community that are invaded are a non-random subset of all communities, the types of organism that are most successful as invaders are a non-random subset of all organisms. Not surprisingly, the vast majority of invading plant species are 'weeds' (an ill-defined perjorative) associated with agriculture, either cultivation or grazing. These plants not only have the opportunity to be transported with agricultural products, but have the even greater advantage of a long period of evolution under the selective pressure of human agricultural activity in the Mediterranean region and Europe. Some weeds have even become specialized as mimics of specific crops (King, 1966; Baker, 1974). Not all invading species are weeds and not all weeds are invaders, since many species become agricultural problems within their original range. Nonetheless, there is a strong correspondence between the traits that contribute to 'weediness' and those of the majority of successful invaders. Baker (1965, 1974) compiled a list of the physiological and life history characteristics of the 'ideal' weed (which fortunately does not exist) (Table 11.3). These characteristics are those that confer competitive success in disturbed environments, and are the kinds of traits that are favored by 'r-selection'.

Biological and Taxonomic Attributes of Invading Species

Among the approximately 8000 plant species that can be classified as weeds, as listed in *A Geographical Atlas of World Weeds* (Holm *et al.*, 1979), the majority are derived from a relatively few groups of plants. The two plant families with the most weed species are the Compositae (also called Asteraceae) with 830 species in 224 genera and the grasses (Gramineae or Poaceae) with 753 species in 166 genera. The legumes (Papilionaceae) rank third with 415 species in 87 genera, and 11 additional families have over 100 species each: the spurges (Euphorbiaceae), mints (Lamiaceae), mustards (Brassicaceae), morning glories (Convolvulaceae), sedges (Cyperaceae), nightshades (Solanaceae), parsley (Apiaceae), roses (Rosaceae), figworts (Scrophulariaceae), buckwheat (Polygonaceae), and mallows (Malvaceae). More than half of the remaining families have ten or fewer weed species.

Clearly, certain groups of plants have traits that predispose them to becoming weeds. These common traits include a herbaceous growth

Table 11.3. *The characteristics of an ideal (?) weed.*

1. Has no special environmental requirements for germination
2. Has discontinuous germination (self-controlled) and great longevity of seed
3. Shows rapid seedling growth
4. Spends only a short period of time in the vegetative condition before beginning to flower
5. Maintains a continuous seed production for as long as growing conditions permit
6. Is self-compatible, but not obligatorily self-pollinated or apomictic
7. When cross-pollinated, this can be achieved by a non-specialised flower visitor or wind
8. Has very high seed output in favorable environmental conditions
9. Can produce some seed in a very wide range of environmental conditions. Has high tolerance of (and often plasticity in face of) climatic and edaphic variation
10. Has special adaptations for both long-distance and short-distance dispersal
11. If a perennial, has vigorous vegetative reproduction
12. If a perennial, has brittleness at the lower nodes or of the rhizomes or rootstocks
13. If a perennial, shows an ability to regenerate from severed portions of the root-stock
14. Has ability to compete by special means: rosette formation, choking growth, exocrine production (but no fouling of soil for itself), etc.

From Baker (1965).

form, rapid growth and early reproduction (often annual or biennial), seeds that are widely dispersed by wind or animals, and the ability to self-fertilize or reproduce asexually (Heywood, 1989; Baker, 1974). Among trees that have become widespread weeds, the same general traits are found, with modification for a woody and relatively long-lived growth form.

Climate and other conditions obviously impose some constraints on what environments a particular species can potentially invade. However, close climatic similarity between the area of origin of an invading species and the ecosystem that is invaded is not a requirement for successful invasion. Apparently successful invaders can be greatly reduced by rare climatic extremes in their new environment that do not occur in their original environment (Quetzal *et al.*, 1989). Successful invaders of the mediterranean climate region of Europe, are generally not from other mediterranean climate regions around the world (Kruger *et al.*, 1989; Quetzal *et al.*, 1989).

The properties of successful invading animal species are more difficult to generalize, and the probability of success of any animal introduction is more difficult to predict (Sharples, 1983; Simberloff, 1986). The intentional introduction of insects or pathogens to control agricultural pests has provided a large number of 'test cases' for theories of invasion, although the details of most of the releases, particularly those that failed, are poorly documented (Clausen, 1978; DeBach, 1974; Laing and Hamai, 1976; Sailer, 1983; Simberloff, 1986, 1989).

The most successful intentional introductions, as well as the most dev-astating accidental introductions, have been of herbivorous insects or plant pathogens. Examples include the chestnut blight (*Cryphonectria parasitica*), white pine blister rust (*Cronartium ribicola*), Dutch elm dis-ease (*Ophiostoma ulmi*), and the pine wood nematode (*Bursaphelenchus xylophilus*) (von Broembsen, 1989). All of these diseases severely dam-aged or in some cases eliminated plant species of the same genus as the pathogen's host in the region where it is indigenous. Insect invaders that have caused major problems, such as the gypsy moth (*Lymantria dispar*) and Japanese beetle (*Popillia japonica*) in North America and the Colorado potato beetle (*Leptinotarsa decemlineata*) and fall web worm (*Hyphantria cunea*) in Europe, are all herbivores with sufficiently gener-alized host preferences that their spread is not limited by the availability of suitable hosts. Successful insect introductions for biological control purposes include *Cactoblastis cactorum*, the moth that successfully con-trolled the introduced prickly pear cactus in Australia (Dodd, 1959), and the beetle *Chrysolina hyperica*, which successfully controlled St. John's wort in western North America (Clausen, 1978; Groves, 1989). The high success of herbivorous insects as invaders is reflected in the fact two insect orders that are exclusively (*Homoptera*) and predominantly (*Thysanoptea*) herbivorous are ten times more abundant among the suc-cessful invaders of North America than their global abundance would predict (Sailer, 1983; Simberloff, 1986).

In general, successful animal invaders tend to be diet generalists, (Ehrlich, 1986) or, if specialists, to have extremely abundant and wide-spread hosts. The successful invasion and establishment of highly spe-cialized species, such as many of the insects introduced for biological control of specific pests, is much more problematical. The failure of introductions for biological control has been attributed to pathogens, predators, and competitive exclusion of later introductions by earlier introductions (Tallamy, 1983; Hall and Ehler, 1979; Ehler and Hall, 1982). Arguments against competition as an explanation for the failure of later introductions include the observation that the first species to be introduced are selected to have a high probability of success and thus later introductions are more likely to fail (Washburn, 1984; Keller, 1984; Dahlsten, 1986). Crawley (1986) found that predators, parasitoids, and diseases were responsible for 41% of the examples of reduced effectiveness of insects introduced for weed control, while climate had a major negative effect in 44% of the cases.

Successful invaders that do not have 'weedy' traits are generally found invading environments very different from those in which weeds are successful (Burdon and Chilvers, 1977; Specht and Moll, 1983; Heywood, 1989; Kruger *et al.*, 1989; Ewel, 1986). Many of the non-weedy invading plant species have traits that are not found in plants of the community that they are invading, such as the ability to fix atmospheric nitrogen through rhizobial symbioses (Allan, 1936; Egler, 1942; Vitousek, 1986). Such invaders represent different plant functional types, at least with regard to some functions, than the plants of the invaded community. Given the strong influence of environmental conditions on species diversity and community structure, it is not surprising that the environment has a strong influence on which types of organisms are likely to invade, as well as whether any invasion is likely to succeed.

Ecosystem Properties Related to Invasibility

Not all types of ecosystem are equally susceptible to being invaded. Part of the resistance of some ecosystems to invasion is clearly a function of the competitive ability and other ecological properties of the species already present. However, in addition to the species attributes that influence invasibility, there seem to be certain combinations of environmental conditions that make some systems particularly susceptible to invasion. Undoubtedly, these environmental conditions interact with the properties of the species already present in the environment and the invading species to determine whether any particular invasion will succeed.

The single most important factor influencing the invasibility of a community is the degree to which it has been disturbed (Elton, 1958), specifically the frequency and intensity of mortality-causing disturbances (i.e., disturbance *sensu* Grime, 1979; Huston, 1979). The disturbance or complete destruction of natural ecosystems is an inseparable component of the same human agricultural activities that have been responsible for the pan-global distribution of weeds and pests. The idea that disturbances play an important role in invasions has a long history (Elton, 1958) although most recent discussions of this idea seem to have been published in 1986 (e.g., Orians, 1986; Fox and Fox, 1986; Crawley, 1986; Kruger *et* al., 1986; McDonald et al., 1986). However, there are numerous situations in which invasion occurs in the absence of disturbance (Mack, 1985; Burdon and Chilvers, 1977; Kruger, 1977). Thus, beyond Elton's original observation about the importance of disturbance for some invasions, there has been little progress in generalizations about which ecosystems are most likely to be invaded (e.g., Orians, 1986; Hobbs, 1989).

Aside from the fact that we identify invading species as being indigenous to some region other than the area being invaded, there is essentially no biological difference between the process of invasion and the process of colonization or recolonization of areas by native plants. Colonization of new substrate or of areas partially or wholly denuded of vegetation by disturbance is a fundamental process of plant succession. The relative abundance of native plants in a community is a result of their success in becoming established and of their ability to compete successfully, or at least grow and reproduce, after they become established.

Just as these processes of population dynamics, inter-specific competition, and other interactions ultimately regulate community structure and species diversity in 'natural' communities, they also regulate the success and failure of invading species. And, just as environmental conditions that affect the dynamic equilibrium between competitive displacement and the effects of disturbance can alter community structure, so can these same environmental conditions influence the invasibility of ecosystems and the likely consequences of the invasions that do occur.

The dynamic equilibrium model makes specific predictions about the types of ecosystem that should be susceptible to invasion. Basically, the same environmental conditions that allow the establishment and growth of native species also allow the establishment and growth of invading species. The key factor that regulates the rate of growth and the 'severity' of invaders is the same factor that determines the maximum potential rate of growth of native species, namely the inherent productivity of the environment, which results from the interaction of available soil nutrients, moisture and temperature. Just as productive environments tend to have high rates of competitive displacement that lead to dominance by a few species and low species diversity, so productive environments are likely to be difficult to invade, but a successful invader will probably become dominant. Of course, the most productive environments have experienced the vast majority of invasions by weedy plants, but these occur only in the presence of the high frequency and intensity of disturbance imposed on the environment by agricultural activities (Baker, 1986; Heywood, 1989).

The fast-growing invaders of productive soils are usually shade-intolerant. Consequently, in the absence of frequent disturbance of productive soils, weedy invaders are usually eliminated rapidly by shading from larger, longer-lived native species (Fig. 11.4). Without additional disturbance, the ecosystem usually proceeds through a successional sequence of native species to a stage dominated by a few species (Groves, 1989). The same dynamic equilibrium of high productivity and low frequency of disturbance that inevitably results in low species diversity among competitors should also be extremely resistant to invasion (cf. Baker, 1986). Only with frequent disturbances can high diversity be maintained in productive environments, as in the case of tall-grass prairies and other fire-maintained grasslands. However, as the frequency and intensity of disturbance increase still further, diversity will be reduced even in productive environments, as in the case of disturbed agricultural areas dominated by annual weeds.

In relatively unproductive environments, competition for light among plants is reduced and the rate of competitive displacement is much lower than in productive environments. These are the situations where the highest diversity among competing plant species is found. Disturbance is relatively less important for the maintenance of species diversity than in productive environments, and may not have much effect on the establishment and growth of either native species or invaders (cf., Hobbs, 1989). In this type of environment, a high frequency or intensity of disturbance can eliminate slowly-growing species and lead to a reduction in diversity.

The predicted invasibility of ecosystems over the range of conditions considered by the dynamic equilibrium model is presented in Fig. 11.5. In Fig. 11.5*a* the three-dimensional (contour) surface represents the probability of successful establishment once an invading species reaches a site. Note that the probability of establishment is highest under the conditions where highest diversity is found. Figure 11.5*b* represents the probable degree of dominance achieved by a successfully established invader. In general, the potential dominance of invading species is inversely correlated to the species diversity of the natural community, just as is the dominance of native species.

The life histories of successful invaders are strongly constrained by the frequency of disturbance. With a high frequency of disturbance, only short-lived plants, or annuals in the most extreme case, can reproduce successfully. However, the enviroment must be sufficiently productive that the plants can complete their life cycle in the intervals between disturbances and produce a sufficient number of propagules to maintain

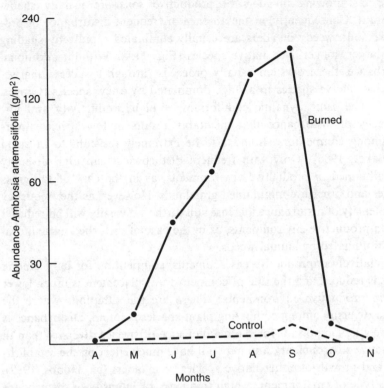

Fig. 11.4 Effect of disturbance (burning) on the abundance of a weedy annual species (*Ambrosia artemesiifolia*) in a four-year-old fallow field on a river flood-plain in Georgia. (From Odum *et al.*, 1974.)

their population. The environments that are invaded by long-lived plants are generally characterized by a low frequency of disturbance and low productivity (Fig. 11.5*b*).

The predictions of the dynamic equilibrium model about the invasibility of ecosystems are consistent with virtually all reported examples of plant invasions. The major contribution of this approach is the clarification of when and why disturbances are likely to be important in the success of invasions. The following discussion briefly reviews the general patterns of plant invasions as they relate to the predictions of the dynamic equilibrium model.

There are few, if any, examples of invading species becoming abundant in undisturbed, productive ecosystems. The only life form of invaders that could dominate in such a situation would be trees, and trees are less likely to be dispersed long distances by man than are herbaceous species

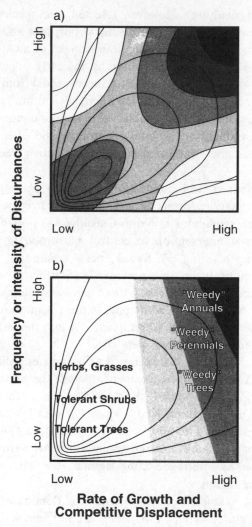

**Rate of Growth and
Competitive Displacement**

Fig. 11.5 Community susceptibility to invasions and the life histories of invaders in relation to the Dynamic Equilibrium Model. (*a*) Susceptibility of communities to invasion. Darker shading indicates higher susceptibility. Note that communities with low diversity, as result of high rates of competitive displacement with low frequencies of disturbance, or of low rates of competitive displacement with high frequencies of disturbance, are least likely to be invaded successfully. The highest probability of successful invasion is in productive environments (high rate of competitive displacement) with very high frequencies of disturbance. (*b*) Predicted life histories of successful invaders under different conditions of dynamic equilibrium. Shading indicates expected dominance of a community by a successful invader. Note that if the invader alters the disturbance regime by increasing frequency or intensity, the community will shift to a dynamic equilibrium with lower diversity and higher dominance by the invader.

associated with agriculture. However, selected tree species are being transported around the world for silvicultural purposes and established in monospecific stands. The successful establishment of most plantations requires substantial and continuous efforts in the early stages to prevent competition from native species (and introduced weeds) from eliminating the plantation species. In the few cases where silviculturally introduced species have successfully become established in natural communities, they form a minor component of the total community and often depend on disturbances to become established (as do most native tree species, as well).

The establishment of white pine (*Pinus strobus*) plantations in the productive hardwood cove forests of the Coweeta Hydrologic Laboratory in the Appalachian mountains of North Carolina was possible only with massive silvicultural intervention to control hardwood growth during the years following planting (W. Swank, pers. commun.). White pine does not occur naturally in this part of the Southern Appalachians, but, once established in plantations, grows vigorously and excludes most other vegetation. Now that the white pine is present in plantations, it is spreading into the 'undisturbed' forest, primarily into the less productive sites on dry ridgetops (M. Huston, pers. obs.).

The examples of invading tree species that become established following disturbance in productive environments include the weeds of the tree world: fast-growing, soft-wooded, short-lived species such as *Ailanthus* and *Paulownia*. The introduced Kudzu vine (*Pueraria lobata*) is capable of overgrowing and killing mature forest in southeastern North America. However, it becomes established and spreads primarily from disturbed areas. In addition, as a nitrogen-fixing legume, it is unique among the vine species of the region.

The only good examples of the invasion of dominant tree species into undisturbed, closed-canopy ecosystems are not actually invasions of exotic species, but rather the recolonization by native species of areas that they probably occupied previously. The northward migration of the dominant tree species of the eastern deciduous forest of North America following glacial retreat has been well documented by palynologists (Wright, 1971, 1972; Watts, 1970, 1973, 1979, 1983; Webb and Bernabo, 1977; Davis 1981; Webb, 1986, 1987; Birks, 1989). These 'invasions' occurred very slowly and some are still taking place. The attributes of these tree species that are becoming established in undisturbed forest ecosystems are virtually the opposite of the attributes of successful weeds. These are the classic 'K-selected' species, with attributes such as large maximum size, slow maximum growth rate, long lifespan, large seeds, and, in many cases, high shade tolerance. In a few cases, these slowly spreading species possess a combination of traits that allows them to almost completely dominate the ecosystem and cause a dramatic reduction in species diversity. This is apparently happening in some African rain forests where monospecific stands of *Gilbertiodendron dewerii* are spreading into mixed forest (Hart *et al.*, 1989; see Chapter 14).

However, the vast majority of successful invasions into productive environments are associated with agricultural activities and other disturbances such as roadbuilding, urban construction, and lawns. The life history of virtually all of these invaders is short-lived, herbaceous plants, and most are quickly eliminated in the absence of disturbances, as has been documented many times in studies of oldfield succession (e.g., Keever, 1950; Odum *et al.*, 1974). The patterns of plant invasions associated with agriculture are well documented (Elton, 1958; Baker, 1974, 1986; Pimentel, 1986; Mack, 1986).

The history of invasions into ecosystems with relatively low productivity is much more interesting because many of these invasions occur with little or no disturbance, and the effects on the ecosystem can be quite dramatic (Vitousek, 1986; Ramakrishnan and Vitousek, 1989; Mack, 1986). The most interesting case studies are invasions of mediterranean ecosystems, which have already been discussed in the context of their high species diversity (see also Chapter 13) and high rates of endemism. These ecosystems have been extensively invaded and altered by introduced species on all five continents where they occur (Mooney *et al.*, 1986; Macdonald and Richardson, 1986; di Castri, 1989; Kruger *et al.*, 1989).

The invasions of mediterranean-climate ecosystems are particularly striking because they are much more extensive than invasions into more productive ecosystems in the same region. In South Africa, the fynbos ecosystems are more extensively invaded than other ecosystems (Macdonald, 1984). Likewise, in Australia the mediterranean heathlands of New South Wales are more severely invaded than other *undisturbed* ecosystems (Fox and Fox, 1986). The Californian mediterranean region, which has been so extensively invaded that some communities have been virtually replaced by introduced species (McNaughton, 1968; Gulmon, 1977), contrasts strongly with the forests of California, which have experienced very few invasions (Mooney *et al.*, 1986).

The types of invasion of mediterranean-climate ecosystems, in terms of the life histories of the successful invaders and the conditions under which they invade, vary between mediterrean environments that differ in productivity and type of disturbance. The variation is consistent with the predictions of the dynamic equilibrium model of species diversity.

Patterns of invasion are similar in California and Chile, the two mediterrenean-climate regions with the most fertile soil. The invasions have been dominated by herbaceous plants, mainly annuals, and have been closely associated with disturbance, both agricultural and natural fires (Heady, 1977; Gulmon, 1977; Keeley and Johnson, 1977; Kruger *et al.*, 1989). Mature chaparral, which is relatively low in species diversity, has virtually no invaders (Kruger *et al.*, 1989). Although some introduced species are often present during the early post-fire succession (Keeley et al., 1981), there have been no successful invasions of shrubs or larger plants into the mature vegetation (Montgomery and Strid, 1976). Invasion of mature vegetation is more extensive in Chile, undoubtedly because of extensive grazing, burning, and cutting of vegetation (Keeley and Johnson, 1977). However, the number of introduced species declines rapidly in post-fire succession (Keeley and Johnson, 1977).

The proportion of invasive species in the California flora is much higher than in South Africa. This is probably a consequence of the higher fertility of the California ecosystems in combination with a high intensity of disturbance. Undisturbed areas in California tend to have a much lower proportion of exotic species than the state as a whole (Baker, 1986; Mooney and Parsons, 1973). Likewise, in South Africa, areas that are relatively more fertile and highly disturbed have a higher proportion of exotic species than the region as a whole (Kruger *et al.*, 1989).

On the poorer soils of the South African and Australian mediterranean-climate ecosystems, the overall proportion of exotic species is much lower than in California, and the proportion of herbaceous annuals is lower as well (Kruger *et al.* 1989). The most prominent invaders of the African fynbos region are trees and shrubs, and their invasions are primarily limited to nutrient-deficient soils (Macdonald and Richardson, 1986) (Table 11.4). A similar pattern is found in South Australia where the more fertile grasslands and *Eucalyptus* woodlands are invaded by herbaceous species from the Mediterranean Basin and South Africa (Specht, 1963, 1972) and the less fertile sclerophyll communities are less invaded, primarily by trees and shrubs.

In the Cape region of South Africa, most of the lowland vegetation on fertile soils (Coastal Renosterveld) has been converted to agriculture (Moll and Bossi, 1984). In contrast, relatively little of the high diversity Mountain Fynbos on very poor soils has been converted to agriculture.

Table 11.4. *List of prominent invasive plant species of four mediterranean-climate regions, with indications of the extent of their invasion. The symbol "I" indicates that the species has been recorded as introduced in the region, but has not been observed to spread; "+" indicates that the species has naturalized but does not spread except in markedly disturbed habitats (e.g., road cuts); and "++" indicates that the species has been observed to spread widely in natural vegetation. The columns headed "low" and "mod" for Australia and the Cape of Africa refer to substrates with low and moderate nutrient supplies, respectively, corresponding to the highly leached and moderately leached categories of Specht and Moll (1983). Information is derived from McNaughton (1968), Specht (1972), Gulmon (1977), Bridewater and Backshall (1981), Macdonald and Jarman (1984), J.A. Vlok (unpublished), and Wells et al., (1986).*

Species	California	Chile	Australia		Cape Africa	
			Low	Mod	Low	Mod
Shrubs and Trees						
Acacia cyclops and congeners	I	I	–	–	++	+
Pinus halepensis	I	–	?	?	+	++
Pinus pinaster	?	–	I	?	++	I
Pinus radiata	–	I	+	++	++	+
Hakea sericea and congeners	I	–	–	–	++	I
Cytisus scoparius	+	?	I	+	I	I
Annual and biennial herbs						
Aira caryophyllea	++	++	I	+	I	++
Anagallis arvensis	+	+	I	++	I	++
Arctotheca calendula	++	++	I	++	–	–
Avena barbata and congeners	++	++	I	++	I	++
Brachypodion distachyon	++	?	I	+	?	?
Briza maxima and *B. minor*	++	++	I	++	I	++
Bromus mollis and congeners	++	++	I	++	I	++
Echium lycopsis	I	I	I	++	I	++
Erodium cicutarium, E. moschatum, and congeners	++	++	I	++	I	++
Hypochoeris glabra	++	++	I	++	I	++
Lolium multiflorum and congeners	++	++	I	++	I	++
Vicia bengalensis	?	?	I	++	I	++

From Kruger et al., 1989.

This repeats a pattern found throughout the world: high-productivity natural ecosystems are virtually eliminated by conversion to agriculture, while low-productivity, higher diversity systems are less affected by agriculture, and often end up as nature preserves.

Experimental work is consistent with the general patterns outlined above. Fertilization of low nutrient heathland habitats in Australia allows mediterranean herbs to invade and grow fast enough to complete their life cycle and reproduce on sites where they could not otherwise survive (Specht, 1963; Heddle and Specht, 1975). In experiments at the Durokoppin Reserve in Western Australia, Hobbs (1989) disturbed the soil and added nutrients in five different types of plant communities. He also added seeds of two common invasive exotic species (a grass and a shrub) to all treatment and control plots. Hobbs found that neither disturbance alone nor fertilizer alone had much effect on the total biomass of the invading species in most of the communities. However, in all cases the combination of disturbance and fertilization together resulted in a significant increase in total biomass of the exotic species in comparison to the controls and other treatments.

Thus, the interaction of productivity and disturbance has a strong influence on the invasibility of ecosystems, as well as on the life history characteristics of successful invaders. Disturbances are critical for invasion of productive habitats, where high rates of competitive displacement result in low diversity in the absence of disturbance. However, disturbances are relatively unimportant in unproductive habitats, where rates of competitive displacement due to competition for light are low and the high diversity communties are much easier to invade.

As discussed earlier, herbivorous insects tend to invade much more successfully than insects at higher trophic levels. This is consistent with ecological studies that demonstrate little or no competition between herbivorous insects (e.g., Rathcke, 1976a,b; Strong, 1982a,b, 1984). The success of invasions by animals at higher trophic levels is more difficult to predict. However, primary productivity must be relatively high to support large herbivores. This may be part of the explanation for the failure of pigs, rabbits, and fallow deer to become established in South Africa in spite of repeated introductions (Brooke *et al.*, 1986).

Even in regions that are relatively unproductive in terms of total productivity averaged over an entire year, a short, highly productive growing season can result in high rates of growth and competitive displacement during the period when plants are able to grow. This is the situation in many of the major grain-producing regions of the world, where most of the crop is produced during a relatively short growing season. In these situations, disturbances are important for allowing invasions by herbaceous plants that must complete their life cycle during the brief period of high productivity. These ecosystems have experienced a high rate of invasion (Mack, 1986).

The evolutionary history of a region and the consequent adaptations of the indigenous species have a strong interaction with the disturbances that are associated with activities such as grazing. The fact that grazing is often a highly species-specific disturbance that occurs at high frequencies and intensities means that grazing can rapidly alter species composition. In many arid grasslands, such as in the western part of North America, as well as tropical grasslands outside of Africa, grasses and herbs do not have many of the adaptations that confer resistance to grazing or the ability to recover quickly from grazing damage, perhaps because they have not had a long history of coevolution with grazing mammals (Milchunas *et al.*, 1988). One of the primary adaptations to grazing is the ability to reproduce vegetatively through the production of tillers that spread from the parent plant and establish new plants where they take root. The ability to tiller is common in grasses that have evolved with grazing animals in Africa and the Mediterranean region of Eurasia, but absent in the 'bunchgrasses' that dominate New World grasslands. Consequently, native grass species have been largely replaced by invaders from the Old World in the semi-arid grasslands of western North America and the tropical grasslands of South America (McNaughton, 1968; Specht, 1972; Moore, 1975; Gulmon, 1977; Milchunas *et al.*, 1988; Mack, 1986, 1989).

Some of the most serious effects of invading plants occur when they are able to alter the environment that they invade (Macdonald *et al.*, 1989; Mack, 1989). There are numerous examples of situations in which invading species have altered the hydrologic or nutrient cycles of ecosystems (cf. Vitousek, 1986; Ramakrishnan and Vitousek, 1989). However, the most dramatic changes wrought by invading species occur when they alter the disturbance regime of the ecosystem, particularly the fire cycle (e.g., Ewel, 1986).

Alteration of natural fire cycles by invading species generally occurs in arid or semi-arid environments where natural fires are relatively infrequent and the indigenous vegetation is not well-adapted to fire. In the absence of any alteration in the disturbance regime, the low rates of growth in these unproductive environments would usually not allow an invading species to become dominant. However, an increase in the fire frequency can quickly lead to dominance by species adapted to the new fire regime. Mack (1989) describes an early example of an invading species that altered the local fire regime and the landscape over vast regions of the Argentine pampas. On the basis of observations made in 1833 during his voyage on the HMS Beagle, Darwin wrote in *The Origin of the Species* that the European cardoon (*Cynara cardunculus*) and a tall thistle (*Silybum marianum*) '... are now the commonest [plants] over the whole plains of La Plata, clothing square leagues of surface almost to the exclusion of every other plant ...' (Darwin, 1872). In southern Uraguay he found that 'very many (probably several hundred) square miles are covered by one mass of these prickly plants, and are impenetrable by man or beast. Over the undulating plains, where these great beds occur, nothing else can now live' (Darwin, 1898). The growth of the giant thistle, which stood six to ten feet tall when in flower, varied from year to year, depending on the weather. During good 'thistle years' the plants were so abundant that they confined horseback travel to narrow cattle trails. By December, the thistle's aboveground growth was dead, and provided a fuel load that resulted in high frequency and intensity of fires. The increased fire frequency, and probably also competition for light, eliminated much of the native vegetation (Mack, 1989). These species were only controlled by extensive plowing of the pampas at the end of the nineteenth century, but both are still considered to be serious problems in Argentina.

More recently, in the sclerophyll woodlands of King's Park, Western Australia, invasion of the African grass *Ehrharta calycina* increased the frequency of fires to an interval of six years (Baird, 1977). This higher frequency (and lower intensity) of fires enhanced the spread of *Ehrharta*. Fire control measures that reduced the frequency of fires decreased the abundance of this invading species (Baird, 1977). Zedler *et al.* (1983) describe how planting the grass *Lolium multiflorum* following a chaparral fire led to increased frequency of fire that reduced the populations of native shrub species and enhanced the persistence of introduced herbs. The spread of *Bromus tectorum*, an exotic annual grass, throughout the intermontane west of North America has potentially disastrous consequences for the arid shrublands and pine woodlands (Mack, 1981; Billings, 1990). This annual grass is able to grow rapidly during the brief spring wet period to build up enough fuel on the ground to carry fires that destroy the native shrubs in the lowlands (e.g., *Artemesia* spp.) and pinyon pines (*Pinus monophylla*) in the highlands (Pickford, 1932; Mack, 1981; Wright and Bailey, 1982; Whisenant, 1990).

Summary

Endemism and invasions represent ecological phenomena at the extremes of several continua of evolutionary and ecological processes. The environmental conditions that allow the evolution and survival of endemic plant species, namely low rates of competitive displacement and a low frequency of major disturbances, also allow high diversity plant communities with many endemic species to be easily invaded by exotic species. Fortunately, the same conditions also prevent the invading species from dominating the community and reducing species diversity *unless* the invasion is accompanied by a change in the disturbance regime of the community. Changes in disturbance regimes can totally alter the dynamic equilibrium of the community and result in a reduction of species diversity and/or a major change in species composition. Examples include the effects of grazing mammals introduced by humans into ecosystems with no evolutionary history of grazing, and alteration of the fire regime that results from altered timing and amounts of biomass production caused by the invading species themselves.

Ecosystems in productive environments are quite resistant to invasion under natural conditions of low disturbance frequencies. However, nearly all of the high productivity terrestrial environments around the world have been severely disturbed by human agricultural activities. Consequently, these environments have experienced the vast majority of invasions of both plants and animals.

The properties of species that influence their ability to resist invasion or survive in the presence of exotic species of competitors or predators are largely the result of their evolutionary history. In particular, the evolutionary history of adaptation to competitors and predators results in a continuum of competitive abilities that range from generally low on islands to highest on large continents. Isolation from competitors over long periods has repeatedly led to mass extinctions of animals on islands and entire continents when competitors from larger and/or less isolated land masses finally invaded. The vulnerability of island species to extinction caused by competition or by predation by introduced predators is responsible for most of the extinctions that have occurred as a result of the phenomenal increase in the transport and dispersal of species by human activities.

12

Case Studies: Species Diversity in Marine Ecosystems

Marine ecosystems cover over two thirds of the Earth's surface and occupy a volume that vastly exceeds that occupied by terrestrial ecosystems. Life presumably appeared in the oceans, and the oceans remain the greatest repository for the diversity of life at the level of phyla. At least 43 of the more than 70 phyla of all life forms are found in the oceans, while only 28 are found on land (Ray, 1988). Our understanding of marine diversity is continuing to increase (e.g., Grassle, 1991), and the question of whether there are more terrestrial species or more marine species remains unresolved. There is no doubt, however, that there are fundamental differences in patterns of species distributions between terrestrial and marine systems, owing largely to the vast size and spatial continuity of the marine environment.

The size of the oceans is complemented by a tremendous range of physical conditions over which marine ecosystems occur. The productivity of marine systems ranges from algal reefs and upwelling areas that support the richest fisheries on earth, to vast nutrient-poor regions in the centers of the largest oceans, where biomass and production are extremely low. Likewise, marine disturbance regimes range from constant pounding by large waves or annual scouring by ice flows, to the nearly constant physical conditions of the deep-ocean benthos. Thus, the marine environment provides an ideal opportunity for comparative evaluation of the predictions of the dynamic equilibrium model.

The general principles that have been discussed in previous chapters and applied to terrestrial ecosystems should also apply to marine ecosystems. Communities of sessile organisms or organisms with low mobility, such as those attached to hard substrates or living in benthic sediments, have important features in common with terrestrial plant communities. Likewise, the highly mobile marine invertebrates and vertebrates utilize

342

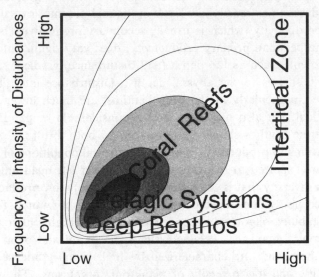

Fig. 12.1 Range of productivity and disturbance conditions encompassed by each of the four marine systems discussed in this chapter. (Based on Huston, 1979, 1985a.)

their environment in much the same ways as terrestrial animals, showing similar patterns of diversity in relation to primary productivity.

The four marine systems that I will discuss in this chapter represent the full range of conditions addressed by the dynamic equilibrium model of species diversity. Some, like the rocky intertidal zone, have been the subject of intensive ecological research, while others, such as the central oceanic gyres, have been little studied. The relative positions of the four systems in relation to the axes of the dynamic equilibrium model are illustrated in Fig. 12.1.

The rocky intertidal zone is characterized by the high primary and secondary productivity that is typical of nearshore waters (Rodin et al., 1975; Ryther, 1969; Mann, 1982). Consequently, rates of competitive displacement are potentially high, and the positive effects on diversity of mortality caused by physical disturbances, predation, or herbivory should be conspicuous. According to the dynamic equilibrium model, variation in diversity among competitors within the intertidal zone should be regulated primarily by the disturbance regime, which varies widely in both frequency and intensity (Fig. 12.1).

In contrast, the productivity of the marine benthos is generally quite low, and drops to extremely low levels in the mid-oceanic abyssal depths. Benthic productivity (which is usually secondary productivity based on detritus, rather than primary productivity) does vary significantly along some gradients, such as the depth (and disturbance) gradient from the continental shelves to the abyssal depths. Disturbance intensities and frequencies, particularly in the deep benthos, are much lower than in the intertidal, but also may vary over a considerable range. Diversity in the marine benthos should be influenced by both disturbance regime and variations in productivity. The environmental conditions of benthic systems are expected to be primarily in the lower left quadrant of Fig. 12.1, with diversity that is often limited by extremely low productivity.

The pelagic systems of the mid-oceanic gyres are noted for their extreme stability, oligotrophic conditions, and remarkably high diversity of planktonic organisms. Pelagic systems of coastal and polar waters, on the other hand, are characterized by high primary and secondary productivity, and low diversity of planktonic organisms. The primary disturbance to pelagic planktonic ecosystems is probably grazing, but there is little indication that grazing regulates their community structure. The primary regulator of species diversity in these systems seems to be nutrient availability, through its effect on growth rates and competitive displacement (Fig. 12.1).

Finally, I discuss tropical coral reefs, the acme of marine species diversity, and undoubtedly the best marine environment for vacationing. Coral reefs share some features of the rocky intertidal zone, since the upper portion of most reefs actually lies in the intertidal zone. However, coral reefs extend to depths far below the intertidal, and consequently have depth gradients of productivity and disturbance not found in the rocky intertidal. The wide range of dynamic equilibria possible over the disturbance and productivity gradients on coral reefs leads to a wide range of species diversity (Fig. 12.1).

In evaluating the diversity of any of these marine ecosystems, as well as in making comparisons between them, it is important to keep in mind the distinction between functional types and functionally analogous species. Because the dynamic equilibrium model specifically addresses the diversity of organisms that are competing or can potentially compete with one another, care must be taken to use appropriate subsets of organisms (i.e., organisms of the same or similar functional types) in testing the predictions of the model. While the diversity of the dominant structural organisms may be strongly influenced by the intensity of competition, the physical heterogeneity provided by these structural organisms can allow the coexistence of many different functional types of interstitial organisms. Thus, it is spatial and resource heterogeneity and resource availability, rather than competitive interactions, that are likely to explain the diversity of the species that live within the structure created by the dominant organisms (Chapters 1 and 4). In many cases where the diversity of the dominant structural organisms is low (e.g., reef crests or mussel beds), the number of interstitial species of a variety of different functional types may be quite high.

Before examining these four marine ecosystems in detail, it is worth considering how the concepts that have been developed primarily in reference to terrestrial ecosystems may differ in their application to marine ecosystems.

Fundamental Differences between Terrestrial and Marine/Aquatic Environments

The dramatic physical contrasts between terrestrial and marine environments result in very different combinations of adaptations among organisms, different types and intensities of disturbance, and consequently communities with different structures and dominant ecological processes. The primary differences between the two environments stem from the physical and chemical properties of water versus air. Although both air and water are fluids, the much higher density of water results in 1. higher thermal inertia and thus buffering capacity with regard to temperature; 2. greater relative buoyancy that counteracts the force of gravity, thus providing three-dimensional mobility and reducing the need for physical supporting structure; and 3. much higher kinetic energy of moving masses of water, and thus more energy transferred, and potentially much more damage done as a result of collisions with organisms and objects.

The physical properties of the marine environment relax some of the major constraints required for survival out of the water, namely some form of thermoregulation, sufficient physical structure to balance the force of gravity, and efficient systems for the conservation and transport of water. However, the physical environment also imposes a major constraint through the force of moving water, which precludes the existence of large, rigid, stationary organisms on the scale of terrestrial woody plants (cf., Denny et al., 1985). Thus the tremendous size-based diversification of sessile organisms, which is found in terrestrial plant communities, is simply not possible in marine environments, and sessile organisms, with a few exceptions such as giant kelp, are limited to a restricted range of relatively small sizes. Consequently, the levels of diversity at higher trophic levels that are generated by the structural and chemical heterogeneity of terrestrial plant communities are simply not found in most marine and aquatic environments, with the exception of the fish communities of tropical coral reefs and seasonally-inundated floodplain forests (Lowe-McConnell, 1987).

The chemical properties of water also result in major contrasts with the terrestrial environment. The effectiveness of water as a solvent for both inorganic and organic molecules means that marine plants can obtain all of their nutritional requirements directly from the same environment, rather than having to obtain some critical resources from the air and some from the soil, as must terrestrial plants. This eliminates a whole suite of physiological and morphological trade-offs involved in the acquisition of carbon, light, water, and nutrients that dominate the ecological processes that structure terrestrial ecosystems. Marine organisms are still subject to many of the same morphological and life history constraints as are terrestrial organisms, such as the trade-offs between growth rate, reproduction, and structural (or chemical) investment that influences resistance to various types of stresses and to herbivory (Hay and Fenical, 1988).

The optical clarity of even the purest water is much less than that of air. Consequently, light is reduced rapidly with depth in marine environments, even in the absence of turbidity resulting from biotic or abiotic material. Thus, light available at the level of the solid substrate on which sessile organisms must grow is limited by water depth, as well as the shading effect of attached organisms or planktonic organisms in the water column. In terrestrial communities, in contrast, light at ground level is limited only by the presence of plants. When plants are removed, the resulting high light levels allow potentially rapid growth rates. Removal of sessile marine organisms does not necessarily result in a significant increase of light at the level of the solid substrate.

In general, the diversity of marine environments at both large and small scales is lower than that of terrestrial environments, primarily as a consequence of the high mobility of the organisms, the wide passive dispersal of their propagules by currents, and the low structural hetero-geneity of the environment. Marine ecosystems are characterized by low standing biomass and very high ratios of production to biomass, while terrestrial ecosystem have standing biomass several orders of magnitude higher, and much low production/biomass ratios (Crisp, 1975). The role of disturbance is much more prominent in some marine environments than in most terrestrial environments. Likewise, the effects of predation and herbivory also tend to be more conspicuous in productive marine systems. At least some of the increased importance of predation and herbivory is a consequence of the greater number of trophic levels that are found in marine environments, possibly as a result of lower energy requirements for maintenance in a chemically and physically buffered environment (cf. Lindemann, 1942; Connell and Orias, 1964; Pimm, 1982; Cohen, 1978). In addition, the low standing biomass of most marine systems makes the effects of herbivory and predation much more conspicuous.

The Intertidal Zone: Pre-eminence of Disturbance, Predation, and Herbivory

Many of the major concepts in ecology, particularly those related to competition, predation, and disturbance, have been experimentally investigated in the marine intertidal zone (e.g., Connell, 1961; Paine, 1966; Paine and Levin, 1981). It is not surprising that the effects of these processes should be particularly dramatic in the intertidal. Along most of the world's coastlines, particularly the north temperate coastlines where most of the research has been done, the intertidal is a highly productive zone, both in terms of primary productivity by plants and secondary productivity by animals (Crisp, 1975).

Nutrient availability in the intertidal zone is generally high, as a result of both terrestrial runoff and nutrient-rich ocean waters (Sverdrup *et al.*, 1942; Mann, 1982). Cool temperatures maintain a favorable balance of production to respiration. Water movement maintains a constantly replenished supply of dissolved nutrients, suspended food particles, and gases necessary for photosynthesis and respiration (Mann, 1982). Consequently, growth rates are high and competition for space and its associated access to resources can be very intense.

Mortality in the Intertidal Zone

Mortality-causing disturbances occur frequently and with high intensity in the intertidal zone. Physiological stresses imposed by extreme tidal variation and damage caused by the powerful physical action of waves and solid objects carried by waves can kill most of the sessile organisms in areas ranging from less than a meter in area to large stretches of coast-line (Wethey, 1985). Likewise, the high productivity of the intertidal zone can support high densities of organisms at higher trophic levels, such as herbivores and predators, and so mortality caused by herbivory and pre-dation can occur at high rates. Consequently, disturbance intensities and frequencies can range from very low in areas protected from damaging wave action or predators, to extremely high. The combination of high rates of growth and competitive displacement with great variation in the frequency and intensity of disturbance allow the effects of disturbance to be manifested rapidly and to produce dramatic differences in community structure in the intertidal zone.

Diversity of the dominant structural organisms of the intertidal zone is relatively low compared with most marine environments, although the organisms that compose intertidal communities are conspicuous, interesting, and often edible. The number of interstitial species may be quite high, even where the diversity of the structural species is low (Hewatt, 1935).

Productive intertidal and shallow subtidal areas generally undergo a rapid succession of organisms (both plants and animals) that are attached directly to the substrate. Menge and Sutherland (1976) report that newly settled mussels can replace newly settled barnacles over the course of a single season at some sites on the New England coast. Marine succession on hard substrates has many similarities to terrestrial plant succession (e.g., Lubchenco and Menge, 1978; Lubchenco, 1978; Breitburg, 1985) and the endpoint of intertidal succession is usually a low diversity (often monospecific) cover of algae or invertebrates such as mussels or oys-ters. Under these conditions of a high rate of competitive displacement and rapid reduction of species diversity during succession, the effect of mortality-causing disturbances (including herbivory and predation) is dramatic and easily observed over short time periods. Herbivory or predation that reduces the abundance of a dominant competitor can pro-duce a significant increase in species diversity over a short time period, as can disturbances such as wave action or ice scour (Paine and Levin, 1981; Sousa, 1979b, 1985; Wethey, 1985).

Because of high algal growth rates in response to the high light and nutrient levels of the intertidal environment, herbivores can play a major role in altering the outcome of algal succession and maintaining different levels of diversity in algal communities subjected to different levels of herbivory (Lubchenco, 1978, 1980). As in terrestrial systems, the relative palatability of rare versus common species determines whether herbivory will increase or decrease species diversity (Lubchenco, 1978; Lubchenco and Cubit, 1980; Lubchenco and Gaines, 1981). The same general patterns of life history and physiological trade-offs that are found in terrestrial plants are also found in marine algae, including an inverse relationship between growth rate and defenses against herbivory (cf. Coley et al., 1985; Hay and Fenicel, 1988). Thus, in the New England rocky intertidal community, early-successional algal species (predominantly green algae) tend to have higher palatability than mid- and late-successional species (predominantly brown and red algae) (Lubchenco and Gaines, 1981). While the specific details of any particular situation may be influenced by accidents of species composition and relative palatibility to particular herbivores, the general patterns of algal species diversity in relation to growth rates and herbivore density clearly fit the predictions of the dynamic equilibrium model (Lubchenco and Gaines, 1981) (Fig. 12.2).

It is interesting to note that wave disturbance, which can be a powerful, non-selective disturbance of all intertidal organisms at high intensities, has a very different effect at moderate intensities. At the moderate wave intensities encountered on many exposed shorelines in New England, the primary effect of waves is to prevent mobile herbivorous and predatory gastropod molluscs from foraging effectively. While this wave intensity does not kill these organisms, it prevents them from having any significant effect on lower trophic levels. As a consequence, intense competition for space among the sessile algae, barnacles, and mussels, leads to rapid establishment of monospecific stands of mussel. In more protected areas, herbivores and predators remove fast-growing algae and mussels, which allows slower-growing (and less palatable) algae to dominate (Lubchenco and Menge, 1978). Thus, low levels of a density- and species-independent disturbance results in an increase in predation and herbivory (e.g., Walde, 1986), which are both highly density- and species-dependent disturbances. Density-dependent disturbances that selectively remove the dominant species are much more effective in maintaining high levels of species diversity than are density-independent disturbances.

The same sorts of physical conditions and ecological interactions that structure marine intertidal communities at small scales may also explain continental-scale differences in intertidal species diversity. Menge and Sutherland (1976) observed that the mid-intertidal zone of the Pacific Northwest of North America has a more diverse animal community than the same zone in New England. This higher diversity is associated with a more complex trophic structure with more trophic levels (Fig.

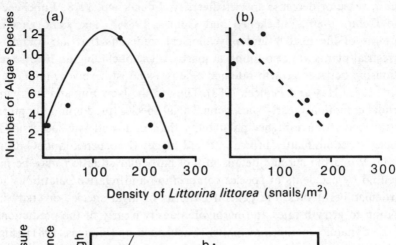

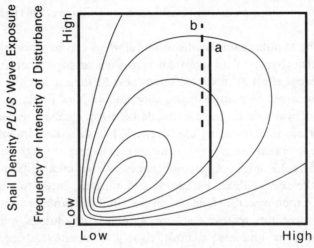

Fig. 12.2 Dynamic equilibrium of algal species diversity in relation to wave disturbance and herbivory by snails in the New England rocky intertidal zone. Exposed substrates (*b*) experience wave disturbance that does not occur in tide pools (*a*). Algal species composition of ungrazed tide pools is dominated by rapidly growing, palatable species (e.g., *Enteromorpha*) that competitively eliminate the slower-growing species that are less palatable, but more resistant to wave damage. The same algal species are present on the exposed sites, but slower growing, wave resistant species (e.g., *Chondrus crispus*) are dominant. An increase in herbivory (snail density) in the low disturbance tide pools where the preferred algal species are competitive dominants produces a unimodal diversity response (*a*). An equivalent increase in herbivory at the exposed sites (*b*), where higher wave disturbance causes the preferred species to be relatively rare, produces a monotonic decrease in species diversity. (Based on Lubchenco, 1978; Lubchenco and Menge, 1978.)

12.3). Menge and Sutherland (1976) note that variability in wave shock, temperature, and precipitation is greater on the East Coast than the West Coast, and severe storms and rapid weather changes are less predictable on the East Coast. A consequence of this variability is a higher frequency and intensity of mortality-causing physical disturbances on the East Coast. This difference in disturbance regime is a probable explanation for the simpler trophic structure and lower diversity of East Coast intertidal communities, where certain types of large, long-lived sessile animals are absent (Wethey, 1985). The disturbances are likely to have a more severe effect on higher trophic levels than on lower levels for several reasons: 1. population size generally decreases with increasing trophic level, and thus a disturbance is more likely to cause local extinction; 2. reproductive rate, and thus the ability to recover between disturbances, generally decreases with increasing trophic level; and 3. periodic low population densities of prey species, caused by intense disturbances, may lead to local extinction of higher trophic levels. This phenomenon is an example of the disturbance axis of the trophic shift in species diversity that is commonly found along productivity gradients (Fig. 5.14). The higher number of herbivorous species on the West Coast is likely to result in higher algal species diversity, particularly at large scales that incorporate between-habitat diversity, but few comparative data are available to evaluate this hypothesis.

Tropical Intertidal Communities

The same general patterns also appear in a comparison of tropical versus temperate subtidal diversity. The relatively constant environmental conditions and food supply of tropical nearshore marine environments allow the survival of an abundant and diverse group of herbivorous fish, which exerts heavy grazing pressure on the algal community (Gaines and Lubchenco, 1982). This intense herbivory may be one cause of the relatively low algal diversity of the tropical inter- and sub-tidal zones. Algal species diversity increases with increasing latitude, reaching a maximum between a latitude of 20° and 40° north, before declining at higher latitudes (Gaines and Lubchenco, 1982). Thus, the latitudinal gradients of these two interacting trophic levels are reversed, with herbivore diversity increasing toward the tropics, and algal diversity decreasing.

In another tropical-temperate comparison, Miller (1974) and Spight (1977) compared gastropod mollusc communities in the rocky intertidal zone on the Pacific Coast of Costa Rica and the Pacific Coast of Wash-

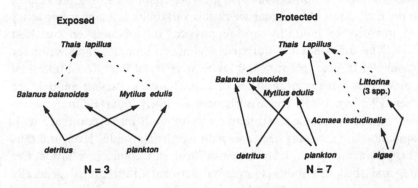

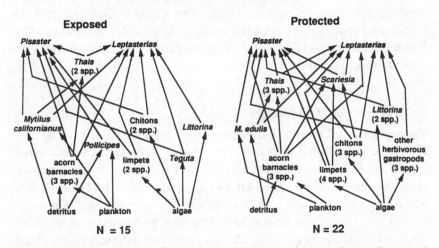

Fig. 12.3 Trophic diagrams for the intertidal communities on the northeast and northwest coasts of North America. Note the simpler trophic structure for the northeast coast communities, where environmental variability is higher and disturbances more frequent. (From Menge and Sutherland, 1976.)

ington. Although the total number of species sampled in Costa Rica was 2.4 times as great as the total number sampled in Oregon (Miller, 1974), the average number of species in a tropical sample was essentially the same as in the temperate samples of the same surface area (9.5 versus 9.0, Spight, 1977).

In the context of the dynamic equilibrium model, the two factors that might vary between the temperate and tropical intertidal sites are the growth rates (rates of competitive displacement and recovery from disturbance) and the disturbance regime. With regard to growth and competitive displacement, the productivity of tropical marine environments is generally lower than temperate marine environments, although this difference is likely to be greater for the open ocean than for the nearshore environment. This difference in productivity may be reflected in the fact that average snail densities (including limpets) are much higher at the temperate site (400 individuals/m^2 versus approximately 160 individuals/m^2, Spight, 1977). However, the different densities might also result from a higher predation rate in the tropics. The number and diversity of snail predators (fish, crabs, octopods, other snails) is higher in Costa Rica than in the Pacific Northwest (Spight, 1977), possibly a result of the larger population sizes that organisms in high trophic levels can maintain in an environment with low variability in food supply. At both locations predation is more intense lower on the shore than in the more frequently exposed upper intertidal, a general phenomenon in the intertidal zone, where upper distributional limits are often set by physical conditions, and lower limits by biotic interactions (cf., Connell, 1961, 1970). Either a lower productivity (and potentially lower rate of recovery from disturbance and predation) or a high predation rate, or both, could contribute to the relatively low diversity of these tropical intertidal zones.

Within the tropical intertidal zone, evidence for a dynamic equilibrium between competition and disturbance can be found in the distribution of species richness of prosobranch molluscs (snails) presented by Spight (1976). Quadrat samples were taken from both rock and cobble areas at Playas del Coco, Costa Rica over a depth range equivalent to the mean tidal range of 2.3 m. Sites were qualitatively classified according to wave exposure. The lowest areas that were exposed at low tide were covered with a dense growth of short, erect red algae and there were few snails in this zone (consequently, no samples were collected). A variety of barnacle species were found over the rocky areas, which supported the snail species that were the major barnacle predators (*Thais melones* and *Chthamalus panamensis*). The most abundant snail species were herbivorous, and there were many less abundant scavengers and predators. Snail density ranged from one individual per m^2 to over 1100 individuals per m^2. If snail density is considered to be an index of the resource availability and intensity of competition for food, the species richness of the snail community can be interpreted as a dynamic equilibrium between rate of competitive displacement and frequency (or intensity) of disturbance as indicated by the exposure index (Fig. 12.4a). At the sites with a high exposure to wave action, the relationship of species richness to organism density forms a typical unimodal curve, with maximum species richness at intermediate densities (and presumably, intermediate rates of competitive displacement). In contrast, sites with a low exposure to wave action show a monotonic decrease of species richness with increasing density, and the maximum levels of species richness are considerably higher than at the exposed sites. This pattern closely resembles the qualitative predictions of the dynamic equilibrium model (Fig. 12.4b).

Succession among sessile organisms growing on hard substrates in the inter- and sub-tidal zone involves both plants (e.g., green, brown, and red algae) and animals (e.g., barnacles and mussels). Various combinations of growth rate, resistance to stress, and susceptibility to herbivory and predation in the presence of different levels of herbivory and predation, determine whether a particular area or tidal zone will be dominated by algae or invertebrates (Menge, 1976; Lubchenco and Menge, 1978).

Building on the ideas of Hairson et al. (1960), Menge and Sutherland (1976) hypothesized that competition would be more important than predation in regulating the number of species within guilds at higher trophic levels because such organisms were more likely to be near the carrying capacity of the environment, and thus competing. Conversely, they hypothesized that predation was likely to be more important than competition in regulating the number of species in guilds at lower trophic levels, and within communities with few trophic levels. They predicted that 'as the number of trophic levels and the number of species per level increase, predation will become relatively more important as an organizing factor.'

While data from the marine intertidal zone generally support the hypotheses of Menge and Sutherland, the dynamic equilibrium hypothesis suggests an alternative explanation based on the effect of productivity and disturbance on the relative influence of predation and competition on the structure of different trophic levels (cf. Fig. 5.14). At any trophic level, predation is likely to have a major influence on structure primarily when rates of growth and competitive displacement are high, and physical disturbances do not cause significant mortality. Under productive conditions that can support high predator biomass, predation can increase the diversity of lower trophic levels. However, in terrestrial and

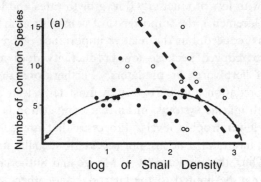

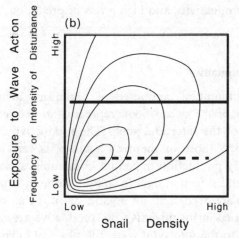

Fig. 12.4 Patterns of snail species richness in the rocky intertidal zone of a tropical beach (Playas del Coco, Costa Rica) in relation to organism density and site exposure. Sites are classified according to exposure to wave action, with 'Z' being the lowest exposure and exposure increasing with reverse alphabetic order. Open circles represent the lower five exposure classes (V through Z), and the closed circles the six highest exposure classes (P through U, plus exposed island sites). (a) Snail species richness decreases with increasing snail density at protected sites (dashed line), and shows a unimodal response to density at exposed sites (solid curve). (b) Hypothesized positions of intertidal snail assemblages on contour diagram of the dynamic equilibrium of species diversity in relation to the rate of competitive displacement (indexed in this data by organism density, individuals/m^2) and the frequency or intensity of disturbance (indexed as exposure to wave action). The shapes of the dashed and solid lines draped across the species diversity surface (b) closely approximate the shapes of the dashed and solid lines in (a). Note that the x axis (organism density) in (a) is a log scale, while the x axis in (b) is linear. Note also that the aggregation of exposure indices into two classes means that the sites are actually scattered above and below the dashed and solid lines in (b). (Data in (a) from Spight, 1976, Table 1; (b) based on Huston, 1979.)

marine systems with low productivity (low growth rates and low rates of competitive displacement), the importance of competition in structuring the community is reduced, but the relative importance of predation and herbivory is also reduced because low productivity does not support high biomass of herbivores or predators. Furthermore, the generally lower levels of predation and herbivory in most terrestrial ecosystems, as compared with marine systems, mean that competition is potentially intense even at lower trophic levels. Processes that are conspicuously important in the intertidal zone are not necessarily equally important in all ecosystems. Thus, the applicability of Menge and Sutherland's (1976) generalizations may be limited to the intertidal and other systems with both high levels of productivity and high levels of predation.

Marine Benthic Communities

Marine benthic communities, composed of organisms living on and within the sediments of the ocean floor, represent an extreme contrast to the communities of the intertidal zone. The productivity of benthic ecosystems, particularly those of the deep benthos, is very low. There is virtually no primary production in benthic communities, because they are below the zone with sufficient light for photosynthesis. Consequently, benthic communities are dependent on organic detritus that drifts down from above or is carried in by currents from productive areas of the sea or land. In contrast to the intertidal zone, the physical environment of the deep benthos is relatively constant and free from the violent action of waves, although the scouring action of bottom currents can affect benthic communities, particularly in shallow areas.

The recent discovery of dense concentrations of organisms around thermal vents in the deep ocean provides a strong contrast with the low biomass and productivity of most of the benthos (Corliss *et al.*, 1979; Edmond and Von Damm, 1983; Rex, 1981; Menzies *et al.*, 1973; Rex *et al.*, 1990). These remarkable organisms have circumvented the light limitation on primary productivity by a food chain based on chemotrophic productivity using the high hydrogen sulfide concentrations of the thermal vents as an energy source (Karl *et al.*, 1980; Edmond and Von Damm, 1985). This productive ecosystem of bacterial mats, tube worms and clams is based on the primary production of bacteria (Jannasch and Wirsen, 1979; Jannasch, 1985; Karl, 1987). Like most highly productive ecosystems, and unlike the rest of the deep-sea benthos, the diversity of these thermal vent communities is very low (Lonsdale, 1977; Enright *et al.*, 1981; Hessler and Smithey, 1983; Grassle, 1985; Arquit, 1990).

Deep Benthos

The marine benthos is conventionally divided into four regions based on depth: 1. the continental shelf (<200 m); 2. the continental slope (200-2000 m); 3. the continental rise (2000-4000 m); and the abyssal plain (>4000 m) (Sanders and Hessler, 1969). Prior to extensive research during the mid-twentieth century, limited data supported the common assumption that the frigid temperatures, extreme hydrostatic pressures, total darkness, and nutrient-poor sediments created a stressful environment where only a depauperate and archaic fauna survived.

One of the most startling results of the first extensive deep-sea investigations was the discovery that the species diversity of deep-sea communities was actually quite high (Hessler and Sanders, 1967; Grassle, 1991). In fact, species diversity increased with depth below the continental shelf and at depths approaching 4000 m reached levels of diversity similar to those found in the tropical subtidal benthos (Sanders, 1968). This surprising result was initially attributed to the equilibrium process of niche differentiation in the extremely stable deep-sea environment (Sanders, 1969). However, as more extensive investigations extended knowledge of deep-sea species diversity out onto the great depths of the abyssal plain (>4000 m), it became apparent that high environmental stability and long evolutionary time periods were not the critical factors, since species diversity began to decline with depth on the abyssal plain (Rex, 1973, 1976, 1981). It is also interesting that the pattern of deep-sea benthic diversity is not related to the total area of the different depth zones (contra Abele and Walters, 1979), as would be the case if the deep-sea diversity patterns were simply a species/area phenomenon based on habitat heterogeneity or evolutionary diversification over large areas. The high diversity upper continental rise (2000-3000 m) covers only 11% of the total Atlantic Ocean area, while the low diversity abyssal plain (4000-5000 m) covers 42% (Rex, 1981).

Accumulating data for many different groups of deep-sea organisms reveal a consistent unimodal relationship between species diversity and depth from the continental shelf to the abyssal plain, with maximum diversity of most groups occurring at depths of 2000 to 4000 meters (Rex, 1973, 1976, 1981). Many different physical environmental factors change monotonically along this depth gradient, including decreasing temperature and temperature variability, increasing pressure and salinity, and decreasing input of organic detritus. Conditions change most rapidly on the contintental slope, where the depth gradient is steepest. The physical instability of the environment would be expected to be greatest on the slope, where sediment slumping and strong currents could create a heterogeneous environment with frequent disturbances. Such conditions would be expected to result in high species diversity. However, most groups of organisms apparently reach their maximum diversity at depths below the continental slope, suggesting that factors other than disturbance are determining the overall pattern of benthic diversity.

One important factor that decreases monotonically with depth and is known to have a strong effect on community structure is productivity. According to Rex (1981), 'our most reliable indication of the food available at any depth is the standing crop of the benthos. Both the density and biomass of the macrofauna [a general size class of small benthic organisms that are retained on a 420 μm sieve, and includes smaller metazoan animals representing most of the typical marine invertebrate phyla] decline exponentially with depth and reach levels of 10-100 individuals and <1 g m^2, respectively, on the abyssal plain (Rowe *et al.*, 1974; Sanders *et al.*, 1965; Thiel, 1975; Rex *et al.*, 1990). Similar reductions with depth are found in the density of the meiofauna (Thiel, 1979), biomass of the bentho-pelagic plankton (Wishner, 1980), catch rates of megafauna [relatively large, mobile vertebrates and invertebrates] (Haedrich and Rowe, 1977; Haedrich *et al.*, 1980), and community respiration (Smith, 1978).'

The unimodal relationship between the species diversity and productivity of deep-sea benthic communities along a depth gradient is consistent with the predictions of the dynamic equilibrium hypothesis (Fig. 5.14). Other processes undoubtedly also contribute to patterns of species diversity along this depth gradient. Predator biomass (and presumably predation pressure) would be expected to be highest in the productive environment of the continental shelf, where humans now dominate as the top of the food chain. However, the potentially high predation rates are apparently not sufficient to prevent a few species from dominating the benthic communities of the continental shelf (Rex, 1973, 1976). It is possible that predation on benthic organisms of the shelf could be lower than suggested above, if higher trophic levels reduce the abundance of those predators that directly affect the benthos, in a manner similar to the 'trophic cascade' suggested by Carpenter *et al.* (1985) for freshwater lakes. In any case, the fact that fewer congeneric species of deep-sea gastropods coexist on the continental shelf than at other depths suggests that competition is more intense in this region (Rex and Warén, 1981). Thus, predator biomass and predation intensity, which are presumably correlated with prey productivity, do not provide a consistent explanation for the diversity pattern over the depth gradient, although predation probably contributes to the high diversity found at intermediate depths. Predator diversity, which is not necessarily correlated with predation intensity, is itself highest at intermediate depths (Rex, 1981).

Spatial heterogeneity, or patchiness, is almost always an important factor in regulating species diversity. Patchiness of benthic species distributions has been documented at scales ranging from 0.01 m² to 100 km (Jumars, 1975a,b, 1976; Hessler and Jumars, 1974; Thistle, 1978). Such patchiness can result from physical differences in the abiotic environment, such as sediment composition and current patterns, from biotic processes such as predation, mortality, and burrowing, and from the unpredictable settling of food resources from the upper ocean. An interesting example of a spatially and temporally unpredictable and very ephemeral resource is wood from drift and sunken ships that settles to the ocean floor and is rapidly consumed by an extremely effective group of colonizing organisms (Turner, 1973).

The difficulties of sampling the ocean benthos without disturbing its structure has meant that the scales of patchiness and the mechanisms by which they influence species diversity are still poorly understood (Rex, 1981). While spatial heterogeneity is clearly an important factor in the regulation of benthic species diversity (Abele and Walters, 1979), there is no evidence that patchiness varies with depth in such a way as to explain the unimodal patterns of diversity.

Perhaps the strongest evidence supporting an interpretation of the unimodal diversity/depth patterns as dynamic equilibria along a productivity gradient is the clear pattern of a trophic shift of maximum diversity to shallower depths for higher trophic levels (Chapter 5, Fig. 12.5). This pattern is only apparent when the productivity of the base of the food chain (detritivores in the case of the marine benthos) is used as the scale for the diversity of higher trophic levels as well. Low levels of primary or secondary productivity that are sufficient to support populations at low trophic levels are apparently insufficient to support stable populations at higher trophic levels. Consequently, the diversity of higher trophic levels may drop to very low levels, or even zero, at relatively low levels of productivity that can support moderate diversity at low trophic levels. Because of the strong inverse correlation between productivity and depth in the marine benthos, the maximum diversity of highest trophic levels is shifted toward shallower depths where productivity is higher.

At the bottom of the food web, the primary producers (or in the benthos, the smallest detritivores) can compete intensely when productivity is high and populations are not reduced by disturbance or predation (Wilson, 1990). Not surprisingly, the highest diversity at low trophic levels is found at very low productivity (Fig. 12.5a). In the deep-sea benthos, the smallest organisms and lowest trophic levels (i.e., the meiofauna) reach their highest diversity at depths of 3000 to 5000 m, and the species richness of these groups is higher than that of the macro- and mega-fauna. The maximum diversity of foraminifera, one of the smallest of the meiofaunal taxa, is found at abyssal depths (>4000 m) (Buzas and Gibson, 1969). Likewise, the diversity of the Protobranchia, a subclass of primitive bivalve molluscs, and of the Cumacea, crustaceans of the subclass Malacostraca that bury themselves in the sediment and feed by filtering or by scraping organic matter off of sand particles, reaches maximum levels at depths of 3000 to 4000 m (Fig. 12.5b and c).

Diversity of the intermediate-sized (and trophically intermediate) organisms peaks at the intermediate depths of 2000 to 3000 meters. The polychaetes, the most diverse of the macrofaunal taxa, include a number of different functional types, such as predators, filter feeders, and sediment gleaners. Highest polychaete diversity is found between 2000 and 3000 m (Fig. 12.5d). The highest diversity of gastropods, many of which are specialized predators on polychaetes, is found between 1000 and 3000 m (Fig. 12.5e). Finally, at the highest trophic levels, the maximum diversity of both vertebrate and invertebrate megafauna reaches its highest levels in the productive waters a few hundred meters deep, although mean diversity peaks at somewhat greater depths (Fig. 12.5f and g). The trophic shift hypothesis is difficult to test quantitatively for the deep-sea benthos because the trophic interactions of the meiofauna and macrofauna are not well understood. Biomass estimates for the two groups are similar (Thiel, 1975), which may reflect trophic independence, or simply a higher turnover rate among the smaller prey organisms. The highest trophic levels of the megafauna may also obtain food by scavenging or preying on pelagic species (see Rex, 1981). In spite of the uncertainty about trophic interactions, the basic pattern is clearly a shift in the maximum diversity of taxa at higher trophic levels to shallower, more productive waters.

The patterns of species replacement, or zonation, along the benthic depth gradient also show evidence of differences in competitive interac-

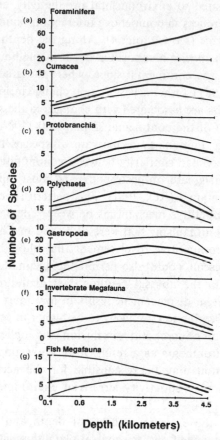

Depth (kilometers)

Fig. 12.5 Trophic shift in the diversity/depth curves toward shallower depths (and higher productivity) at higher trophic levels. Upper figures (*a,b*) represent smaller organisms and presumably lowest trophic levels (meiofauna) and lower figures (*f,g*) represent larger organisms and higher trophic levels. (megafauna). In each figure the thick line represents the means for all sites within a depth class and the thin lines outline the ranges of all sites. (*a*) Foraminifera, protozoans with chitenous or calcareous multichambered shells (Buzas and Givson, 1969); (*b*) Cumacea, small filter-feeding and scraping crustaceans (Jones and Sanders, 1972); (*c*) Protobranchia, primitive bivalve molluscs (H.L. Sanders, unpublished data cited in Rex, 1981); (*d*) Polychaeta, filter-feeding, gleaning, and predatory worms (Hartman, 1965); (*e*) Gastropoda, scraping and predatory snails (Rex, 1976, 1977); (*f*) invertebrate megafauna, primarily large echinoderms, giant amphipods, decapods, and pycnogonids (Haedrich *et al.*, 1980); (*g*) fish megafauna (Haedrich *et al.*, 1980). Note that variability in species diversity increases in the shallow, productive environments, where the variation in rates of disturbance and predation is much higher and more important in maintaining species diversity. For all groups except the Foraminifera, the diversity values are the expected number of species per sample of 50 individuals, calculated using the rarification technique of Hurlbert (1971). (Based on Rex, 1981.)

tions that may be related to environmental productivity, environmental stability, and the differences in competitive interactions among sedentary versus mobile organisms (see Chapter 4). Along the depth gradient the most rapid changes in species composition occur at the boundary of the continental shelf and the continental slope, where the gradient changes most abruptly (Rex, 1981). Other areas of rapid change in species composition at greater depths are associated with changes in the steepness and sediment composition of the continental slope (1000-1100 m) (Haedrich *et al.*, 1975). Among different taxa, the rate of species turnover with depth (based on percentage similarity between communities) is least in polychaetes, somewhat greater in gastropods, and most rapid in the fish and invertebrate megafauna (Rex, 1981). This pattern is consistent with the differences in competitive mechanisms between sedentary organisms and mobile perceptive organisms that were discussed in Chapter 4.

The low rates of competitive displacement that allow high diversity on an ecological timescale could also permit the accumulation of large numbers of species in the abyssal benthos over evolutionary time, if low rates of competitive displacement result in low rates of extinction (Chapter 4). Rates of extinction are likely to be higher in very productive environments as a result of competitive exclusion, and higher in extremely low productivity environments as a result of very low population sizes. This same phenomenon may be responsible for the accumulation of endemic species in low productivity terrestrial environments (Chapter 11).

In summary, over the extreme gradient of depth and other factors between the continental shelf and the abyssal plain, the factor that seems to best explain the overall pattern of benthic community structure is productivity. While other conditions in addition to productivity also vary along the depth gradient, the dominant patterns are consistent with those expected along a strong gradient in the rate of competitive displacement with relatively low disturbance intensities (Fig. 5.5). The low rates of competitive displacement that are associated with low productivity environments of the continental slope and rise allow the maintenance of high species diversity, undoubtedly augmented by other processes such as predation and environmental heterogeneity. However, in high-productivity shallow (0-200 m) benthic environments, the presumably high rates of competitive displacement result in low species diversity, in spite of the diversity-increasing effects of predation and environmental heterogeneity. Finally, in the extremely low productivity environment of the abyssal depths, population growth rates are apparently so low that few species can maintain viable populations, and diversity drops to low levels. If we were able to study these communities in more detail, it is likely that we would find smaller-scale patterns in which predation, disturbance, or environmental heterogeneity had a greater effect on species diversity than productivity and the rate of competitive displacement. Unfortunately, such data are not yet available (Grassle, 1991), and we are only able to detect relatively coarse patterns at the scale of the entire gradient of the deep benthos.

Shallow Benthos

Shallow benthic communities are relatively easier to study than the abyssal benthos, and their productivity can be quite high (and diversity low), particularly in polar regions where they receive organic matter from the highly productive surface waters (Walsh *et al.*, 1989). In the shallow, sandy-bottomed Chirikov Basin in the northern Bering Sea, a large (27 mm) amphipid, *Ampelisca macrocephala,* forms dense mats of tubes with densities of up to 22,000 individuals/m^2 and biomass of 941 g/m^2 (Nerini and Oliver, 1983). This highly productive benthic area is the primary feeding ground of the California gray whales, which migrate annually from their breeding grounds off Baja California to calve and feed in the productive waters of the Bering and Chukchi Seas (Nelson and Johnson, 1987). In the mud and gravel sediments of the region surrounding the Chirikov Basin, productive clam beds support a permanent population of 200,000 Pacific walruses. Both whales and walruses cause significant physical disturbance to the benthic sediments (Nelson and Johnson, 1987). Although these animals feed on the dominant benthic organisms, and initiate a successional sequence among the invertebrates that recolonize the disturbed areas (Nerini and Oliver, 1983), the disturbances are apparently insufficient to increase significantly the species diversity of these productive communities.

Recent investigations in the Bering and Chukchi Seas provide interesting insights into the smaller-scale dynamics of shallow (20-50 m) arctic marine benthos. Grebmeier *et al.*, (1988, 1989) report on a spatially complex system in which different water masses and current patterns create sharp discontinuities in species composition and diversity. The study area spans the waters to the north (Chukchi Sea) and south (Bering Sea) of the Bering Strait between eastern Siberia and western Alaska (Fig. 12.6). The relationship between benthic biomass and the amount of organic matter that drifts down from shallower waters is well established for the shallow benthos (Rowe, 1971; Grebmeier *et al.*, 1988). In this Arctic system, surface water productivity varies strongly between different water masses, with dramatic effects on benthic biomass and community structure.

Two distinct water masses flow northward through the Bering Strait and into the Chukchi Sea. Flowing along the east side is the Alaska Coastal Water, which is low in nutrients and has low primary productivity (50 g C m^{-2}, Sambrotto *et al.*, 1984; Walsh *et al.*, 1989). Water from the Anadyr Strait merges with Bering Shelf water, forming a gradient from the high-productivity Bering Shelf-Anadyr Water in the west to the low-productivity Alaska Coastal Water in the east. These Bering Shelf-Anadyr waters have an estimated annual primary production of 250-300 g C m^{-2} (Walsh *et al.*, 1989; Grebmeier *et al.*, 1988) with peak productivity during the short Arctic summer. Both the total organic carbon in surface sediments and the benthic biomass are closely related to the productivity of the overlying waters (Fig. 12.7)

In these relatively shallow benthic communities, scouring by deep currents is a physical disturbance that is largely independent of productivity conditions. Spatial variation in current intensity is reflected in sediment composition, with fine sediments dominating in areas with weak currents, and coarse sediments predominant in areas with stronger currents. If sediment composition is used as an index of the intensity of disturbance due to current scour, a complex pattern of species diversity in relation to biomass is revealed that is consistent with the dynamic equilibrium hypothesis. For sites with predominantly fine sediments, reflecting a low intensity of current scour, benthic species diversity decreases with increasing biomass (Fig. 12.8*a*). This pattern is the expected consequence of an increased rate of competitive displacement that should result from increasing productivity at low disturbance intensities (Fig. 12.8*b*). In contrast, for sites with predominantly coarse sediments, reflecting a high intensity of current scour, species diversity increases with increasing biomass (Fig. 12.8*a*). This pattern is consistent with more rapid population recovery from disturbance under conditions of higher productivity (Fig. 12.8*b*). In some shallow and potentially productive areas, extremely high intensity disturbances resulting from seasonal ice gouging are associated with low density, low diversity communities (J. Grebmeier, personal communication).

These patterns in the arctic benthos are consistent with the results of experiments in coastal soft-sediment communities, where there is evidence that both competition and predation can influence species diversity

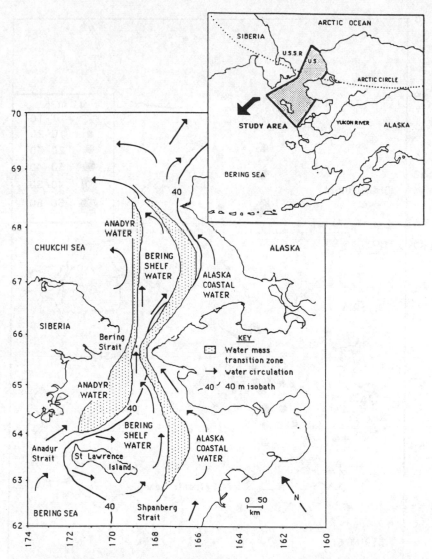

Fig. 12.6 Region of the northern Bering and Chukchi Seas showing local water circulation, water masses, and bathymetry. The nutrient-poor Alaska Coastal Water flows along the east, and the nutrient-rich Bering Shelf-Anadyr Water flows along the west. The frontal zone between the Alaska Coastal Water and the Bering Shelf Water is very distinct, while the contact between the Anadyr and Bering Shelf Water forms a more gradual gradient. (From Grebmeier *et al.*, 1988.)

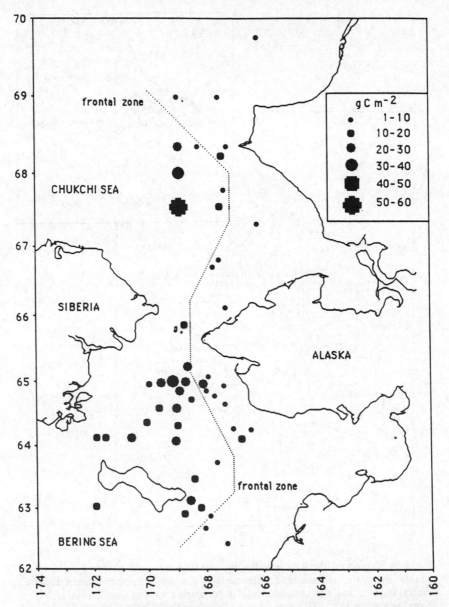

Fig. 12.7 Distribution of benthic biomass in the northern Bering and Chukchi Seas in relation to the frontal zone between the nutrient rich Bering Sea-Anadyr Water and the nutrient-poor Alaska Coastal Water. (From Grebmeier *et al.*, 1988.)

(Wilson, 1990). It is interesting that exclusion of epibenthic predators ususally results in an increase in density and biomass, but an increase or little change in species diversity (Wilson, 1990). This pattern contradicts the predictions of the intermediate disturbance hypothesis, but is the pattern predicted by the dynamic equilibrium model for conditions with low rates of growth and competitive displacement (e.g., Fig. 5.5*a*).

A curious variant of the high productivity benthic communities of shallow polar waters are the 'inverted benthos' of the underside of sea ice. These communities at the ice-sea interface are supported by high algal productivity using light that penetrates through the snow and ice (Dunbar, 1968; Alexander, 1974; Tremblay *et al.*, 1989). Algal productivity is inversely correlated with the depth of water below the ice (Pike and Welch, 1990), suggesting that nutrients may be limiting the productivity of this ecosystem. The high algal productivity is associated with a very high density of macroinvertebrates, primarily gammarid amphipods, which support fish, seals, and sea birds. As expected, the diversity of the algae and amphipods is quite low, with one or two species often contributing 90% of the amphipod biomass (Carey, 1985; Pike and Welch, 1990). The structure of the meiofauna (primarily nematodes and copepods) and the microfauna (protozoa) communities is not yet well understood (Carey, 1985).

It is interesting to note that although the diversity of both primary producers (phytoplankton) and the secondary producers of the benthos is quite low in polar waters, the high primary productivity of these systems provides another dramatic example of the trophic shift of diversity in response to productivity (Fig. 5.14). The regional biomass and species richness of the marine birds and mammals at the highest trophic levels are greater in polar regions than anywhere else, and are closely tied to the densities of their prey species (Brown, 1980; Hunt and Schneider, 1987; Springer *et al.*, 1986, 1987; Heinemann *et al.*, 1989). The high densities of Arctic and Antarctic seabirds (e.g, puffins, murres, penguins, etc.) and of marine mammals such as seals and whales, provide some of the most spectacular aggregations of large animals anywhere on earth (Gambell, 1985; Laws, 1985).

In summary, marine benthic ecosystems are based on organic matter that drifts down from surface waters. Consequently, the productivity of benthic ecosystems is dependent on the primary productivity of the overlying waters, or other sources of organic matter. Because of their depth below the surface, these ecosystems tend to be well-buffered from environmental variation and disturbances due to storms and wave ac-

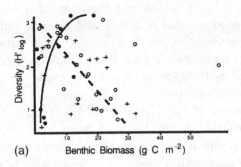

(a)   Benthic Biomass (g C m^{-2})

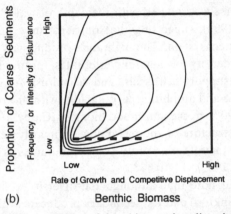

(b)   Benthic Biomass

Fig. 12.8 Dynamic equilibria of benthic species diversity in the Bering and Chukchi Seas in relation to biomass and intensity of current scour (indexed by sediment composition). (a) Solid circles and solid line indicate sites with the sum of medium sand, coarse sand, and gravel composing more than 15% of the sediments. Open circles and dashed line indicate sites with coarse sediments composing less than 15% of the sediments, and most sediment in the classes of silt and clay, very fine sand, and fine sand. Crosses indicate sites for which sediment composition was not reported (based on data from Grebmeier et al., 1989). The three outlier open circles in the upper right portion of the figure are sites in the transition zone between the two water masses. These sites are subjected to seasonal variation in carbon input ranging from high to low, and may also experience greater variability in current scour than other sites (J. Grebmeier, personal communication). (b) Hypothesized locations of benthic sites on the contour surface of the level of species diversity expected to result from the dynamic equilibrium between disturbance and the rate of competitive displacement (based on Huston, 1979). The dotted line corresponds to those sites in (a) with a low intensity of current scour, so diversity is expected to decrease as growth rates and rate of competitive displacement increase with increasing productivity (indexed by biomass). The solid line corresponds to those sites in (a) with a high intensity of current scour, where diversity is expected to increase with increasing biomass as higher growth rates allow more populations to survive the disturbances and thus increase the number of species present.

tion. Thus, both coastal and deepwater benthic ecosystems represent an extreme contrast to the marine intertidal zone, which is characterized by both high productivity and a high frequency and intensity of disturbances. Among benthic communities, variation in the rate of competitive displacement (directly related to productivity) seems to be the predominant factor that determines the large-scale patterns of species diversity, such as those along depth gradients (productivity generally decreases with depth) and horizontal gradients that result from variation in the productivity of surface waters. In benthic systems where both the rate of competitive displacement and the disturbance frequency or intensity vary sufficiently, species diversity can be expected to reflect the dynamic equilibrium between these opposing processes.

Marine Open Water Ecosystems

Patterns of Productivity in the Open Ocean

The productivity of marine open-water systems is dominated by phytoplankton in the upper 200 meters of the ocean. Productivity is highly dependent on nutrient levels, and ranges from very high (up to 10 g C m^{-2} d^{-1}) in upwelling areas to very low (less than 0.1 g C m^{-2} d^{-1}) in oligotrophic regions in mid-ocean (Russell-Hunter, 1970; Sumich, 1976; Vinogradov, 1983). The vast majority of the ocean surface has very low productivity (Table 12.1, Fig. 12.9). The species diversity of the planktonic organisms of the world's oceans shows the same inverse relationship to productivity that is found in most other ecosystems. However, the distribution of productivity and diversity in the open oceans is not a simple gradient, but rather a complex pattern based on currents and upwellings (Vinogradov, 1983).

Although monotonous and apparently uniform when viewed from the surface, the open ocean is actually a complex system of currents and countercurrents, convergences and divergences, and upwellings and downwellings, driven by the prevailing winds and the rotation of the earth. Vast areas of uniform conditions, the central gyres of the major oceans, are separated by zones where adjacent currents flow in opposite directions (Fig. 12.10). To both the north and south of the equator, major currents flow westward in all the world's major oceans (the North Equatorial and South Equatorial currents). Opposing the westward-flowing equatorial currents is the Equatorial Counter Current, which flows eastward, displaced somewhat to the north of the equator. The

Table 12.1. *Rates of net algal primary productivity in the major regions of the world's oceans.*

Region	Percentage of Ocean Area	Normal Range (Average) (g C m^{-2} yr^{-1})	Total Production (10^9 g C yr^{-1})
Open Ocean	91.9	1-187 (57)	18.9
Continental Shelf	7.4	94-281 (164)	4.4
Upwelling Areas	0.1	187-469 (227)	0.09
Algal Beds and Reefs	0.2	234-1875 (1136)	0.7
Estuaries (excluding marshes)	0.4	94-1875 (682)	0.95

From Whittle (1977)

general westward movement of wind toward the equator in both the northern and southern hemispheres tends to produce a convergence of currents along the eastern margin of continents. Associated with these convergences is an accumulation and downwelling of nutrient-poor surface water. In contrast, along the western margins of the continents, the equatorial currents pull water away from the continents, leading to a divergence of surface water and an upwelling of nutrient-rich deep water. Thus, the world's most productive fisheries are found along the western margins of continents, where northward flowing currents in the southern hemisphere and southward flowing currents in the northern hemisphere mix cold polar waters with cold nutrient-rich upwellings (Sverdrup *et al.*, 1942; Vinogradov, 1983). It is interesting to note that, in addition to supporting an incredibly high productivity of marine organisms, these cold currents with nutrient-rich upwellings tend to produce deserts and semi-arid mediterranean climate conditions on the adjacent western margins of continents, which contrast starkly with the high productivity of the nearby ocean.

The same pattern of high plant species diversity associated with the low primary productivity of the mediterranean climate regions (Chapter 11), is also found in the various low-productivity regions of the open ocean. Perhaps the most striking pattern is the remarkably high planktonic species diversity of the central 'anti-cyclonic' oceanic gyres, vast areas of slowly circulating and downwelling nutrient-poor water that are found in the central northern and southern regions of both the Atlantic and Pacific Oceans, as well as in the Indian Ocean.

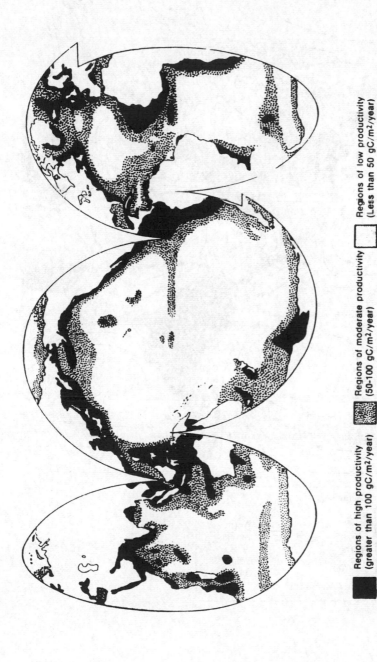

Fig. 12.9 Geographic distribution of marine primary production. Note the high productivity of coast, polar, and upwelling areas (including the equatorial upwelling), and the low productivity of the open ocean areas of the central oceanic gyres, which are distributed symmetrically about the equator in both the Atlantic and Pacific Oceans. (Based on FAO 1972, from Sumich, 1976.)

■ Regions of high productivity (greater than 100 gC/m²/year)

▨ Regions of moderate productivity (50-100 gC/m²/year)

☐ Regions of low productivity (Less than 50 gC/m²/year)

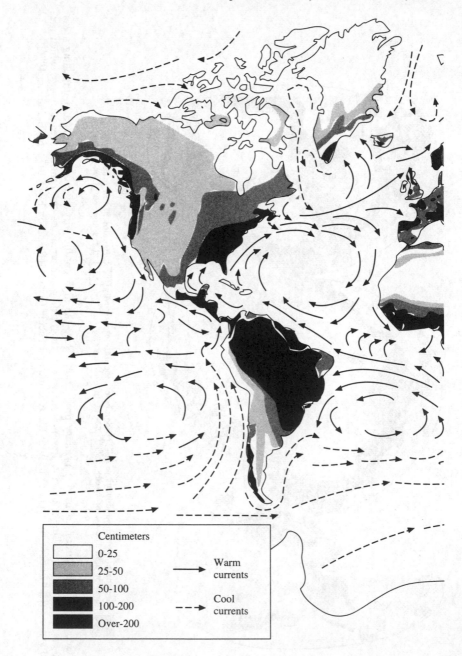

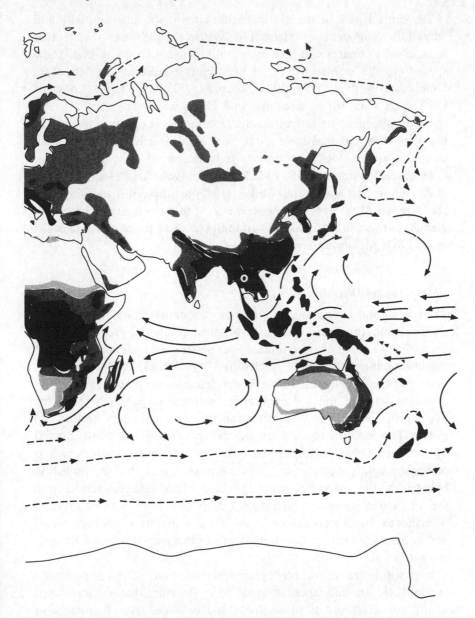

Fig. 12.10 Major currents of the world's oceans, showing upwelling and down-wellings areas, cold and warm currents, and the distribution of precipitation on land. Note the locations of major deserts in relation to cold currents.

The central gyre of the North Pacific Ocean is a huge spatially and temporally homogeneous oligotrophic system that has been investigated in an effort to understand how its incredibly high diversity of planktonic species can be maintained in an apparently equilibrium system (McGowan and Walker, 1979, 1985; Venrick 1982). This large (1.5×10^7 km^2) water mass slowly circulates in a clockwise direction, and is relatively uninfluenced by lateral mixing from adjacent current systems. In the upper few hundred meters, the horizontal distribution of physical, chemical, and biological properties is homogeneous in comparison to other current systems (NORPAC Committee, 1960; Reid 1962; Hayward *et al.*, 1983). McGowan and Walker (1985) attribute this uniformity to the large size of the gyre and the weakness of the lateral mixing processes. Neither is there much seasonal variability in these properties (Hayward *et al.*, 1983; McGowan and Walker, 1979, 1985).

Phytoplankton Diversity

The spatial and temporal uniformity of conditions, as well as the physiological and functional similarity of most of the co-existing species, particularly among the phytoplankton, present great difficulties for an equilibrium interpretation of species diversity (Venrick, 1982). Venrick reports that the phytoplankton can be divided into two groups (Fig. 12.11), a shallow stratum and a deep stratum, with the boundary between the two varying from 75 to 90 meters, and the lower boundary around 180 meters. The productivity of the upper group, which is apparently limited by nutrients, is much higher than that of the lower group, which is limited by light (Eppley *et al.*, 1973). Venrick argues that the transition between the shallow and deep groups is a dynamic zone that results from the interaction of light, which decreases exponentially with depth (and is unaffected by phytoplankton at the low densities at which they occur) and nutrients, which increase with depth as nutrients diffuse up through the nutricline.

Even within the diverse phytoplankton community of this oligotrophic region, there are differences in community structure that are associated with depth gradients of productivity and biomass. The dominance of the most-abundant species is higher in the more productive shallow stratum than in the deeper stratum, where abundances are much more evenly distributed (Fig. 12.12, upper panel). This is true both for large-scale averages, as well as single samples (Fig. 12.12, lower panel). The larger total number of species sampled in the shallow stratum (178

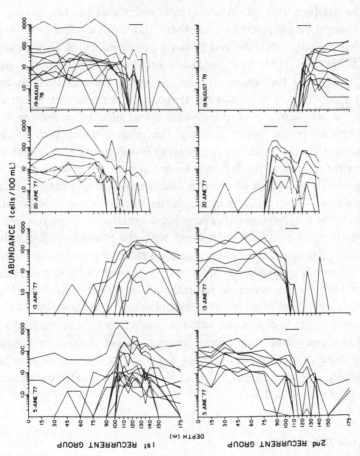

Fig. 12.11 Depth distributions of individual species in the shallow and deep phytoplankton assemblages ('recurrent groups') at four locations (indicated by different dates on the sampling voyage) in the central gyre of the North Pacific Ocean. Each vertical pair of graphs shows the two groups at a particular station, ordered by the number of species in the group. Species compositions of groups at different stations do not necessarily correspond. Vertical bars indicate the depth range of the transition zone. (From Venrick, 1982.)

species versus 139 in the deeper stratum) is primarily a consequence of the greater volume of the shallow stratum, which is three to four times thicker than the deeper stratum. Thus, the local diversity of the phytoplankton increases with depth (i.e., decreasing productivity), even in this highly oligotrophic system.

Positive correlations between the abundances of different species within groups identified by Venrick (1982) suggests that these individuals are responding similarly to environmental conditions and are not having a significant negative effect on one another. Based on evidence on zooplankton community structure and biomass presented by McGowan and Walker (1979), it is likely that predation on the phytoplankton is significantly greater in the upper stratum, as would be expected on the basis of phytoplankton productivity. In spite of the potential effect of predation by zooplankton in preventing dominance by a few phytoplankton species in the upper stratum, the relative dominance of the most-abundant species in the upper stratum is still much higher than in the lower stratum, where predation rates are apparently much lower.

This pattern is consistent with the predictions of the dynamic equilibrium hypothesis for a situation in which the rate of population growth and competitive displacement (presumably correlated with phytoplankton productivity) is positively correlated with the rate of predation by zooplankton. The low rate of competitive displacement in the low light, low productivity deep stratum is apparently sufficient to allow the maintenence of high species diversity, in spite of lower predation rates (Fig. 12.13). The high diversity of similar phytoplankton species in the North Pacific Central Gyre is consistent with the prediction of non-equilibrium competition theory that very similar species should coexist for longer periods of time than dissimilar species, since slight differences in competitive ability will take longer to be expressed than greater differences (Huston, 1979; Caswell, 1982).

Zooplankton Diversity

McGowan and Walker (1979, 1985) describe a similar situation for the copepods, which are a major component of the zooplankton and feed primarily on phytoplankton. Copepods are found over a much greater depth range (0-600 m) than the phytoplankton and some species have vertical diurnal migrations of 200 meters or more. McGowan and Walker (1979) separated the copepods into seven groups of similar species, with each group distinguished by a different combination or variability of

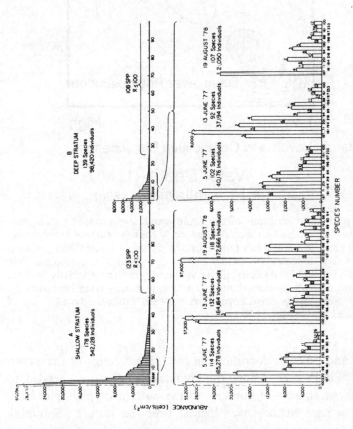

Fig. 12.12 Rank order of species abundances in the shallow and deep phytoplankton assemblages in the central gyre of the North Pacific Ocean. Upper panel (A and B) shows the total ranked abundances for the three stations presented separately in the lower panel. The lower panels present the ranked abundances of the 15 most abundant species at each station, with the species identification numbers indicated along the x axis, and the rank at that station at the top of each bar. Note the higher degree of dominance by the most abundant species in the shallow assemblage (stratum), both on average and by separate stations. (From Venrick, 1982.)

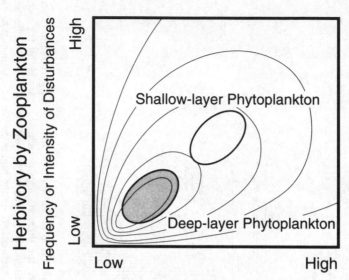

Fig. 12.13 Hypothesized positions of the shallow and deep phytoplankton communities of the North Pacific Central Gyre on a contour surface representing the level of species diversity maintained by the dynamic equilibrium between rate of competitive displacement (which is very low overall as a result of low nutrient availability and is regulated along the depth gradient primarily by light availability) and the rate of disturbance (resulting primarily from herbivory by zooplankton). Note the positive correlation between productivity and rate of predation. (Based on Huston, 1979.)

prey species and physical conditions in their depth range of maximum abundance. Within each group, the abundances of species tended to be positively correlated. In contrast to the pattern of phytoplankton, zooplankton species richness was highest near the surface. Summing those groups that were most abundant in the upper 100 m (Groups I, II, IV, and V), the total number of species was 50, while for those groups most abundant below 100 m (III, VI, and VII), the total number of species was only 20. Comparing zooplankton to phytoplankton, there is a shift of maximum diversity toward higher productivity environments with increasing trophic level.

There was little evidence for spatial or temporal variability that could explain the high copepod species diversity. Species abundance patterns were similar at scales of several hundred meters to 1000 km (Fig. 12.14), suggesting that there was insufficient spatial patchiness to explain the high diversity on the basis of either disturbances or variation in competitive conditions (McGowan and Walker, 1985). Although temporal and spatial variability in physical conditions tended to be greater in the winter, there were no significant differences in community structure between winter and summer, nor were there seasonal patterns of plant-nutrient concentrations, primary productivity, or zooplankton biomass (McGowan and Walker, 1985). Even a large-scale change in sea surface temperature from a 1.4 °C warm anomaly in 1968 to a −1.0 °C cool anomaly in 1969, which resulted in a doubling of primary production and zooplankton biomass, resulted in no significant change in the proportions of the dominant species. Both vertebrate and invertebrate predators are present and have a community structure that is as constant as that of the copepods (Clarke, 1978; Barnett, 1983; Hayward and McGowan, 1979). While these predators may well help maintain the diversity of the zooplankton community, the low productivity of the system suggests that the total predator biomass, and the regulatory effect of higher trophic levels, is likely to be low (Ryther, 1967; Crisp, 1975).

Thus, for the copepods as well as the phytoplankton, the high species diversity of the North Pacific Central Gyre is difficult to explain by any mechanism other than the extremely slow rates of competitive displacement that are expected in this oligotrophic, low productivity system. Furthermore, the maintenance of diversity by other processes, such as low levels of predation, and whatever spatial and temporal heterogeneity does exist, is likely to be much more effective under conditions of very low productivity than it would be at high levels of productivity with high rates of competitive displacement.

In contrast to the stable oligotrophic conditions of the North Pacific Central Gyre, the California current to the east and the sub-Arctic water mass to the north differ dramatically in species composition and diversity. Sub-Arctic copepods (the most thoroughly studied component of the zooplankton) have a much higher biomass and dramatically lower species diversity, with fewer total species and much greater dominance by the most abundant species (McGowan and Walker, 1985; Vinogradov, 1983; Springer et al., 1989) than in the central gyre (Fig. 12.15). Further to the north, in the Bering and Chukchi Seas, extremely high primary productivity (Sambrotto et al., 1984; Walsh et al., 1989) supports both a productive benthic ecosystem (see previous section) and a high biomass

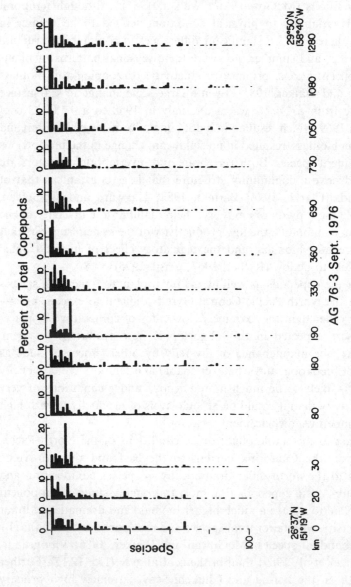

Fig. 12.14 Constancy in the relative abundances of copepod species along an east–west transect in the central gyre of the North Pacific Ocean. Each histogram represents the copepod community structure at a station, with increasing distance east of 26°37'N, 151°19'W. Note that species distribution changes little over a distance of 1200 km. (From McGowan and Walker, 1985.)

of zooplankton, both with low diversity and a high degree of dominance by a few species (Springer *et al.*, 1989). To the east, the California current has much higher productivity and also high spatial and temporal variability in physical conditions and copepod community composition (McGowan and Walker, 1985). Although the copepod species diversity in the California Current is much higher than in the sub-Arctic and Bering-Chukchi Seas, it is not as high as in the Central Gyre.

In summary, the species diversity of phytoplankton and zooplankton in the world's oceans has a strong, inverse relationship with primary productivity. This widespread pattern is most likely the result of variation in the rate of competitive displacement along the productivity gradient, with disturbance, predation, and mixing acting primarily to keep the system in a state of small-scale non-equilibrium. This mechanism results in a spatially complex pattern of oceanic diversity that is determined by the effects of currents on the upwelling of nutrient-rich deep ocean waters, or on other sources of nutrients to support high primary productivity. In addition to upwellings and the polar regions, continental shelf areas tend to have high primary productivity and consequently low species diversity of phytoplankton and zooplankton. As expected, the maximum diversity of the highest trophic levels is found in the most productive areas, a fact that is most dramatically reflected in the high biomass and diversity of marine mammals and birds in polar regions.

Species Diversity on Coral Reefs

Coral reefs are probably the most beautiful ecosystems on earth. The amazing variety of shapes and myriad colors of the corals, sponges, and gorgonians are animated by the pulse of wave action. Yet the complex framework of the reef is only the backdrop for a spectacular three-dimensional display of shifting galaxies of fish of all sizes and hues, punctuated by individuals or small groups of fish of fantastic shapes and colors. It is no wonder that these systems have attracted the interest of researchers, as well as the public at large.

The reef crest and shallow portions of the reef (down to 5 meters) have received most of the popular and scientific attention, largely because they are easily accessible even without specialized diving equipment, and because it is here that much of the animal life is concentrated and the spectacular colors are most easily observed. In fact, the remarkable colors of the reef are only visible under natural light in shallow water where all wavelengths of light can penetrate. Only the shorter wavelengths are able

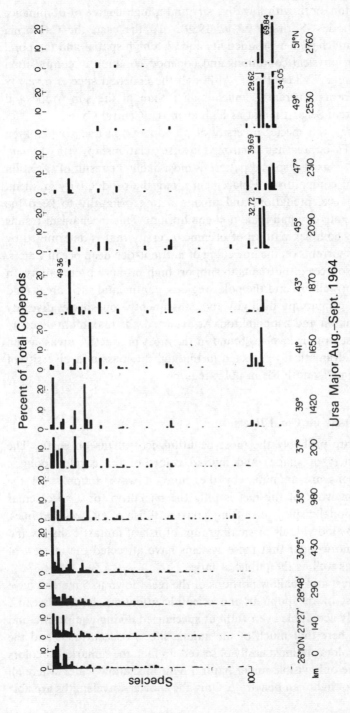

Fig. 12.15 Number and relative abundance of copepod species along a transect from near the center of the North Pacific Central Gyre (26°10' N) northward into the sub-Arctic Water Mass (~44-56° N). The upper 100 species are those most abundant in the Central Gyre, and the lower species are those encountered outside the Central Gyre. Data collected August–September, 1964. (From McGowan and Walker, 1985.)

to penetrate to the deep reef, where everything appears in muted shades of bluish green except in the artificial light of a camera flash. Most of the animal life is also concentrated in the shallow portions of the reef, where productivity is highest. The intense biological activity and high spatial heterogeneity of the shallow reef provide an infinite supply of visual and biological curiosities.

Basic Patterns of Coral Diversity

Many different gradients of environmental conditions are found on coral reefs. Because of the massive physical structure of reefs, they create their own gradient of exposure to the open ocean, ranging from high exposure on the reef crest and slope facing the open ocean, to much lower exposure on the protected side. Obviously, the relative location of a particular reef in relation to islands, other reefs, and prevailing winds and currents results in great differences in the conditions experienced by different reefs. In addition, the decrease in light availability and wave energy with depth create strong vertical gradients of productivity and disturbance. All of these conditions interact to regulate the species diversity of coral and other sessile reef organisms in a dynamic equilibrium between growth conditions and disturbance properties.

Just as in other productive ecosystems, such as the rocky intertidal zone, species diversity at the reef crest is highly variable and strongly influenced by predation, storms, and other mortality-causing disturbances. Although the reef crest is often dominated by one or two species of corals, high diversity can be found on reef crests and shallow reef flats that are periodically exposed at low tides (Ditlev, 1978; Fishelson, 1973; Glynn, 1976) or subjected to other disturbances that cause mortality. High diversity has also been found under conditions apparently unfavorable for coral growth. Ohlhorst (1980) recorded the highest diversity at Discovery Bay ($H' = 2.34$) in a very turbid 18-m site within the bay; diversity was also high at 6 m in the back reef, where high temperatures and reduced water circulation may negatively affect coral growth. Notwithstanding the dramatic variability of the reef crest, the complex structure and life of coral reefs extend far below the surface, and the depth zonation of dominant species and growth forms on coral reefs is one of the most striking patterns found in any natural community (Goreau, 1959a; Goreau and Goreau, 1973; Stoddart, 1969).

Prior to the deep-reef studies made possible by the development of SCUBA, biologists believed that coral species diversity decreased monotonically with depth, following the gradient of decreasing light availability. This pattern was consistent with available data from sampling with dredges (e.g., Wells, 1957). However, the first systematic study of species coexisting at different depths on a well-developed reef found the opposite pattern (Loya and Slobodkin, 1971; Loya, 1972). The reefs near Eilat in the Red Sea extend from a reef flat at the surface to a depth of 40 to 50 m. In linear transect samples over a 0 to 30-m-depth range Loya (1972) found that the number of species per 10-m transect increased from 13 on the reef flat to a maximum of 30 at the deepest site. Diversity increased from the surface to a depth of 8 to 12 m and then remained relatively constant to 30 m (Fig. 12.16a). Identical patterns of increasing diversity with depth have been found on other reefs as well.

At Discovery Bay, Jamaica, the reef extends from the surface to a depth of 60 to 70 m, where a vertical escarpment occurs (Liddell and Ohlhorst, 1981). Huston (1985b) sampled from 0 to 30 m in depth, using the same 10-m linear transect technique as Loya. The study was made in March 1977, prior to the destruction of much of the upper reef by Hurricane Allen in 1980 (Woodley, 1980). Coral species richness was much lower than that found at Eilat, but the pattern of increasing diversity with depth was similar. Species richness ranged from two species encountered in 30 m of line transect on the reef crest (dominated by *Acropora palmata*) to 17 species per 30 m of transect at a depth of 20 m. The same general trends were found in transects on the east and west ends of the fore reef, (Fig. 12.16b) and in all groups of organisms.

In the Indian Ocean, Sheppard (1980) sampled coral species diversity from the surface to a depth of 60 m on two atolls in the Chagos Archipelago. Species richness increased from the surface to a maximum at around 20 m and then declined gradually to the deepest sites. It is of interest that the same pattern was found on both the inner slopes of the lagoon (0 to 40 m) as was found on the outer seaward slopes (Fig. 12.16c), where disturbance intensities are much greater. As at Eilat and Discovery Bay there were striking zones of species dominance along the depth gradient, but species richness increased and decreased independently of the zones.

It is not surprising that Sheppard found that the increase in diversity with depth does not continue indefinitely. The decreasing light levels with increasing depth must inevitably lead to a decrease in the number of photosynthetic species that are able to survive. Diversity of photosynthetic organisms begins to decline near 30 m and presumably continues decreasing to the lower limit of photosynthesis. At Discovery Bay, stud-

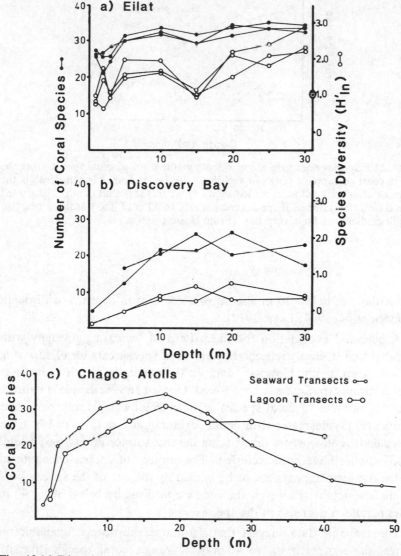

Fig. 12.16 Diversity patterns with depth on three coral reefs. Solid circles represent number of coral species and open circles the *H'* diversity index (based on natural logarithms). (*a*) Eilat, Red Sea. Data are means based on three 10-m linear transects from each depth along three depth gradients (Loya, 1972); (*b*) Discovery Bay, Jamaica. Data were collected as in Loya's study and are from two separate depth gradients (Huston, 1985c); (*c*) Chagos Atolls, Indian Ocean. Data are means based on 10-m transects from 19 seaward depth gradients and approximately 19 lagoon depth gradients (Sheppard, 1980). (From Huston, 1985a.)

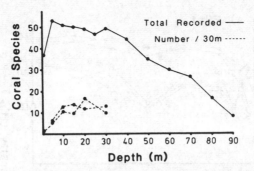

Fig. 12.17 Species pool sizes along a depth gradient for 62 coral species from the north coast of Jamaica (Goreau and Wells, 1967). Pool size at each depth is the number of species whose recorded depth range includes that depth. The dotted line is the total number of species encountered in 30 m of line transect along two depth gradients at Discovery Bay. (From Huston, 1985a.)

ies of the reef below 30 m also show a decrease in diversity with depth (Liddell *et al.*, 1984; Loya, 1972).

A potential explanation for the pattern of increasing diversity with depth is that it simply reflects the number of species capable of surviving at any given depth. However, data on the maximum depth range over which various species have been found (Goreau and Wells, 1967) indicate that the number of coral species capable of growing at any particular depth actually decreases with depth beginning at 5 m (Fig. 12.17). Even at shallow depths, where diversity on the reef surface is very low, a total of 37 species have been recorded. The number of species encountered in the transects appears not to be limited by the size of the species pool at shallow depths. However, the decrease in diversity below 30 or 40 m does parallel a decrease in the species pool.

The available data suggest that if suitable substrate is available on a reef slope over the 0- to 40-m-depth range, coral species diversity increases from the surface to 10 to 15 m, is highest at 15 to 30 m, and begins to decline between 25 and 40 m. This pattern of low diversity near the surface and highest diversity at intermediate depths cannot be explained by any simple physical gradient. Oxygen, temperature, and salinity do not vary with depth sufficiently to be ecologically significant on a single reef (Dana, 1976; Ohlhorst, 1980; Ott, 1975) but may be important on a larger scale (Glynn, 1977; Glynn and Stewart, 1973).

Components of the Dynamic Equilibrium on Coral Reefs

In contrast to the rocky intertidal zone where nearly all biological interactions take place in an environment with high productivity and high growth rates, the competitive environment on coral reefs ranges from high levels of productivity near the surface to very low productivity deep in the reef. As in the rocky intertidal, there is great variation in the frequency and intensity of disturbances in shallow water. However, nearly all well-developed coral reefs have a strong gradient of disturbance intensity and frequency that decreases with depth, paralleling the decrease in growth rates with depth. Thus, coral reefs provide an ideal system in which to test the complex predictions of the dynamic equilibrium hypothesis.

Abiotic Disturbances

Tidal Exposure. The consequences of coral mortality resulting from occasional extreme low tides are well documented. Fishelson (1973) reports that sporadic, extremely low tides that occur in the Red Sea near Eilat kill all or part of the coral population growing on the reef flat. The rapidly growing (10-15 cm yr^{-1}) branching species (*Acropora*, *Stylophora*, *Pocillopora*) are most sensitive to exposure, but their fast growth compensates for the higher survival of the slowly growing (a few mm yr^{-1}) massive forms. The occasional mass mortality that occurs at Eilat apparently maintains the high reef flat diversity by preventing the monopolization of space by the rapidly growing species (Loya and Slobodkin, 1971). In the absence of disturbances, these same species form monospecific stands over the subtidal reef flats of the Gulf of Suez, where extremely low tides do not occur. Intermediate levels of diversity were found on reef flats at the southern end of the Gulf of Eilat, which experiences extreme tides less frequently than does Eilat at the northern end. Similar observations have been made in other reef systems (Glynn, 1976; Ditlev, 1978; Connell, 1978).

Wave Action. Wave energy decreases rapidly with depth. Most wave energy is released when the wave breaks and so is concentrated near the reef crest. Water movement is reduced to 4% of the surface movement (diameter of particle orbit at the surface equals wave height) at a depth of half the wavelength. The decreasing depth gradient of wave energy is reflected in the effect of Hurricane Allen on the reef at Discovery Bay. Although storm effects were visible to a depth of 30 m, the damage was concentrated on two species (*Acropora cervicornis* and *A. palmata*) in the shallower reef zones (0 to 15 m, S.L. Ohlhorst and W. D. Liddell, personal communication).

Wave damage can influence reef diversity, through both the chronic effects of regular wave action and the extreme effects of destructive storms. Grigg and Maragos (1974) found that the degree of exposure to waves and swells affected coral cover and diversity at their 8-m depth study sites in Hawaii. In general, diversity was highest and cover lowest at the most exposed stations, indicating that disturbance from chronic wave action can also act to maintain diversity. Connell (1978) emphasized the diversifying effect of occasional severe hurricane damage on Australian reefs. Although heavy wave action may cause considerable destruction (for example, see Tunnicliffe, 1983; Woodley, 1980) some water movement is necessary to maintain conditions conducive to coral growth (Hubbard, 1974; Jokiel, 1978; Stoddart, 1969; Wells, 1957).

At Discovery Bay in Jamaica, the frequency of mortality-causing disturbances is apparently too low to maintain high species diversity at the crest. The devastating Hurricane Allen of 1980 was the first major storm to follow this course in 63 years (Neumann *et al.*, 1978). Thus the patterns of diversity recorded by Huston in 1977 (Fig. 12.16*b*) had developed during a 60 yr period of freedom from major storm disturbance. This relative freedom from disturbance may partially explain the lower coral diversity of the reef crest at Discovery Bay in comparison with more exposed reefs elsewhere in the Caribbean or at Eilat, in the Red Sea.

Although some disturbance is apparently necessary to maintain high coral species diversity, diversity may be reduced by extremely frequent or severe disturbances that prevent coral species from surviving. Reef crests subjected to extreme wave action tend to be devoid of corals and are characterized by algal ridges, which are smooth, massive structures covered by encrusting calcareous red algae. Such ridges are well developed in the Indo-Pacific on windward crests (Wells, 1957) and also in the Caribbean (Glynn, 1973). Under conditions of less severe wave action, a community of increasing diversity and structural complexity can develop. Wells (1957) reports a 'richer growth of reef corals' on the crests of leeward reefs where algal ridges are poorly developed or absent.

For the Caribbean, Geister (1977) defines six 'breaker zones' of decreasing wave strength (Fig. 12.18). The first five breaker zones encompass a progression from low diversity caused by extreme abiotic conditions (Zones a and b), through relatively high diversity where abiotic conditions (wave disturbance) prevent the expression of competitive dominance (Zone c), to low diversity where competitive dominance is expressed rapidly under conditions of rapid growth and low levels of disturbance (Zones d and e). Zone f, which has very little wave action, does not fit this hypothetical sequence, possibly because the extremely low water circulation or high temperatures often found in backreef areas create conditions unfavorable for coral growth, reducing growth rates and consequently increasing diversity. Grigg (1983) reports a similar pattern related to wave exposure on 14 islands across the Hawaiian Archepelago. Diversity was lowest at the extremes of the disturbance spectrum and highest at intermediate levels, as predicted by the intermediate disturbance hypothesis. However, unexpectedly low diversity values were found on some islands where seasonally low water temperatures apparently slowed coral growth.

Sedimentation. Both tidal exposure and wave action have their main effect at or near the surface. Sedimentation is another physical factor that affects coral and can be important at any depth. Calcium carbonate produced by coral and calcareous algae is continuously broken up by physical and biological action and transported off the reef by gravity aided by water movement (Goreau and Goreau, 1973). Sedimentation can affect coral by (a) settling directly on the coral and thus interfering with feeding and/or photosynthesis, (b) reducing the amount of substrate suitable for coral growth by covering hard substrate with shifting calcium carbonate sand, and (c) reducing light availability through turbidity caused by fine suspended sediment. Reefs may also be affected by sediment discharged from rivers. Sedimentation may act both as a disturbance (episodic or chronic) and as a regulator of coral growth rates (Dodge *et al.*, 1974).

Coral species differ in their ability to remove sediment. Hubbard and Pocock (1972) analyzed 26 species of coral for their ability to remove sediment particles of different sizes and found strong differences that correlated with polyp size and morphology. Large particles are removed by controlled distention, which is most efficient in species with large, complex calices (e.g., *Montastrea cavernosa*). However, some corals with small polyps are extremely efficient at removing sediment particles with their tentacles (e.g., *Porites* spp.). Fine sediment is generally removed by ciliary action, and some species that are poor at removing large particles are efficient at removing silt (e.g., *Acropora* and *Agaricia*). As a consequence of these differing abilities, sediment load can have a major

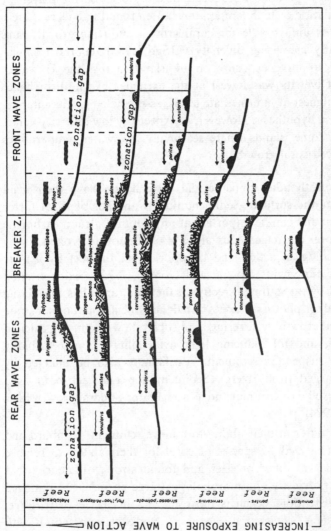

Fig. 12.18 Idealized wave zonation of the six basic Caribbean reef types, showing arrangement of breaker zones and wave zones. Complete zonal sequences, as illustrated, represent 'climax zonations' of Caribbean coral reefs. Relative degree of wave exposure is indicated by arrows. In order of decreasing wave action, the reef crests are characterized by (*a*) the algal ridge, (*b*) interlocking growth of the hydrocoral *Millepora*, with the matlike colonial zooanthid *Palythoa* and some coral, (*c*) thickets of massive branching *Acropora palmata*, with *Diploria strigosa* and a few other species, (*d*) nearly monospecific stands of *A. cervicornis*, (*e*) nearly monospecific stands of *Porites porites*; and (*f*) a mixture of backreef and deepwater species. (From Geister, 1977.)

effect on the distribution of coral species and, thus, the composition of the reef community.

Sediment load is highly variable on reefs and depends at least partially on such microsite differences as proximity to resuspendible sediment. Both Ott (1975) and Ohlhorst (1980) found highest sediment deposition near the surface, where water turbulence is highest, and Ohlhorst reports a significant inverse correlation between sediment deposition and depth. In spite of high resuspension and deposition of sediment near the surface, the largest accumulations of sediment occur deeper in the reef, where hard substrate suitable for coral growth may be covered by sediment. Coral diversity can be reduced by extreme scarcity of suitable substrate caused by large amounts of sediment (e.g., Mergner and Scheer, 1974) or increased by moderate amounts of sedimentation (e.g., Ohlhorst, 1980). However, at both Discovery Bay (Huston, 1985c; Liddell *et al.*, 1984) and Eilat (Loya, 1972), coral species diversity showed no relation to the amount of coral cover (or, inversely, the cover of sand and debris). Highest levels of diversity in particular transects were found at coral covers of 30 to 95%.

Biotic Disturbance

Herbivores. On coral reefs, the effects of herbivores and corallivores are much more subtle than the dramatic changes caused by tides or storms at the reef crest, but may be much more important to the structure of the reef. One reason the effect of herbivores is so difficult to detect on a natural reef is that it is so pervasive there are few natural controls to demonstrate the effect (Hatcher and Larkum, 1983; Carpenter, 1986; Lewis, 1986). Only when herbivores are prevented from grazing the reef's surface by experimental removal (Sammarco, 1982a, b; Sammarco *et al.*, 1974) or by natural causes such as turbulence or mass mortality (Hughes *et al.*, 1987; Liddell and Ohlhorst, 1986; Morrison, 1988; de Ruyter van Steveninck, 1986; Carpenter, 1985; Lessios, 1988), does it become clear that without herbivores the coral would be rapidly overgrown and killed by heavy growth of algae.

The dominant grazers on coral reefs are sea urchins and fish, particularly parrotfish. All urchins and most species of coral-grazing fishes feed primarily on algae, although in the course of continually scraping algae from the surface of living coral the grazers may damage the coral. Only a few fish species actually feed on coral (Ogden and Lobel, 1978;

Randall, 1974), and these corallivores compose a relatively small portion of the total fish biomass.

Herbivorous fishes and urchins have a direct effect on algae and, thus, an indirect effect on coral. The usual effect of grazers is a great reduction in the biomass of algae, which may have a high rate of productivity in spite of very low biomass under intense grazing pressure (cf. Lessios, 1988). Sea urchins, particularly the genus *Diadema*, are conspicuous and important algal grazers on reefs throughout the world. In the Virgin Islands, removal of all the *Diadema antillarum* from an isolated patch reef resulted in a more than tenfold increase in the algal biomass over a four-month period, in comparison with results on naturally grazed controls (Sammarco *et al.*, 1974). In the Red Sea, *Diadema setosum* can prevent algae from monopolizing space, and there is a strong inverse correlation between *Diadema* density and algal cover (Benayahu and Loya, 1977; Dart, 1972). Sammarco (1980, 1982b) found the same inverse correlation between algal cover and urchin density in cages with experimentally manipulated *Diadema* densities.

Moderate urchin grazing may have a diversifying effect on tropical algal communities similar to that reported by Paine and Vadas (1969) for temperate algae and urchins. Higher algal diversity was found in grazed as compared with ungrazed patch reefs in the Virgin Islands (Sammarco *et al.*, 1974), where a single algal species dominated the ungrazed system. The opposite result was found at Discovery Bay, where no algal species was dominant under ungrazed conditions (Sammarco, 1982b).

The interaction of algal palatability and grazing pressure can result in conspicuous shifts in algal species composition (Bryan, 1975; Lubchenco, 1978; Tsuda and Bryan, 1973; Tsuda and Kami, 1973). A clear example can be seen in the increase in encrusting calcareous (coralline) algae under heavy grazing, contrasted with elimination of crustose algae by rapidly growing filamentous or foliose species under reduced grazing pressure. This shift in species composition has been reported in cases of urchin grazing (Benayahu and Loya, 1977; Sammarco *et al.*, 1974) and in grazing by parrotfish (Brock, 1979). The calcareous algae are a suitable substrate for coral settlement and growth.

While the effect of herbivores on algae is clear, their effect on corals is more complex. In addition to the positive effect of allowing the survival of coral which would otherwise be smothered by more rapidly growing algae, herbivores may have a negative effect by killing recently settled coral recruits in the course of thoroughly scraping off algae. There are reports that some herbivorous fish avoid eating even very small coral colonies (Birkeland, 1977; Brock, 1979), but sea urchins are not so particular and destroy coral spats while grazing (Sammarco, 1980, 1982a; Schuhmacher, 1974). On a surface with little physical relief to provide protected refuges for coral spats, optimal conditions for coral survival were found at intermediate urchin densities, where enough algae were removed to prevent corals from being eliminated by competition, but where scraping damage was not severe enough to damage most corals (Sammarco, 1980). Where coral recruits are protected from physical damage by spatial refuges such as crevices, coral survival is greater under highest (fish) grazing pressure (Brock, 1979).

Although heavy algal growth can kill coral, moderately high algal densities may reduce the amount of interspecific contact between corals and thus allow survival of coral species that would otherwise be eliminated by competition with other corals. At two 30-m sites at Discovery Bay, Jamaica, coral diversity was highest at the site with high algal cover (46% algal cover, no *Diadema* urchins) compared with the site having a low cover of fleshy algae (10% cover, *Diadema* present, M. A. Huston unpublished data). On shallow patch reefs at Discovery Bay, where algal growth is very rapid, coral diversity was highest where urchins were present and algal biomass was low (Sammarco, 1982a).

The importance of urchins as algal grazers may be abnormally high for reefs with heavy fishing pressure, which is where most of the experimental urchin research has been conducted. Because fishing removes both urchin predators and herbivorous fish, the relative importance of urchins is much lower for unfished reefs (Hay, 1984). Urchin densities may fluctuate drastically, for reasons not fully understood, as evidenced by the recent massive die-off of *Diadema* in the Caribbean (Lessios *et al.*, 1984; Lessios, 1988).

A special case of algal grazers is the damselfish (Pomacentridae), which create and defend algal 'gardens' growing on dead coral. The damselfish may kill living coral by nipping away live tissue (Kaufman, 1977; Vine, 1974) and may actually 'weed' their gardens to control the algal species composition (Ogden and Lobel, 1978). Potts (1977) found that the algal growth alone is sufficient to kill coral by overgrowing and shading it, causing exhaustion of the coral's metabolic reserves. The algae that grow on the damaged coral are protected from other herbivores by the aggressive damselfish, although occasionally schools of herbivorous fish may enter the territory and graze heavily, overwhelming the damselfish by their numbers (Vine, 1974).

These fish gardens may cover a substantial portion of the reef surface (up to 60% on reef flats, Wellington, 1982b) and significantly affect coral species composition. Massive smooth-surfaced corals grow and recover from fish damage more slowly than the branching species. In the Caribbean, damselfish tend to eliminate the massive *Montastrea annularis* from the upper parts of the reef, which are dominated by the fast-growing branching *Acropora cervicornis* (Kaufman, 1977).

In the eastern Pacific the same phenomenon occurs involving the massive species *Pavona gigantea* and the branching species *Pocillopora damicornis* (Wellington, 1982b). Wellington further observed that *Pocillopora* recruitment and growth are favored in damselfish territories because damselfish protect the area from grazing corallivorous fishes. Damselfish activity is greatest in the upper part of the reef (above 12 m) where algal productivity is highest (Vine, 1974). Damselfish may thus contribute to reduced diversity in the upper reef by increasing the dominance of the branching species at the expense of massive species.

Corallivores. A few species of fish feed primarily on coral. These include the pufferfishes (Tetraodontidae), triggerfishes (Balistidae), butterflyfishes (Chaetodontidae), filefishes (Monacanthidae), and some of the larger parrotfishes (Scaridae) (Randall, 1974). Some of these species browse only the polyps, which allows the coral to regenerate quickly. Corallivores that break off and ingest portions of the coral skeleton feed primarily on branching species of coral, which are clearly more susceptible to this sort of feeding. Thus, in the Atlantic, the branching species *Acropora* and *Porites* are particularly affected, while in the Pacific, such genera as *Pocillopora, Acropora*, and *Montipora* are most affected (Neudecker,1977; Wellington, 1982b).

In the upper parts of the reef corallivores may increase diversity by selectively damaging the dominant branching species, while in the deeper parts of the reef where the same branching species are less abundant, corallivores could actually decrease diversity. The only data available indicate that there is no difference in corallivorous fish activity between 15 and 30 m, so it is difficult to determine whether there is a gradient in this biotic disturbance, or simply a reduction near the surface.

A potentially devastating coral predator with a long history of dramatic population fluctuations (Walbran *et al.*, 1989) is the crown-of-thorns starfish (*Acanthaster planci*). On reefs off the Pacific coast of Panama, *Acanthaster* selects the less-abundant non-branching coral species and avoids the abundant *Pocillopora*, which harbors symbiotic crustaceans that repulse the starfish (Glynn, 1976). Selective destruction of the rarer species results in a significant reduction in species diversity, an effect opposite that of density-independent mortality or of a predator that preys selectively on the most abundant species.

Summary of Disturbance Effects on Coral Reefs

The effect of physical disturbance from wave action and tidal exposure is concentrated near the surface and decreases monotonically with depth. While the relationship between diversity and the depth gradient of abiotic disturbance is consistent with the intermediate disturbance hypothesis, the effect of wave action is so small at 20 to 30 m, where diversity is highest, that the correlation is probably spurious. This conclusion is reinforced by the observation that diversity increases with depth, even on reefs where surface wave action is insufficient to maintain high diversity on the reef crest, such as Discovery Bay. Disturbance levels vary greatly among reefs, resulting in a strong depth gradient where wave energy is high, or virtually no gradient on protected reefs. Similar levels and patterns of diversity were found on both the inner and outer slopes of a barrier reef (Ott, 1975) and on both the seaward and lagoon slopes of an atoll (Sheppard, 1980). Thus, coral diversity increases with depth regardless of the strength of the disturbance gradient.

The potential effects of biotic disturbance on coral diversity are more complex, because corallivores and herbivorous damselfish can have opposing effects on coral species composition. Since the expected effects of corallivores are opposite those actually found on reefs, they apparently do not exert a controlling influence. While damselfish may contribute to the dominance of branching coral in shallow waters on some reefs, the same depth gradients of diversity occur regardless of the presence of damselfish. More work is needed on the interacting effects of corallivores and damselfish and the possibility that human intervention through fishing has altered the natural pattern. Hay (1984) found that herbivory by urchins and fish on the sea grass *Thalassia* increased along the depth gradient on heavily fished reefs but decreased on unfished reefs. Based on available information, one must conclude that there is no depth gradient of either biotic or abiotic disturbance that can explain the general increase in coral diversity with depth.

Coral Growth and Competition

There is no doubt that competition for space can be intense on the reef surface. In shallow water the growth of fleshy algae is rapid, and exclusion experiments indicate that unless grazed by fish and urchins, algae would overgrow and kill the coral (Benayahu and Loya, 1977; Birkeland, 1977; Brock, 1979; Sammarco, 1980; Sammarco et al., 1974). In a very basic sense, the existence and growth of coral reefs depend on the removal of competing algae by herbivores. Kinzie (1973) observes, 'the strong light in shallow water excludes the deep water species [of gorgonians] by favoring the growth of algae which smother the gorgonians'. Corals also grow most rapidly near the surface (Huston, 1985b), and competitive interactions among corals occur by several different mechanisms.

The amount of light reaching the reef decreases along the depth gradient and produces a strong gradient in the productivity and growth of the reef organisms. Even in the clear tropical waters where coral reefs are found, light intensity is reduced 60 to 80% in the top 10 m of water (Kinzie, 1973; Ott, 1975). Below 10 m the decrease in light intensity is much more gradual, reaching 1% at 30 m. It is inevitable that total energy available for all life processes of autotrophs is also reduced with depth, although the presence of photosynthetic algae and coral at depths of 60 m or more indicates that even extremely low light levels are sufficient to support some growth.

Because reef-building corals are basically photosynthetic organisms, their primary source of energy is light, although most coral species also capture planktonic prey with their tentacles. All species of reef-building corals have algal endosymbionts (zooxanthellae) whose photosynthesis has been shown to increase calcification rates (Kawaguti and Sakumoto, 1948; Goreau, 1959b; Goreau and Goreau, 1959). Many other reef organisms also depend on light for growth. Algal endosymbionts are also found in shallow-water gorgonians (Kinzie, 1973), colonial anthozoans, such as *Zooanthus*, and species of such diverse phyla as molluscs, flatworms, and sponges. Various types of macro-algae are also important components of the reef community. All of these organisms depend directly on light for energy, although some may also be partially heterotrophic.

The presence of annual density variation in the calcium carbonate skeleton produced by coral (analogous to growth rings in temperate trees) allows determination of the rate of coral growth (linear extension of the $CaCO_3$ skeleton) by X-radiography (Buddemeier et al., 1974; Dodge and Thompson, 1974; Macintyre and Smith, 1974). The annual

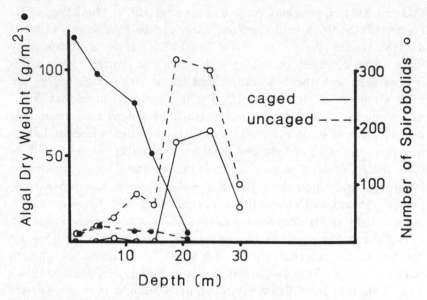

Fig. 12.19 Relationship between algal biomass and the settlement of feather-duster worms (Spirobolids) along a depth gradient. (Based on data in Vine, 1974.)

nature of these bands is well established (Buddemeier, 1974; Buddemeier *et al.*, 1974; Highsmith, 1979; Weber *et al.*, 1975), although not all species have detectable bands, and even within a species that has bands, not all colonies show distinct banding (Weber and White, 1974; Huston, 1985b). The measurement of coral growth is reviewed by Buddemeier and Kinzie (1976) and Bak (1976).

Results from numerous studies (Baker and Weber, 1975; Barnes and Taylor, 1973; Buddemeier *et al.*, 1974; Huston, 1985b; Weber and White, 1974; Woodhead, 1971) support the generalization that coral growth rates decrease with depth within a species, although there are species that show no such trend. When the growth differences with depth are compared between species rather than within a species, the pattern is much stronger. The fastest growing coral species predominate in shallow water, and slower-growing species increase in abundance with depth. Decreasing growth rates resulting from the decrease in light with depth affect not only reef-building corals, but also all of the organisms dependent on photosynthesis, including algae (Fig. 12.19) and gorgonians (Kinzie, 1973).

Most species of reef-building corals are partially heterotrophic. Corals feed primarily on zooplankton (Johannes *et al.*, 1970; Muscatine, 1973; Porter, 1974b, 1976), and plankton levels can be significantly reduced as water passes over a reef (Glynn, 1973). The degree of heterotrophy seems to be positively correlated with polyp size (Porter, 1976), which is consistent with the observation that species with large polyps are more abundant in the deep reef. The fast-growing corals that domi- nate the upper parts of most reefs have small polyps and presumably depend much more on photosynthesis than on plankton feeding. Exper- iments on the effect of light and plankton availability on the growth of three species of coral at two depths (1 to 2 m and 7 to 10 m) support Porter's generalization about the relationship of polyp size to heterotro- phy (Wellington, 1982a). For the three species that Wellington examined, reduced light at the deep sites resulted in a greater decrease in growth than did reduced plankton, suggesting that plankton feeding does not compensate for reduced light at deep sites. The available information indicates that plankton availability decreases with depth (Ohlhorst, 1980; S. L. Ohlhorst, unpublished information) and might thus complement the effects of the light gradient.

In summary, there is a strong gradient in light energy, and thus in autotrophic productivity, on all coral reefs. Slight variation may occur with cloud cover or water clarity, but the overall pattern of the light extinction curve is the same on all reefs. This nearly constant gradient of light contrasts with the disturbance gradient, which is highly variable because it depends on the strength of wave action impinging on the reef.

Mechanisms of Competition among Coral

Near the reef surface, the cover of coral or other organisms such as encrusting algae often approaches 100%, and intense exploitative com- petition for resources apparently occurs. Rapidly growing branching coral species achieve dominance by overtopping other corals, reducing light availability and water circulation. This mechanism has been sug- gested as an explanation of the dominance of pocilloporid corals at shallow depths in the Pacific (Glynn *et al.*, 1972; Porter, 1974a) and of *Acropora* in the Caribbean (Shinn, 1972). When free of herbivores, algae can overgrow and 'smother' coral as well as other organisms, such as gor- gonians (Kinzie, 1973). Sediment trapped in algal growth or at the base of branching corals may contribute to mortality caused by overtopping.

Another mechanism of interaction among coral that has received con- siderable attention is extra-coelenteric digestion, which could be classified as interference competition. Lang (1973) reported that corals were able to extrude mesenterial filaments capable of digesting the surface of their neighbors, thus winning competitive interactions for space. Lang's experiments showed that a hierarchy of 'digestive dominance' exists among Caribbean corals, with position in the hierarchy correlated with polyp size. This hierarchy is inversely correlated with coral growth rate and the ability to overgrow and smother competitors. Thus, these inversely related competitive abilities offered an apparent mechanism that could help explain the high diversity on coral reefs in the context of equilibrium competition theory. However, field observations in the Pacific showed a reversal of this hierarchy, with *Pocillopora* being both fast-growing and digestively dominant (Glynn *et al.*, 1972; Porter, 1974a). Laboratory experiments on the Pacific corals contradicted the field observations and indicated a digestive hierarchy consistent with the results from the Atlantic (Glynn, 1974).

These discrepancies were finally resolved by Wellington's (1980) long-term experiments, which demonstrated that the previously reported hierarchy based on extension of mesenterial filaments was a short-term response. The short-term results were reversed when the fast-growing species (*Pocillopora*) recovered from the initial damage and produced sweeper tentacles that killed the tissue of the slower-growing species (*Pavona*) and established a buffer zone that was ultimately covered by encrusting coralline algae. No experimental results have been reported for an equivalent species pair from the Caribbean (e.g., fast-growing *Acropora* versus digestively dominant *Montastrea annularis*), although Richardson *et al.* (1979) report that *Montastrea cavernosa* can produce sweeper tentacles to defend itself against *Montastrea annularis*. Thus, at least for some Pacific corals, the fast-growing species can achieve dominance by both of the major mechanisms, overtopping and digestion. This reduces the probability that diversity can be maintained by an equilibrium balance of competitive abilities.

Under conditions where competitive dominance is expressed slowly, changes in environmental conditions may cause shifts from one type of competitive mechanism to another. When this occurs, relative competitive abilities may be reversed as the system shifts to a competitive hierarchy based on a different mechanism, but the net effect would still be a slower reduction in diversity. Such alteration of competitive hierarchies by changing environmental conditions is an alternative mechanism that could produce the non-hierarchical 'circular competitive networks' on the underside of coral shelves, as reported by Buss and Jackson (1979).

Such variability in the ranking of competitive ability may explain the results of a one and a half year study of interphyletic interactions at 15 m on the reef at Discovery Bay, Jamaica. Ohlhorst (1980) found that *Montrastrea annularis* lost space in 50% of all encounters with fleshy and encrusting sponges. However, *Montastrea* also lost space in 19 of 40 cases in which there was no apparent competitor. Space was maintained in at least half of the encounters with other classes of organisms, including other corals, non-coral coelenterates, algae, and encrusting foraminifera.

Spatial patterns on the reef surface may provide an important insight into competitive interactions. Ohlhorst's (1980) data show a perfect correlation ($r_S = 1.00$, $p < 0.01$, $n = 5$, Spearman's Rank Correlation) between abundance on the reef surface at 15 m depth (transect data) and percentage of perimeter contact with *Montastrea* for algae, sponges, non-coral coelenterates, miscellaneous organisms, and bare substrate or sand. The fact that interactions (physical contacts) are proportional to abundance suggests that coral contacts are essentially random for these organisms, and there are no positive or negative spatial interactions. However, for two classes of organisms this relationship is significantly non-random. For other corals the frequency of perimeter contact (at 0 to 2 cm) is much lower (6.5%) than their abundance on the reef would indicate (48.1%). For the encrusting foraminiferan *Gypsina*, the contacts are much more frequent than would be expected based on their abundance (25.6% as compared to 3.2%). These data, together with Wellington's (1980) observation that coral border zones killed by sweeper tentacles are colonized by encrusting calcareous algae (ecological equivalents in shallow water of the encrusting foraminifera), indicate that coral may avoid interspecific interactions by establishing border zones maintained by encrusting organisms. In a sample of 101 coral-*Gypsina* contacts over $1\frac{1}{2}$ years, 60% of the contacts showed no change, while coral lost space in 20% and gained space in 19% (Ohlhorst, 1980). Thus, once space is occupied, inter-coral competitive interactions may be greatly reduced and stable spatial relationships established.

Changes in coral colony size that result from competitive interactions (primarily digestive) occur slowly at 15 m; therefore, a low frequency of disturbance would be adequate to prevent competitive equilibrium and maintain species diversity. Such spatial stability and slow rates of change are less likely to be found at shallower depths, where rapid growth and overtopping provide an alternative mechanism of competitive interactions.

Consequences of Reduced Growth Rates in the Deep Reef

Reduced competition for space under low-light conditions apparently allows the survival of slow-growing, 'shade-tolerant' species that are unable to compete successfully in the high productivity environment of the upper reef. Most of these species are not physiologically limited to a particular depth and can survive in shallow water under non-competitive conditions, such as in turbid areas or in caves.

The dramatic reduction of algal growth and biomass with depth has consequences for all sessile organisms, which are very small when they settle out of the plankton and must compete with algae and other organisms for space. Vine (1974) found that the number of featherduster worms (Serpulidae) settling on unprotected plates remained close to zero from the surface to a depth of 15 m. Below 15 m, where algal productivity decreased rapidly, the number of serpulids which settled increased sharply until it began to decrease because of the sedimentation below 25 m (Fig. 12.19).

Birkeland (1977) studied coral recruitment on artificial substrates at depths of 9 to 34 m on both the Caribbean and Pacific coasts of Panama. As in Vine's study, the biomass accumulation of filamentous algae dropped dramatically with depth. Associated with this reduction in algal biomass was a fivefold increase in the number of surviving coral recruits between 9 and 20 m and an increase in the number of identified coral genera from 7 to 12 m. Although the coral grew fastest at the 9-m site, their survival was greatest at 20 m, where algal growth was less. At both the 9- and 20-m sites coral survival and growth were greatest on the vertical sides of the settlement blocks, where light availability (and algal biomass) was lower than on the upper surface. At 34 m coral grew only on the upper surface.

Birkeland (1977) suggests that diversity differences between reefs show the same relationship to the productivity of algae (and other 'fouling community' organisms) as is found along the depth gradient. The lowest rates of biomass accumulation (at a 9-m depth) among five different reef sites were found on the high-diversity reefs of the San Blas Islands on the Caribbean coast of Panama. The extremely high productivity of the fouling community (including barnacles, bryozoans, and tunicates) in nutrient upwelling areas on the Pacific coast effectively prevented the survival of any coral recruits and led to low coral diversity (Birkeland, 1977). The artificial substrates used in Birkeland's experiments were not protected from grazing; thus, the biomass differences may reflect differential herbivore pressure as well as different rates of productivity. This does not, however, affect the conclusion that coral recruitment, growth, and diversity can be greatly reduced by competition from dense growth of algae and other organisms.

Reduced light levels caused by turbidity in shallow parts of a reef are sometimes associated with species composition and morphology more typical of the deep reef (Bak, 1975; Bonem and Stanley, 1977), although very high turbidity can eliminate coral and reduce coral diversity (Loya, 1976; Roy and Smith, 1971). Out of ten sites studied near Discovery Bay, Jamaica, Ohlhorst (1980) found the highest coral diversity ($H' = 2.34$, $J' = 0.81$) at an 18-m site which was 'very turbid, with visibility often less than 5 m'. At this site, coral cover was the lowest (18%) and the diversity of both fleshy and encrusting sponges was the highest recorded at any of the ten sites. Over the ten sites, sedimentation was positively correlated with species diversity, and Ohlhorst suggests that sediment may have a diversifying effect similar to that of predators. Although the 18-m site is within the depth range where highest diversity is generally found, these results indicate that certain levels of light reduction and other stresses associated with turbidity and sedimentation do not necessarily reduce coral species diversity, but may actually increase it.

The effect of light on the species composition (and ultimately diversity) of the reef community is demonstrated by the fact that in several taxonomic groups the non-photosynthetic species, which are independent of light availability and hence should be able to grow anywhere, are rarely found in the upper parts of the reef. They increase in abundance and diversity only in the deep reef, although deepwater and non-symbiotic species are often found in caves and other low light situations at shallow depths (Faure, 1974; Jaubert and Vasseur, 1974; Vasseur, 1974). Kinzie (1973) found that gorgonians with symbiotic algae dominated the gorgonian fauna to a depth of 55 m, where non-symbiotic species became predominant. Goldberg (1973) reports the same pattern on a Florida reef, although the non-symbiotic forms became abundant at a shallower depth, perhaps because of reduced water clarity. Sponges, most of which are non-symbiotic, also increase in abundance and diversity with depth, as do non-photosynthetic corals (Goreau and Goreau, 1973).

The same pattern of diversity that occurs along the depth gradient of light availability is also found along light gradients at much smaller scales. On the Great Barrier Reef, several species of *Acropora* have a table-shaped growth form, with a strong light gradient from the edge to the center of the shadow cast by the table. The number of coral species found in the shadow is low at the center and edge and highest in the zone of intermediate light levels (Sheppard, 1981).

All of these patterns are consistent with Slobodkin's laboratory experiments (1964) with freshwater hydra, which showed that at high light levels a species with photosynthetic algal symbionts outcompeted a non-symbiotic species, but both coexisted at low light levels. Thus it appears that rapidly-growing photosynthetic species may exclude non-photosynthetic and slow-growing species from the upper parts of the reef, while slow growth conditions in the deeper reef and turbid shallows allow the coexistence of many species.

The Interaction of Competition and Disturbance

Reef diversity patterns in relation to the dynamic equilibrium between growth rates and disturbance frequencies are summarized in Fig. 12.20. The pair of thick lines that converge at the lower left corner of the figure encompass the range of physical conditions encountered at different depths on the forereef slope. Near the surface (0 m) the growth rates are high and the range of disturbance frequencies (and intensities) is great, reflecting variation in the degree to which the reef is exposed to or protected from storms and normal wave and tidal action. Growth rates are reduced with depth, and the frequency (and intensity) of wave action is reduced to low levels in the deep reef. The ellipse represents the range of conditions found at shallow turbid sites (generally in backreef areas or lagoons), where turbidity reduces light levels and, presumably, coral growth rates (which may also be reduced by the detrimental effects of sedimentation). Because these sites are close to the surface, the range of disturbance frequencies is greater than at deeper sites on the forereef having the same light availability. Both shallow turbid sites and deep clearwater (forereef) sites may have the appropriate combinations of growth rates and disturbance frequencies to allow high levels of diversity.

Competitive interactions in the low-disturbance environment of the deep reef are clearly on the equilibrium end of the spectrum of non-equilibrium conditions. In fact, it would be difficult to distinguish a true equilibrium situation from a slowly changing non-equilibrium one without detailed long-term study. Very slow growth rates at these depths slow the rate of approach to equilibrium, which could ultimately be attained in the absence of disturbance. Even in a system that would ultimately reach equilibrium, non-equilibrium coexistence can be prolonged by slow rates

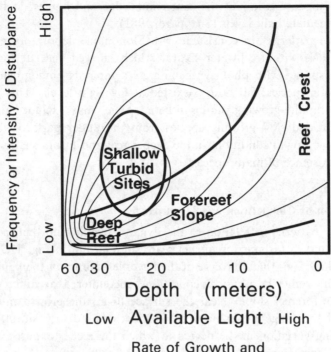

Fig. 12.20 Predicted relationship of coral reef physical conditions to species diversity that results from the non-equilibrium interaction of growth rate and frequency of disturbance. (Based on Huston, 1979). See text for details.

of change and high similarity between species (Huston, 1979; Caswell, 1982). Extensive monospecific stands in the deep reef of *Montastrea* or *Agaricia* indicate that a low-diversity equilibrium can occur at depths where higher diversity also occurs.

Equilibrium can be approached much more rapidly in the upper parts of the reef, where a higher frequency of disturbance is necessary to prevent equilibrium from occurring. Not surprisingly, reef spatial patterns are much less constant near the surface than deeper in the reef (Bak and Luckhurst, 1980). The low-diversity reef crests of *Porites porites* or *Acropora cervicornis* sometimes found on protected reefs may represent competitive equilibrium achieved in the absence of disturbance.

Evolutionary History and Biogeography of Coral Reefs

Because living corals create highly characteristic stone (calcium carbonate) skeletons that form the structure of reefs, the history of reef structure and diversity is extremely well-preserved in the fossil record (Liddell *et al.*, 1984; Fagerstrom, 1987). It is possible to collect data on the diversity and structure of the calcified sessile organisms of fossil reefs that are directly analogous to data collected from living reefs. Only completely soft-bodied organisms are missing from the record.

Recently summarized paleontological research (Fagerstrom, 1987; Kauffman and Fagerstrom, in press) demonstrates that the remarkable species diversity of coral reefs is not the result of a long history of evolution and coevolution in a stable environment. Rather, the geological history of reefs is one of massive extinctions and rapid evolution of organisms to form new coral reef communities. Coral reef systems, perhaps because of their limitation to a specific range of water temperatures and an inability to keep pace with rapid changes in water levels, have undergone massive extinctions with a greater frequency than other marine communities. Based on the fossil record, calcified reefs have been completely eliminated by mass extinctions every 26-30 million years, with each extinction followed by a delay of 3 to 10 million years before reefs reappeared in the world's oceans. However, once reefs reappeared, species diversity of corals and other hard-bodied organisms rapidly increased to levels equal to or greater than those found on contemporary reefs (Fig. 12.21.)

The fossil record also makes it clear that the center of coral diversification has been the eastern Pacific Ocean throughout much of geologic history. The complex array of islands and archepelagos of the Indo-Pacific provides a classic 'species-pump' for the evolutionary diversification of organisms through alternating cycles of isolation (allopatric speciation) and reunification (competition, niche diversification, and sympatric speciation), as well as periodic local extinctions and the recolonization of previously occupied or completely new environments (e.g., Moore, 1954). This environment has repeatedly lead to the formation of a diverse coral reef community that has subsequently spread throughout the world. The Caribbean region has relatively low species diversity in comparison with reefs of the Indo-Pacific and Mediterranean regions, which were the source of Caribbean coral species (Stehli and Wells, 1971; Fig. 12.22).

The peculiar geography of the Indo-Pacific region is also the probable explanation for the high diversity of other groups of organisms, including

DIVERSITY IN REEF COMMUNITIES
THROUGH TIME

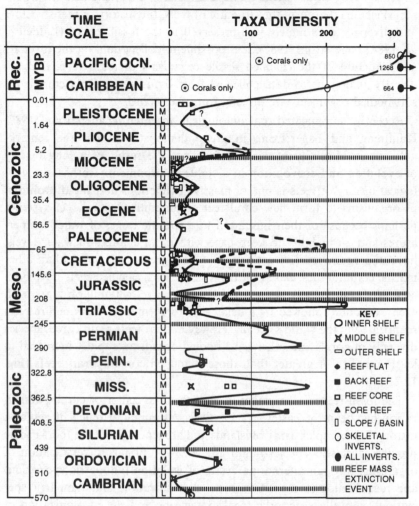

Fig. 12.21 Fluctuations in the occurrence and taxonomic diversity of reef communities over geologic time. Solid line indicates the number of taxa with hard skeletons from shallow reef environments (crest and upper forereef slope). Dotted line indicates the number of taxa from deep reef environments (middle to lower slope and basin habitats). The horizontal bars indicate global mass extinction (From Kauffman and Fagerstrom, in press).

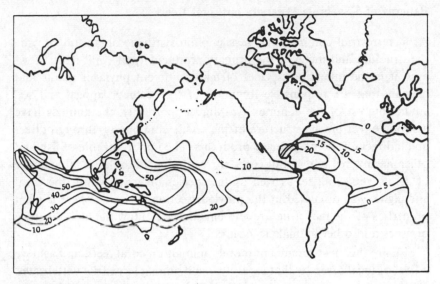

Fig. 12.22 Distribution of the number of genera of hermatypic corals. (After Stehli and Wells, 1971.)

mangroves, which are much more diverse in southeast Asia than in the New World, and the rainforest trees of the family Dipterocarpaceae, which have achieved a remarkable diversity on the islands of southeast Asia (Ashton, 1989). The repeated isolation and reunification of these islands as sea level has risen and fallen over geologic time is similar to the repeated expansion and contraction (with isolation of core regions) of the African grasslands that has apparently contributed to the remarkable ungulate diversity now found there (see Chapter 13).

The fossil record demonstrates that the increase in species diversity with depth on coral reefs is a consistent phenomenon that has appeared repeatedly throughout geologic history during the periods of evolutionary radiation between the massive coral reef extinctions (Fig. 12.21). The highest diversity in fossil reefs is consistently found in the deep-reef communities (i.e., 20-40m). Thus, as in the case of many other high diversity communities, environmental conditions on the deep reef are not only conducive to the contemporary coexistence of many species, but have contributed to the accumulation of species over evolutionary time, either through enhanced survival of species that immigrate or evolve there, or through increased rates of speciation, or both.

Patterns of Vertebrate Diversity on Coral Reefs

As in terrestrial communities, sessile, photosynthetic organisms provide the structure and much of the primary productivity of coral reefs. However, it is the interstitial species of many different phyla of plants and animals that are responsible for much of the aesthetic appeal and scientific interest of the community. More importantly, the animals have a strong effect on the structure of the sessile community, through direct and indirect effects on primary producers. The important role of fish and other herbivores and corallivores in regulating the relative abundance of corals versus algae, as well as the species diversity within the coral and algae, was discussed in the previous section. The issue here is what properties of the reef structure and environment allow the high diversity of fishes and other animals to occur.

Fish are the most prominent mobile animals on coral reefs, and achieve a level of local diversity that is rarely found among terrestrial vertebrates. The high fish diversity is unusual in that it occurs along with high total densities of individuals and high total biomass. Another unusual feature of the high fish diversity is the large number of closely related species found on most reefs. For example, in the Capricorn group of reefs at the southern end of Australia's Great Barrier Reef, there are around 850 species of fish, representing 84 families and 297 genera. A number of genera have over a dozen species, including *Chaetodon*, 25; *Scarus*, 24; *Apogon*, 23; *Pomacentrus*, 19; *Acanthurus*, 13; and *Halichoeres*, 12 (Ehrlich, 1975). Fifty or more species commonly coexist on patch reefs only three meters in diameter, and even more species can be found coexisting within a similarly sized area at the northern end of the Great Barrier Reef, where fish diversity is even higher (Sale, 1978). The coexistence of large numbers of closely related species is contrary to the predictions of equilibrium competition theory, and suggests that some factors are preventing competitive equilibrium from occurring. A similar phenomenon is found in the great rift lakes of Africa (Malawi, Tanganyika, Victoria) (Lowe-McConnell, 1987).

There is a remarkable variety of specialized feeding morphologies and behaviors among coral reef fishes (Hiatt and Strasburg, 1960; Ehrlich, 1975; Goldman and Talbot, 1976). However, there is also considerable overlap in the resources used by species coexisting on a single reef (Sale, 1977; Williams, 1982). Motta (1988) conducted a detailed ecomorphological analysis of ten species of Pacific butterfly fishes and found a great deal of overlap in what the species actually ate. He concluded that their specialized morphologies were more important in determining how and where they ate, rather than what they ate. Given an abundant food source, such as zooplankton, both the highly specialized species as well as the generalized species fed opportunistically on the same resource. In fact, there was even evidence for evolutionary *convergence* in the feeding morphology of the highly specialized species. These observations are consistent with conclusions based on underwater observations in the Bahamas (Collette and Earle, 1972), where 'time-sharing' mechanisms and competition for space were more important than food in limiting the number and kind of fish species.

Unlike vertebrates in most terrestrial communities, many coral reef fishes tend to be extremely sedentary, limited to small areas of the reef where they may be resident for long periods of time (Bardach, 1958; Springer and McErlean, 1962; Ehrlich, 1975). In relation to body size, the home ranges of coral reef fish are an order of magnitude smaller than those of other vertebrates (Sale, 1978). Some species are highly territorial, vigorously defending their small patch of reef against interspecific and intraspecific intruders (e.g., the damselfish). However, the degree of territoriality varies greatly among species as well as between different life stages of the same species, and many species with restricted home ranges do not actually defend them. The high degree of site fidelity, as well as the fact that many grazing species form multispecies schools, suggests that predation pressure may exert a strong effect on the structure of the reef fish community (Ehrlich and Ehrlich, 1973).

Peter Sale (1977, 1978, 1979, 1980a,b; Sale and Dybdahl, 1975) has found high variability in the fish species composition of isolated reef patches and little evidence for microhabitat specialization among the species he examined. Based on evidence that fish populations are limited by the availability of appropriate spaces on the reef surface that provide protection from predators and/or access to food, Sale hypothesized that the regulation of species diversity on reefs is essentially a random process. Most species of reef fish have planktonic larvae that are passively carried by ocean currents and must settle in an unoccupied suitable site on a reef in order to mature. The fact that suitable spaces on reefs are almost always occupied, and are made available by mortality caused by predation or other factors that are unpredictable in time and space, means that the successful establishment of an individual fish is an unpredictable, random process that is largely independent of the local abundance, competitive ability or other properties of that species (Talbot *et al.*, 1978; Williams and Sale, 1981). Success in obtaining space results simply from arriving first, which provides a major competitive advantage over later arrivals. Thus, the species composition of the reef fish community is not determined by species-specific competitive interactions and niche partitioning, but rather represents a dynamic equilibrium between the rate at which space is made available and a random colonization process, analogous to a lottery (Sale, 1978).

Sale (1978) argued that the appropriate life history strategy for organisms in an environment where space is limiting and the availability of space is completely unpredictable is: 1. stay put; 2. breed often; and 3. disperse the resulting offspring. The fact that most coral reef species produce planktonic larvae means that the competitive success or local abundance of adult fish of a particular species do not determine the future size of the local population. This results in a non-deterministic, non-equilibrium regulation of population size (see also Shulman and Odgen, 1987; Hughes, 1990).

However, there are also large-scale patterns that have been interpreted in terms of a more deterministic, competition-based regulation of reef fish communities. Anderson *et al.* (1981) surveyed the patterns of coexistence among Pacific butterfly fishes (Chaetodontidae) along a 50 mile transect on the Great Barrier Reef, in the same area where they had previously found stochastic patterns of community structure on small artificial reefs (Talbot *et al.*, 1978). Along the transect they found that there were conspicuous differences in the niches of many of the chaetodontids that were coexisting locally, and that there was a spatial replacement among species occupying the same niche along the 50 mile transect. Sale and Williams (1982) reanalyzed the data of Anderson *et al.* (1981) and argued that there was no convincing evidence for an equilibrial, niche partitioning interpretation of the local community structure of reef fishes. Although some chaetodontids were associated with particular conditions across a broad range of environments, the number and identy of locally coexisting species was highly variable, and was independent of feeding guilds. The apparent pattern and stability of reef fish communities at large scales (e.g., Anderson *et al.*, 1981; Molles, 1978), may be simply a large-scale stochastic equilibrium of many non-equilibrium patches (cf., DeAngelis and Waterhouse, 1987).

The open ocean adjacent to nearly all coral reefs has an important effect on the structure and species diversity of the reef. Moving, oxygenated water is critical to the survival of sessile coral and most other reef organisms as well. In addition, water from the open ocean also brings in photoplankton and zooplankton that support a wide variety of reef organisms. The ocean serves as a reservoir for the mixing and dispersal of the planktonic larvae not only of reef fishes, but of nearly all reef organisms. Furthermore, the open ocean and landward lagoons serve as reservoirs for predators that feed on the reef and may help regulate the structure of the reef fish community without in turn being strongly regulated by the reef community. This superimposition of predators that are relatively independent of the productivity of the reef community is probably a major explanation for the high predation pressure that leads to a space-limited fish community and the non-equilibrium maintenance of high species diversity.

Summary

Marine systems span a broad range of productivity and disturbance conditions that produce many different dynamic equilibria of species diversity, ranging from very high to very low. The high productivity intertidal zone has historically been the most thoroughly studied marine system, which has led to a strong emphasis on processes such as disturbance and predation that have a dramatic effect on community structure under productive conditions where competition is potentially intense. The highest diversity in most marine ecosystems is found under conditions of very low (although not the lowest) productivity, and the strongest gradients of marine species diversity are along gradients of productivity. Non-equilibrium processes in low productivity environments apparently regulate diversity over most of the ocean. In general, those marine systems that are most stable also tend to have extremely low productivity. However, environmental stability is not a common feature of most marine ecosystems, either on ecological or geological time scales, and non-equilibrium processes associated with predation, disturbance, catastrophic extinction episodes, and low rates of growth and competitive displacement seem to explain most of the variability in marine species diversity.

The most diverse marine ecosystems are coral reefs. Reefs are influenced by a variety of physical and biological processes, which may have opposing effects on species diversity. High growth rates of coral and algae in shallow water lead to intense competition for space and reduced species diversity, resulting from dominance by competitively superior species. A variety of physical disturbances can increase diversity on the reef crest by reducing the population size of the dominant species, but even when physical conditions maintain high diversity on the reef crest, diversity increases with depth. The ubiquity of the depth gradient of species diversity on coral reefs, and in the open ocean as well, is probably associated with the single physical gradient that has the same relative shape on all reefs, in all waters, and throughout geologic time, that is, the gradient of decreasing light availability with depth.

13

Case Studies: Species Diversity in Fire-influenced Ecosystems

Certain dramatic adaptations to fire, such as serotinous pine cones that open and release their seeds *only* when exposed to the temperatures reached during major fires (LeBarron and Eyre, 1939), led to the recognition by early ecologists that fire was extremely important in some natural communities (e.g., Clements, 1910; Cooper, 1913). However, only recently has the ubiquity and significance of fire in virtually all terrestrial ecosystems been recognized (e.g., Walter, 1973; Wright, 1974; Wright and Bailey, 1982). Recent work on reconstructing the fire history of vegetation in North America before settlement by Europeans has demonstrated that fire was an important process that maintained a forested landscape very different from current forests (Spurr, 1954; Heinselman, 1973; Swain, 1973; Pyne, 1982; Cronan, 1983; Clark, 1990; Harris, 1984). It is now known that even tropical rain forests may burn during occasional severe droughts (Sanford *et al.*, 1985; Uhl *et al.*, 1988; Kauffman *et al.*, 1988; Leighton and Wirawan, 1986; Goldammer, 1987), and many tropical forests are currently burning with much greater frequency as a result of human land-clearing practices and alteration of vegetation structure (Uhl and Buschbacher, 1985; Periera and Wetzer, 1986; Booth, 1989; Uhl and Kauffman, 1990). Arctic tundra, and even semi-aquatic environments, such as bogs and marshes, can become dry enough to burn during severe droughts (Lutz, 1956; Cochran and Rowe, 1969; Viereck, 1973).

High species diversity, particularly of grasses, forbs, and shrubs, is often found in vegetation that is regularly burned. Increasing recognition of the importance of fire for maintaining the populations of certain species, for increasing overall species diversity, and for preserving landscape pattern in general has led to a revolution in ideas about the use of fire in managing forests and other natural ecosystems (Loucks, 1970; Wright and Bailey, 1982; Pyne, 1984). 'Controlled' use of fire for agricultural and silvicultural management dates back to the earliest records of human activities and, after a relatively brief period of management by fire suppression, is rapidly regaining an important management role.

The effect of fire on the biological diversity of communities and landscapes demonstrates virtually all of the concepts addressed in this book and provides an illustration (and perhaps a comparative test of the qualitative predictions) of the dynamic equilibrium model of species diversity across the full range of disturbance regimes and growth conditions found in terrestrial plant communities. The central feature of dynamic equilibria involving fire as the primary disturbance is that both the frequency and intensity of the disturbance, as well as the rate and pattern of recovery following the disturbance (i.e., succession), are regulated by the availability of a single resource: water.

Understanding the effect of fire on the biological diversity of a landscape requires an understanding of the pattern and regulation of plant succession on that landscape. The structure and diversity of any particular community is influenced not only by the characteristic disturbance regime of that landscape, but also by the length of time since the most recent fire, and the specific characteristics of that fire. Thus, it is sometimes difficult to distinguish between different seres and different stages of the same sere on landscapes where fire is a prominent process. The landscape spatial pattern caused by fire is strongly influenced by the degree of heterogeneity that is established under a particular dynamic equilibrium (e.g., Fig. 8.1).

Because of the central role of water availability in dynamic equilibria involving fire, the ecosystems to be discussed in this chapter will be organized in order of increasing water availability, beginning with deserts and ending with rain forests (Fig. 13.1). The location of each community type in relation to the two axes indicates the relative level of diversity among competing species in each community, as well as the direction of change in diversity that would be expected with a change in disturbance frequency or resource availability (water, and in some cases, soil nutrients). Thus, an increasing frequency of fire would be expected to reduce species diversity in deserts, but increase diversity in productive grasslands.

The distinction between functional types and functional analogues is particularly important in fire-influenced communities, since changes in diversity over succession result from the accumulation of additional functional types (often 'interstitial' species) as well as changes in the number of species within the structural functional types that define the

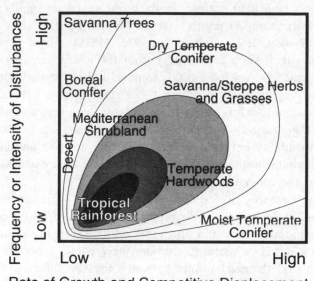

Fig. 13.1 Approximate locations of terrestrial plant communities in relation to the axes of the dynamic equilibrium model of species diversity. (Based on Huston, 1979.)

community. Depending on environmental conditions, frequent fires may increase the diversity among competing species, while reducing the total number of species through loss of interstitial functional types. The interaction of plant succession and fire was the basis of one of the first descriptions of the effect of 'intermediate' disturbance rates on species diversity (Loucks, 1970).

The Predictable and Unpredictable Effects of Fire

Fire has a dramatic and significant effect on many ecosystem properties, including species diversity. In some environments, the frequency and effects of fire are quite predictable, while in others the frequency and effects are much more difficult to predict. Under dry and windy conditions, fire can act as a completely density- and species-independent disturbance, while under other conditions the mortality caused by fire is extremely selective. Sometimes fire cannot even be considered a disturbance, because no living biomass is destroyed.

The critical variables that determine the characteristics of any particular fire regime are 1. Fire type and intensity (e.g., crown fires, severe surface fires, and light surface fires); 2. mean size of significant fires; and 3. fire frequency or return intervals for specific land areas (Heinselman and Wright, 1973; Heinselman 1973, 1981a). These variables are not independent, since each results from the interaction of vegetation properties such as fuel load, total biomass, and the spatial pattern and amount of vegetation with different flammabilities, as well as weather patterns such as monthly precipitation, relative humidity, and wind speed and direction. Various combinations of these factors can produce an incredible variability between the effects of one fire and another, which makes it extremely difficult to predict the effects of a particular fire on vegetation and ecosystem properties.

The consequences of fire, like those of other disturbances, can be understood in the context of the productivity and competitive conditions of the specific environment in which they occur. Before discussing specific ecosystems, it will be useful to consider the general properties of fire in vegetation and the mechanisms by which fire can influence landscape pattern and species diversity.

Predictable Properties: The Occurrence and Frequency of Fire

Given the right conditions of temperature and pressure, nearly every known element will combine with oxygen, generally with a release of energy. According to Lovelock (1988) an increase of only 4% from the atmosphere's current oxygen content of 21%, would lead to the uncontrolled spontaneous combustion of most plants and wooden structures on earth. Oxidation is an unavoidable natural phenomenon, occurring as the result of enzymatic activity within the cells of all organisms, as well as abiotically under the range of temperatures and pressures found on Earth. The only uncertainty involved in oxidation is its rate, not its occurrence.

At the present atmospheric oxygen content, wood will burn spontaneously when it reaches a temperature of approximately 350 °C (Anderson, 1970; Wright and Bailey, 1982). The primary natural sources of temperatures high enough for ignition are lightning strikes and, under rare conditions, heterotrophic respiration (microbial 'spontaneous combustion'). Wet plant material burns more slowly than dry plant material simply because the heat lost as the water evaporates prevents the material from reaching a temperature high enough for combustion until nearly all the water has evaporated. Once organic matter is sufficiently dry, the heat released by burning is more than enough to evaporate what little water remains and produce a chain reaction of combustion (Fransden, 1971, 1973; Rothermel, 1972).

With the right conditions of dryness, amount of fuel, and wind to provide oxygenated air and move the fire, fires in vegetation are inevitable and virtually impossible to stop, as demonstrated by the fires in western North America in 1988, most notably in Yellowstone National Park. Only a change in weather conditions associated with the onset of winter extinguished the fires (Christensen *et al.*, 1989; Schullery, 1989). A much larger fire, the Black Dragon Fire, which occurred in the boreal forest region of China and the Soviet Union in 1987 may have been the largest single fire on earth in the past 500 to 1000 years (Salisbury, 1989).

The probability of fire in vegetation is related in a complex nonlinear, but predictable way to precipitation (Fig. 13.2). The complexity arises because precipitation has a positive influence on plant productivity and thus the amount of living and dead material present to burn (i.e., the fuel load), but a negative influence on the flammability of the plant material that is present. Precipitation influences fire on temporal scales ranging from 1000-year or longer climatic cycles to hourly changes in humidity and precipitation. Regional climate determines what forms of plants are present and thus the quantity and flammability of the plant material, as well as the maximum biomass that can be accumulated. Seasonal or annual variability of precipitation determines the relative abundance of particularly flammable fuel types such as grasses and herbs and the moisture conditions of the living and dead plant material. Weekly, daily, and even hourly patterns of precipitation determine whether fuel moisture conditions will become dry enough to allow fires and whether a fire that gets started will be extinguished by a rain or snowstorm, or continue to grow.

In desert areas with very low precipitation, natural vegetation fires are rare, because plant productivity is so low that sufficient fuel to support a fire rarely accumulates (Humphrey, 1962, 1974; Wright and Bailey, 1982). However, in regions of high precipitation, plants grow rapidly and achieve large sizes, so high amounts of fuel accumulate rapidly. However, the high precipitation prevents the vegetation and dead plant material from becoming sufficiently dry to burn. In temperate and tropical regions with high precipitation distributed evenly throughout the year, fires are extremely rare. Normal year-to-year variation in precipitation, which can produce 'droughts' that noticeably affect vegetation and agriculture, is generally insufficient to dry the vegetation enough to burn. Fires occur only during the extreme droughts that have occurred at intervals

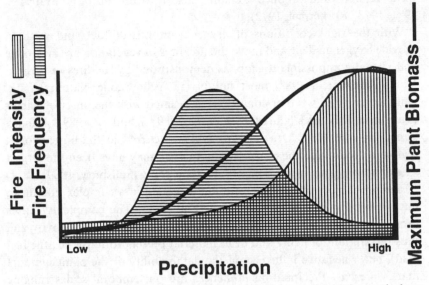

Fig. 13.2 Relationship of fire frequency and intensity to precipitation and plant biomass.

of several hundred to a thousand years (Sanford *et al.*, 1985), or when humans alter the structure of the vegetation to make it burn more readily (Uhl and Kauffman, 1990).

In regions with levels of precipitation intermediate between deserts and wet forests, fire frequency is determined by either the rate of accumulation of fuel (if precipitation and productivity are low) or the frequency of periods that are sufficiently dry to allow fuel to burn (if precipitation and productivity are high) (Heinselman and Wright, 1973). In many coniferous forests of western North America, frequency of fire at any location is determined primarily by the low but predictable rate at which fuel of sufficient amount and spatial arrangement to support fire accumulates (Heinselman, 1973; Loope and Gruell, 1973; Romme, 1982; Romme and Despain, 1989). In more productive areas, frequent small fires can reduce the fuel load and result in a lower frequency of major fires (Kilgore, 1973). The deciduous forests of the North American southeast experience major fires at a lower frequency than the western coniferous forests, and the occurrence of fire tends to be much less predictable because it is determined by the occurrence of severe droughts during the summer and fall (Keetch and Byran, 1968; Lorimer and Gough, 1988; Martin, 1989). Consequently, the proportion of fire-adapted species in southeastern deciduous forests is much lower than in the western and northen coniferous forests.

Throughout most of the world where there is enough precipitation to produce sufficient fuel to support a fire, the predictable seasonality of precipitation in relation to plant growth is the key that determines the frequency of fire. The alternation of wet and dry seasons produces ideal conditions for fire, and fire is a prominent and largely inevitable landscape-scale process wherever a dry season predictably follows the growing season. Such a pattern occurs in most of the grassland regions and all of the mediterranean climate regions around the world, where shrub communities accumulate sufficient biomass over the course of three to ten years to produce a highly predictable fire cycle in a community of plant species that exhibit classic adaptations to fire (Shantz, 1947; Naveh, 1973, 1974; Le Houerou, 1974; Gill and Groves, 1981; Gill, 1981; Specht, 1981; Westman *et al.*, 1981; Christensen, 1985). Even if total annual precipitation were insufficient to produce enough fuel when distributed evenly throughout the year, concentration of precipitation into a wet season can result in a pulse of plant productivity that produces sufficient fuel to burn during a dry season.

Variation in resource availability that influences the rate of accumulation of living and dead biomass is often associated with predictable variation in fire frequency and intensity. In the New Jersey Pine Barrens, where fires occur frequently in all forest types, the fires that occur in the more productive lowland pine communities tend to be much more intense and cause much higher tree mortality than the fires that occur on the drier uplands (Little, 1979).

There is often a strong inverse correlation between the frequency and intensity (and size) of fires (Fig. 13.2), particularly in regions where the fire frequency is determined primarily by the rate of fuel accumulation (Rogers, 1961; Loucks, 1970; Mutch, 1970; Heinselman and Wright, 1973). Even if plant and fuel moisture conditions and weather are ideal for fire, fires may not be able to sustain themselves and spread if insufficient fuel biomass has accumulated. Consequently, the probability of fire is not constant, even though the frequency of lightning strikes and other sources of ignition may be relatively constant (Heineslman, 1981a). The probability of fire increases through time as fuel accumulates, and so is most appropriately modeled as a renewal process, rather then as a constant probability (Clark, 1989a). The probable intensity of fire also increases through time (with increasing fuel load) under these conditions (Wright and Bailey, 1982; Hobbs and Gimingham, 1984a).

For example, in the coniferous forests of Yellowstone National Park, vegetation does not become susceptible to major fires until the later stages of succession, when the first cohort of trees has died and fallen to the forest floor, and an understory has developed a continuous fuel load from the forest floor to the canopy (Romme, 1982; Romme and Despain, 1989). Fires that occur before maximum fuel load and fuel continuity have been accumulated are likely to be surface fires of small extent and low intensity, except under extreme weather conditions, such as those in 1988 that produced the crown fires that burned areas of all successional stages and even such vegetation types as willow thickets and wet meadows.

This inverse correlation between fire frequency and intensity is the basis of the use of fire as a silvicultural management tool. In the extensive southeastern pine forests and plantations of loblolly and longleaf pine, fires are intentionally set annually or biannually to prevent the accumulation of dead wood, leaf litter, and shrub biomass that could support fires of sufficient intensity to damage the pines or even lead to crown fires (Van Lear and Waldrop, 1989). The herbaceous communities in these fire-maintained pine savannahs are among the most diverse herbaceous plant communities anywhere in the world (Walker and Peet, 1983).

Fire frequency in one vegetation type can be strongly influenced by the fire frequency in an adjacent plant community (Heinselman and Wright, 1973; Heinselman, 1973, 1981b). In fact, even within a single vegetation type, fire frequency is influenced more strongly by the probability of fire spreading from an adjacent area, than by variation in the relatively low probability of ignition (e.g., lightning strike) at any particular location (Heinselmann, 1981b). In chaparral, the proximity of flammable vegatation types increases the frequency of fire in less flammable types (Radke *et al.*, 1982), while in Arctic heaths the fire frequency is higher in those areas near the boreal forest (Rowe *et al.*, 1974).

Unpredictable Properties: Spatial Variation in Fire Intensity

Given the virtually certainty that fire will occur in most ecosystems, at intervals ranging from one year to 1000 years, the primary uncertainty involved in fires is the effect the fire will have when it does occur. The maximum intensity of a fire is determined by the amount of fuel that has accumulated since the previous fire. The instantaneous effect of a fire on a plant community (as well as on the animals and the full suite of ecosystem processes), has two primary, and interdependent, components:

1. the intensity of the fire, which results from a combination of the maximum temperature that the fire reaches, and the length of time the maximum temperature is maintained; and 2. the scale of the spatial heterogeneity in the intensity of the fire.

The most significant effects of fires on vegetation, and thus on landscape pattern, result from the effects of fire on the last and the first stages of the life cycle of plants. Plant mortality produces an instantaneous change in the relative abundance of plant species in the community, and sometimes in the total species richness as well. Fire also dramatically alters the conditions for seed germination and plant regeneration. Removal of plant biomass completely changes the light environment at the soil surface, and the effects of fire on organic and mineral soils may affect the availability of resources such as mineral nutrients and water. The effects of fire on the soil are strongly dependent on pre-fire amounts of organic matter and soil nutrients, as well as the fire temperature (Stark, 1974). Variation in burn intensity often creates a high degree of spatial heterogeneity in plant survival and conditions for germination and growth (Schullery, 1989; Christensen *et al.*, 1989).

Temperatures in fires can range from 100 °C or less at the ground surface to over 1000 °C several feet above the ground, with significant variation in biological and geochemical effects over that range (Wright and Bailey, 1982; Christensen, 1985). Temperatures that are too low may fail to stimulate the germination of seeds that have a heat requirement for germination, while temperatures that are too high may kill all the seeds in the surface soil (see reviews in Heinselman, 1981a; Christensen, 1985). Heinselman (1981a) identified the extent to which the soil organic layer was consumed as important to the regeneration of northern coniferous forests for a variety of reasons.

The predictability of fire intensity (energy released per unit length of burning front per unit time, Byram, 1959; Brown and Davis, 1973) tends to be greater when the fire frequency is high than when fire frequency is low. This is simply a consequence of the fact that the potential variability in intensity is lower when the total fuel availability is low than when it is high. In ecosystems with high maximum fuel availability, such as in a coniferous forest with an average between-fire interval of 200-300 years, potential fire intensity ranges from small surface fires and incomplete patchy burns when the fuel load is low or when fuel moisture and relative humidity are high, to crown fires that kill all plant material and consume most of the organic matter of the system, when fuel loads are near maximum and weather conditions are suitable. In addition, in systems where the fire frequency is low as a result of the length of time needed to accumulate a sufficient fuel load, the particular weather conditions when a fire finally does occur have a tremendous influence on the intensity and size of the fire (Wright and Heinselman, 1973).

The scale of spatial variability almost always increases with increasing intensity of fire. High intensity fires tend to produce large areas that are uniformly burned, since the effect of small-scale variability in fuel amount and moisture is overwhelmed by the effect of high temperatures produced by the fire. In contrast, low intensity fires tend to produce a high degree of small-scale heterogeneity, since slight variability in fuel amount and moisture is sufficient to determine whether a particular area will or will not burn. It is evident that variation in fire intensity and thus in the amount and spatial pattern of burning of the organic layer can produce incredible spatial variability in the effects of a single fire, as well as dramatic differences between fires. Nonetheless, there are some predictable aspects of fire effects on ecosystems, vegetation structure, and landscape patterns.

The intensity of a fire has a predictable effect on the number and type of plants that are killed or damaged by the fire. In general, plant susceptibility to damage or death from fire increases with decreasing plant size, so that herbs, seedlings, and small saplings can be killed by a low intensity fire that has little or no effect on trees. Fire damage results from the interactions of maximum temperature and the length of time for which a high temperature is maintained (Nelson, 1952; Hare 1965). The cambium or meristem can be killed by temperatures above 60 °C for greater than 60 seconds (Wright and Bailey, 1982). Bark thickness, which varies among species and increases with increasing plant size, determines how well the cambium is insulated from heat at the plant surface and thus whether the plant is likely to be injured or killed by a fire of a given intensity and duration (Martin, 1963; Siren, 1973; Uhl and Kauffman, 1990) (Fig. 13.3).

Fires of low to intermediate intensity are capable of producing mortality that is highly selective with regard to plant size and species, and may be extremely heterogeneous at small scales; high intensity fires generally kill all plants regardless of size or identity and produce homogeneity at small scales and heterogeneity at larger scales. Thus, under some conditions the properties and effects of fire are highly predictable, while under other conditions the effects are unpredictable.

The Yellowstone fires of 1988 provide a clear example of the high

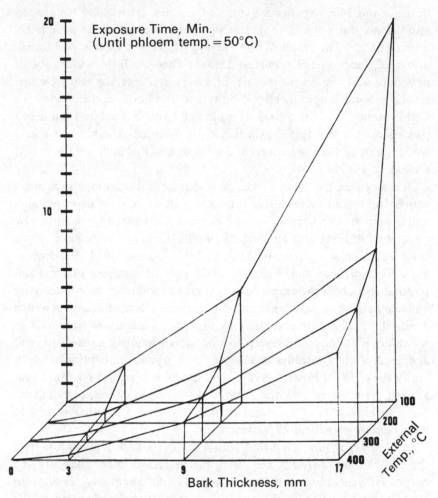

Fig. 13.3 Interactive effects of fire temperature and bark thickness on the time needed for the phloem to reach 50 °C. (From Siren, 1973.)

variability in the effects of forest fires. Although the total area within fire perimeters was 562,310 ha in the Greater Yellowstone Area (GYA, 4.8 million ha in Wyoming, Montana, and Idaho), the effect of the fire within the overall burn perimeter was extremely variable. In Yellowstone National Park, 50% of the total park area was included within burn perimeters within which 45% of the park actually experienced some level of burn (Guth and Cohen, 1989). 'In the GYA, 61% of the burned area experienced canopy burn (trunk, limbs, and needles or leaves of trees burned), and 34% experienced surface burn (fire crept along the ground and did not burn the canopy, leaving many trees partly or completely unburned). ... The result of the varying burns is a mosaic of burned, unburned, and partially burned areas. There is little uniformity of pattern or scale across the mosaic; fires were influenced by many factors, including wind, slope, fuel availability and condition, and humidity, so that in some areas the mosaic is quite fine (with burned and unburned patches only a few feet or yards across), whereas in other areas it is coarse (with all the vegetation for dozens of hectares uniformly burned).' (Schullery, 1989).

In most cases, the effect of fires on resources and on competitive interactions influenced by resources is relatively short-lived, as plant biomass and organic matter accumulate and any lost nutrients are replenished by mineral weathering and atmospheric inputs. Fires generally have only a small effect on soil nutrients (McKee, 1982; Raison, 1979; Waldrop *et al.*, 1987; van Lear and Waldrop, 1987) and the influence of local soil physical and chemical properties has a greater influence on productivity and competitive interactions than the transient effect of fire on resource availability. A major exception to this generalization is situations in which fire is followed by severe erosion, which removes most of the soil and totally alters resource availability and physical conditions (Klock and Helvey, 1974; Foster, 1976). This situation occurred on Mt. Monadnock, New Hampshire, the summit of which was changed from dense forest to bare granite by severe erosion following a fire that is said to have been set to get rid of wolves.

In most vegetation, even that with a relatively high frequency of fire, the predominant patterns are often due to factors other than fire. In particular, vegetation patterns on a landscape commonly result from topographic variation in the availability of nutrients and water, which influences the rate of succession and the intensity and outcome of competitive interactions, as well as the frequency and intensity of fire. Where the frequency of fire is low but predictable, pattern imposed by fire may be a prominent feature of the landscape. Where fire is very infrequent and unpredictable, its influence on landscape patterns is likely to be very dramatic, but highly localized in time and space. The following sections on the role of fire on different types of plant communities will emphasize the spatial and temporal scales at which fire produces predictable effects.

Deserts

The production of plant biomass and thus fuel for fires is limited by water availability in deserts. Deserts with an annual rainfall of less than 17 cm burn rarely, if at all (Wright and Bailey, 1982). The decomposition of plant litter, and the consumption of leaves and wood by animals (e.g., Whitford *et al.*, 1982) is generally sufficient to prevent the low productivity from leading to an accumulation of fuel. Plant cover is usually intermittent over the surface so there is rarely a continuous bed of fuel to support the spread of fire, even if it does get started in an area with a local concentration of fuel.

With increasing precipitation and plant production, the probability of fire in deserts increases. Desert grasslands and shrublands may be more likely to burn after a period of abnormally high precipitation, when grasses and herbaceous plants can achieve much higher biomass than normal. However, hot and dry conditions are more consistently associated with desert fires than are preceeding wet growing seasons (Swetnam and Betancourt, 1990).

Most species of desert shrubs and grasses are harmed by fire, although frequent fires (every 10 to 20 years) tend to shift community composition toward grasses (Wright and Bailey, 1982), with elimination of many sensitive species of shrubs, grasses, and most cacti. Shrubs that can be greatly reduced or eliminated by burning include mesquite (*Prosopis glandulosa*), creosote bush (*Larrea tridentata*), and big sagebrush (*Artemesia tridentata*), while rabbitbrush (*Chrysothamnus* spp.) is often enhanced by fire (Cottam and Stewart, 1940; Young and Evans, 1974; Wright and Bailey, 1982). Many grasses, including the important range species, black grama (*Bouteloua eriopoda*), are severely affected by fire.

The spread of an exotic species of annual grass, *Bromus tectorum*, into the arid shrublands and woodlands of the North American west is apparently increasing the probability of fires as a result of its high productivity during the brief early-spring wet periods (Pickford, 1932; Stewart and Young, 1939; Stewart and Hull, 1949; Piemeisel, 1951; Mack, 1981; Young and Tipton, 1990; Whisenant, 1990). *Bromus* produces sufficient biomass in the open spaces between shrubs and trees to allow the spread of fires during the dry summer and fall, after the *Bromus* goes to seed and dies. Once *Bromus* invades an area, a fire cycle may be initiated that can totally eliminate the tree and shrub species that have never become adapted to fire (Blaisdell, 1953; Wright and Bailey, 1982; Sparks *et al.*, 1990; Billings, 1990).

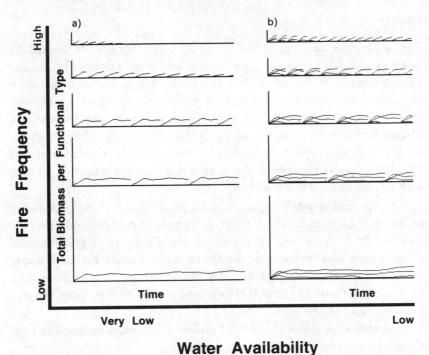

Water Availability

Fig. 13.4 Interaction of fire frequency with the successional dynamics of desert ecosystems. Under dry conditions (column *a*), fire frequency has little effect on species composition or diversity. Under slightly wetter conditions (column *b*) the effect of disturbance is more pronounced.

Because of the simple 'auto-succession' that occurs under arid conditions (West, 1982; Fig. 13.4), fire does not have a dramatic effect on landscape pattern in deserts. The low total biomass and short stature of the mature vegetation does not result in much contrast between burned and unburned areas. Severe droughts, which may occur relatively frequently in deserts, can cause mortality of shrubs and grasses just as effectively as fires (Nelson, 1934; Wright and Bailey, 1982). Fires and droughts generally decrease plant species diversity, as predicted by the dynamic equilibrium model of species diversity. Deserts usually do not have much spatial heterogeneity caused by temporal asynchrony of disturbances, although responses to gradients of soil moisture or other resources (or regulators) can be quite dramatic over the large expanses of land visible in many desert areas.

As with all types of vegetation, deserts represent a continuum of plant community composition and biomass. With increasing soil water availability, the height of desert vegetation increases, grading into shrublands and open montane woodlands. The higher biomass that can be accumulated in shrublands allows a higher frequency and intensity of fire, and results in some of the most diverse fire-maintained plant communities found anywhere.

Fire-dependent Shrub Communities of Mediterranean Climate Zones

Shrub communities with continuous cover of fire-adapted woody vegetation from one to four meters tall are found throughout the world in areas that receive from 200 to 1000 mm of annual precipitation. The woody shrubs are characterized by a variety of adaptations to fire and generally have thick 'sclerophyllous' leaves. The most distinctive and best known of these shrublands are found in five widely separated regions of the world (Emberger, 1930, 1955; Walter, 1964/1968; di Castri and Mooney, 1973; di Castri *et al.*, 1981). These regions have in common poor soils and what has come to be called a mediterranean climate, characterized by wet, cool (rarely cold) winters and summers that are mild to hot, with drought periods ranging from 1 to 12 months. The pronounced annual dry seasons result in a high fire frequency, although variation in fire frequency and time since a fire can result in striking differences in vegetation structure and species diversity (Fig. 13.5).

In spite of vastly different taxonomic origins, the flora and fauna of all five mediterranean regioms have some striking similarities in their appearance, adaptations to fire, and ecology (Mooney and Dunn, 1970; Cody and Mooney, 1978; Mooney, 1977; see however, Barbour and Minnich, 1990). Similar shrublands are also found outside of the mediterranean climate zones (Barbour and Minnich, 1990), including several areas in the interior mountains of western North America (Keeley and Keeley, 1988).

The five major mediterranean vegetation regions are, in order of total land area, located in the Mediterranean basin, Australia, California, Chile, and South Africa (Fig. 13.6). Each of these areas is located on the western side of a continent and to the east of a major cold oceanic current, which produces the winter rainfall. The total area of mediterranean climate covers 4.6% of the earth's surface, but only 2.3% when the wettest and driest extremes, the perhumid (1-2 month summer drought) and the perarid (11-12 month summer drought), are excluded. Of the 2.3%, only approximately half is covered by shrubland, including both the typical evergreen sclerophyll shrubland, as well as shrublands

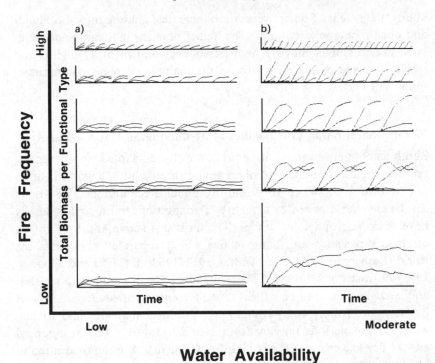

Water Availability

Fig. 13.5 Interaction of fire frequency with successional dynamics in mediterranean climate regions, where vegetation ranges from desert through the charactistic shrublands to closed forest. With increasingly moister conditions, the effect of fire frequency and time since fire on the structure and diversity of the vegetation becomes more pronounced.

with a variable proportion of drought-deciduous shrubs, savannah, and open woodland. Depending on precipitation and soils, the other 50% of the mediterranean climate area is covered by closed woodland or forest and by drier, shorter shrubby heaths that grade into deserts. Obviously, there is a great deal of variability in mediterranean vegetation, and because of the generally high topographic relief of these areas, several different vegetation types can often be found within a small area (di Castri, 1981).

The mediterranean shrublands result from a very particular combination of environmental conditions that allows dominance of shrubs, rather than either woodlands or grasslands and savannahs. While precipitation amount and pattern define the climatic region, an equally important environmental axis is soil fertility. Shrublands exist only on nutrient-poor soils, decreasing in stature as nutrient levels decrease so that the poorest

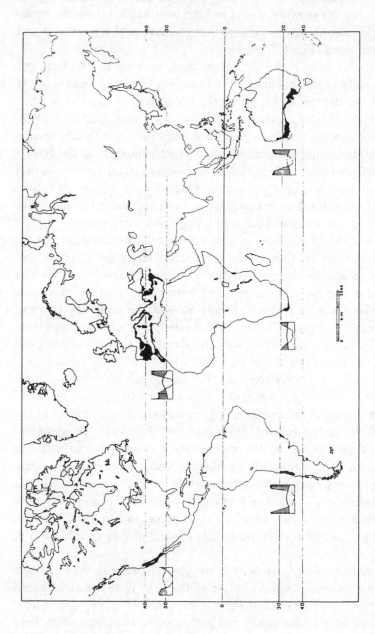

Fig. 13.6 World map of the five areas with mediterranean-type regions, a Gaussen-Walter climate diagram (temperature curves with dotted lines; precipitation curves with solid lines) illustrates the main climatic patterns. These climate diagrams are from Long Beach, California; Valparaiso, Chile; Rabat, Morocco; Dasseneiland, South Africa; and Cape Naturaliste, Australia. (From di Castri, 1981.)

'oligotrophic' soils support a dwarf shrubland generally called heathland. On soils of substantially higher fertility, grasses can accumulate enough biomass during the growing season to form continuous grasslands or savannahs, eliminating shrubs through a combination of competition and high fire frequency (Fig. 13.7).

Soil nutrients are critical in determining vegetation structure and species diversity of shrublands at both the continental and local scale. At the local scale, the stature of the vegetation, as well as species diversity, are closely related to the productivity of the vegetation (Milewski, 1983). The classic unimodal relationhip between plant productivity and species diversity is also found in mediterranean vegetation such as the fynbos of South Africa (refer to Fig. 11.2). The environmental conditions that influence spatial variability in plant productivity on a local scale tend to be highly correlated with topography: both soil nutrients and water accumulate in low areas and areas of topographic convergence.

The interaction of the rate of growth and competitive displacement with the frequency of fires makes the evaluation of the effect of soil conditions on species diversity somewhat complex. Since diversity can be reduced over time in the absence of disturbance even on poor soils, care must be taken to compare communites of the same age to avoid confounding the effects of the rate of successional processes (presumably correlated with soil conditions) with the length of time these processes have been occurring (e.g., Cowling 1983a, b).

Specht and Moll (1983) conclude that 'in the mediterranean climate regions of the world, the availability of soil nutrients appears to be the major environmental variable affecting the nature and distribution of the vegetation.' Ravinovitch-Vin (1983) concluded that rock type was more important than climate in determining plant community structure in areas receiving greater than 600 mm annual precipitation in Isreal. While strong evidence supports the differentiation of vegetation in response to soil nutrients *within* the mediterranean shrublands, it is equally clear that the major transitions from forest to woodland to shrubland are controlled primarily by water availability (Specht and Moll, 1983; Beard, 1983) (Fig. 13.8).

At the continental scale, soils differ significantly between the five major mediterranean climate regions. Most of the soils in the mediterranean climate regions of Australia and South Africa have developed on nutrient-poor bedrock or siliceous sands, and prolonged weathering and leaching have further reduced their nutrient status (Rundel 1979; Specht 1979; Milewski, 1983). In contrast, soils in the other three regions are relatively

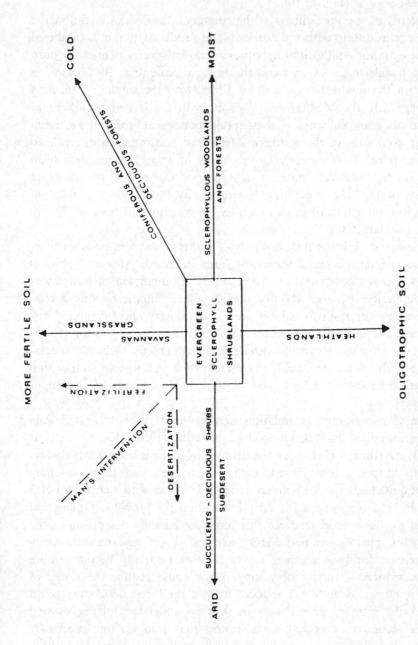

Fig. 13.7 Dynamics of evergreen sclerophyll shrublands along environmental gradients determined by increased or decreased aridity, increased or decreased soil fertility, and increased control by low temperatures. (di Castri, 1981).

more fertile (di Castri, 1981; Rundel, 1983) (Fig. 13.9). These soil differences are reflected in the nutrient content of plant material from different regions (Fig. 13.10).

Patterns of species richness at the continental scale closely follow the inverse relationship with soil nutrients and productivity that is predicted by the dynamic equilibrium hypothesis. Lowest species richness at both the plot scale (e.g., 100 m^2) and the regional scale (e.g., 10,000 km^2) is found in the relatively rich soils of Chile (with the smallest total area of chaparral), the Mediterranean Basin (with the largest total area of chaparral), and California. Highest species richness at both scales is found on the poor soils of the southern Africa and Australian mediterranean regions, both of which are relatively small in area. The mediterranean zone on the relatively nutrient-rich soils of Chile differs from other regions in that the vegetation is structurally more complex, with an extensive understory of grass and often an overstory of trees (di Castri, 1981; Rundel, 1981).

In contrast with the high diversity of plants in nutrient-poor mediterranean vegetation, the diversity of birds in South African fynbos is relatively low (Siegfried and Crowe, 1983). Comparison of bird diversity differences between the five mediterranean climate regions is confounded by variation in habitat area, evolutionary and taxonomic history, and the species composition and structure of the vegetation (Cody, 1981). Nonetheless, structural variation in mediterranean-climate vegetation provides some interesting insights into the well-known relationship between bird species richness and vegetation structure (or foliage height diversity).

The classic positive correlation between bird diversity and foliage height diversity (FHD) first noted by MacArthur (1958, 1964; MacArthur and MacArthur, 1961), does not always hold in mediterranean-climate regions. In fact, in some closed forest areas, where foliage height diversity can be quite high, species richness actually decreases with increasing FHD (Ralph, 1985) (cf. Fig. 2.18). In this situation the probable explanation is that available food resources for birds are actually decreasing as the forest becomes denser and FHD increases. Food resources are likely to decrease for two reasons: 1. as the forest canopy becomes more dense, reduced diversity of woody plants could reduce the range of food resources available; 2. reduced light in the forest understory could reduce the primary productivity of fruit by understory plants, as well as the secondary productivity of insects that feed on the understory plants. It is important to note that canopy density is not necessarily

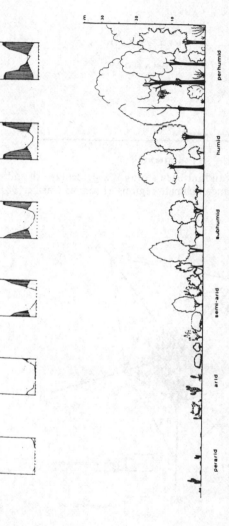

Fig. 13.8 Schematic latitudinal gradient of mediterranean climatic types and corresponding vegetation types in Chile. The climatic trends are represented by means of Gaussen-Walter climate diagrams (temperature curves with dotted lines; precipitation curves with solid lines). The plant formations are depicted schematizing some of their dominant plants.
Perarid Type (Copiapó, 27°21'S): subdesert with *Copiapoa, Trichocereus,* and chamaephytes. Arid Type (La Serena, 29°54'S): open shrubland with *Puya, Baccharis, Trichocereus, Lithraea caustica,* and *Adesmia.* Semi-arid Type (Llay Llay, 32°50'S): matorral with *Retanilla ephedra, Lithraea caustica, Cryptocarya alba, Trevoa trinervis, Colletia spinosissima, Satureja gilliesii* and *Kageneckia oblonga.* Subhumid Type (Talca, 35°26'S): Sclerophyllous woodland with *Cryptocarya alba, Lithraea caustica, Aristotelia chilensis, Quillaja saponaria* and *Azara petiolaris.* Humid Type (Chillán, 36°36'S): mesophilous woodland with *Laurelia, Myrceugenella, Quillaja saponaria* and *Nothofagus obliqua.* Perhumid Type (Traiguén, 38°15'): hygrophilous forest with *Nothofagus dombeyi, Persea lingue, Drimys winteri, Chusquea* and epiphytes. (From di Castri, 1981.)

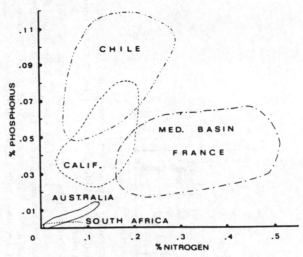

Fig. 13.9 Levels of phosphorus and nitrogen (total as a percentage of soil mass) in the soils of the five mediterranean climate regions. (From di Castri, 1981.)

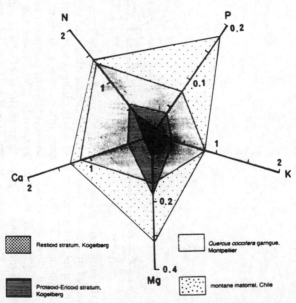

Fig. 13.10 Foliar analysis (% dry weight) of four representative mediterranean shrublands. At the site near Kogelberg, South Africa, the canopy Proteoid-Ericoid stratum has higher nutrient concentrations than the subcanopy Restioid stratum. Note the much higher nutrient concentrations from the Mediterranean region and Chile. (From Specht and Moll, 1983.)

correlated with forest productivity. Canopy density is a consequence of shade tolerance, and thus is strongly influenced by water availability, as well as soil nutrients.

In the South African fynbos, bird diversity shows little relation to vegetation structural diversity. However, at equivalent levels of structural diversity, the more productive lowland plant communities have two to three times as many bird species as the low productivity mountain fynbos (Siegfried and Crowe, 1981). Cody's (1975) data on bird species richness along gradients of vegetation height diversity in mediterranean-climate regions on three continents seems to show a change from a positive to a negative correlation at higher levels of foliage height diversity (Fig. 13.11). In fact, the data suggest that the shift to a negative correlation occurs at lower levels of vegetation height diversity on sites with poor soils (South Africa) in comparison to sites with richer soils (California). Maximum bird species richness increases with increasing soil nutrients (and presumably plant productivity), reaching its maximum in California.

Small mammal diversity in mediterranean climate regions is correlated with habitat heterogeneity (Fox, 1981). Insects tend to be highly specialized on host plants, and insect diversity can be quite high in Australian mediterranean vegetation (Morrow, 1981). According to Milewski (1983), cold-blooded animals (which are presumably more energetically efficient) are the predominant predators in vegetation on poor soil, while on more fertile sites, mammals predominate. In general, animal diversity, particularly of large animals such as vertebrates, would be expected to be lower in mediterranean vegetation than in more productive plant communities.

The evolutionary history of the mediterranean-type ecosystems is moderately well known. The oft-noted phenotypic convergence (Mooney and Dunn, 1970; Mooney, 1977; see also Barbour and Minnich, 1990) of distantly related taxa in the mediterranean regions suggests that the dominant force driving the structure and diversity of these ecosystems is strong natural selection within a similar, stressful environment, rather than accidents of evolutionary history or phytogeography. The two northern mediterranean-climate regions became physically separated from the three southern regions in the mid- to late-Cretaceous, at which time the three southern continents with present-day mediterranean floras began to separate, while the two northern continents remained connected for a longer period, initially across the Atlantic, and later across the Pacific (Deacon, 1983). The general phylogenetic relationships of the floras reflect the well-known similarities and differences in the mammalian faunas of these regions (Keast, 1972).

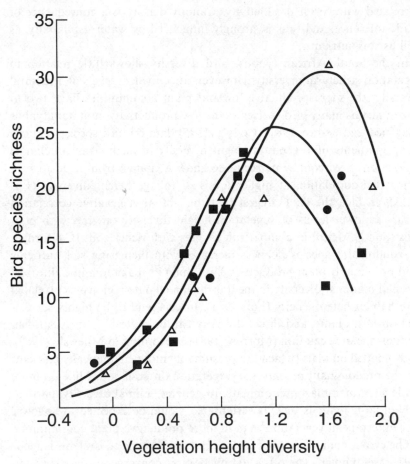

Fig. 13.11 Bird species richness along gradients of increasing vegetation height diversity in three mediterranean-climate regions, showing a decrease in bird species richness at high levels of foliage height diversity. Solid squares indicate sites in south-west Africa, dots indicate Chile, and triangles indicate California (after Cody, 1975).

The most significant differences in the geological (and presumably evolutionary) histories of the five regions result from the fact that two of the regions, in Africa and Australia, are on rifted or sheared continental margins, while the other three are on subduction zones that produced significant mountain building during the Cenozoic, when most of the flora evolved (Deacon, 1983). Major climatic changes were globally synchronous during the Cenozoic, with a trend toward cooler, drier climates culminating in the glacial cycles of the Pleistocene. In North

America the first fossils of sclerophyllous taxa are found at the Eocene-Oligocene boundary (38 my BP) (Axelrod, 1973, 1975), associated with an apparent shift from a tropical to a seasonal or semi-arid climate (Peterson and Abbott, 1979). In contrast, the present mediterranean areas of Africa and Australia were dominated by forest taxa until the late Miocene (15 my BP), with an increase in schlerophyllous vegetation thereafter (Deacon, 1983). Thus, it is apparently not evolutionary time *per se* that explains the greater species diversity of Australia and South Africa. Nor is it simply the relative surface areas of the different regions that are associated with the diversity patterns. The fact that the highest species diversity is found within the smallest region (Southern Africa) and that no correlation exists between species richness and area suggests that ecological and evolutionary factors acting on a smaller, more local scale are the probable explanation for the patterns.

One key to the diversity differences between the regions seems to be soil conditions. The ancient landscapes of the rift zones have soils that are extremely low in nutrients, while the more geologically active subduction zones (California, Chile, and the Mediterranean) have younger, more fertile soils. There are also differences in current climate that may be significant. Dry season (summer) precipitation is highest ($\sim$15% of annual total) in the South African Cape and Australia, and lowest in Chile ($\sim$1-5%) (Barbour and Minnich, 1990). Reduced dry season water stress may allow the survival of plant types that are more shade-tolerant than those that can survive in the other mediterranean regions. This would allow a more complex successional sequence, with a better developed understory, in areas that escaped fire for longer periods. These conditions should create a more heterogeneous landscape as a result of temporal asynchrony and physical heterogeneity, and might result in a larger regional species pool.

Low rates of competitive displacement that result from low soil nutrients and arid conditions, coupled with a moderately high rate of disturbance caused by frequent fires, are the conditions under which the dynamic equilibrium theory would predict the highest species diversity. It should be noted that the high species richness of mediterranean regions is the result of both high numbers of ancient taxa (paleoendemics) as well as high number of recently evolved taxa (neoendemics). This suggests that the conditions that allow the survival of ancient species are the same conditions that promote the evolution and survival of new species (see discussion of endemism in Chapter 11).

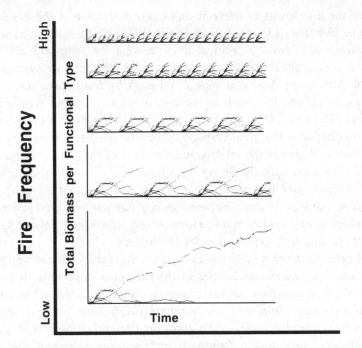

Fig. 13.12 Interaction of fire frequency and successional dynamics in grasslands. With low fire frequencies, woody plants overgrow and shade out the grasses and herbs.

Grasslands

A predictable fire cycle occurs in climates that are sufficiently dry that grasses and herbs are the primary life form. Classic examples of this pattern are found in temperate grasslands (particularly tall-grass steppe or prairies) and tropical grasslands (Curtis, 1959; Jackson, 1965; Daubenmire, 1968). Grasslands are found in regions with insufficient annual precipitation to support well-developed closed canopy forest (generally less than 100 cm, Sims, 1988), but sufficient rainfall during the spring and/or early summer growing season to allow high productivity. Fires play a major role in maintaining most grasslands, eliminating trees and shrubs that would outcompete and replace the grasses in the absence of fires (Gleason, 1913, 1923; Sauer, 1950; Curtis, 1962; Hulbert, 1973; Bragg and Hulbert, 1976) (Fig. 13.12).

Plant productivity in most grasslands is limited not by soil nutrients, but by water availability (Walter, 1939). So, during years with abundant water during the growing season, plant productivity and fuel accumulation can be extremely high. These ecosystems tend to have very high agricultural productivity, both in terms of harvested crops and grazing. Consequently, they are often converted to agriculture and the natural ecosystem and its fire cycle eliminated. The managed ecosystems (and formerly the natural ecosytems) usually support high densities of grazing mammals, paricularly ungulates, which are often able to consume a sufficient proportion of the plant biomass that the frequency and intensity of fires is reduced (Chew and Chew, 1965; Martin, 1975). Many of the plants in these ecosystems show the ultimate adaptation to frequent fires: no permanent aboveground parts.

On a global scale, grasslands (broadly defined) are the largest of the four major natural vegetation types, covering 24% of the earth's vegetated area or 4.6 billion ha (Gould, 1968; Shantz, 1954; Olson *et al.*, 1983). Grasslands are variously called prairies (North America), steppes (Russia), velds (South Africa), and pampas (South America). In North America, the distribution of the major grassland formations (tall-grass, mixed grass, and short-grass prairies) roughly parallels the north-south precipitation zones (Collins, 1969) (Fig. 13.13), although other factors that influence soil moisture availability, such as evapotranspiration, soil physical properties, and temperature, also influence the distribution of grasslands.

While grasslands occur intermixed with trees as part of the continuum of conditions found in savannah regions, extensive areas of grassland are generally found in areas too dry to support woody vegetation except along streams, such as much of central North America east of the Rocky Mountains and west of the Mississippi River, or with a high enough fire frequency to suppress woody vegetation. Soil nutrient levels must be sufficiently high that enough aboveground biomass can be produced in a single season to support an extensive root system and allow reproduction. On sites with nutrient levels too low to support grasslands, shrublands are usually found.

No permanent aboveground biomass is accumulated by the grasses and forbs, and fires regularly remove all aboveground living and dead organic matter. Variation in both the frequency of fires (which do not always act as a disturbance, but virtually always affect community structure in some way) and in the rate of competitive displacement (which is strongly correlated with plant productivity) make grasslands an ideal ecosystem in which to observe the consequences of the dynamic equilibrium between competition and disturbance.

Grasslands occur over a broad range of climatic and soil conditions,

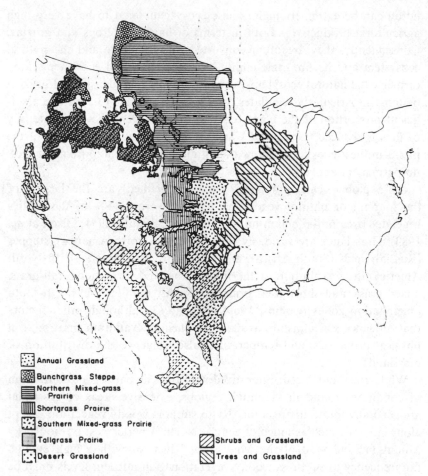

Fig. 13.13 North-south zonation of major grassland types of North America. (From Risser *et al.*, 1981.)

and consequently productivity and the rate of competitive displacement vary significantly over a wide range of temporal and spatial scales. Across the longitudinal gradient of precipitation from the tall-grass prairies to the short-grass prairies of North America, plant productivity, as well as plant height (Weaver and Fitzpatrick, 1934), are strongly correlated with average annual precipitation, and a similar pattern is found in African grasslands and savannahs (Fig. 13.25). At any particular location, grassland annual productivity is strongly influenced by annual precipitation, particularly precipitation that occurs prior to and during the growing season (Rogler and Haas, 1947; Smoliak, 1956; Dahl, 1963; Shiflet and

Dietz, 1974). Similarly, within any location, spatial variation in grassland productivity is strongly correlated with topographic variation in water availability (Albertson, 1937; Weaver and Fitzpatrick, 1934; Barnes *et al.*, 1983; Abrams *et al.*, 1986). Other soil properties, such as organic matter, total nutrient pools, and available nutrients, also vary with topography, generally being higher in swales and other areas of runoff convergence. In some areas, particularly ridge and slope positions, mineral nutrients may be so limiting that additional water has little stimulatory effect on plant growth. In an experiment on short-grass prairie in Colorado, Lauenroth *et al.* (1978; Dodd and Lauenroth, 1979) found that the addition of both nitrogen and water produced an increase in the biomass of the dominant plant types (warm season grasses and half-shrubs) that was generally two to ten times greater than the response to either nitrogen or water alone.

Moist, but well-drained lower slopes and level bottoms are generally dominated by tall, dense, low diversity stands of big bluestem, while wet, poorly-drained areas are dominated by grasses such as cordgrass (*Spartina pectinata*) or species of *Calamagrostis* (Weaver and Fitzpatrick, 1934). In areas with sufficient precipitation and suitable soil properties, tall grasses can dominate all topographic positions. Species richness, and particularly evenness, on a meter square basis is often relatively low in these areas dominated by tall grasses, although the total species pool in big bluestem stands can be quite high (Fig 13.14). In disturbed or eroded stream courses, rank growth of sunflowers or other forbs can form dense monocultures.

In the tall-grass prairie zone, drier upland sites and upper slope positions tend to have higher species diversity at both small (e.g., 1 m^2) and larger (e.g., 100 m^2) spatial scales than the stands dominated by tall grasses (Barnes *et al.*, 1983; Abrams and Hulbert, 1987; Gibson and Hulbert, 1987). The grasses are smaller in stature than those that dominate the moist, fertile areas (Weaver and Fitzpatrick, 1934; Albertson, 1937), and the degree of dominance by any species is generally less than in the moist, fertile areas (Fig. 13.15). Major grass species of drier sites in the North American prairies include little bluestem, *Stipa spartea*, prairie dropseed (*Sporobolus* sp.) and in drier, western areas, gramma grass (*Bouteloua* spp.).

On the driest sites, particularly those with thin, poor soils, plant cover is often less than 100%, and both species richness and evenness drop. Long-term climatic fluctuations, such as prolonged droughts, can have the same effects on grassland species diversity and composition as are seen along spatial moisture gradients (Weaver and Albertson,

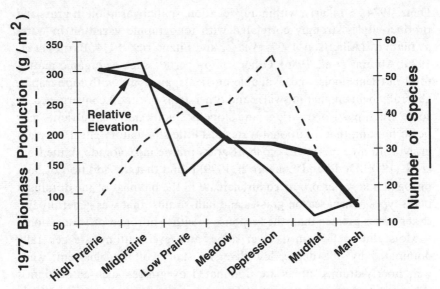

Fig. 13.14 Relationship of species richness to plant productivity and relative moisture availability along a topographic gradient on the Ordway Prairie, South Dakota. The thickest line indicates the relative elevation of the seven sites. (Based on Barnes *et al.*, 1983.)

1936, 1939; Albertson, 1937). It is interesting to note that, although little bluestem dominates many relatively dry upland sites under normal climatic conditions, it is more severely affected by drought than any other abundant grass species, and was dramatically reduced during the droughts of the 1930s (Albertson, 1937; Weaver and Albertson, 1936).

Thus, grasslands show the classic unimodal relationship of plant species diversity to plant productivity (Grime, 1973) that is predicted by the dynamic equilibrium model. Because of the great variation in the rate of competitive displacement (which is correlated with productivity) on prairies, the effect of a particular intensity or frequency of grazing, fire, or other disturbances can vary greatly from one area to another. In general, a particular level of disturbance would be expected to produce a much greater increase in diversity in productive sites than in high-diversity, unproductive sites (Milchunas et al., 1988), the least productive of which might decrease in diversity with an increase in disturbance frequency or intensity (Fig. 13.16). Woody plants, which grow at a much lower rate than grasses and herbs, decrease strongly in diversity and abundance with increasing fire frequency on prairies.

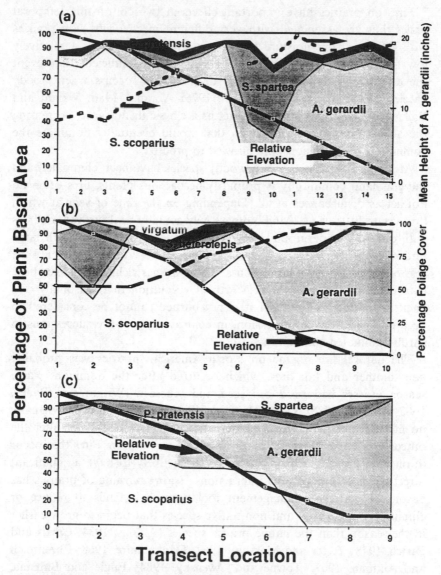

Fig. 13.15 Topographic gradients of species composition in three prairies, illustrating the shift in dominance from little bluestem (*Schizachrium scoparius*) on the dry upland areas to big bluestem (*Andropogon gerardii*) in the more productive lowlands. Note that dominance is highest (and evenness lowest) in the lowland prairies. (Based on Tables 1, 6, and 10 in Weaver and Fitzpatrick, 1934.)

Fires on prairies have important effects at two contrasting temporal scales that are associated with very different mechanisms. On prairies with sufficient soil moisture to support woody vegetation, a relatively low fire frequency (fire every 10 to 30) years is sufficient to prevent the dramatic reduction of species diversity that occurs when woody species invade (Bragg and Hulbert, 1976; Abrams, 1986; Wright and Bailey, 1982). At this scale, fires act as a classic disturbance, decreasing the abundance of woody species that would eventually dominate the community if succession were allowed to proceed.

At higher frequencies, when woody species have been eliminated and the grassland community is primarily herbaceous plants, fires often do not act as disturbances at all. Depending on the time of year at which fires occur, little or no living biomass may actually be destroyed. Grasses with extensive underground root systems, and forbs or trees with large underground root systems and storage organs survive fire with only the loss of their aboveground biomass. Thus, species richness changes little as a result of fire, and the regrowth of vegetation that seasonally dies back to ground level even if it is not burned cannot be considered to be a post-disturbance succession, in contrast to post-fire successions in shrub-dominated communities.

The natural fire regime on North American prairies was probably late summer and fall fires, which occurred after the dominant warm season grasses had produced seed and senesced (Wright and Bailey, 1982; Axelrod, 1985). Under these conditions, fire would cause virtually no mortality of living grass or herb tissue, and so could not alter the outcome of competitive interactions by this mechanism. Fires that occur during the growing season are much more likely to have a significant direct effect on competitive interactions. Spring burning of prairies has become a preferred management tool because it tends to reduce or eliminate many weedy and non-native species that begin to grow earlier in the season than the native prairie species (Aldous, 1934; Curtis and Partch, 1948; Tester and Marshall 1962; Daubenmire, 1968; Ehrenreich and Aikman, 1963; Towne and Owensby, 1984; Engle and Bultsma, 1984).

Even though fire often does not act as a true disturbance on prairies, it does have important effects on the community, primarily through altering resource availability by mineralizing nutrients tied up in organic matter and by reducing shading through removal of both dead and living aboveground biomass. In the absence of fire, a thatch-like layer of dead leaf material accumulates rapidly on productive prairies. This thatch has

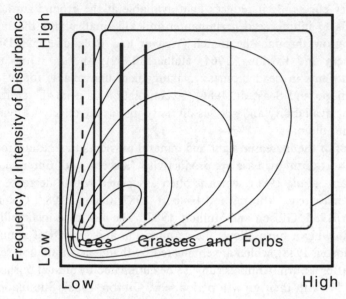

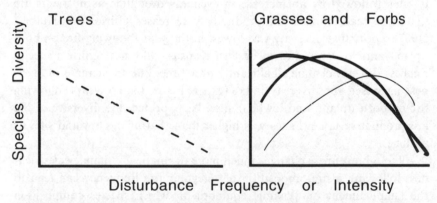

Fig. 13.16 Range of diversity responses to increasing disturbance frequency/
intensity on prairies. Trees and woody plants have low rates of growth and
competitive displacement in relation to grasses and forbs. Consequently, woody
plant diversity decreases monotonically with increasing disturbance on prairies,
which are near the physiological limit of woody plant growth in relation to
moisture availability. Grasses and forbs have a broad range of growth rates, in
response to variation in moisture and nutrient availability. Depending on the
effect of local environmental conditions on growth rates, the response of grass and
forb diversity in response to the same increase in disturbance frequency/intensity
can be either increasing, unimodal, or decreasing. (Based on Huston, 1979.)

two primary effects: 1. it reduces light available at the ground surface, which tends to inhibit seed germination and kill small plants that are unable to grow through the thatch (Weaver and Roland, 1952; Dix, 1960; Kucera and Koelling, 1964; Hulbert, 1969); and 2. it ties up mineral nutrients in dead biomass, making them unavailable for plant growth (Knapp and Seastedt, 1986). Accumulation of dead thatch can reduce light availability and species diversity just as effectively as living aboveground biomass.

The effect of the increased light and mineral nutrients that result from fire is a short-term increase in productivity (Kucera and Ehrenreich, 1962; Abrams et al., 1986), which is often associated with a decrease in species diversity (e.g., Abrams and Hulbert, 1987; Collins, 1987; Collins and Barber, 1985; Gibson and Hulbert, 1987). The increased availability of light following a burn allows the germination and growth of annual species (Gibson, 1988), which are eliminated by thatch accumulation several years after a burn. Annuals can also be eliminated by annual burning that coincides with their growth period (e.g., Gibson, 1988). Stimulation of growth by increased nutrient availability, particularly nitrogen (Hobbs and Schimel, 1984), can lead to reduced diversity, primarily in terms of evenness, immediately following a fire. The initial reduction in evenness is often followed by an increase in evenness over time as nutrients are tied up in dead biomass and productivity decreases. Gibson and Hulbert (1987) report that diversity was lowest in the year following fire on both upland and lowland sites in eastern Kansas tall-grass prairie. Species richness did not change significantly over time, but an increase in evenness increased the diversity index (exp H') over the six years following fire on both upland and lowland sites. As expected, the diversity of the less productive upland sites was higher than that of the lowland sites at all times.

Although evenness changes much more dramatically than species richness following a fire, imposition of a regular fire frequency can lead to the establishment of dynamic equilibria in which species composition and richness vary predictably with fire frequency. In general, annual fires lead to a reduction in broadleaf forbs, and a productive, lower diversity grassland dominated by the warm-season grasses (which use the C_4 photosynthetic pathway) such as big and little bluestem, Indian grass (Sorghastrum nutans), and switchgrass (Panicum virgatum) (Gibson and Hulbert, 1987). A somewhat lower frequency of burning (e.g., 3-8 years, depending on the rate of competitive displacement) allows maintenance of a higher diversity community with more forbs, woody plants, and additional grass species. A very low frequency of burning results in a dynamic equilibrium with lower species diversity as a result of thatch accumulation and eventually, invasion and dominance by woody species.

Thus, in many cases, the effect of fire on species diversity in grasslands is not through action as a mortality-causing disturbance that increases diversity, but rather as a stimulus to growth that increases the rate of competitive exclusion and actually results in a decrease in diversity, at least for the few years following the fire. The positive effect of fire on grassland diversity is not so much the maintenance of species diversity among grasses and herbaceous plants, as the prevention of replacement of the high diversity herbaceous community by a much lower diversity community of woody plants.

Grazing of plants by herbivores is also an important process on most grasslands. Unlike fire, which in some cases removes only dead biomass, grazing almost always acts as a disturbance by removing living biomass. Grazers tend to be selective in the species that they consume (Schwartz and Ellis, 1981; Senft *et al.*, 1985), so their effect on species diversity depends on the relative abundance and growth rates of the preferred food species. Grazing generally results in an increase in the abundance of unpalatable species, and a decrease in the abundance of palatable species (Dyksterhuis, 1958). Because the dominant tall-grass species of the North American prairies are highly palatable (Dyketerhuis, 1958), grazing generally results in increased diversity on productive sites. In contrast, grazing often reduces diversity in the less productive short-grass prairie region (Weaver and Albertson, 1940; Milchunas *et al.*, 1988)

Collins (1987) conducted an experimental study of the effects and interaction of fire (three years of annual mid-April burns) and grazing by cattle (three years of mid-May to September grazing at 0.4 ha per animal) on tall-grass prairie in Oklahoma. He found that grazing alone had the strongest effect in reducing the relative area covered by the dominant grasses (from 70.1% in the control plots to 56.4%) and increasing the cover of forbs and minor grasses (from 38.1% in the controls to 43.1%), while annual burning alone had the greatest effect in reducing the cover of minor species (from 38.1% in the controls to 11.9%) and increasing the cover of the dominant grasses (from 70.1% in the controls to 93.6%). Average species richness (sampled with 15 0.25 m^2 quadrats in each of three 0.1 ha replicates per treatment) increased from 24.3 in the ungrazed, unburned treatments to 29.0 in the grazed, burned treatments. Species richness was slightly higher in the burned treatment (27.7) than in the grazed treatment (24.7), but evenness was much lower in the burned treatment.

Milchunas *et al.* (1988, 1991) discuss the interaction of evolutionary history with the effect of grazing intensity on grassland community structure along a moisture/productivity gradient from semi-arid to subhumid grasslands. They postulate that on grasslands with a long evolutionary history of grazing, plant species have acquired adaptations that reduce the effect of grazing as a disturbance and increase the rate of recovery from disturbance (analogous to an increase along the growth rate/competitive displacement axis, Fig. 13.17). Their generalizations about the response of species diversity to grazing under different moisture conditions, based on an extensive review of the literature, are consistent with the predictions of the dynamic equilibrium model. In semi-arid grasslands an increase in grazing intensity results in either no change in diversity or a monotonic decrease (Fig. 13.17*a*, *b*). In productive subhumid prairies, competition for light and relative high rates of competition displacement result in a classic 'intermediate disturbance' unimodel response of diversity along a gradient of increasing grazing intensity (Fig. 13.17*c*). On sub-humid prairies with a long evolutionary history of grazing, diversity increases more gradually with grazing pressure and high diversity is maintained at higher grazing levels than on similar prairies with a short history of grazing (Fig. 13.17*d*).

Any disturbances that create spatial heterogeneity within an area are likely to increase total species richness within the area by increasing between-habitat (beta) diversity, even thought the small-scale, within-habitat (alpha) diversity may be unchanged over most of the area, and decreased at the actual sites of the disturbance. This is the effect of many soil-disturbing processes on prairies, such as buffalo wallows, gopher and badger mounds, and prairie dog towns (Platt, 1975; Bonham and Lerwick, 1976; Coppock *et al.*, 1983; Collins and Uno, 1983), as well as patchiness that results from spatial heterogeneity of grazing intensity (Milchunas *et al.*, 1988; Bakker *et al.*, 1983). Measurement of grassland diversity at large scales (e.g., 0.5 ha, Collins and Barber, 1985) is likely to obscure the small-scale mechanisms that influence species diversity by including both the effects of processes that influence within-habitat diversity and processes that influence between-habitat diversity.

Spatial heterogeneity is clearly important in maintaining grassland diversity, as it is in virtually all ecosystems. However, it is also important to consider functional-type diversity as well. Even though the stature of grassland vegetation is short in relation to forests, grassland vegetation is

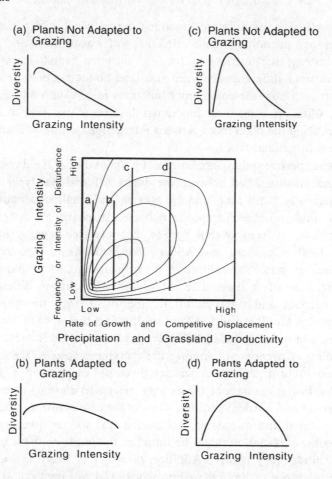

Fig. 13.17 Hypothesized dynamic equilibrium values of grassland species diversity in relation to grazing along a moisture/productivity gradient, illustrating the effect of evolutionary adaptation to grazing on the dynamic equilibrium of species diversity. (*a*) Strong monotonic decrease in species diversity with increased grazing intensity on semi-arid short-grass grasslands with little evolutionary adaptation to grazing pressure. (*b*) Slight decrease in diversity with increased grazing intensity on semi-arid short-grass grasslands with a long evolutionary history of grazing. Plant adaptation to grazing allows more rapid recovery of aboveground biomass following grazing, which effectively shifts the community toward higher rates of growth and competitive displacement. (*c*) Unimodal response of diversity to grazing intensity on productive subhumid grasslands with short evolutionary history of grazing. (*d*) Unimodal response of diversity to increased grazing intensity on productive subhumid grasslands with a long evolutionary history of grazing. Adaptation to grazing shifts the community to higher values on the competitive displacement axis, which has the effect of shifting the diversity maximum to higher intensities of grazing, and allowing the maintenance of diversity at grazing intensities higher than would be possible without adaptation to grazing (e.g., (*c*)). (Based on Huston, 1979, and Milchunas *et al.*, 1988.)

composed of many different functional types, which differ in maximum size, growth phenology, shade tolerance, water acquisition and conservation strategy, and nutrient source (e.g., nitrogen fixers). Recognition of the functional differences between grassland plants can provide important insights into how the community functions (e.g., Coffin and Lauenroth, 1990), while ignoring these functional differences can lead to confusion by obscuring the differences between interactions within and interactions between functional types.

Most experimental manipulations of grassland or oldfield communities have not distinguished between the different functional types of plants, and generally lump large canopy species and small understory species into a single number for species richness. In nearly all cases, addition of nutrients reduces overall species diversity (see reviews in Huston, 1979, 1980; Goldberg and Miller, 1990). However, the addition of nutrients or water, or subjection to a disturbance such as mowing or grazing, can affect these different groups of plants in different ways. In a nutrient and water addition experiment on a first-year oldfield community, Goldberg and Miller (1990) found that a single large annual species, common ragweed (*Ambrosia artemesiafolia*), dominated all plots regardless of treatment. Among the remaining species, the treatments resulted in little change in species diversity, in spite of a significant increase in the biomass of the canopy species (*Ambrosia*) in response to both water and nutrient addition. Only in the case of nitrogen addition (in both watered and unwatered treatments) was species diversity reduced. A possible explanation might be found in the photosynthetic physiology of the understory plants. Addition of water is likely to increase the shade tolerance of the understory plants, and so might compensate for the reduction in light that penetrates the denser overstory. Although addition of nitrogen would be expected to increase the rate of gross photosynthesis of all the plants, increased nitrogen (and thus protein) levels could also lead to higher respiration rates, and thus no net gain in shade tolerance for the understory plants.

In summary, spatial variability in species richness and community structure on grasslands results from variation in water and nutrient availability, which can produce high growth rates that lead to competitive displacement and low diversity. Species diversity on North American prairies is closely related to resource availability, producing a unimodal response of diversity in relation to productivity. High productivity sites in fertile, moist lowlands are usually dominated by a few species. Highest species richness is generally found on sites with relatively low productivity, such as dry ridges and hillslopes. Such patterns are found in a wide varietiy of grasslands, including those on serpentine soils, where biomass is generally lower and species diversity higher than on adjacent non-serpentine soils (Spence, 1959; Whittaker, 1960; McNaughton, 1968: Hart, 1980; Huenneke *et al.*, 1990), and on other rocky substrates (Silvertown, 1983; Mahdi *et al.*, 1989). Experimental additions of nutrients to grasslands and oldfields almost always result in a decrease in species diversity (Bakelaar and Odum, 1978; Willem, 1980; see reviews in Huston 1979, 1980). Depending on timing, fire may or may not act as a disturbance, and often has the effect of temporarily increasing the availability of light and mineral nutrients. The effect of grazing is highly variable and depends on the productivity and species composition of the vegetation.

Savannahs: Fire and Grazing in Grass/Tree Systems

The plant communities dominated by grass and trees clearly reflect a peculiar balance of environmental conditions that favors these contrasting plant growth forms, but disfavors plants of intermediate stature, such as shrubs. These types of ecosystems are generally called savannahs, although the term lacks a precise definition (Pratt *et al.*, 1966; Bourlière and Hadley, 1983). The term savannah has been used to describe vegetation types ranging from open coniferous woodlands to certain arid shrublands (Bourlière and Hadley, 1983). True savannahs are characterized by scattered trees in a more or less continuous matrix of grassland. Frequently savannahs are interspersed with treeless patches of pure grassland and by areas of closed canopy forest with few grasses.

The significant features of savannahs are: 1. alternating wet and dry seasons; 2. structure determined primarily by competition for soil moisture between grasses and trees; and 3. strong modification by fire, herbivores, and soil nutrients (Walker and Noy-Meir, 1982). In the context of the dynamic equilibrium model, savannahs experience a broad range of disturbance frequencies (and intensities) that is bounded by frequencies too high to allow the survival of woody vegetation, which produce a pure grassland, and frequencies so low that a closed canopy forest can form and eliminate the grasses (Fig. 13.18). In few other ecosystems does the balance between the frequency of disturbance and the rate of competitive displacement have a more dramatic effect on the structure, species diversity, and visual heterogeneity of the landscape.

Savannahs are primarily a tropical phenomenon (Walter, 1964/68,

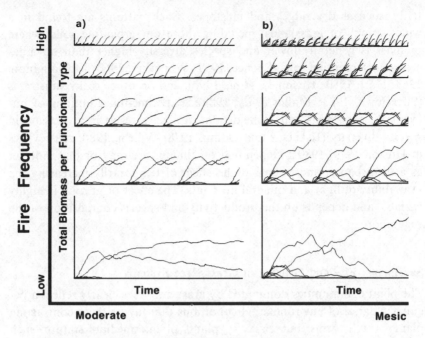

Water Availability

Fig. 13.18 Effect of variation in disturbance frequency and plant growth rates on successional dynamics in a savannah region. Vegetation ranging from pure grassland to closed canopy forest can be represented, depending on the growth conditions, average fire frequency, and time since last fire. The coexistence of grass and trees in the same location results from a particular balance of moisture conditions and disturbance regime (see text).

1985), although areas of interspersed grass and trees are also found in the temperate zone, particularly in mediterranean climate regions and at the boundary between grasslands and forests.

Vegetation that can be classified as savannah covers approximately 15% of the land surface (Olson *et al.*, 1983; Atjay *et al.*, 1979), with most being found in Africa (Fig. 13.19), and lesser amounts in South America to the north and south of the Amazon Basin rain forest. Walter (1985) distinguishes four types of tropical savannahs:

'(1) *Fossil savannahs,* which formed under different climatic conditions but are maintained under present climate unsuitable for the formation of savannas.

(2) *Climatic savannahs,* at the boundary between tropical deciduous forests and deserts, with precipitation between 300 and 500 mm per year.

(3) *Edaphic savannahs,* where soil conditions result in excessive water during the rainy season, reduced water availability as a result of impermeable barriers, or nutrient levels too low to support a forest.

(4) *Secondary savannahs,* which are created and maintained by fires, grazing, or various human interventions.'

The low total amount and high variability of precipitation in savannah regions create a precarious balance of plant community structure that can be shifted toward treeless grassland or toward closed forest by subtle variation in any one of a number of environmental conditions (Walter, 1971; Sarmiento and Monasterio, 1975; Walker 1981; Walker and Noy-Meir, 1982). Consequently, the spatial pattern of vegetation in savannah regions tends to be very heterogeneous. Vegetation patterns in savannahs are influenced by: 1. spatial variation in soil water caused by variation in topography and soil drainage properties; 2. temporal variation in water resulting from climatic variability; 3. spatial variability in soil mineral nutrients; 4. the intensity and timing of grazing and browsing by mammals and herbivory by insects; and 5. the frequency and timing of natural and anthropogenic fires. As the transitional zone between grasslands and closed woodland, savannahs can exhibit dramatic shifts in composition as the consequence of subtle variations in environmental conditions (e.g., Markham and Babbedge, 1979; Scholes, 1990), differences in fire frequency or history, or simply time since the most recent fire.

Savannahs also provide an example of how the number and diversity of different functional types within a plant community changes through time and in response to environmental variation in conditions (Fig. 13.20). In relation to the frequency of fires or the intensity of grazing and browsing, trees have very low growth rates and rates of competitive displacement, while grasses have high growth rates and competitive displacement rates. Savannahs generally have a very low diversity of trees, often extensive stands of a single species, and moderate or high diversities of grasses and herbaceous plants. An increase in the level of resources such as water or mineral nutrients usually results in a decrease in the diversity of grasses and herbaceous plants as a result of increased competitive displacement (competition for light). The same increase in resources usually leads to an increase in the diversity of woody plants, as a consequence of increased ability to recover between disturbances, as well as the addition of more shade-tolerant functional types to the community. Similarly, a decrease in the frequency of disturbance results in an increase in the diversity of

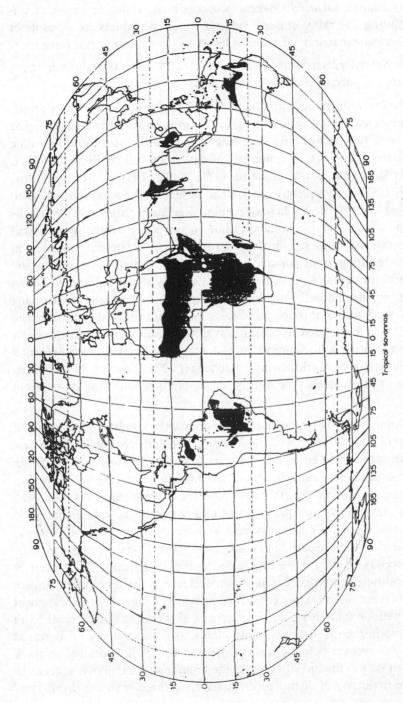

Fig. 13.19 Global distribution of tropical savannas. (From Bourlière, 1983c.)

woody plants and a decrease in the diversity of grasses and herbs (Fig. 13.20).

The low annual total and high seasonality of precipitation make savannah regions marginal for the survival of woody plants. The coexistence of a mixture of grass and woody vegetation may also result from temporal and spatial resource partitioning through a two-layer rooting system, in which grasses use water that is seasonally available in high amounts near the surface, and the woody vegetation uses both surface water and deeper subsoil water that is available over a longer time period (Walter, 1939, 1971, 1985). A stable mixture of grass and trees is possible if 1. there is sufficient rainfall (300-500 mm annual precipitation concentrated in a single rainy season, Walter, 1985) that some is able to penetrate through the grass roots to the subsoil; and 2. there is insufficient rainfall for the maintenance of a closed woodland (Walker and Noy-Meir, 1982). While these conditions could conceivably allow a situation of coexistence at competitive equilibrium to occur, in most cases savannahs are clearly non-equilibrium systems, in which periodic droughts or disturbances by fire and animals interrupt the course of plant succession that would eventually lead to dominance by woody vegetation and elimination of grasses (Medina and Silva, 1990). It should be noted that trees can also be eliminated by too much water, which leads to anoxia and root death. Some grasslands and palm savannahs in the wet tropics result from this mechanism (Walter, 1985; Sarmiento, 1983; Sarmiento and Monasterio, 1974; Menaut, 1983; Pires and Prance, 1985), as do wet seasonal grasslands that occur on very shallow soils in dry savannah regions (Walker, 1981).

Plant succession on savannahs leads to closed forest whenever there is sufficient moisture availability to support woody species that are more shade tolerant than the savannah dominants. Suitable moisture conditions are often found in topographic depressions, on particular soil types, or may result from climatic variation in annual precipitation or seasonality. The gradual invasion and growth of shade-tolerant tree species under the dominant species of open savannahs occurs more rapidly under conditions of higher moisture availability (Blaisdell *et al.*, 1973), but does occur even under relatively dry conditions in the absence of disturbances such as fire (Smith and Goodman, 1987). Succession in woody vegetation tends to eliminate open-savannah species, but leads to an increase in woody species diversity as functional types of increasing shade tolerance are added to the community. However, these shade-tolerant functional types tend to be intolerant of dry conditions, so periodic severe droughts

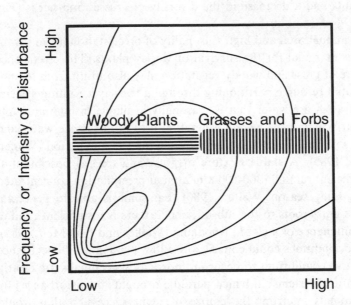

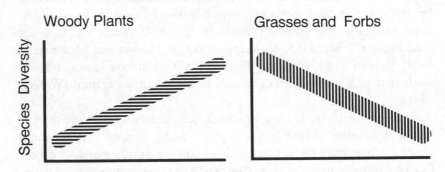

Fig. 13.20 Differences in the response of the dynamic equilibria of the woody species versus the grasses and forbs in a savannah community. For a given frequency of reduction (by fire) the woody vegetation has a very low growth rate as a result of low water availability. An increase in growth rates (resulting in an increase in the rate of recovery from disturbance) will increase the species diversity of woody plants, while a similar increase in the growth rates of grasses and forbs will lead to a decrease in diversity as result of competitive displacement. Along the disturbance (frequency of reduction) axis, an increase in fire frequency will lead to a reduction of woody plant diversity, and an increase in the diversity of the grasses and forbs. (Based on Huston, 1979.)

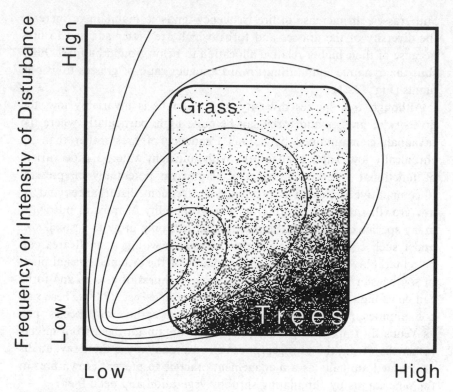

Fig. 13.21 Dynamic equilibrium of the balance between grasses and trees in savannahs. Trees increase in relative abundance with low frequencies of disturbance (fire, grazing) and high growth rates, while grasses increase in relative abundance with high frequencies of disturbance and low growth rates. (Based on Huston, 1979.)

can slow or reverse woody succession and reduce the diversity of woody species. Disturbances such as fire or grazing are also important in slowing or stopping this successional pattern (Walker *et al.*, 1981; Walker and Noy-Meir, 1982).

The effect of fire on woody plants in savannahs is generally to reduce diversity. This occurs because smaller individuals of shade-tolerant species are more susceptible to fire mortality than the larger trees in whose shade they become established, and because only a small subset of large trees are able to survive fires by virtue of adaptations such as thick, corky bark (Walker, 1981). Because fires have a negative effect on the woody vegetation that would eventually outcompete and eliminate forbs and grasses, an increase in fire frequency tends to maintain or increase the diversity of the grasses and forbs, which are better adapted to fires because of their higher relative allocation to belowground biomass. Fires shift the dynamic equilibrium toward a higher ratio of grasses to woody plants (Fig. 13.21).

Although tree species diversity on savannahs is invariably low, the diversity of grasses and forbs can be quite high, particularly where the savannah is maintained by a high frequency of fires rather than by chronically low water availability. With sufficient water, a wide variety of functional types of herbaceous plants can potentially be present. If competitive exclusion is prevented by frequent disturbances and/or low growth rates that result from low availability of mineral nutrients, many species can coexist. Because plants can only grow to a small size under such conditions, the number of species within a small area can be remarkably high. In the pine savannahs of the Atlantic coastal plain of southeastern North America, the species richness of grasses and forbs can be as high as 42 species/0.25 m^2 (Walker and Peet, 1983). These rich communities include a significant proportion of carnivorous species such as Venus fly traps and pitcher plants that obtain nitrogen from insects rather than from the nitrogen-poor soils. In many areas these savannahs are burned annually as a management practice to prevent crown fires in the pine stands by eliminating shrubby vegetation and dead fuel.

Animal Diversity on Savannahs

Africa, with by far the most extensive area of savannahs in the world, is epitomized by the vast herds of large mammals found on its savannahs and grasslands. These large herbivores are all ungulates, from three orders: the Proboscidea, with a single species, the elephant; the Perissodactyla, with six species, including rhinoceros and zebras; and the Artiodactyla, with many species in five families. Of the five Artiodactyl families, the Bovidae, with about 78 species of antelopelike animals of a range of sizes, is by far the most diverse and abundant; the Suidae, or pig family, is ranked next with three species. The high diversity of the Bovids is a relatively recent phenomenon, with most of the speciation occuring in the Pliocene and Pleistocene, apparently as a consequence of the evolution of the divided ruminant stomach, which is highly efficient for digesting cellulose (Langer, 1974).

The high diversity of the Bovids and other large and small mammals on savannahs is associated with the highest biomass of mammals found in any terrestrial ecosystem (Table 13.1). This remarkable diversity is the consequence of a variety of ecological factors and an evolutionary history unique to Africa. Ecological factors that contribute to the coexistence of many closely related and ecologically similar species include: 1. a high degree of spatial habitat heterogeneity, with great variability in the structure and productivity of vegetation resulting from seasonal and spatial variability in soil moisture; 2. high productivity of digestible plant material resulting from the dominance of herbaceous plants, particularly grasses, and the relatively low proportion of woody vegetation; 3. high seasonal and annual variability in the productivity and thus the carrying capacity of the environment, resulting from variation in precipitation and leading to density-independent population reduction (which can also have a strong density-dependent component)(Corfield, 1973; Phillipson, 1975); and 4. strong density-dependent population regulation resulting from parasites, diseases, and predators (Bourlière, 1983b).

In spite of the basic ecological similarity of many African ungulates, and the high degree of overlap in the habitats and resources that they utilize (Ferrar and Walker, 1974; Sinclair, 1983), there is substantial resource partitioning on the basis of size (Bovid weights range from <4 kg to >800 kg), and size-dependent requirements of food quality and quantity (Gwynne and Bell, 1968; Sinclair, 1983); habitat (Lamprey, 1963; Ferrar and Walker, 1974); and seasonal use of different resources and different habitats, including long-distance migration (Pennycuick, 1975; Maddock, 1979) and local shifts in habitat use, sometimes called the 'grazing succession' (Vesy-Fitzgerald, 1960, 1965; Bell, 1970, 1971) (Fig. 13.22).

The abundance of large predators is one of the most dramatic aspects of the plains and savannahs of Africa, and attests to the high productivity of primary producers and herbivores. Lions, leopards, cheetahs, wild dogs, and hyenas represent a variety of different hunting strageties, and have exerted a selective pressure that has led to a variety of behavioral strategies to avoid predation (Sinclair, 1983; Hamilton, 1971; Jarman, 1974; De Vos and Dowsett, 1966; Sinclair, 1977). Predator-avoidance strategy is influenced both by environment (e.g., open grasslands versus closed woodlands), and by prey size, and may be similar or opposite to the selective pressure exerted by competition. Among small ungulates in complex habitats with areas of shrub and woodland, both predation and competition select for small size and solitary existence (Sinclair, 1983). In open grasslands and savannahs, predation selects for large size and large herds, while competition selects for smaller size and smaller

Table 13.1. *Ungulate biomass and species richness in some national parks and other protected areas.*

Location	Habitat Type	Number of Ungulate Species	Live Biomass (t km^{-2})	Domestic Stock (t km^{-2})
Africa				
Tarangire Game Reserve, Tanzania[a]	Open *Acacia* Savanna	14	1.1	–
Kafue National Park, Zambia[b]	Tree Savanna	19	1.3	–
East Tsavo National Park, Kenya[c]	Open *Commiphora Acacia* woodland	13	4.4	–
Nairobi National Park, Kenya[d]	Open Savannah	17	5.7	–
Serengeti National Park, Tanzania[e]	Open and Tree Savanna	20	8.2	–
Ruwenzori National Park, Uganda[f]	Open Savanna and Thickets	11	12.0	–
Ruwenzori National Park, Uganda[g]	Same Habitat, Overgrazed	11	27.8-31.5	–
Virunga National Park, Zaire[h]	Open Savanna and Thickets, Overgrazed	11	23.6-24.8	–
South Asia				
Gir Forest,Gujarat, India[i]	Dry Deciduous Woodland and Tree Savanna	6	0.4	6.2
Wilpattu National Park, Sri Lanka[j]	Open Forest and Scrub	7	0.7	?
Kanha National Park, Madhya Pradesh, India[k]	Open *Shorea robusta* Forest and Grass Meadows	10	0.9-1.2	2.9-3.0
Karnali-Bardia National Park, Terai, Nepal[l]	Open *Shorea robusta* Forest and Grass Flood Plain	6	2.8-3.1	47.9
Kaziranga Wildlife Sanctuary, Assam, India[m]	Grass Flood Plain	9	3.8	?
Chitawan National Park, Tarai, Nepal[n]	Tall Grass and Riverine Forest	6	18.5	41.8

groups (Jarman, 1974; Sinclair, 1983). Over evolutionary time, predation pressure in this spatially heterogeneous environment has contributed to the remarkable radiation of grazing and browsing mammals.

The role of predation in regulating ungulate population size is highly

Table 13.1. *Continued.*

Location	Habitat Type	Number of Ungulate Species	Live Biomass (t km^{-2})	Domestic Stock (t km^{-2})
South America				
Estacion Biologica de los Llanos, Masaguaral, Venezuela[o]	Mosaic of Savanna Types	2	0.3[1]	–

[1]To which a biomass of 0.3 t km^{-2} must be added, representing the population of capybaras *Hydrochoerus capybara*), the large rodent which is the ecological counterpart of a small ungulate

[a]Lamprey (1964)
[b]Dowsett (1966)
[c]Leuthold and Leuthold (1976)
[d]Foster and Coe (1968)
[e]Sinclair and Norton-Griffiths (1979)
[f]Petrides and Swank (1966)
[g]Petrides and Swank (1965)
[h]Bourlière (1965)
From Bourlière (1983a).

[i]Berwick (1974)
[j]Eisenberg and Lockhart (1972)
[k]Schaller (1967)
[l]Dinerstein (1980)
[m]Spillett, in Seidensticker (1976)
[n]Seidensticker (1976)
[o]Eisenberg *et al.* (1979)

variable, having virtually no effect in some situations and a strongly limiting effect in other situations (Sinclair, 1983), depending on the productivity and seasonality of the ecosystem. This variability was illustrated by the elimination in 1964 of rinderpest, an exotic disease that had imposed high juvenile mortality on many of the most abundant ungulates. The elimination of rinderpest had dramatically different effects in the Ngorongoro Crater, which receives sufficient rain throughout the year to support grass growth, than in the adjacent Serengeti ecosystem, where the highly seasonal precipitation has led to the long-distance seasonal migrations of wildebeest, the most abundant ungulate. In the Serengeti, the population size of the predators (lions and hyenas) is limited by the abundance of the resident prey (topi, hartebeest, and warthog), which suppport the predators when the wildebeest migrate to other areas (Hanby and Bygot, 1979). Consequently, the predators do not regulate the size of the wildebeest population, removing only about 1% of the wildebeest per year (Elliott and Cowan, 1978). Elimination of rinderpest resulted in a fivefold increase in the wildebeest population (to 1.3 million) in the Serengeti, and a similar increase in buffalo (Sinclair, 1983). In the Ngorongoro Crater in contrast, where the wildebeest are permanent res-

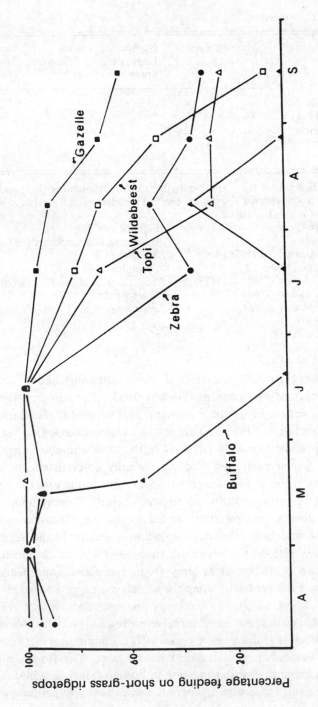

Fig. 13.22 Seasonal grazing succession in the Serengeti Plains of Africa. Large ungulates, such as the buffalo, leave the dry, short-grass ridgetops and move to wetter, tall-grass areas earlier in the season than do smaller ungulates, such as wildebeest and gazelles. The species using the short-grass ridgetops follow a progression of decreasing size through the season. (From Bell, 1970.)

idents and do not migrate, the lions and hyaenas remove about 14% of the wildebeest population annually (Elliott and Cowan, 1978). Removal of rinderpest had no effect on the population size of wildebeest, which fluctuated between 10,000 and 15,000 between 1958 and 1978, suggesting that predators, rather than food supply, were regulating population size (Sinclair, 1983).

The availability of food, along with the spatial heterogeneity of the habitat, are the critical factors supporting the high diversity of animals on savannahs. Savannahs, along with grasslands and tundras, have the highest ratio of plant productivity to plant biomass of any terrestrial ecosystem (Fig. 13.23), and nearly all the productivity is palatable and accessible to a wide size range of animals, without need for specialization for climbing trees (small size is one of the primary requirements for an arboreal lifestyle). The dominant role of plant productivity in regulating the biomass and diversity of savannah animal communities is illustrated by the fact that seasonal patterns in the biomass and diversity of insects, amphibians, birds, small and large mammals are strongly correlated with seasonal variation in precipitation and plant productivity (Gillon, 1983; Lamotte, 1983; Fry, 1983; Happold, 1983; Sinclair, 1983).

The combination of high plant productivity and a variety of different plant resources that allows specialization has led to high diversity of small mammals (Happold, 1983) and invertebrates (Gillon, 1983; Anderson and Lonsdale, 1990) as well as large mammals. Animals, ranging from lions to spiders, that are specialized to prey on other animals make an important contribution to the overall animal diversity of savannahs (Gillon, 1983; Bourlière, 1983d). As in other ecosystems, the diversity of higher trophic levels increases with increasing productivity of prey species, which is closely tied to the amount of the plant productivity that can be used by animals. The large number of carnivore species found in Serengeti National Park (27, Bourlière, 1983d) is undoubtedly related to the relatively high prey biomass and the high ratio of prey to predator biomass (Schaller, 1972).

Bird diversity is also extremely high on savannahs, as a result of the structural heterogeneity and high availability of plant seeds and insect prey (Fry, 1983). Savannah bird diversity is surpassed only by the bird diversity of tropical rain forests. However, the size diversity and morphological diversity of savannah birds, which include ostriches and large ground-dwelling storks and cranes, are actually higher than in rain forests, where the average size and the total size range of birds are much smaller. Savannah bird species comprise approximately half of the total

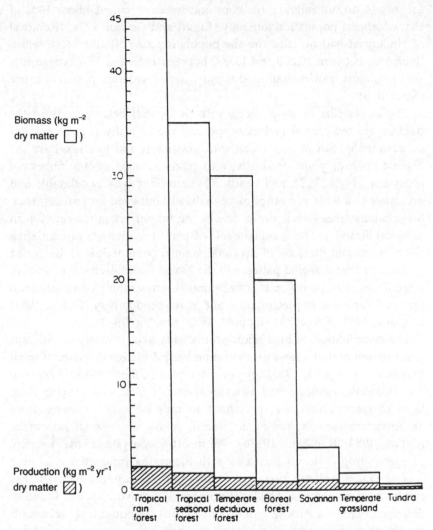

Fig. 13.23 Ratio of plant productivity to plant biomass for major ecosystems (Based on data collated by Lieth and Whittaker, 1975, from Cox and Moore, 1985.)

bird faunas of Australia and of Africa south of 20 °N (227 of 460, and 708 of 1500, respectively) but only 16% of the neotropical bird fauna (521 of 3165) (Fry, 1983), where the relative and absolute area of savannah in relation to rain forest is much smaller.

In addition to the ecological factors that allow the maintenance of high animal diversity on savannahs, it is clear that evolutionary processes have had a major role in producing the high species diversity of savannahs. In particular, the high diversity of nearly all animal taxa on African savannahs in relation to other tropical savannahs requires an evolutionary or historical explanation. The large size of Africa and the fact that it symmetrically spans the equator have allowed the persistence of three geographically separated arid refugia during periods of extreme wet climate (Leuthold, 1977) (Fig. 13.24), in contrast to the New World, where transformation of savannahs to steppe during the Miocene resulted in the extinction of most large savannah mammals (Webb, 1978). During dry climatic periods, the semi-arid African savannah could expand to form a continuous band around the wetter region that extends eastward along the equator from the western rain forest area. During periods of cooler, wetter climate, the areas of mesic vegetation were much more extensive. This climatically-driven pattern of vegetation change, fragmentation, and coalescence has been described as an 'evolutionary pump' of adaptive radiation resulting from alternating allopatric and sympatric evolutionary processes (Morton, 1972). Present-day patterns of abundance and distribution support this interpretation (Sinclair, 1983). A similar phenomenon of repeated geographic fragmentation and coalescence may be responsible for the high species diversity of rain forest trees, mangroves, and marine organisms such as coral in the island complex of the Indo-Pacific (see Chapter 12).

Animal Effects on Landscape Pattern in Savannahs

Grazing animals can interact strongly with fire, since both fire and grazers consume the same resource. High levels of grazing can reduce the frequency of fires, and the intensity of those fires that do occur, by reducing the accumulation of flammable biomass (Sinclair, 1979). Because grazers are much more selective in the biomass that they consume than are fires, community structure differs in response to whether grazing or fire is the predominant disturbance to the vegetation. Where there is sufficient moisture to support woody vegetation, grazing is likely to result in reduced diversity as a result of dominance by unpalatable woody vegetation.

The short-term effect of fire or human activity on plant nutrient content can induce a positive feedback interaction with herbivores that may introduce persistent heterogeneity into the landscape (e.g., Scholes, 1990; Blackmore *et al.*, 1990; Holt and Coventry, 1990). In African savannahs on poor sandy soils, the woody vegetation is composed of several

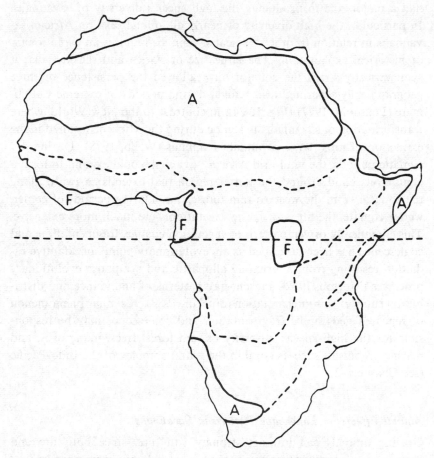

Fig. 13.24 Location of arid refugia (A) and forest refugia (F) during periods of extreme wet and dry climate. A drought corridor, indicated by dotted lines, currently links the arid savannah regions into a continuous band around the central forest region. (From Sinclair, 1983, based on Leuthold, 1977.)

species, dominated by *Burkea africana* with a mixed species herbaceous layer dominated by the coarse grass *Eragrostis pallens*. Nutrient levels are increased by dung and urine in cattle pen areas, and following abandonment, the vegetation shifts to a low diversity savannah dominated by *Acacia* trees and a palatable grass, *Cenchrus ciliaris*. Because *Cenchrus* is heavily preferred by most wild and domestic herbivores, nutrients continue to be concentrated in these areas, which maintains islands of low diversity, and palatable high-nutrient vegetation (Walker, 1981). The response of herbivores to the increased nutrient content of post-fire vegetation is a well-known phenomenon (Weaver, 1967; Komarek, 1969; Nelson, 1974) and can lead to high concentrations of herbivores on a small proportion of the total landscape.

Although plant productivity in savannahs increases with precipitation (Fig. 13.25), the plant productivity that is available to support the secondary production of animals actually decreases at high levels of precipitation, as the grass-dominated savannah shifts to a tree-dominated closed forest, with less digestible and less accessible plant biomass. Consequently, the number of species of ungulate grazers and browsers tends to decrease at high levels of precipitation, as their primary food sources are displaced by trees that produce much less vegetative biomass that can be used by animals (Fig. 13.26). In areas with sufficient precipitation, open savannah vegetation represents a successional stage that would eventually change to closed woodland in the absence of disturbance. In these ecosystems, elephants can be important in maintaining the structural and compositional diversity of the plant community, which allows the coexistence of a high diversity of grazing ungulates. When elephants are removed from these systems, the forest canopy closes, and the diversity of grazing and browsing mammals can drop dramatically, as suggested by the habitat preferences of major species (Fig. 13.27).

The effect of increasing human populations in Africa, and in particular of the extensive trade in now-illegal elephant ivory, has been to reduce elephant populations drastically throughout most of Africa (Weisburd, 1988; Booth, 1989; Perlez, 1990; Morell, 1990) and to concentrate remaining groups of elephants in refuge areas where their populations occasionally reached abnormally high levels.

A dramatic effect of elephant overcrowding that was noted in the past was extreme damage to the forests. In the course of obtaining food, and in some other activities, elephants damage and knock over trees. At normal elephant population levels, the killing of a few trees has the positive effect of maintaining open savannah and preventing canopy closure. However, at high population densities, the elephants may actually destroy the food base on which they and other animals depend (Eltringham, 1979). This dramatic, and abnormal, effect has to some degree distracted attention from the critical positive role that elephants play in maintaining an open, savannah vegetation structure in certain types of African forests.

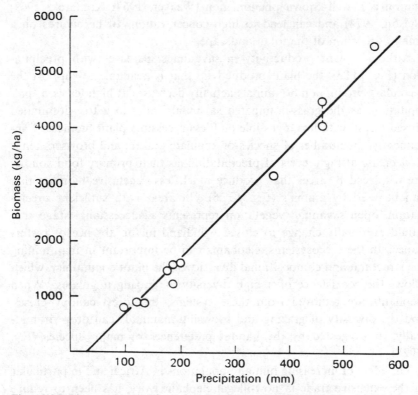

Fig. 13.25 Biomass production of grasslands in Southwest Africa in relation to annual precipitation. (From Walter, 1939, 1985.)

Fire in Forests

While the role of fire in certain forest types, particularly those dominated by pines, has been recognized for some time (Clements, 1910; Maxwell, 1910; Braun, 1950), it is only relatively recently that forest ecologists have recognized that fires play an important role, at least periodically, in all forests (e.g., Goldammer, 1990). In eastern North America, where fire suppression has been a dominant forest management policy, it is now known that fires have influenced forest structure throughout evolutionary and prehistoric time (Niering, 1981; Delcourt and Delcourt, 1987; Clark, 1989; Heinselman, 1973), probably reaching their maximum frequency and effect on vegetation within the past 200 years (Cronon, 1983; Martin, 1989), and decreasing only in the twentieth century. For forests, the dynamic equilibrium is bounded by disturbance frequencies too high to allow the survival of trees given the local growth conditions and by rates

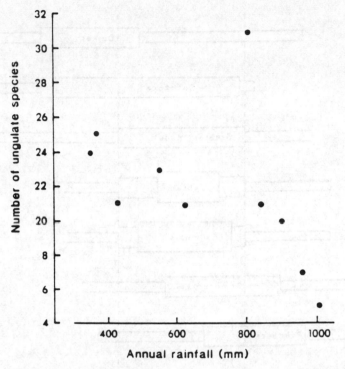

Fig. 13.26 Relationship of ungulate species richness to annual precipitation in East African savannas. (From Deshmukh, 1986, based on data from Western and Ssemakula, 1981 and Coe *et al.*, 1976.)

of growth (and the associated rate of competitive displacement) too low to allow the survival of trees under a particular disturbance regime (Fig. 13.28).

Fire-dependent Tree Communities

The ultimate plant adaptation to fire is the production of seeds that are released or are able to germinate only after being exposed to temperatures so high that they only occur during fires that are intensive enough to cause the mortality of most trees (Stone and Juren, 1951; Went *et al.*, 1952; Lotan, 1974). Serotinous (late-opening) cones that open and release seeds only after exposure to high temperatures are found in several species of pine that tend to occur in monospecific stands that are regularly destroyed by fires and then regenerate from seed. North American examples include the lodgepole pine (*Pinus contorta*) of the intermontane west (Clements,

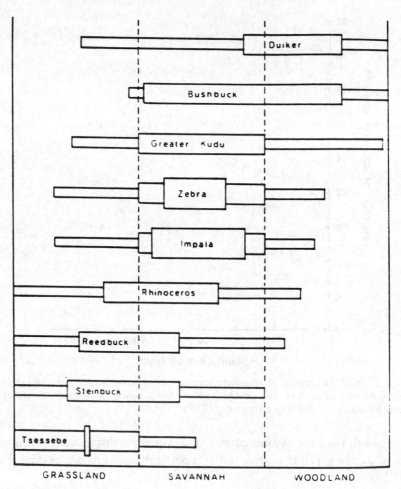

Fig. 13.27 Habitat segregation among African savanna ungulates illustrating the decrease in species richness between open-canopy woodland and closed-canopy gallery forest. (From Deshmukh, 1986.)

1910), the jack pine (*Pinus banksiana*) of the Boreal region, the table mountain pine (*Pinus pungens*) of the southeast (Barden, 1977; Zoebel, 1969), and the knobcone pine (*Pinus attenuata*) of the Pacific coast (Vogl, 1967). Chronically dry conditions, along with the flammable resins of pine needles, contribute to the inevitable occurrence of major fires in these ecosystems.

Such adaptations to fire are found in situations where fires cause nearly complete mortality of all plants, and most regeneration is from

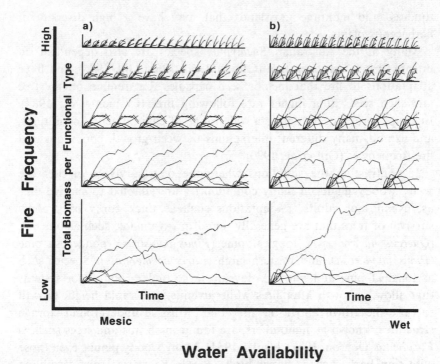

Fig. 13.28 Interaction of fire frequency with successional dynamics in forests.

seed. Low productivity and relatively frequent disturbance contribute to very low species diversity in these forests, as does the low water availability, which reduces the number of possible functional types with regard to shade tolerance. The forests are often either monocultures or dominated by only a few tree species. Some shrubs may be present, but herbaceous cover is generally very low as a result of the dry conditions (Olsvig *et al.*, 1979). Because of the low productivity, the probability of fire increases gradually through time as dead wood slowly accumulates during succession (e.g., Romme and Despain, 1989), although young stands may be particularly susceptible to fires because the flammable crowns are close to the ground (Little, 1946, 1979).

The number of functional types of plants in these low productivity regions should be higher on sites with a lower disturbance frequency or with higher productivity resulting from increased water and nutrients. Although tree diversity is very low under these conditions, the diversity of herbaceous perennials, which are little affected by fire, can be quite high. Such low diversity conifer stands are often interspersed with meadows, tundras, and montane grasslands that may have a high diversity of herbaceous plants.

Many of the Australian *Eucalyptus*. species are also dependent on intense fires for regeneration (Gilbert 1960, 1963; Gill, 1975), and have adaptations to fire that include seed capsules that remain on the tree and shed seeds at a higher rate following fires (Cremer *et al.*, 1984). Eucalypt forests occur across a wide range of environmental conditions, and exhibit many different interactions between growth conditions and fire frequencies (Gill *et al.*, 1990).

In contrast to the situations where the only survivors of fires are seeds, woody plants in many communities are adapted to survive fires as juveniles or adults. Adaptations such as thick corky bark allow survival of trees that are generally found in savannahs, such as bur oak (*Quercus macrocarpa*), longleaf pine (*Pinus palustrus*), ponderosa pine (*Pinus ponderosa*), and western larch (*Larix occidentalis*). Species such as longleaf pine and bur oak develop large underground root systems that allow regrowth after fires while juvenile, and rapid height growth to raise the crown above the fire zone. Massive underground storage structures, known as lignotubers, are found in shrubs and trees such as *Eucalyptus* (Jepson, 1931; Jacobs, 1954). Many woody plants, even those with thin bark such as paper birch (*Betula papyrifera*) and trembling aspen (*Populus tremuloides*), resprout from adventitious buds at the base of the stem or on shallow roots.

Many of the western coniferous forests, although not dependent on fire to the extent of cone serotiny, do exhibit strong adaptation to fire. In the relatively dry, open mixed-coniferous forests, frequent low intensity ground fires are important for preventing the fuel buildup that could lead to a crownfire conflagration. In addition, the seedlings of many species, including the giant sequoia (*Sequoiadendron giganteum*), germinate most successfully after intense groundfires (Harvey *et al.*, 1980; Witherspoon et al., 1985). Some of these forests include trees of spectacular dimensions, and a diversity of dominant species that is high in comparison to most coniferous forests. Major species include sugar pine (*Pinus lambertiana*), white fir (*Abies concolor*), Douglas fir (*Pseudotsuga menziesii*), giant sequoia, ponderosa pine, and incense-cedar (*Libocedrus decurrens*) (Kilgore, 1973). Fire-maintained pine forests are also found throughout the tropics (Goldammer and Pénafiel, 1990; Stott *et al.*, 1990; Koonce and González-Cabán, 1990).

The coniferous forests of the Pacific Northwest of North America are unique in the world in terms of the total living and dead biomass of the forest, as well as the massive size of trees of several different species, most notably the coast redwood (*Sequoia sempervirens*) (Waring and Franklin, 1979). The huge sizes (diameters of 2 to 5 meters) and great ages (commonly 500 to over 1000 years) attest to the low frequency of major fires. The massive accumulations of dead wood found in many of these forests (Franklin and Waring, 1981; Harmon *et al.*, 1986) suggest that groundfires are infrequent as well. In spite of the apparent low frequency of fire (a natural fire rotation of 434 years, Hemstrom and Franklin, 1982) in these ecosystems, infrequent massive fires may be essential for the successful regeneration of many of the species (Franklin and Hemstrom, 1981).

Fire-influenced Tree Communities

When sufficient moisture is available to form a closed canopy forest, the frequency of fires drops to a much lower level than in savannahs or shrublands. Fire has some influence on the vegetation of virtually all forests of North America (Lotan *et al.*, 1981), and acts as a mortality-causing disturbance of variable (and inversely correlated) frequency and intensity with both a size-selective and species-selective effect (Komarek, 1981). With increasing annual precipitation across geographic gradients, or increasing soil moisture across topographic gradients, the proportion of trees that are adapted to survive fires decreases (Lotan *et al.*, 1981; Martin, 1989). Consequently, the damage and mortality caused by fires tends to increase along these same gradients as the frequency of fire decreases (Garren, 1943; Martin, 1989). Adaptations to occasional fires include thick bark and the ability to resprout, either from the roots or from the main stem (Flint, 1925; Starker, 1934; Hodgkins, 1958; Lyon and Stickney, 1976).

Surface fires, which primarily kill shrubs and small trees, were probably much more common in prehistoric times, when they were set by indigenous people (Pyne, 1984; Day, 1953; Loucks, 1970; Komarek, 1974; Barden and Woods, 1976; Lorimer, 1977; Russell, 1983). Such fires helped maintain the open, cathedral-like stands with grass underneath that were frequently noted by early travelers (see references in Cronon, 1983; Van Lear and Waldrop, 1989). Natural lightning fires tend to be smaller and less intense than human-caused fires because they usually occur when humidity is high and there is some precipitation (Barden and Woods, 1973).

Given the right combination of weather, fuel load, and fuel condition, even mesic forests can burn (Keetch and Byram, 1968). Thin-barked tree species of mesic areas, such as spruce and fir species in the North American west or American beech (*Fagus grandifolia*) and sycamore (*Platanus occidentalis*) in the east, are more likely to be killed by occassional large ground fires than are thicker-barked species such as redwoods, Douglas fir, oaks, or the formerly dominant American chestnut (Martin, 1989). Depending on its intensity, fire can act to increase or reduce tree species diversity, and the species composition and degree of adaptation to fire can influence the change in diversity in a manner analogous to that described for grazing on grasslands by Milchunas *et al.* (1988) (cf. Fig. 13.16). The effects of fire can range from a slight change in species composition, to nearly complete mortality that initiates a post-fire succession (Lutz, 1956; Lyon and Stickney, 1976).

Fire almost always reduces the species richness of understory trees and shrubs (Keetch, 1944; Loomis, 1977; DeSelm *et al.*, 1973; Vogl, 1964) and often increases the richness of the herbaceous component, which may result in an increase in total species richness (DeSelm *et al.*, 1973). The most complete study of the effect of fire on forest vegetation is the 40-year experiment on the "Santee Fire Plots" conducted by the USDA Forest Service in Francis Marion National Forest in South Carolina (Lewis and Harshbarger, 1976; Waldrop *et al.*, 1987). The treatments (with three replicate 32 × 32 m plots per treatment) were 1. no burn; 2. periodic summer burn (average interval was 5 years); 3. periodic winter burn (average interval was 5 years); 4. biennial summer burn; 5. annual summer burn; and 6. annual winter burn, established in 1946 in 42-year-old loblolly pine stands that had regenerated naturally following logging.

The results of this remarkable long-term experiment clearly demonstrate the contrasting effects of fire on different components of the plant community. The experiments were designed to study the effects of fire on the development of the understory below a canopy of loblolly pines. The fire treatments had no significant effect on the growth of the overstory pines, although the data suggest that mortality of the pines increased with increasing fire frequency/intensity (Waldrop *et al.*, 1987). Although there was no treatment effect on total species richness of the understory, when the community was separated into woody and herbaceous species, clear patterns emerged (White *et al.*, 1991), that can be understood in the context of a dynamic equilibrium between disturbance frequency and changing growth conditions (Fig. 13.29).

The understory of these loblolly pine stands exists in relatively light shade cast by the overstory pines. Since the fire had little detectable effect on the overstory pines, the light environment at the top of the understory was unaffected by the treatments. For the tallest component of the understory, the woody plants, the primary effect of the treatment was the increasing mortality and stress caused by the fires. In the unburned plots, a dense understory of hardwood trees and shrubs developed. The species richness of woody plants decreased monotonically with increasing frequency and intensity of burning (Fig. 13.29). The annual summer burn resulted in the lowest species richness of woody plants. This treatment caused the greatest mortality and reduction of woody plant growth rate because it not only removed physiologically active biomass prior to retranslocation of carbon and nutrients, but reduced the leaf area available for photosynthesis during the summer growing season (Waldrop *et al.*, 1987).

In contrast to the woody vegetation, the light environment of the herbaceous understory was affected by the treatments, as a result of the response of the taller woody understory species to fire. Increasing fire frequency/intensity allowed more light to reach the forest floor, and this resulted in increased herbaceous growth rates that led to higher diversity (Fig. 13.29). With the exception of the annual summer fires, the treatments had little or no direct effect on the herbaceous vegetation. In the unburned treatment, herbaceous vegetation was nearly eliminated by shading by the forest canopy. Increasing frequency and intensity of fire (summer burns caused greater mortality of both woody and herbaceous plants than winter burns), opened up the understory canopy and removed understory vines and shrubs that would compete with the herbaceous plants (Fig. 13.30). Consequently, the biomass and number of species of herbaceous plants increased with increasing fire frequency/intensity. These results demonstrate the different dynamic equilibria of species composition and diversity that occur with different disturbance regimes, as well as contrasting patterns of diversity in different functional groups of plants.

The interaction of fire frequency and successional dynamics can produce a landscape with patches of vegetation that differ dramatically in structure and appearance, but are actually different stages of the same sere. These contrasting patches may result from different fire regimes, or simply from different lengths of time since the last fire. Myers (1985) reports such a phenomenon on the sandy ridges of northern and central Florida. Two 'distinct' vegetation types have long been known from this region: 1. sandhill, which is composed of open, savannah-like stands of

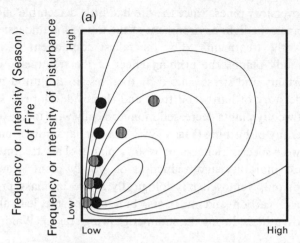

Rate of Growth and Competitive Displacement

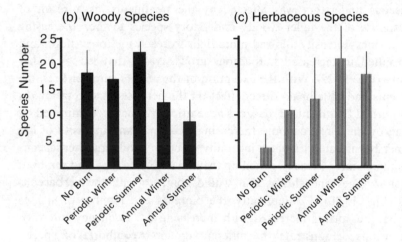

Fire Frequency or Intensity (Season)

Fig. 13.29 Hypothesized dynamic equilibria of the species diversity of woody and herbaceous understory species in the Santee Fire Plots. (a) The maximum growth rates and rate of competitive exclusion of the woody plants (solid circles) is largely unaffected by treatment. Note that at this relatively early stage of forest succession on dry soils, there is little evidence of competitive displacement among the woody understory species and the primary effect of the treatment is a reduction in diversity resulting from the inability of species to recover between disturbances. The maximum growth rates of the herbaceous understory plants (shaded circles) increase in response to increased light that penetrates the woody understory canopy at higher disturbance frequency/intensity. The primary treatment effect is an increase in diversity that results from more species being able to survive with higher light availability (Based on Huston, 1979). Species richness of (b) understory woody and (c) herbaceous plants. (From White et al., 1991.)

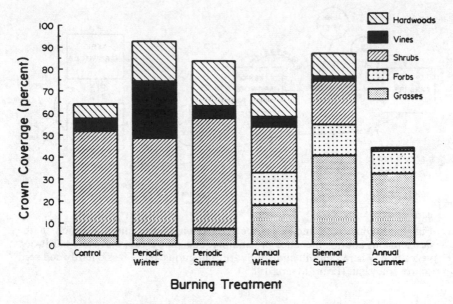

Fig. 13.30 Percentage crown cover of all understory plants less than 1.5 m tall after 30 years of prescribed burning. Note the decrease in hardwoods, vines, and shrubs, and the increase in forbs and grasses with increased frequency and intensity (based on season) of burning. (From Waldrop *et al.*, 1987.)

longleaf pine (*Pinus palustris*) or slash pine (*Pinus elliottii* var. *densa*), with grass ground cover; and 2. sand pine scrub, with a closed canopy of sand pine (*Pinus clausa*), and an understory of evergreen hardwoods (primarily oaks and hickories). Sandhill is characterized by a high frequency of low intensity fires, while sand pine scrub has less frequent, but more severe fires. Myers presents evidence that, in the absence of fires, both of these vegetation types develop into xeric hardwoods dominated by evergreen oaks and hickories (Fig. 13.31). A similar fire-maintained landscape mosaic in Zambia was described by Lawton (1978).

A widespread phenomenon in North America that seems to be related to past patterns of fire is the presence of oak-dominated forests in which oaks are not regenerating (McGee, 1979; Van Lear and Waldrop, 1989; Carvell and Tyron, 1961; Carvell and Maxey, 1969; Swan, 1970; Abrams, 1992). One plausible hypothesis is that these forests represent a mid-successional stage following post-settlement suppression of fires on prairies and savannahs, and prior to domination by more shade-tolerant species (such as sugar maple). Domination by shade-tolerant species is occurring most rapidly on mesic sites (Nowacki *et al.*, 1990). The

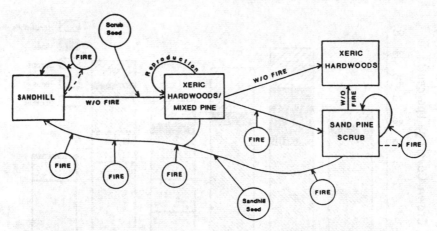

Fig. 13.31 Qualitative model of the interaction of fire and vegetation in the development and maintenance of sandhill, sand pine scrub, and xeric hardwood vegetation, where sandhill and scrub are contiguous and a mesic hardwood seed source is absent. (From Myers, 1985.)

post-settlement disturbance regime of clear-cutting and fires may have contributed to oak dominance, when repeated fires killed the saplings of most species, and allowed the oaks to dominate by virtue of their large root systems as seedlings and rapid development of thick bark (Abrams and Nowacki, 1992). Martin (1989) suggests that oak forest has replaced mesic forest over much of the Cumberland Plateau as a result of frequent burning over the past 100-150 years. Tulip poplar (*Liriodendron tulipifera*) regeneration also seems to be related to fires (Sims, 1932; Shearin *et al.*, 1972), although it regenerates well on almost any disturbed, bare area. Tulip poplar seeds are wind-dispersed and remain viable in the soil for up to 8 years (Sanders and Clark, 1971), and germinate rapidly following fire (McCarthy, 1933).

The species diversity of broad-leaved deciduous forest is influenced by many processes that may have opposing effects on different components of the community (Fig. 13.32). Relatively dry forests are likely to have high species diversity among the dominant canopy species, both as a result of low rates of competitive displacement (e.g., Huston and Smith, 1987), and periodic fires. Understory diversity is likely to be low, both because of the fires, and because dry conditions preclude the survival of shade-tolerant species. In contrast, the diversity of canopy species in productive mesic areas is likely to be reduced by a higher rate of competitive displacement, and a low frequency of fires, although spa-

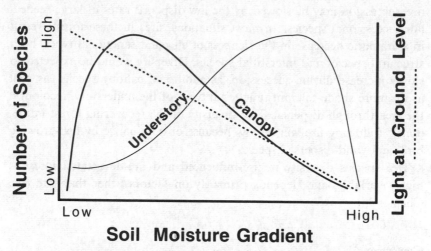

Fig. 13.32 Hypothesized pattern of species diversity among canopy and sub-canopy plants along a moisture gradient. Canopy species are responding soley to the moisture gradient, while subcanopy species must respond to both the moisture gradient and the light gradient caused by variation in the biomass and leaf area of the canopy species. (Based on Fig. 10.9.)

tial and temporal heterogeneity caused by other disturbances, such as treefalls, can help maintain high diversity. In mesic sites, the diversity of shade-tolerant subcanopy and understory trees and shrubs is likely to be quite high in comparison to dry sites, so the total number of woody species is generally higher, even though the evenness of canopy dominants is lower than on many dry sites. Thus, along a gradient of increasing moisture/productivity, one major component of the forest community increases in diversity while another major component decreases in diversity.

Fire-devastated Tree Communities

Evergreen wet and rain forests experience fires only under extreme climatic conditions that occur every several hundred years (Leighton and Wirawan, 1986; Sanford *et al.*, 1985; Goldammer and Siebert, 1990) or because of human modification of vegetation structure (Uhl and Buschbacher, 1985; Kaufmann and Uhl, 1990; Soares, 1990; Fearnside, 1990b). Fire is catastrophic when it occurs and causes almost complete mortality because virtually none of the species are adapted to fire (Uhl and Kaufmann, 1990). Depending on the size of the area affected by the fire, succession may be slowed by the low dispersal rates of large-seeded, late-successional species. In most situations, fires in these forests result in a dramatic decrease in the richness of all plant functional types, both structural species and interstitial species. Diversity would be expected to slowly increase during succession, as additional canopy species, as well as the more shade-tolerant functional types of the understory, recolonize the area through dispersal. Tree diversity of the recovering forest immediately following fire will be low because of dominance by fast-growing bird- and wind-dispersed species.

Tree species diversity in fire-influenced and fire-devastated forest is highly variable, and depends primarily on factors other than fire (see Chapter 14).

Conclusion

The productivity of an ecosystem, the frequency of fire in that ecosystem, the intensity of the fires that do occur, and the physiology and life history characteristics of the organisms in that ecosystem, are closely interrelated in complex network of positive and negative feedbacks. Fires occur periodically in all terrestrial ecosystems. However, those ecosystems with the most frequent fires tend to exhibit the least conspicuous ecological responses to fires. In such ecosystems, the predominant patterns of species composition and diversity on the landscape result not from variation in the frequency or intensity of fire, but from variation in environmental conditions that influence plant productivity, and thus the rate of competitive displacement and recovery from fire. Likewise, when fires are infrequent, the overall diversity and pattern of the environment is determined by factors related to plant productivity, rather than the effects of fire, although the transient effect of fire is spectacular when it does occur.

In the context of the dynamic equilibrium between succession and disturbance, the temporal and spatial variability in diversity and species composition found on a landscape result from the general pattern of plant succession under the moisture conditions of the landscape, and the frequency with which succession is reinitiated by fire (or some types of vegetation are eliminated by fire). Plant productivity (which is also influenced by nutrients when sufficient water is available) sets the rate at which burnable biomass accumulates, and the patterns of climate and weather determine how often the ecosystem becomes dry enough to burn. Figure 13.33 illustrates the range of plant community structure

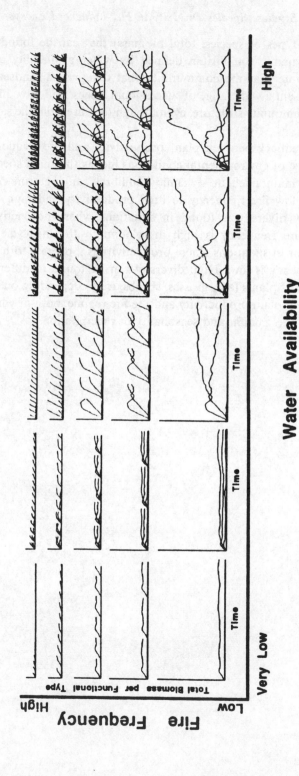

Fig. 13.33 Summary of temporal dynamics of vegetation expected under different combinations of water availability and fire frequency. The successional sequences for different moisture conditions are based on Fig. 9.9. Note that the temporal variability is an index of spatial variability on a mosaic landscape with asynchronous successional dynamics.

(species richness, types of species, total biomass) that can be found at any particular dynamic equilibrium defined by water availability and fire frequency. Note that the communities that comprise a landscape mosaic may represent a wide range of successional stages, and thus differ dramatically in community structure, even though they are from the same sere.

The positive feedback between plant productivity and fire frequency over a broad range of environmental conditions means that high species diversity can be maintained at dynamic equilibria ranging from low productivity and low fire frequency to high productivity and high fire frequencies. Low diversity is found in situations where productivity is low and the fire frequency is high in relation to the life cycle of most organisms, or in situations where productivity is moderate to high, but the fire frequency is low. High diversity can be found in different functional groups of plants (e.g., grasses, shrubs, trees) depending on the interaction between plant productivity and the fire regime, both of which are influenced by the amount and seasonality of precipitation.

14

Case Studies: Species Diversity in Tropical Rain Forests

Tropical rain forests, which cover only 7% of the Earth's surface, are estimated to contain more than half of all the species of living organisms on Earth (Wilson, 1988). Depending on which of the many estimates of the total number of species is used, the number of species in tropical rain forests is somewhere between 2.5 and 15 million (Parker, 1982; Arnett, 1985, Erwin, 1983, Wilson, 1988). One of the perplexing aspects of tropical diversity is the morphological and genetic similarity of many of the coexisting organisms. Not only are many of the plants similar in appearance (Richards, 1952), but closely related species often have broadly overlapping geographic ranges (Ashton, 1977; Whitmore, 1984). For virtually all types of organisms, at almost any scale, diversity is higher in the tropics than anywhere else on earth. This diversity of diversities suggests that many different mechanisms influence the number of species in tropical rain forests.

Explaining the latitudinal gradient of species diversity, and in particular the high species diversity of tropical rain forests has long been and still remains the 'Holy Grail' of ecology and evolutionary biology. I believe that the failure to attain the grail has been caused by two fundamental misconceptions about tropical rain forests, one of which has been gradually abandoned by most ecologists over the past ten years, and the other of which I hope to lay to rest with this chapter. The first misconception is that the tropics are a stable and unchanging environment, and have been so for most of their evolutionary history. The second misconception is that tropical rain forests are highly productive ecosystems where rapidly growing plants provide an abundant supply of resources for the animals that they support.

I will argue that the diversity of tropical rain forests is the evolutionary and ecological result of the same type of dynamic equilibrium that I have postulated for all of the other ecosystems discussed in this book. As with virtually all ecosystems (except those with high disturbance frequencies), the maximum diversity in tropical rain forests is found under conditions of low resource availability and plant growth rates. Hopefully, this 'counter-intuitive' statement should no longer be counter-intuitive. I believe that this relationship, which is the key to understanding the diversity of tropical rain forests, is also the key to saving the rain forests (Chapter 15).

The high precipitation that is the signal feature of rain forests affects biological diversity by a variety of mechanisms. As discussed in Chapter 7, wetter conditions allow plants to be more shade tolerant, so a greater number of plant functional types can survive. This phenomenon contributes to the high diversity of understory plants found in some tropical rain forests. High precipitation also contributes directly and indirectly to the high diversity by other mechanisms that will be discussed in the following sections.

The Distribution and Phytogeography of Tropical Rain Forests

Between the latitudinal limits of the Tropic of Cancer (23° 27′ N) and the Tropic of Capricorn (23° 27′ S), tropical rain forest occurs wherever precipitation is sufficient, generally above 1500 - 2000 mm/yr (Walter, 1973; Holdridge, 1967). The distribution of rainfall throughout the year is at least as important as the total amount (Walter, 1973; Stephenson, 1990) (Fig. 14.1). Tropical rain forest (sensu Schimper 1898, 1903) is the evergreen forest of the humid lowlands where rainfall is distributed evenly throughout the year. With increasing seasonal variation in precipitation, most trees in the forests drop their leaves for part of the year. These forests are variously called monsoon forests or tropical dry forests. Many other types of tropical (and subtropical) forest can be defined, based on variation in precipitation, elevation, etc. There are many systems for classifying tropical forests, one of which, the Holdridge Life Zone System (Holdridge, 1947, 1967) is illustrated in Fig. 14.2. With the exception of the driest forests, most tropical forests have many features in common with the true evergreen rain forest. Most forests of the moist tropics have relatively high tree species diversity and a similar appearance in terms of leaf sizes and shapes and buttressed trunks (Richards, 1952).

In spite of the similar appearance of rain forest plants on different continents, there is virtually no overlap in species composition between the three major rain forest regions. However, at the higher taxonomic levels

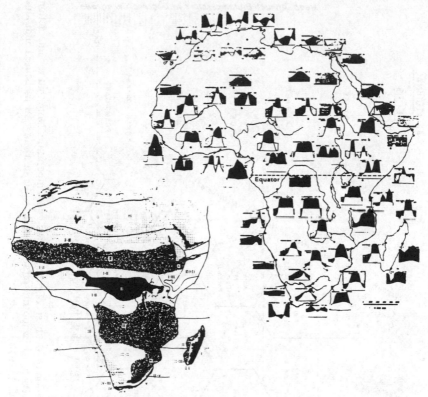

Fig. 14.1 Seasonal distribution of precipitation in Africa and its effect on vegetation zones. Climate-diagram map of Africa showing 66 stations. Black indicates an excess of precipitation over evapotranspiration and grey indicates an excess of evapotranspiration over precipitation. Note the high precipitation (truncated at the top of the diagram) and the even seasonal distribution of precipitation along the equator with increasing seasonal variability and decreasing precipitation (particularly in the Sahara) away from the equator. The inset illustrates the distribution of vegetation types or Zonobiomes. I. Equatorial Humid Climate with Evergreen Tropical Rain Forest. II. Tropical Summer-Rain Region with Deciduous Forest. III. Subtropical Arid Climate with Deserts. IV. Winter-Rain Region with Sclerophyll Woodlands. V. Warm-Temperate Humid Climate. (Modified from Walter, 1985.)

of genera and families the three regions are quite similar, presumably as a consequence of their common origin as parts of the supercontinent Gondwanaland (Chaloner and Lacey, 1975; Lacey, 1975). Forests that are dominated by unusual groups of plants with restricted distributions can be found, particularly on islands and in other isolated situations, yet

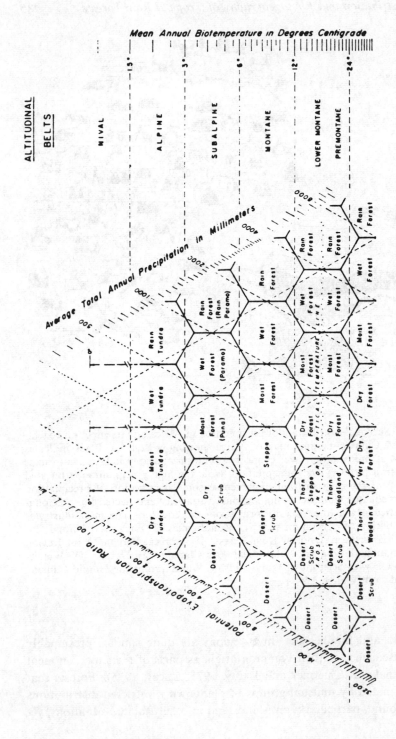

Fig. 14.2 Holdridge life zone system for classifying vegetation on the basis of precipitation and temperature. Note that the interaction between precipitation and temperature (indicated by the Potential Evapotranspiration axis) results in a range of moisture conditions for any given annual precipitation. (From Holdridge et al., 1971.)

tropical forests around the world are notable for their taxonomic and morphological similarities rather then their differences.

Recently, Gentry (1982, 1988) compiled a large set of comparable diversity samples from a range of tropical forest types on several continents. The basic sampling area is 0.1 ha, generally composed of smaller subsamples, with all rooted plants recorded which are larger than 2.5 cm in diameter at breast height (DBH). Thus, these samples include not only trees, but vines, shrubs, and some large herbs. This large comparative study has provided new insights into both the species diversity and the taxonomic composition of tropical forests (Gentry, 1988).

American Tropics

The dominant plant family of neotropical lowland forests is the Leguminosae. Of 43 neotropical lowland sites in Gentry's (1988) compilation, the legume family had the highest number of species at 39 sites. This degree of dominance is comparable to the well-known dominance of the Dipterocarps in Southeast Asia (Whitmore, 1984; Richards, 1952). Among trees over 10 cm DBH, Dipterocarps comprised 12-15% of the species in two samples from Southeast Asia, while legumes ranged from 11 to 21% of the tree species at nine neotropical sites (Gentry, 1988).

The second most dominant family among neotropical trees is the Moraceae, which includes the figs. Of the nine neotropical sites mentioned above, the Moraceae had the second greatest number of species at six sites, and on fertile soils may have as many species as the Leguminosae (Gentry, 1986b, 1990b). Other prominent tree families are the Lauraceae, which is often the family with the most species at mid-elevation sites and includes such species as the avocado, and the Annonaceae, including the neotemperate pawpaws (*Asimina triloba*) and the tropical custard fruit (*Annona* spp.).

When all plant forms are considered, including shrubs and lianas, families such as the Rubiaceae, Palmae, Melastomaceae, and Bignoniaceae are notable for their large contribution to species richness. According to Gentry (1988) the same 11 families (Leguminosae, Lauraceae, Annonaceae, Rubiaceae, Moraceae, Myristicaceae, Sapotaceae, Meliaceae, Palmae, Euphorbiaceae, and Bignoniaceae) compose about half of the species richness of any 0.1 ha sample of neotropical lowland forest (38% - 73%, $\bar{x} = 52\%$). Many of the same families are also among the ten most species-rich in African and Asian forests as well.

One of the characteristic features of neotropical lowland forests is the presence of large palms as components of the forest canopy. The only other places where large palms are so prominent are on a few islands such as Madagascar and New Caledonia (Gentry, 1988).

African Tropics

As in the Neotropics, the legume family has the highest pecentage of species in most lowland African forests. Several families that are represented by a only a few species at sites on other continents are consistently among the ten most species-rich at African sites. These include the Olacaceae, Sterculiaceae, Dichapetalaceae, Apocyanceae, Sapindaceae, and the Ebenaceae (ebony family), which is also species-rich in Asia, but rare in the neotropics.

Two notable structural features of African lowland tropical forests are a greater abundance and diversity of lianas than in either Neotropical or Asian forests, and a higher total basal area (an average of 70 m^2/ha versus 35-45 m^2/ha in Neotropical and Asian forests), reflecting a higher number of large trees for the sites reported by Gentry (1988).

Indo-Malayan Tropics

The most notable feature of these lowland tropical rain forests is the dominance of trees of the family Dipterocarpaceae (386 species), which is found nowhere else in the world. In addition, two genera, *Eugenia* (Myrtaceae) and *Ficus* (Moraceae) have approximately 500 species each (Whitmore, 1984). The family Myrtaceae is always among the most species rich in the Asian tropics, but is rarely species-rich in the Neotropics, and virtually absent from Africa. Whitmore (1984) notes that many of the speciose genera are characterized by large numbers of species with overlapping geographical ranges and apparently identical ecological characteristics, a situation that is incompatible with explanations of species diversity based on competitive equilibrium.

Gentry (1988) concludes that the taxonomic composition of lowland tropical forests around the world is remarkably similar at the family level, with the notable exception of the Dipterocarpaceae in Southeast Asia. Three families, the Rubiaceae, Annonaceae, and Euphorbiaceae, are always among the ten most species-rich families on all continents. The rest of the species-rich families of the Neotropics are also present in Africa and Southeast Asia, and are even among the most species-rich families at certain sites on these continents.

Even at the generic level there is strong taxonomic similarity among tropical forests. Gentry reports a generic overlap of about 30 percent between some African and Neotropical sites, which does not seem surprising given their plate-tectonic history. However, the overlap between the Neotropics and Asia is nearly as great (23%), as is the overlap between the Neotropics and Australasia (25%). Surprisingly, the greatest overlap at the generic level is between the Neotropics and a site on Madagascar (Gentry, 1988).

The Structure and Diversity of Tropical Forests

The physical structure and complex microclimate of tropical forests are provided by the trees, which are overwhelming in their taxonomic and structural variety, although not necessarily in their size. Few tropical forests have the high density of massive trees that is found in the most productive temperate forests, such as the Douglas fir forests of the Pacific Northwest, or the redwoods of California (Waring and Franklin 1979; Hart *et al.*, 1989) (Table 14.1).

Even though the size of rain forest trees and the biomass of the forests are not remarkable in a global context, in terms of the number of species per unit area, the diversity of tropical rain forests is unmatched anywhere on earth. The number of tree species greater than 10 cm in diameter at breast height ($\sim$ 1.5 m) found in 0.1 ha can be over 200, as compared to 15-20 for the most diverse temperate forests, such as those of the Great Smoky Mountains of North America (Whittaker, 1956, 1966; Gentry, 1988). This dramatic difference is in within-habitat, or alpha, diversity, which refers to the number of species that are coexisting within an area sufficiently small that the individuals could interact with one another. Even more striking than the high number of tree species found in 1.0 ha or even 0.1 ha, is the fact that as sample area is increased, the number of tree species at some sites continues to rise, with no sign of leveling off (e.g. Ashton, 1964; 1977; Gentry, 1988; Fig. 14.3).

High local diversity in the tropics is not the result of many widespread species that coexist over a large geographic area. Among most taxa in tropical forests, the increase in diversity with increasing sample size seems to be a different type of phenomenon, called geographical, or gamma, diversity. Geographical diversity is measured in terms of the taxonomic differences between the groups of species that occupy the same habitat in different geographic areas (cf. Chapters 2 and 3). If many of the species in a habitat in one area are replaced by taxonomically different

Table 14.1. *Comparison of biomass and size of temperate and tropical forests and trees.*

	Basal Area (m²/ha)	Stem Volume (m³/ha)	Biomass (t/ha)	NPP (t/ha/yr)	Diameter (max.) (cm)	Height (m)
Pacific Coast of North America[1]						
Western Hemlock/Sitka Spruce, 110 yrs.						
Western Hemlock	98	1,987	871	10.3	90-120 (260)	50-65
Sitka Spruce					180-230 (525)	70-75
Coast Redwood Stand, 1000 years						
Coast Redwood	338	10,817	3,461		150-380 (501)	75-100
Coast Redwood Stand, "old growth"	247	9,000	3,200	14.3		
Douglas Fir and Western Hemlock Stand						
Douglas Fir	127	3,600	1,590		150-220 (434)	70-80
Appalachian Mountains, S.E. United States[2]						
Cove Forest, Poplar Creek Flats (site 18)	53	764	500	11.5		
Tulip Poplar Forest, Greenbrier Cove (22)	34	346	220	24.0		
Hemlock-mixed Forest, Surrey Fork (24)	64	890	610	11.5		
Tropical						
Amazon Caatinga[4] San Carlos, Venezuela	36³		268	8.9		
Lower Montane Rain Forest[5] El Verde, Puerto Rico			228	10.3		
Evergreen Forest[6] Banco, Ivory Coast	55-75³		513	12.2		
Dipterocarp Forest[7] Pasoh, Malaysia			475	12.7		
Lowland Rain Forest[8] Le Selva, Costa Rica	30		382			
Riverine Forest[9] Panama			1174			
Rain Forest[10] N.E. Australia	60					40-45

1 Waring and Franklin, 1979
2 Whittaker, 1966
3 Brunig, 1983
4 Jordan, 1985

5 Odum et al., 1970
6 Huttel and Bernhard-Reversat, 1975
7 Kato et al, 1978

8 Holdridge et al., 1971
9 Golley et al, 1975
10 Stocker and Unwin, 1990

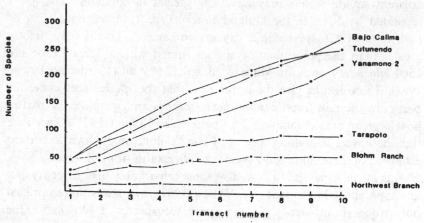

Fig. 14.3 Typical species area curves for tropical forests. Curves are based on 100 m² subsamples of 0.1 ha samples from high and low diversity forests, and include all plants greater than or equal to 2.5 cm in diameter at breast height. (From Gentry, 1988.)

(but ecologically similar) species in the same habitat in another area, and vice versa, the geographical diversity of the region that encompasses both areas is high. In tropical forests, geographical diversity is higher than almost anywhere else in the world. Consequently, high local diversity in the tropics aggregates to astronomically high regional diversity.

For example, Gentry (1982) reports on two 1000 m² samples from the same 0.8 km² area of mature rain forest vegetation at Rio Palenque, Ecuador. Each sample was composed of ten randomly located transects 50 m long and 2 m wide, in which all rooted plants greater than 2.5 cm in diameter at breast height were counted. The first 1000 m² sample included 119 species and the second sample also included 119 species. However, in spite of the identical number of species, the taxonomic composition of two samples was quite different. Only half of the species in each sample were shared with the other sample. This suggests an amazing constancy in the number of species per unit area, but remarkable variability in which species are present in different areas of the same habitat.

A similar phenomenon is reported by Erwin (1983, 1988) who has collected beetles from rain forest canopies by insecticidal fogging near Manaus, Brazil and Tambopata, Peru. At Tambopata, over one million individual beetles have been collected from fifteen 12 × 12 m plots, with three replicate plots in each of five forest types. Within a single plot of 144 m², the species/area curve increases linearly, indicating that even adjacent square meters may have few species in common. Two plots separated by 50 m in the Upland Forest Type 1 shared only 8.7% of their total of 126 species in a dry season sample. This is only slightly greater than the percentage of species shared with a Terra Firme site 1500 *kilometers* away at Manaus, Brazil (2.6% of 113 species, Erwin, 1988). These results give the impression that the species/area curve for beetles in the rain forest canopy never reaches an asymptote, no matter how large an area is sampled. This pattern led Erwin (1982) to speculate that there were 30 million species of insects alone, in contrast to current estimates of 2.5 million total species of all taxa in tropical forests.

Ants at the same site in Peru also show remarkably high diversity, but less apparent habitat specificity (Wilson, 1987). In a study that examined only arboreal ant species, Wilson found 135 species of 40 genera. One tree yielded 43 species of 26 arboreal genera, which is nearly as many ant species and more genera than are found in all of the British Isles (48 species of 16 genera, Barrett, 1977). The habitat specificity of the ants with regard to the four forest types at Tambopata, Peru, was not as great as the habitat specificity of beetles sampled at Manaus, Brazil. 53.5% of the ants occurred in only one forest type (versus 84% of the beetle species) and 13.1% occurred in all four forest types that were sampled (versus 1% of the beetle species).

A fundamental question is whether this heterogeneity in local species composition reflects greater habitat specificity (particularly in comparison to habitat specificity in the temperate zone) or whether there are other reasons unrelated to habitat specificity and the competitive resource partitioning that it implies. The incredibly high diversity of plants and invertebrates in tropical rain forests has a number of potential explanations, each with different implications for understanding the theoretical basis for tropical diversity.

The initial focus of this chapter will be on understanding patterns of plant diversity at several scales, particularly the diversity of trees. If the diversity of trees can be understood, most of the diversity of other organisms can be understood as a consequence of the resulting structural and resource heterogeneity and their effect on interstitial species. Specific examples of animal diversity patterns are discussed in later sections of this chapter.

Table 14.2. *Characteristics of the Three Major Regions of Tropical Rainforest*

Region	Total Area ($\times 10^6$ ha)	Vascular Plant Species (estimate)	Species/km^2
American (Neotropical)	400	90,000	0.025
Indo-Malayan	250	35,000	0.014
Malaya	13	7,900	0.061
African	180	30,000	0.017
Temperate Comparisons			
British Isles	30	1438	0.0048
United States	737	20,000	0.0027

Continental-scale Patterns of Total Plant Species Richness

In terms of the total number of plant species in the three major regions of tropical rain forest (Fig. 14.4), species richness at the continental scale is correlated with total area of the rain forest in each region (Table 14.2). The greatest number of plant species (estimated at 90,000, Raven, 1976), as well as the highest density of species per area, are found in the Neotropics, which include nearly half of the total rain forest area on the globe. Africa, with the smallest total rain forest area, has the smallest total number of plant species (estimated at 30,000). The relatively high number of species per unit area found in parts of the the Indo-Malaysian region is undoubtedly related to the isolation of the many islands and peninsular areas that facilitate genetic isolation (analogous to the 'species pump' phenomenon of expansion and contraction of savannah areas in Africa, cf. Chapter 13). This total regional diversity is a combination of within-habitat, between-habitat, and geographical diversity. Geographical diversity, in particular, would be expected to be greater in large regions than in small regions, but one of the unusual features of tropical diversity is high geographical diversity within relatively small areas. The principal locations in the tropics where species diversity of both plants and animals is generally low are on islands, although the proportion of endemic species is often very high (see Chapter 11).

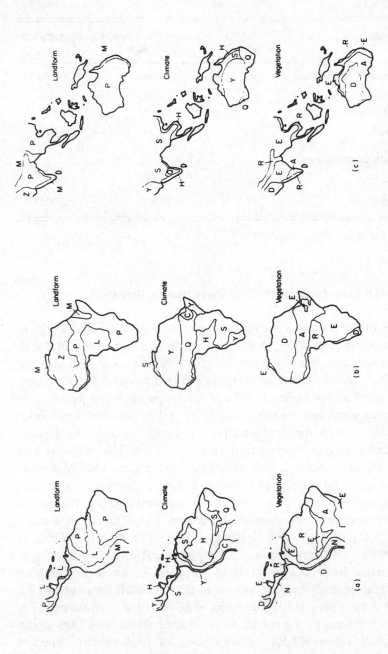

Fig. 14.4 General patterns of landforms, climate, and vegetation in the three major regions of tropical rain forest. *Landforms*: L, lowland; P, plateau; Z, desert; M, mountain; *Climate*: H, humid; S, seasonal or monsoonal; Q, semi-arid; Y, dry; C, cool. *Vegetation*: R, rain forest; E, seasonal or monsoonal forest; A, savannah; D, desert; N, montane. (From Jordan, 1985b.)

Intercontinental Patterns of Within-habitat Species Richness

Not only is there great similarity in the taxonomic composition of tropical forests around the world, but the species richness of sites with similar environmental conditions is remarkably consistent around the world (Gentry, 1988; Anderson and Benson, 1981). In spite of the large differences in the total number of plant species at the continental level (Table 14.2), sites with high within-habitat richness are found in all three regions. However, as more data from the Neotropics have been collected, the traditional ranking of tree species within-habitat (alpha) richness as highest in Southeast Asia has not held up.

In terms of the number of tree species found in a small sample (e.g., 0.1 or 1.0 ha) at a particular site, the highest richness is apparently found in the Neotropics, where up to 300 species of trees over 10 cm DBH have been found in a one hectare sample which included 606 individuals (Yanamono, Peru, Gentry, 1988). The maximum richness found in Southeast Asian forests is 225 per hectare (Sarawak, Whitmore, 1984), while a maximum reported for Africa is approximately 60 (Huttel, 1975; Bernhard-Reversat *et al.*, 1978), although Hladik (1986) reports numbers equivalent to the neotropics for rain forests in Gabon, Africa. This ranking based on tree species richness per hectare is maintained with 0.1 ha samples that include all rooted plants over 2.5 cm DBH (Gentry, 1988). While there is a great range in the number of species found in different types of tropical forests (e.g., those with different amounts and seasonality of precipitation), the species diversity of virtually any tropical forest is higher than that of the most diverse temperate forests.

Intra-continental Variation in Species Richness

Within each of the three major regions of tropical rain forest there is great variation in the within-habitat richness found in different parts of the region. Much of this difference is between-habitat diversity associated with distinctly different types of forests that are found under different conditions of soil or rainfall. However, evidence for past variation in vegetation and climate in the tropics has led some researchers to suggest that present patterns of species diversity and distributions are the result of past climatic conditions when rain forest was restricted to a few relatively small areas within the zones that are now covered by rain forest. This hypothesis has been called the 'refuge theory' because it suggests that current distribution patterns of rain forest organisms are associated with the locations of prehistoric high rainfall areas presumed to be refuges for rain forest plants and animals during dry periods (Haffer, 1969; Simpson and Haffer, 1978).

All of the postulated refuges are in areas that currently receive high amounts of rainfall, primarily because of their relatively higher elevation. These areas presumably also received more precipitation than the regional average during the cooler and drier climates that occurred during past glacial periods. During those times, much of the Amazon basin is thought to have been covered by grassland and savannah vegetation (Bigarella and Andrade-Lima, 1982; Ab'Saber, 1982; van der Hammen, 1982; Haffer, 1981), and the evergreen rain forest biota may have been restricted to a number of wet refuges that were much smaller in total area than the current area of evergreen rain forest. However, there is little direct evidence for the vegetation cycles on which the refuge hypothesis is predicated, and the data that do exist are consistent with other hypotheses as well (Colinvaux, 1987; Bush and Colinvaux, 1990).

The areas thought to represent these refuges are currently the centers of distribution for a variety of taxa, with many species apparently in the process of spreading outward from these centers. Some of the refuges that have been independently described for birds, trees, and butterflies have a significant degree of overlap in their locations (Haffer, 1981; Prance, 1981; Brown, 1982; see also Beven et al., 1984; Fig. 14.5). For some taxa, diversity is highest within the refuges, while for other taxa higher diversity is found in the zones of overlap of species spreading outward from two or more centers. Similar centers of endemism and diversity are found in African rain forests (Hamilton, 1989). It remains unknown whether the allopatric speciation that may have occurred in the isolated refuges over the series of Pleistocene glaciations was sufficient to compensate for the extinctions that probably occurred when large areas of rain forest were replaced by dry savannahs and grasslands.

Most of the explanations for the variety of patterns of large-scale variation in diversity within the tropics do not provide any insight into the fundamental questions of the small-scale ecological regulation of contemporary patterns of species diversity. Of the estimated 90,000 species of plants found in the Amazon basin, only a very small percentage actually live together, coexisting and possibly competing in the same location. Nonetheless, this small percentage represents a number that is 10 to 100 times greater than the number of species that can be found coexisting in temperate forest communities.

The key to understanding the high biological diversity of the tropics,

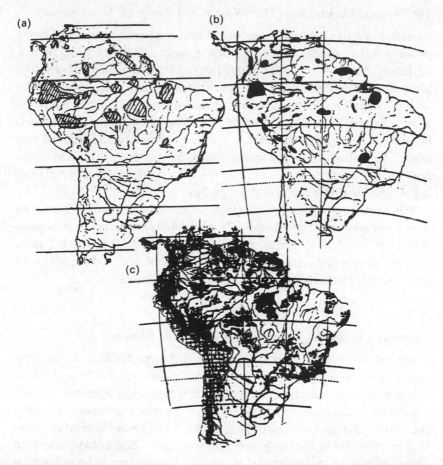

Fig. 14.5 Locations of postulated Pleistocene forest refuges for a) trees, b) birds, and c) butterflies in South America. Map for trees and woody angiosperms is from Prance (1982, 1985). Map for butterflies from Brown (1982). Map for toucans and lizards from Haffer (1969) and Vanzolini (1970).

as well as the causes of variation in diversity within both the tropics and the temperate zone, is explaining the patterns of species coexistence in habitats that differ in diversity. Whether the explanation turns out to be an equilbrium one, a non-equilibrium one, or part of both, the starting point is to discover if there are physical and ecological characteristics that are commonly associated with high within-habitat diversity that differ from those associated with low within-habitat diversity.

The Tropical Environment: High Spatial and Temporal Heterogeneity

Variation in environmental heterogeneity has long been recognized as one of the major explanations for variation in species diversity. Environmental heterogeneity over spatial scales ranging from that of the individual plant to subcontinents reaches its maximum in the tropics, where snow-covered mountains may rise from lowlands of savannah or rain forest. As a consequence of the many types of environmental heterogeneity, there is no such thing as a typical tropical forest, or even a typical tropical rain forest. Notwithstanding the much remarked convergence of leaf shape and tree structure, even adjacent, superficially similar tropical forests can differ significantly in important properties.

The tropical environment has extreme spatial variation in nearly all major environmental properties. Tropical climates range from oppressively hot and humid lowlands to cold, snow-covered mountains, from hot, dry deserts to cold, dry deserts, from extreme seasonal variability of precipitation to nearly constant year-round conditions.

Temporal and Spatial Variation in Weather and Climate

Temporal variation in rainfall is higher in the tropics than in the temperate zone, and contrasts sharply with the low seasonal variation in temperature that has given the tropics their false reputation for climatic constancy. Seasonal temperature variation decreases monotonically from the boreal zone to the equator (Fig. 14.6a) while seasonal rainfall variation increases from the temperate zones to about 25° latitude, and then decreases toward the equator (Fig. 14.6b). Seasonal variation in rainfall in the tropics is equivalent to or greater than in most of the temperate zone.

Higher spatial variation in total precipitation in the tropics as compared to temperate climates is a consequence of the greater evaporation from warm tropical oceans. This produces much greater maximum amounts of rainfall, as well as more extreme rainfall gradients across continental margins and mountain ranges. Spatial variation in seasonal patterns of precipitation results from both the latitudinal interactions of the trade winds with topography and from the intertropical convergence zone (Fig. 14.7).

Extremely low seasonal precipitation affects vegetation just as dramatically as low temperature. During the dry season in much of the tropics, trees lose their leaves and the grass turns brown, giving the landscape the

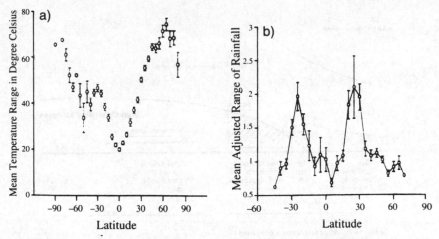

Fig. 14.6 Latitudinal patterns of seasonal variation in a) temperature and b) precipitation. Data are from 1056 stations compiled by Müller (1982). Negative latitudes are in the Southern Hemisphere. Temperature data represent record highs and lows over an approximately 30 year period. Rainfall data are the difference between the annual maximum and annual minimum divided by the mean annual rainfall. (From Stevens, 1989.)

appearance of a temperate winter, an illusion spoiled only by the high temperature, lack of snow, and the occasional palm tree. Length of the dry season has an effect on vegetation that is just as significant as the total annual rainfall (Fig. 14.8).

In spite of, or perhaps because of, the low seasonal variability in temperature in the tropics, spatial differences in temperature may present more of a barrier to tropical organisms than to organisms in temperate zones. Elevational variation in temperature produces greater biological differences between elevational zones in the tropics than in temperate regions because of the low annual variation in temperature at any particular elevation in the tropics. In temperate mountain ranges, organisms that live at any specific location experience a wide range of temperatures between summer and winter, and the annual temperature range overlaps with the seasonal extremes of temperature at much higher or lower elevations, even though the averages will be much different. In contrast, in a tropical mountain range, organisms experience only a narrow variation about the average temperature at any particular elevation, which does not overlap with the narrow temperature range at higher or lower elevations. For this reason, it has been suggested that mountain ranges in the tropics represent much greater barriers to the migration or dispersal of

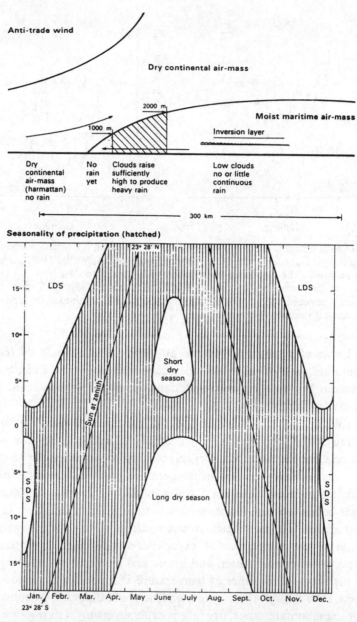

Fig. 14.7 Latitudinal distribution of seasonality of precipitation. Above: cross-sectional view of the tropical convergence zone, which shifts to the north or south following the zenithal position of the sun. Below: resulting distribution of wet and dry seasons. Note the continuously wet equatorial rain belt and the asymmetry about the equator. (From UNESCO, 1978.)

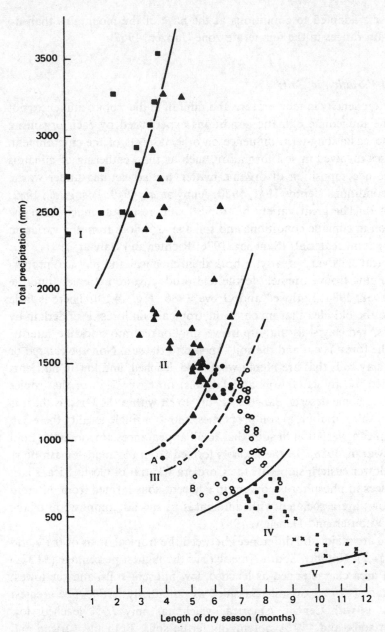

Fig. 14.8 Interactive effect of length of dry season and total precipitation on forest vegetation in India. Vegetation types are: I, Evergreen tropical rain forest; II Semievergreen tropical rain forest; III, Monsoon forest (A, moist; B, dry); IV, Savanna (thornbush forest); V, Desert. (Modified from Walter, 1985.)

organisms adapted to conditions at the base of the mountains than do mountain ranges in the temperate zone (Janzen, 1967).

Spatial Variation in Soils

The heterogeneity in temperature and rainfall in the tropics influences not only the immediate climatic conditions experienced by each organism, but also has a long-term influence on other aspects of the environment. Processes involved in soil formation, such as the weathering of minerals and the decomposition of organic matter, are accelerated under warm, moist conditions (Jenny, 1941, 1980; Jenny et al., 1949; Post et al., 1982; Fig. 14.9). The great variety of tropical soils results as much from the variation in climatic conditions and soil age as it does from the variation in soil parent materials (Sanchez, 1976; Richter and Babbar, 1991).

Tropical soils vary greatly in both their chemical and physical properties, ranging from extremely fertile and productive to extremely infertile (Sombroek, 1984; Lathwell and Grove, 1986; Fig. 14.10). There is little truth to the old idea that much of the tropical rain forest is underlain by 'latosols,' red clay soils that are irreversibly baked into bricklike 'laterite' when the forest is cut and the soil exposed to the sun. Nonetheless, red or yellow clay soils that are highly weathered, leached, and low in nutrients (classified mostly as Oxisols and Ultisols) do cover 51% of the tropics (including some deserts, Sanchez, 1976). Even within the Oxisols there is considerable variation in soil properties, and at a single locality there can be significant variation in soils that to all appearances are identical (Buol and Eswaran, 1988). In the Oxisols formed from old mud- or ash-flows on the lower eastern slopes of the Cordiera Central of Costa Rica, slight differences in phosphorus content in adjacent soils formed from different mud flows are associated with differences in species composition of the forest (Vitousek and Denslow, 1987).

There are major soil differences between the tropical areas of the world (Table 14.3). Notably, South America has the highest percentage (52.3%) of land area characterized as leached, low nutrient soils, and the lowest percentage (13.7%) of soils classified as potentially fertile. The greatest contrast is with Central America, which has only 7.9% leached, low nutrient soils, and 44.1% potentially fertile soils. Both the African and Asian tropics have approximately half the percentage of low nutrient soils and twice the percentage of potentially fertile soils as South America. Thus, the largest and most species rich of the tropical rain forest areas has, on average, the poorest soils.

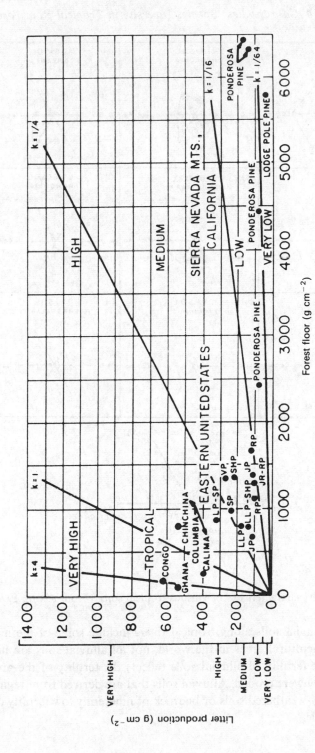

Fig. 14.9 Estimates of leaf litter decomposition rates from tropical to temperate evergreen forests. The decomposition rate factor, k, used in the equation, $C_t = C_0 e^{-kt}$, is calculated from the ratio of annual litter production to the approximate steady state mass of the forest floor. (From Olson, 1963.)

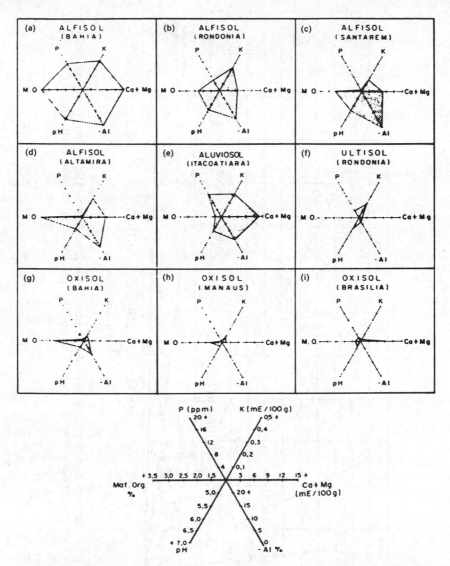

Fig. 14.10 Fertility characteristics of various tropical soils. (From Alvim, 1978.)

Although alluvial soils along tropical rivers include some of the most productive agricultural areas in the world, not all alluvial soils are high in fertility. The fertility of alluvial soils reflects the fertility of the areas from which they were eroded. Alluvial soils that are derived from regions with old, highly-weathered soils or bedrock of inherently low fertility can be quite infertile.

Table 14.3. *Distribution of major soil mapping units in the tropics, based on 1:5,000,000 FAO/UNESCO Soil Map of the World for nations mainly within the Tropics of Capricorn and Cancer. Soil names in parentheses are from Soil Taxonomy (1975, 1987). Values are percentages unless otherwise noted. Percentages do not sum to to 100 because minor soil mapping units have been omitted from table.*

Soil Type	Description	Central America	Africa	South America	Asia	Global Tropics	United States
Leached, Low Nutrient soils							
Ferrasol	Sesquioxide-rich clays; (Oxisols)	0.2	18.5	40.0	1.8	21.2	0
Acrisol	Acid, argillic (Ultisols)	7.7	3.8	12.3	29.1	11.1	19.8
Total Percentage		7.9	24.3	52.3	30.9	32.3	19.8
Young, Poorly-developed Soils, Fertile or Infertile							
Cambisols	Incipient change in structure (Inceptisols)	11.1	4.5	1.6	10.2	4.9	5.9
Potentially Fertile Soils							
Luvisol	Nonacid, argillic; (Alfisols)	12.8	9.6	7.7	11.7	9.6	15.8
Vertisol	Shrink-swell, clayey (Vertisols)	5.9	4.0	0.5	7.8	3.7	1.9

Table 14.3. (*continued*).

Soil Type	Description	Central America	Africa	South America	Asia	Global Tropics	United States
Fluvisols	Alluvial soils (Aquents)	1.1	2.6	1.6	5.6	2.7	0
Kastanozem	Organic-rich with low acidity (Mollisols)	13.0	tr	1.2	0	1.1	23.9
Andosol	Volcanic ash; high organic; amorphous (Andepts)	7.0	0.2	0.7	1.0	0.9	0
Nitosol	Low CEC in argillic (Alfisol, Ultisol)	3.9	5.1	1.3	5.3	3.9	0
Phaeozem	Thick, base-rich organic horizon (Mollisols)	0.4	tr	0.7	0.2	0.3	6.0
Total Percentage		44.1	21.5	13.7	31.6	22.2	47.6
Dry, Rocky, or Sandy Soils							
Lithosol	Rocky, shallow (lithic subgroups)	8.9	10.4	9.5	9.6	9.9	0
Arenosol	Sand (Psamments)	0	16.4	4.7	4.3	9.7	0

Yermosol	Desert soils (Aridisols)	8.9	9.6	0.2	3.3	5.5	12.4
Xerosol	Dry soils of semi-arid regions (Aridisols)	5.4	4.7	0.9	0	2.7	2.7
Total Percentage		23.2	41.1	15.3	17.2	27.8	15.1
Wet or Poorly-drained Soils							
Gleysol	Reduced horizons due to wetness (Aquepts, Aquents)	2.3	2.1	3.7	3.7	2.9	5.8
Planosol	Poorly-drained with abrupt A-B boundary (Alfisols, Ultisols)	0.5	0.4	2.6	1.3	1.2	0
Histosols	Organic soils (Histosols)	0.9	0.1	0.2	2.8	0.6	0
Total Percentage		3.7	2.6	6.5	7.8	4.7	5.8
Total Land Area (x 10000 ha)		27,708	225,892	153,865	86,063	493,527	77,047

Modified from Richter and Babbar (1991).

Table 14.4. Soil mapping units of the Brazilian Amazon River Basin, based on GIS-interpreted Mapo do Solos do Brazil. Distróphico means low in nutrients. Eutróphico means high in nutrients.

Primary Mapping Unit	Area 1000 km²	Percent of Total	Secondary Mapping Unit	Area 1000 km²	Percent of Total
Latosolos	1805.8	39.14	Amarelo Distrófico	787.3	17.06
			Vermelho-Amarelo Distrófico	923.4	20.01
			Roxo Distrófico e Eutrófico	1.0	0.02
			Vermelho-Escuro Distrófico	87.6	1.19
			Vermelho-Escuro Distrófico e Eutrófico	6.6	0.14
Podzólicos	1489.6	32.28	Vermelho-Amarelo Distrófico	1118.4	24.24
			Vermelho-Amarelo Eutrófico	160.8	3.48
			Plintico Distrófico	210.4	4.56
Lateritas Hidromórficas	322.2	6.98	Distrófica (Tb)	273.6	5.93
			Distrófica e Eutrófica (Ta)	3.0	0.06
			Indiscriminadas	45.7	0.99
Solos Gley	292.4	6.34	Distróficos	212.8	4.61
			Distróficos e Eutróficos	79.6	1.72
Solos Litolicos	207.9	4.51	Distróficos	148.6	3.22
			Distróficos e Eutróficos	56.9	1.23
			Humicos Distróficos	2.4	0.05
Solos Arenoquartosos	203.3	4.41	Distróficos	178.1	3.86
			Hidromórficos Distróficos	25.2	0.55
Podzol	130.2	2.82		130.2	2.82

	Area	%		Area	%
Cambissolos	62.0	1.34	Distrófico (Tb)	19.0	0.41
			Eutróficos (Tb, Ta)	42.1	0.91
			Humico Distrófico	0.9	0.02
Solos Aluviais	35.5	0.77	Distróficos e Eutróficos	4.4	0.10
			Eutróficos	31.1	0.68
Grupamento Indiviso de Solos	32.9	0.71	Eutróficos	32.9	0.71
Terras Roxas Estructuradas	25.0	0.54	Eutróficas	8.5	0.18
			Similar Distrófica e Eutrófica	2.4	0.05
			Similar Eutrófica	14.1	0.30
Solos Salinos	5.5	0.12	Solochak	2.9	0.06
			Indiscriminad Costeiros	2.6	0.06
Planossolos	1.8	0.04	Solodico	1.8	0.04
TOTAL PRIMARY UNITS	4614.1	100.00	TOTAL SECONDARY UNITS	4614.1	100.00
Total Classified as Distrófico				3962.5	85.88
Total Classified as Eutrófico				242.5	5.26

From Richter and Babbar (1990).

In the Amazon basin, dramatic differences in the characteristics of rivers draining different regions led Sioli (1950) to classify the rivers into three types according to color: whitewater, clearwater, and blackwater. Whitewater rivers are turbid with high loads of suspended inorganic solids. Alluvial soils are relatively fertile along these rivers, which include the Amazon itself, the Madeira, Purus, Juruá, Manú, and Jutai. Clearwater rivers are greenish and clear, with few suspended solids. Alluvial soils are generally infertile along such clearwater rivers as the Tapajos, Trompbetas, Xingú, and Curuá Una. Blackwater rivers are clear and dark with dissolved organic compounds. They usually drain swampy areas or areas with sandy soils that have few clays to bind and remove organics. Fertility of alluvial soils is low along such blackwater rivers as the Rio Negro and the Urubú (Junk and Furch, 1985).

Fittkau (1971) related the hydrochemical conditions in Amazonian rivers and streams to geochemical conditions of the basin. He divided the basin into three basic geochemical provinces (Fig. 14.11).

1. The Andean and pre-Andean region (western peripheric region);
2. The archaic shields of Guiana and Central Brazil (northern and southern peripheric regions);
3. Central Amazonia.

These three provinces differ considerably in respect to their geochemistry and geomorphology. The Andes are young mountains which started to rise in the Tertiary. The intensive erosion processes affect former marine sediments rich in mineral elements, especially calcium and magnesium. Consequently, rivers with their catchments areas in the Andes and pre-Andean region contain high sediments loads and high concentrations in mineral salts with a large percentage of alkali-earth metals and neutral pH.

The archaic shields of Guiana and Central Brazil belong to the oldest geological formations on earth. In comparison to the Andes the relief is low, erosion processes being relatively minor. However, the weathering of the rocks delivers small quantities of mineral nutrients. Water derived from these areas has little sediment load and is poor in mineral elements, especially in alkali-earth metals, thus having low pH value.

Central Amazonia is covered with Tertiary sediments of fluviatile and lacustrine origin. These soils have high fractions of

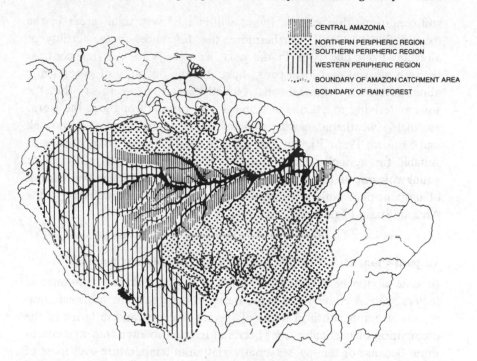

Fig. 14.11 Geochemical regions of the Amazon basin. (According to Fittkau, 1971.)

sandy and kaolinitic material. The availability of mineral nutrients and the pH values are extremely low. The relief is low and gradients in elevation are small. Dense vegetation cover inhibits erosion processes. Rivers with catchment areas in Central Amazonia have transparent and very acid water with extremely low concentrations of dissolved minerals.

... Total phosphorus may be used as an additional criterion for the description of water types. There seems to be a gradient in the concentration of this element, whitewater being rich in total phosphorus, blackwater, extremely poor, and clearwater intermediate. Considering the fact that phosphorus can be a limiting factor for plant growth, this parameter is important in respect to the evaluation of the fertility of Amazonian waters.' (Junk and Furch, 1985).

The juxtaposition of fertile alluvial soils from the Andes (the 20-100 km wide 'várzea' floodplain) with the infertile upland ('terra firme') soils of the central Amazon basin is one example of the heterogeneity of soil conditions that can be found within relatively small areas in the tropics (Sioli, 1975b). Furthermore, the differences in the fertility of alluvium from one river to the next, as well as the varying contrast between alluvial and local soils, illustrates the difficulty of generalizing about patterns of tropical soils. Nonetheless, it is safe to say that the soils underlying much of the tropical rain forest throughout the world are highly weathered, acidic, and low in available nutrients (Vitousek and Sanford, 1986; Richter and Babbar, 1991). Highly productive soils suitable for agriculture are found in areas areas of rich alluvium or young volcanic soils, but these are the exceptions to the general pattern of nutrient-poor soils in tropical rain forest regions, particularly in the Amazon basin (Tables 14.3 and 14.4).

Tropical Plant Diversity: The Influence of Precipitation and Soils

In spite of the heterogeneity of tropical forests and of environmental factors such as climate and soils, there do exist patterns of forest community structure, both in terms of species diversity, and in terms of the distribution of particular tree species in relation to environmental conditions. Because of the low seasonal variation in temperature over most of the tropics, temperature-related patterns of community structure are not apparent, except in the case of altitudinal gradients, which are analogous to latitudinal gradients of temperature. However, as discussed above, precipitation has extremely high variability, and is a critical resource that affects plant growth and community structure (Kramer, 1983; Wigley *et al.*, 1984; Woodward, 1987; Smith and Huston, 1989).

Recent work has shown that precipitation is strongly correlated with species diversity in Costa Rican forests for trees over 10 cm dbh (Huston, 1980a), and for all vascular plants, including trees, understory plants, and epiphytes over a range of Neotropical (Gentry, 1982) and African (Hall and Swaine, 1976) forest sites, probably as a consequence of several very different mechanisms. Some potential mechanisms would be expected to cause species diversity to increase with increasing precipitation, while others would cause it to decrease. In the studies cited above, tree species diversity increased with increasing precipitation.

The first explanation that comes to mind to explain the positive correlation between precipitation and species diversity is that plants need water to grow and should grow better with more water (up to the point where the soil becomes anaerobic). This explanation is a specific case of the 'productivity/diversity hypothesis,' discussed in Chapter 4. While some components of total diversity, such as the number of functional types, are expected to increase with increasing productivity, the number of competing species within a functional type may either increase or decrease, depending on the level of productivity and the trophic level being considered (cf. Chapter 5). For primary producers competing for light, diversity tends to decrease with increasing productivity, since increasing productivity is associated with increasingly intense competition for light. However, at very low levels of productivity, diversity increases with increasing productivity, producing the unimodal 'hump-backed' curve of diversity in relation to productivity (Grime, 1973; Huston, 1979).

Precipitation, Productivity, and Soils – Spurious Correlations

If the relationship between precipitation and species diversity in tropical forests were based solely on a positive effect of precipitation on productivity, the relationship of precipitation to the diversity of potentially competing species would be expected to be negative, rather than the positive relationship that has been actually found. However, precipitation influences plant growth not only directly, but indirectly through its effect on soil formation and nutrient leaching. Increased precipitation results in increased rates of weathering of soil parent material (dissolution and chemical alteration of primary minerals), which makes nutrients available to plants, but precipitation also leads to the loss of nutrients through leaching from the rooting zone (Craig and Hallais, 1934; Sanchez, 1976). Both of these processes are accelerated under high temperatures (Jenny, 1941). It should be noted that the sites being discussed are *forest* sites. Water availability severely limits plant productivity throughout much of the tropics, but such dry environments do not support forest vegetation. Although, water availability can also limit annual productivity in seasonally dry forests, water is generally not a limiting resource in tropical rain forests.

The negative effect of precipitation on soil nutrient availability in the warm tropics has been documented in several areas (Craig and Halais, 1934; Hall and Swaine, 1976; Huston, 1980a) (Fig. 14.12). In an analysis of data from Costa Rica published by Holdridge *et al.* (1971), Huston (1980a) found a negative correlation between precipitation and both total base cations ($r_{Spearman} = -0.51$, $P < 0.01$, $N = 39$) and a fertility index based on P, K, and Ca ($r_S = -0.62$, $P < 0.01$, $N = 36$). An important aspect of forest structure was consistent with this negative correlation between precipitation and low fertility: the number of stems/0.1 ha was

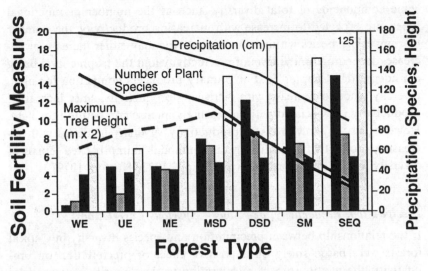

Fig. 14.12 Effect of precipitation on species richness, tree height and properties of rain forest soils in Ghana. The forest types are presented along a rainfall gradient, with the wettest types (WE) on the left and the driest (SEO) on the right. Soil fertility measure are indicated as shaded bars, from left to right: black bars are total exchangeable bases (milliequivalents / 100g soil); gray bars are base saturation (% / 10); dark gray bars are pH; pale gray bars are available phosphorous (ppm) – note that the rightmost value (125) is off the scale. Precipitation is the midpoint of the range for each forest. Number of plant species is the average number of vascular plants per 0.0625 ha plot. Maximum tree height (doubled to scale more clearly) is the mean height of the tallest tree in or near each plot. The seven forest categories (and the number of plots ("A" samples) in each) are: WE, Wet Evergreen (13); ME, Moist Evergreen (27); UE, Upland Evergreen (altitudinally distinct from the other types) (6); MSD, Moist Semi-Deciduous (45); DSD, Dry Semi-Deciduous (41); SM, Southern Marginal (17); and SEO, South-East Outlier (6). (Based on Hall and Swaine, 1976.)

positively correlated with precipitation ($r_S = 0.43$, $P < 0.01$, $N = 39$). A high density of small stems in a mature forest is a good indicator of a low productivity site (Spurr and Barnes, 1973).

Tree species diversity in these Costa Rican forests was negatively correlated with soil nutrient availability (and presumably, productivity), as measured in several different ways (Fig. 14.13a; Table 14.5). The postulated negative relationship between tree species richness and site productivity is further supported by the positive correlation of richness with the number of stems/0.1 ha ($r_S = 0.63$, $P < 0.01$, $N = 35$) and by the fact that the highest species richness was found in forests of only intermediate height (Fig. 14.14). Similar relationships of tree species

Table 14.5. *Stepwise regression of soil nutrients on species richness for forty-six sites in Costa Rica using forward selection. Although Na entered as the final significant variable at step 7, it is not included because its relationship with species richness is probably spurious, based on its high correlation with K.*

Order of entry into regression

Step	Variable	R^2	Std. Error	Partial	Signif.
1	K	0.28	8.28	-0.53	<0.0002
2	Ca	0.40	7.65	-0.41	<0.006
3	Mn	0.51	7.03	-0.42	<0.005
4	Exchange Capacity	0.55	6.76	0.31	<0.05
5	Organic Matter	0.60	6.47	-0.33	<0.04
6	Avail. N	0.71	5.60	0.52	<0.0006

Analysis at Step 6

Source	df	Sum of Square	Mean Square	F-statistic	Signif.
Regression	6	2981.0	496.8	15.86	<0.00005
Error	39	1221.5	31.3		
Total	45	4202.5			

Multiple $R = 0.84$ $R^2 = 0.71$ SE = 5.60

From Huston (1980).

richness to soil nutrients, precipitation, and tree height were reported by Hall and Swaine (1976) for 155 forest sites in Ghana (Fig. 14.12).

It is clear that the inverse correlations between soil nutrients and tree species richness found in several comparative studies of tropical tree diversity are a consequence of being on the descending limb (i.e., the high productivity portion of the gradient) of the unimodal relationship between productivity and diversity (see Fig. 5.5). For sites on the ascending limb, the correlation will be positive, and for sites that cover the entire range of fertility (productivity) conditions there will not be any linear correlation (Fig. 14.13c, d). Throughout the tropics, extremely nutrient deficient sites, such as white sand forests, are characterized by low species diversity and vegetation with thick, sclerophyllous leaves that are high in secondary compounds (McKey *et al.*, 1978; Anderson, 1981). That a study in Costa Rica or Africa (Fig. 14.13a, b) should find a negative

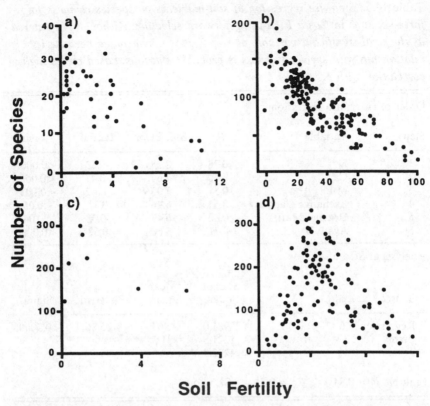

Fig. 14.13 Relationship between tree species richness and soil fertility in four tropical forest areas. *a*) Costa Rican forests. Soil fertility index is the sum of the percentage values for phosphorus, potassium, and calcium obtained by dividing the value for each nutrient by the mean value of that nutrient for all 46 sites. Number of species is tree species greater than 10 cm dbh per 0.1 ha. (Based on data from Holdridge *et al.*, 1971, from Huston, 1980.); *b*) Ghana closed forest. Soil fertility index is a Principal Components axis (Axis 1), which is positively correlated with precipitation and negatively correlated with total exchangeable bases and other measures of soil nutrient availability. Number of species is total number of vascular plant species in 0.0625 ha. (From Hall and Swaine, 1976.); *c*) West Malesian forests. Soil fertility indexed as in (*a*), but using only P and K. Number of species is tree species over 10 cm dbh per 1000 individuals. (Based on Ashton, 1977.); *d*) Amazon Basin. Hypothesized pattern of tree species richness based on greater range and lower average soil fertility in the Amazon basin. Number of species is tree species over 10 cm dbh per ha.

correlation between tree species richness and site fertility, while a survey of Amazonian forests would probably find no significant correlation (as a consequence of a unimodal pattern) is not a surprise given the rela-

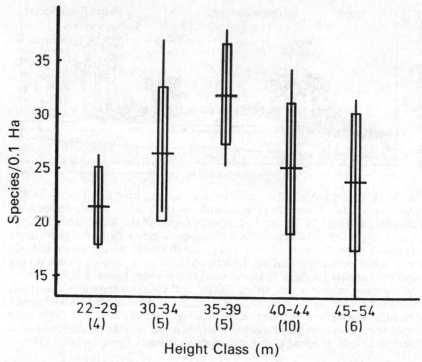

Fig. 14.14 Relationship between tree species richness and tree height in Costa Rican lowland rain forests. Height classes are based on the mean of the heights of the tallest trees in 0.1 ha plots at each site. Horizontal line indicates mean, rectangle indicates 1 SD about the mean, and vertical line indicates the range of values for each height class. The number in parentheses below each height class is the number of sites in that class. Sites used in this figure include only those in the moist, wet, and rain forest life zones of Holdridge, with all swamp forests, tidally inundated forests, dry forests, and sites above 2000m eliminated. (Based on data from Holdridge *et al.*, 1971, from Huston, 1980.)

tively high fertility of soils in Central America and Africa (Table 14.3) in relation to South America.

The general pattern of tree species richness in relation to both precipitation and soil nutrients that was detected in Costa Rica, Ghana, and Malesia, and South America (for precipitation only, soil nutrients were not measured, Gentry, 1982), can also be found on a much smaller scale (Fig. 14.15). Near San Carlos de Rio Negro, there is strong zonation of forest structure and composition that is associated with variation in soil nutrients and water availability. Jordan (1985a) reports that the tallest and most productive forest in the region is on well-drained *tierra*

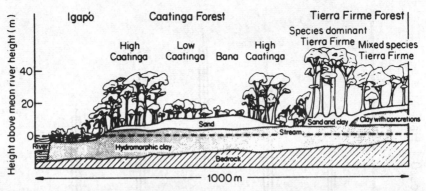

Fig. 14.15 Vegetation zonation and species richness near San Carlos de Rio Negro, Venezuela. Schematic representation of the relation between hydrology, vegetation, and soils in the upper Rio Negro region of the Amazon basin. Areas close to rivers and streams contain hydromorphic gley soils and support igapó (seasonally flooded) forest. Moving away from the rivers, where surface soils are sandy, igapó grades into caatinga forests, which occur on Spodosols and do not experience annual flooding. 'High caatinga' occurs where the sand is fine-grained and the water table remains within 100 cm of the surface throughout the year; 'low caatinga' and 'bana' occur on slightly higher, better drained sites with coarse sand soils, which experience alternating standing water and droughts. Tierra firme forests occur where the clay content of the soil is higher, with legume-dominated forest on ultisols or mixed dominant stands on oxisols. (From Jordan, 1985.)

firme soils near the stream. As expected, this productive forest is low in tree species richness, being dominated by the species *Eperua purpurea* or 'Yevaro.' Adjacent to the 'Yevaro' is a mixed forest of smaller stature that occurs where the nutrient-poor clay soils are near the surface. This forest has the highest diversity in the region, over 80 species per hectare (Uhl and Murphy, 1981). Associated with deep sandy soils on a slight elevational gradient away from the stream is a distinct zonation of decreasing forest height and species diversity. Soil water availability, and possibly nutrients as well, decrease with distance from the stream. Nearest to the stream is 'high caatinga', followed by low caatinga (or 'campina'), and finally the scrubby 'bana' vegetation (Klinge and Medina, 1978). Bordering the streams are the low stature 'igapó' forests, which are flooded for six months or longer each year.

Additional data from the same site are provided by Brunig (1983), which shows a general pattern of highest tree species richness at low (but not lowest) stem volume and mean crown cross section. The pattern of tree species richness along an elevational gradient in Puerto Rico shows

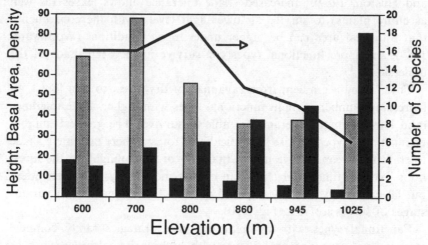

Fig. 14.16 Forest structure along an elevational gradient in Puerto Rico. Note that the highest species richness is found at intermediate levels of tree height (in metres, indicated by black bars), total basal area (m^2/ha, pale gray bars), and stem density (number/ha, dark gray bars) The highest elevation forests are elfin cloud forests. (Based on White, 1963).

a similar relationship to forest height and basal area (White, 1963, Fig. 14.16).

For canopy trees the positive correlation between precipitation and species diversity is probably an indirect consequence of the negative relationship between soil nutrient availability and species diversity. This reflects the negative relationship between forest productivity and tree species diversity, rather than a positive relationship suggested by the precipitation pattern alone (e.g., Gentry, 1982). The hypothesized mechanism is that the rate of competitive displacement is reduced under conditions of low productivity because trees grow more slowly and do not attain large size.

Increased Precipitation Allows More Plant Functional Types

Although, favorable conditions of nutrient availability (with adequate water) are expected to lead to intense competition for light and reduce diversity among competing plant species, increased water availability allows plants to survive at lower light intensities. As discussed in Chapter 7, 9, and 10, the increasing shade tolerance that is possible with higher soil water availability allows more plant functional types to coexist (Smith and Huston, 1989). Increased shade tolerance allows leaves (as well as entire plants) to survive at lower light levels, with the consequence that total leaf area can be higher under moist conditions (Woodward, 1987), and more functional types can survive in the deep shade of the understory.

Thus, along a gradient from savannah or dry forest to rain forest, an increasing number of plant functional types with higher shade tolerance (and lower drought tolerance) is able to survive. The greatest number of functional types coexist in tropical rain forest, where extremely shade tolerant types can survive in less than 1% of full sunlight on the forest floor. Not all functional types can coexist indefinitely with each other, but all can survive within a spatial matrix of forest patches in various stages of succession (e.g., Fig. 5.10).

Functional types with different light-use strategies rarely compete strongly with one another. Plants with high shade tolerance, such as understory herbs, live in the moist, deep shade created by larger, less shade-tolerant plants, such as trees (i.e., they are interstitial species). While it is true that competition for light between tree seedlings and herbaceous plants can be intense, extremely *shade-tolerant* herbs rarely occur in the same high light environments where shade intolerant tree seedlings germinate and grow. Competition between species with different growth forms and maximum sizes is likely to occur for relatively brief periods of time and on average will be less intense than competition between species of similar growth form and maximum size. Nonetheless, competition for light between trees of different functional types (e.g., similar in maximum size but differing in shade tolerance) can be severe.

Endemism and Tropical Species Diversity

Endemic species (species with spatial ranges below a certain size or confined to a limited area) are often major contributors to high species diversity in tropical forests, both in terms of within-habitat diversity and small-scale geographic diversity. In the Chocó region of Panama, Gentry (1982) reports that 'veritable swarms of endemic species in such evolutionarily plastic genera as *Anthurium* (59 species), *Piper* (80), *Cavendishia* (34), *Calathea* (23), *Miconia* (63), *Psychotria* (59), *Peperomia* (42), *Clidemia* (31). ... account for a very high proportion of the endemism and species richness of the region.'

The tropics have an important feature in common with another environment that has high levels of endemism. Soils of the mediterranean-climate regions, which have extremely high levels of endemism, are generally quite poor, and there seems to be an inverse correlation between soil fertility and endemism in these regions. The low fertility of most tropical rainforest soils, and, in particular, the correlation between high rainfall and low nutrient availability were discussed earlier in this chapter. Low productivity may also result from reduced light availability, and many of the genera mentioned by Gentry (1982, 1986) as having 'species swarms' are primarily understory shrubs and herbs. Low productivity resulting from low soil fertility and low light availability may contribute to reduced pollen and seed dispersal, and thus be an indirect cause of high levels of endemism and high geographical diversity (see Chapter 11).

Epiphytes: Coexistence in a Wet and Dangerous World

Much of the high total diversity of vascular plants in some tropical rain forests is a result of the many species of epiphytic plants that live on the trunks, branches, and leaves of the trees. Gentry and Dodson (1987a) report that epiphytes comprised 35% of the species and 50% of the individuals in a 0.1 ha sample in the extremely high diversity wet forest at Rio Palenque, Ecuador. Epiphytes are plants with a strategy that assures that they will exist in a high light environment without investing the resources to become very large. Many rain forest epiphytes have strong functional and taxonomic similarity to the plants that dominate arid environmments, having a high tolerance (or effective avoidance) for low moisture conditions and a low tolerance for shade (Benzing, 1990). Epiphytes are totally dependent upon trees to provide the physical structure that allows them to avoid heavily shaded conditions. Telephone poles, wires, rooftops, and research towers also provide suitable habitat for many epiphyte species.

Epiphytes are the quintessential interstitial species, dependent on structure and microclimate provided by the trees, but having relatively little effect on the trees (see however, Strong, 1977). Because epiphytes live in an environment that is very different from the environment experienced by the trees on which they live, they can be expected to have different patterns of diversity than trees and to exist under very different conditions of growth rates and disturbance frequency.

The two key features of the epiphytes' environment result from its physical separation from the soil: 1. it is detached from the soil as a source of water and nutrients, and thus epiphytes must depend on small amounts of nutrients released or leached from the trees or obtained from rainfall; and 2. it does not experience the physical buffering provided by the soil, and thus epiphytes are subjected to extreme variation in temperature, moisture availability, and nutrient availability. The epiphytes' environment is characterized by high light availability, which is favorable for plant growth, but also by stress caused by heat and low levels of water and nutrients, which can slow growth and reduce productivity.

Because nearly all epiphytes depend completely on precipitation (including dew and fog) for the moisture they need to survive, they are most abundant in humid regions and are limited in their distribution by the length of dry season and occurence of frost (Gentry and Dodson, 1987a,b; Sanford, 1969; Madison, 1977; Benzing, 1990). The abundance and diversity of epiphytes are positively correlated with precipitation, and presumably with the productivity of epiphytes.

Epiphyte diversity would be expected to be positively correlated with epiphyte productivity and biomass because epiphytes rarely reach densities at which they compete for resources. Epiphyte densities are low in relation to the resources they use because their dependence on trees for structural support limits them to a very small proportion of the total three-dimensional space of their environment. In addition, wind damage, branch breakage, and bark shedding by trees impose a high frequency of disturbances on epiphytes.

When an epiphyte grows large enough to effectively compete for light with other epiphytes, it is likely to be too heavy for its structural support, which results in high mortality rates for the larger size classes of epiphytes. The light environment of epiphytes is controlled by the trees on which they grow, rather than by the epiphytes themselves. The primary exceptions are large branches, tree forks, and other stable situations where organic matter can accumulate. In these situations, which are much more similar to the soil environment than to the typical canopy environment, epiphytes can achieve sufficiently high biomass to compete for light and reduce the diversity of that particular epiphyte assemblage.

The epiphytes' environment is also extremely heterogeneous, which certainly plays some role in the high diversity of epiphytes. Few epiphytes seem to be restricted to particular species of hosts *per se*, although bark physical properties are important (Went, 1940; ter Steege and Cornelissen, 1990; Benzing, 1990). Variation in light availability, exposure to drying winds and sunlight, accumulation of organic matter, branch structure, bark texture, and other factors provide a highly diverse physical environment to which epiphytes show a wide variety of adaptations.

However, epiphytes show few adaptations for competing for resources such as light, mineral nutrients, and water, as would be expected in a environment that is both stressful and subject to frequent disturbances (Grime, 1973, 1979). Rather, adaptations for conserving and storing water and nutrients are common (Benzing, 1990). Epiphytes store water in specialized stems, roots, and cups or tanks formed by leaves (Schimper, 1888; Benzing, 1990). Some epiphytes obtain scarce nutrients such as nitrogen by attracting insects to live on and in them (Kleinfeldt, 1978; Huxley, 1980). A photosynthetic mechanism specialized for conserving water (CAM - Crassulacean Acid Metabolism) was first discovered in epiphytes of the family Crassulaceae (Osmond, 1978). There are numerous species of epiphytic cacti in tropical rain forests, as well as other families common in deserts (e.g., Bromeliaceae, Gnetaceae).

Many of the most dramatic examples of coevolution and highly specialized adaptations are found in the reproductive systems of epiphytes. Where resources for growth and reproduction are extremely limited and suitable sites for germination and growth are widely scattered, epiphytes have evolved efficient dispersal mechanisms such as sticky, bird-dispersed seeds that adhere to branches after they are defecated (van der Pijl, 1969, 1972; Baker *et al.*, 1983)

In summary, epiphytes are a major component of tropical plant species richness, but their diversity is influenced by mechanisms very different from those that influence diversity of the trees upon which they grow. Epiphytes exist in an environment where intense competition with other epiphytes occurs only infrequently, as a result of frequent disturbance (e.g., dislodgement or branch breakage) and low growth rates caused by low water and nutrient availability. Unlike the diversity of canopy trees, epiphyte diversity is positively correlated with epiphyte productivity and biomass, which are regulated by the amount and seasonality of precipitation. Limitation of diversity by competitive exclusion is expected to occur very rarely, under conditions where organic matter and high epiphyte biomass can accumulate in tree forks and on large branches.

Understory Diversity: Coexistence in Wet, Low Light Conditions

The coexistence of many interstitial plant functional types (characterized by different degrees of shade and drought tolerance) is a major explanation for the high *total* plant species diversity of tropical rain forests, which includes many types of epiphytes, as well as highly diverse genera of understory plants in such families as Rubiaceae, Melastomaceae, and Araceae. The number of species within a particular functional type (e.g., highly shade-tolerant and drought-intolerant understory shrubs) can vary greatly, depending on many factors including biogeography and competition within the functional type. Diversity within some functional types may be inversely related to the diversity within other functional types.

For example, shade-tolerant understory herbs live in a light environment created by trees of the forest canopy. As the productivity of the canopy trees increases, as a result of increasing water and/or nutrient availability, canopy density increases, competition for light among canopy trees may become more intense, and tree diversity is likely to decrease. Along the same gradient of increasing tree productivity and canopy density, light available to understory plants will decrease, with a decrease in productivity of the understory plants.

Observations and experiments in a wide range of systems demonstrate that the diversity of understory plants, which compete among themselves for light and other resources, increases with decreasing light availability to quite low levels, below which it decreases. In an experiment on plant community regeneration in an open agricultural field in the tropical lowlands of Costa Rica, light available to the plants was reduced with neutral density shade cloth to 28% and 8% of full sunlight in plots that were 15 × 15 m. An adjacent unshaded area provided a 100% sunlight treatment. After six months of growth from bare ground, biomass was clipped from ten 0.25 m^2 areas in each treatment, sorted by species, dried, and weighed. In the full sunlight, the total biomass was 2.5 times the biomass of the 28% treatment, and the plots were dominated by a single species (*Borreria sp.*, Rubiaceae) that comprised 82%(± 8 SE) of the biomass. In the 28% treatment, the same dominant species comprised 62%(± 8) of the biomass, and there were a total of 16 additional species present (in comparison with only five additional species in the full sun treatment). At the lowest light level, 8%, a different dominant species comprised only 22%(± 8) of the biomass, and there were 14 additional species (Fig. 14.17). Clearly, diversity in these experiments increased with decreasing light availability, although it is equally obvious that the diversity of photosynthetic plants must decrease to zero in total darkness.

Plant diversity is also correlated with light availability along natural light gradients, such as the shadow cast by a tree in a lawn. Fig. 14.18 illustrates the increase in the species richness and evenness among the grasses and forbs in a mowed lawn in Tennessee along a gradient of decreasing light from open lawn into the deep shade near the trunk of a large spreading tree (Huston, unpublished data). Similar patterns can be found in all terrestrial systems, although they may be complicated by the

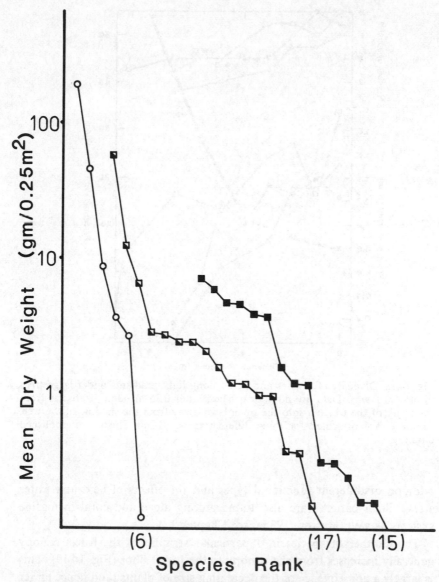

Fig. 14.17 Dominance/diversity curves for three light levels in tropical herbaceous understory experiment. Open circles represent species from plots in full sunlight. Half-shaded squares represent species from plots with 28% full sunlight, and shaded squares represent plots in the deepest shade, 8% of full sunlight. Number in parentheses indicate total number of species found in each treatment. Note the dramatic decrease in dominance of the most abundant species between full sun and the deepest shade. (From Huston, unpublished data.)

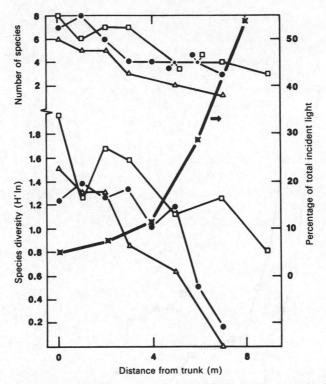

Fig. 14.18 Diversity of grasses and forbs along light gradients under large trees in mowed lawns. Data are number of species per 0.25 m^2 along transects from the trunk of the tree out into the open lawn away from the shadow of the tree. Transects are presented for three different trees. (From Huston, unpublished data.)

addition of different functional types and the effects of heat and water stress. Such patterns are also found among algae and coral in marine systems as well (Huston, 1985a, see Chapter 12).

The number of species in a particular stratum of the forest canopy generally increases from the canopy to the forest floor (Fig. 14.19). This is largely a consequence of the decreasing size of plants (and hence larger number of individuals per unit area) as one moves from the canopy (i.e., trees) to the ground surface, along with the fact that small individuals of canopy and subcanopy trees may coexist with herbaceous plants on the forest floor. Calculating the number of species for a specific number of similarly-sized individuals would compensate for the bias introduced by sampling plants of different sizes (e.g., trees and herbs) using the same plot

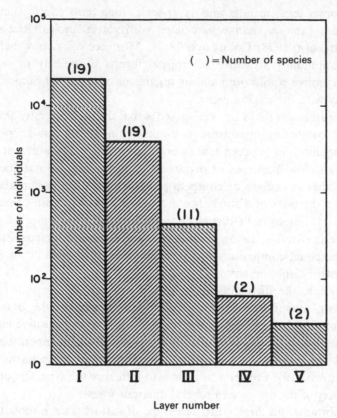

Fig. 14.19 The number of individuals and species of woody plants per hectare in different layers of the canopy of a Hawaiian rain forest. I = herbaceous plant layer (0-0.5 m); II = tree fern layer (>0.5-5 m); III = low-stature tree layer (>5-10 m); IV = intermediate-stature tree layer (>10-15 m); V = emergent layer (>15 m). (From Mueller- Dombois, *et al.*, 1981.)

size. This approach to comparing diversity patterns of different canopy layers has rarely been taken, so few data are available to appropriately address this interesting comparison.

Precipitation and Disturbance

Although tropical rain forests were initially thought to be stable environments where the organisms were likely to be at equilibrium with each other and with their environment, it is now recognized that tropical forests are extremely dynamic, subject to a variety of disturbances and environmental fluctuations, both small-scale, short-term effects such as windstorms and treefalls and large-scale, long-term effects such as drought and climatic change associated with glacial cycles (Richards, 1952; Hartshorn, 1978; Colinvaux, 1987). Most ecologists now believe that the community structure of tropical forests is unlikely to reflect either competitive equilibrium among organisms or an equilibrium of the community with the environment.

The interpretation that I have proposed is that the high diversity among potentially competing organisms in tropical rain forests results from a dynamic equilibrium between low rates of competitive displacement and a moderate to low frequency of disturbances and environmental fluctuations that alter the effects of competition (Huston, 1979). Accumulating evidence on the rates of disturbance in tropical forests indicates that the frequency and intensity of disturbances differ little from those of temperate forests (Runkle, 1985). Therefore the explanation for coexistence of many potential competitors must be sought in differences in the rates of competitive displacement.

Precipitation, the distinguishing physical property of the rain forest environment, may influence both major components of the dynamic equilibrium model of species diversity. The negative and positive effects of precipitation on plant productivity, the intensity of competition for light, and the number of coexisting plant functional types have been discussed previously. Precipitation can also influence the type, frequency, and intensity of disturbance in several different ways.

An important, and largely intrinsic, type of disturbance in most tropical rain forests is treefalls and branchfalls, which damage or kill a few large trees and result in increased light, water, and nutrient availability to many smaller plants. Treefalls provide much of the spatial and temporal heterogeneity of vegetation dynamics in tropical rain forests (e.g., Denslow, 1987), and are essential for both the prevention of competitive equilibrium at a local scale and the maintenance of the dynamic equilibrium of the forest mosaic at large scales. Precipitation tends to be positively correlated with the frequency and magnitude of treefalls for a number of reasons.

1. High precipitation is often associated with shallow root systems in trees, which increases the susceptibility of trees to windthrow. Shallow rooting can be caused by anoxic conditions in the soil that continually or periodically kill roots below a certain depth, or it may result from preferential root growth on or near the soil surface where nutrient levels are higher than in deeper soil (Odum, 1970; Stark and Jordan, 1978; Klinge and Herrera, 1978; Vitousek and Sanford, 1986; Fig. 14.20).

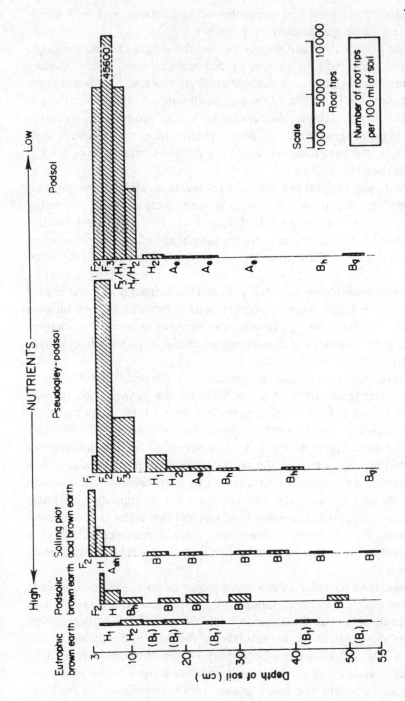

Fig. 14.20 Increase in relative and absolute amount of surface roots with decreasing soil nutrients in European soils. Soils are arranged in order from the most nutrient rich on the left to the most nutrient poor on the right. (Based on Meyer and Göttsche (1971), from Jordan, 1985.)

2. High precipitation can soften the soil and subsoil, making it structurally less stable, particularly on hillslopes.

3. High precipitation can greatly increase the weight of a tree's canopy, because of water held on leaf and branch surfaces, water held in cuplike epiphytes, and water held in mats of epiphyte roots and organic matter. Such an increase in weight, often associated with an altered distribution of weight, can destabilize the balance of a tree, resulting in uprooting or breakage. Where periods of intense precipitation are associated with high winds, the probability of treefall is further increased (e.g., Strong, 1977; Brokaw, 1982).

4. The height or total biomass of trees increases along a precipitation gradient from dry to wet forests, and then decreases with increasing precipitation in rain forests (Holdridge et al., 1971; Hall and Swaine, 1976; Huston, 1980a), presumably because of the effect of precipitation on soil nutrients. Where trees are larger, treefall gaps are likely to be larger.

5. The contrast between treefall gaps and the surrounding forest matrix is likely to be larger where precipitation is sufficiently high to allow a high leaf area. Increasing precipitation is therefore likely to be associated with an increasing relative importance of treefall gaps for allowing light to penetrate to the forest floor.

The meandering of Amazonian rivers may provide a disturbance of 'intermediate' frequency that contributes to the maintenance of high diversity in some Amazonian forests (Salo et al., 1986; Rasanen et al., 1987). Colinvaux (1987) presents evidence that erosional activity by rivers may have been higher in the past, as a result of regional increases in precipitation associated with glacial cycles in the higher latitudes. While it is true that river meandering does act as a disturbance to floodplain forests, as well as providing new substrate for the initiation of forest succession, only a small proportion of tropical rain forest occurs in river floodplains. There is no evidence that such disturbance occurs at a significantly greater rate in tropical rain forest areas than in other parts of the world.

Because river meandering is a consequence of flow rate, bed load, and geomorphology, there is variation in the rate of meandering by rivers in different regions of the Amazon basin. Such variation in disturbance can alter the dynamic equilibrium level of diversity in floodplain forests, which is also influenced by the fertility of the alluvium deposited by the river. Meandering of rivers does produce a high degree of habitat heterogeneity within the flood plains, which contributes to between-habitat (beta) diversity, but the within-habitat diversity depends on other factors as well. The local diversity of many riverine forests is low, as a consequence of several factors including high nutrient levels of alluvial soils and early successional status. The highest diversity forests are generally found on upland areas away from active river floodplains (Foster, 1990).

Fire is another major disturbance whose role in tropical forests has recently been recognized. While ecologists have long been aware of the role of fires in maintaining savannahs, it is only recently that fire has been found to be an important periodic disturbance in rain forest areas (Sanford *et al.*, 1985; Goldammer, 1990) as well as dry forests. In high rainfall areas, natural fires occur only during infrequent severe droughts, at intervals of 300 to 600 years in parts of the Amazon basin (Sanford *et al.*, 1985). However, throughout virtually all of the tropics, fires set to trees and brush during dry seasons are a major tool of traditional 'slash and burn' agriculture, as well as the more extensive land clearing that is currently occurring throughout the tropics.

Floods and landslides are other disturbances that occur with increasing frequency and intensity as precipitation increases (Veblen and Ashton, 1978; Garwood *et al.*, 1979; Colinvaux, 1987).

Low-Diversity Tropical Forests

Not all tropical forests have high tree species diversity. Any theory of species diversity must be able to explain not only high diversity, but why some forests have only a few dominant species. There is clearly more than one explanation for the different types of low diversity plant communities found in the tropics.

Early Successional Plant Communities

The early stages of plant succession are sometimes dominated by one or a few plant species. In many cases, particularly secondary successions where soil conditions are suitable for most species, dominance results when a single species is very successful in dispersing into the area or becoming established by other means. Where disturbances are large, species dispersed by wind or birds are often the first to invade, and may form nearly monospecific stands. In the temperate zone, wind-dispersed species such as pines, aspens, or birches often dominate in early succession. In the tropics, bird-dispersed species such as *Cecropia* in the neotropics or the closely related *Musanga* in Africa, can dominate large disturbances. Some low diversity forests, such as the so-called 'storm forests' of Kelantan in Malaysia, are clearly early successional stands (Wyatt-Smith, 1954).

Experiments on early succession in Costa Rica illustrate the importance of nutrients for community structure and diversity even in the earliest stages of succession. Harcombe (1977b) studied fertilized (70 kg N, 87.5 kg P, and 166 kg K ha^{-1} yr^{-1}) and unfertilized plots for 12 months following clearcutting. In the fertilized plots, a single rapidly growing herbaceous species, *Phytolacca*, achieved a biomass double that which it achieved in the unfertilized plots. In the fertilized plots *Phytolacca* suppressed all the other species of grass, shrubs, trees, and vines for a full year. Diversity was much higher in the unfertilized plots, where other plant types achieved a cover value equal to or greater than that of *Phytolacca* after 12 months (Fig. 14.21). In a similar experiment that focused on competition among trees found in light gaps in the primary forest, Huston (1982) found that a single fast-growing early successional species, *Hampea* (Malvaceae) quickly dominated the fertilized (83 kg N, 250 kg P, 50 kg K/ha/yr) plots, while *Hampea* shared dominance with a second species in the unfertilized plots.

Early successional plant communities in the fertile alluvial plain of the Manu river in the Peruvian Amazon form a temporal sequence (and spatial zonation) of monospecific stands of increasing height until the *Cecropia* monoculture breaks up into a mixed species forest that is eventually dominated by large species of the fig and mahogany families. High diversity forests form after the death of the dominant figs and mahoganies, and when the river moves a sufficient distance away that the sites become well drained and there is no longer a frequent input of fresh alluvium (Gentry and Terborgh, 1990; Foster, 1990). A similar pattern of lower diversity in the frequently flooded *Várzea* forests than in adjacent *terre firme* forests on unflooded and less fertile soils is found throughout the Amazon basin (Campbell *et al.*, 1986, 1992).

Extreme Soil Conditions

Nutrient deficient soils such as white sand soils, or soils that are extremely wet or dry often support very low diversity forests (Richards, 1952; Ashton, 1971; Brunig, 1973; Janzen, 1974; Anderson, 1981). Mangrove forests, existing in the stressful saline (but high nutrient) environment at the margin of the sea, are typically dominated by a single species,

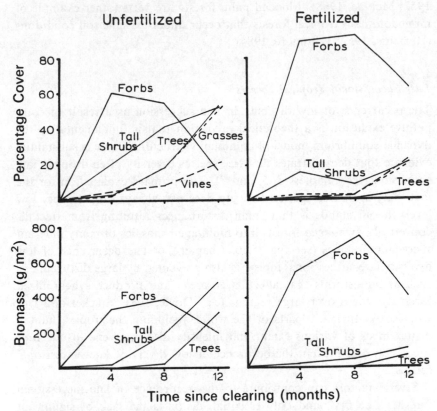

Fig. 14.21 Biomass of major plant groups over 12 months of secondary succession on fertilized and unfertilized plots in Costa Rica. (Based on Harcombe, 1977b.)

although there may be several monodominant zones along the gradient from land to the sea (Steenis, 1958; Chapman, 1976). The relative importance of stress tolerance versus competition or other mechanisms (e.g., propagule dispersal; Rabinowitz, 1978a,b) in determining patterns of mangrove zonation has not been resolved (cf. Semeniuk, 1983). However, mangrove forests are often very productive, which suggests that among this group of salt-adapted species, competition may have an important influence on community structure (Walter, 1971). In comparison with the Neotropics, African and Southeast Asian mangrove communities have higher species richness. This higher richness results in more complex and finer-grained patterns of zonation and distribution than in Neotropical mangroves, but the zones are still characterized by single-species dominance (Watson, 1928; Schimper and Faber, 1935; Richards,

1952; McNae, 1968). Flooded palm forests are yet another example of monodominant tropical forests that occur under extreme soil conditions (Richards, 1952; Whitmore, 1984).

Late Successional Tropical Forests

The occurrence of low diversity in late succession as a result of competitive exclusion is a theoretical prediction that is fundamental to the dynamic equilibrium model of community structure. There is substantial evidence that demonstrates that tree species diversity often drops in late succession (see Chapters 3, 5, and 7). The primary explanation for the fact that forest tree species diversity does not always drop to very low levels in old stands is that small disturbances resulting from treefalls convert old even-aged forests into multi-aged mosaics of many different successional stages (see Fig. 5.10, Chapter 7). The occurrence of low diversity late-successional forests is also prevented by large disturbances, such as windstorms that affect large areas and produce synchronized forest dynamics over large regions (e.g., Doyle, 1981; Stocker and Unwin, 1990). In many parts of the world, including the tropics, human disturbances of various extent and intensity also intervene to interrupt succession and prevent the occurence of low diversity, late-successional forests.

Several factors can contribute to the occurrence of late-successional forests. The first, and only essential, factor is the lack of significant disturbance over the time interval required for competitive exclusion to occur. Second, favorable growth conditions may accelerate the rate of competitive exclusion, and allow a low diversity, late-successional stage to be reached more quickly (Huston, 1979; Huston and Smith, 1987). Third, the presence of appropriate late-successional species, that have such properties as large size, shade tolerance, and possibly resistance to some types of disturbance, can both accelerate the reduction in diversity, and also cause the almost complete elimination of other species from the forest canopy. In North American forests, such species as sugar maple, American beech, and hemlock are often late-successional dominant species.

In the tropics, the most enigmatic low diversity forests are certain monodominant stands that occupy large areas (several hectares to several thousand km^2), occur on a variety of soils, and are nearly identical to adjacent high diversity forests with regard to soil conditions, and the number and identity of species present in the subcanopy and understory

(Table 14.6). Hart *et al.* (1989) studied a forest dominated by the Caesalpinioid legume *Gilbertiodendron dewevrei* in central Africa and compared it to adjacent mixed forest that had much higher diversity. Few significant differences in soils, seed predation, or treefall gap occurrence and area were found between the monodominant forest and the adjacent mixed forest. Virtually all the tree species found in the high diversity mixed forest were also present in the understory of the monodominant forest. However, the dominant *Gilbertiodendron* was completely absent from the mixed forest.

Based on their studies in Central Africa and a review of other monodominant tropical forests (Table 14.6), Hart *et al.* (1989) observed that the dominant tree species of these enigmatic forests share a number of important characteristics. First, they are all shade tolerant and persist as seedlings on the forest floor. This has important consequences including the ability to regenerate beneath their own canopy, and the ability to maintain a deep crown with a high leaf area index that reduces subcanopy light to very low levels. Second, they have very large, poorly dispersed seeds. This contributes to seedling survival (if the seeds avoid predation), but means that these species disperse into adjacent forest at very low rates following large disturbances. Third, they tend to be large trees, and monodominant stands are characterized by a higher density of large individuals than are adjacent mixed stands. These are the classic characteristics of late successional dominant species.

Hart *et al.* (1989) conclude that these forests represent a late, near-equilibrium successional stage where competitive exclusion has occurred as a result of infrequent disturbances and the presence of a strongly dominant species. In contrast, adjacent high diversity mixed forests represent a non-equilibrium intermediate stage where more frequent disturbances and/or the absence of potentially dominant species allow the coexistence of many species.

The potential effect of individual species with the appropriate physiological and life history characteristics that allow dominance in late succession is emphasized by the results of Hart *et al.* (1989). The extreme low diversity of the monodominant *Gilbertiodendron* forests depends on the presence of this single large, shade-tolerant, long-lived species. Apparently, only the slow dispersal of this species prevents larger areas from being reduced to a low diversity state. In most tropical forests, the late-successional species do not dominate so completely and various types of disturbance intervene to prevent dominance. The occurrence of late-successional species that are shorter-lived and/or less resistant to

Table 14.6. *Monodominant tropical forests of well-drained soils.*

Location	Dominant Species (Family)	Rainfall (mm/yr)	Size of Stands	Degree of Dominance
Malaya	*Shorea curtisii*[a,b] (Dipterocarpaceae)	>2000	groves on hills	excludes other dominants
Malaya, Sumatra	*Dryobalanops aromatica*[a,c] (Dipterocarpaceae)	>2500	1000s km^2	60-90% of timber trees
Borneo, Sumatra	*Eusideroxylon zwageri*[d] (Lauraceae)	>2000	1000s km^2	pure stands
Trinidad	*Mora excelsa*[e] (Caesalpiniaceae)	>2000	largest stand >210 km^2	85%-95% of canopy
India	*Poeciloceuron pauci-florum*[f] (Guttiferae)	>5000	2-4 km^2	almost pure
East Africa	*Cynometra alexandri*[g] (Caesalpiniaceae)	~1800	>110 km^2	>75% of canopy
West Africa	*Talbotiella gentii*[h] (Caesalpiniaceae)	~1000	1-2 ha	pure stands
Central Africa	*Gilbertiodendron dewevrei*[i,j] (Caesalpiniaceae)	~1800	largest stand >100 km^2	>90% of canopy
Central Africa	*Julbernardia seretii*[i] (Caesalpiniaceae)	~1800	?	>90% of canopy

[a]Whitmore, 1984   [e]Beard, 1946
[b]Burgess, 1969   [f]Kadambi, 1942
[c]Lee, 1967   [g]Eggeling, 1947
[d]Koopman and Verhoef, 1938

From Hart *et al.* (1989).

breakage or windthrow allows the establishment of the asynchronous suc-cessional mosaics that help maintain the high diversity (both within-and between-habitat) of tropical forests.

Organisms that Depend on Trees - The Regulation of Animal Diversity

The entire biotic community of tropical rain forests depends on the resources and physical structure provided by trees. The structurally minor,

Table 14.6. *(continued.)*

Substrate Specificity	Associated Species	Dispersal	Shade Tolerance
hill crests	similar to lowland dipterocarp forest	poor	seedlings persistent
wide range of soils and topography	similar to adjacent dipterocarp forest	poor	shade-tolerant
wide range of soils	?	seeds large and heavy	shade-tolerant
wide range of soils and topography	similar to adjacent mixed forest	seeds fall beneath parent	shade-tolerant
similar to adjacent forest types	similar to adjacent forest types	?	seedlings persistent
variable soils	shared with mixed forest	poor	shade-tolerant as saplings and poles
variable soils and parent material	few associated canopy species	?	shade-tolerant
wide range of soils and topography	shared with adjacent mixed forest	very poor	shade-tolerant
shallow soils	shared with adjacent mixed forest	?	shade-tolerant

[h]Swaine and Hall, 1981 [i]Gérard, 1960 [j]Louis, 1947

but still important, groups of plants such as epiphytes, epiphylls, parasites, and lianas, find their physical support and their needed microclimatic conditions on the trees. Likewise, the animal food web depends on the leaves, flowers, and fruits of the trees and their associated epiphytes and understory. Even the decomposer food web, in all its unstudied complexity, depends on the primary production of the trees, and may be influenced by the same leaf chemistry that affects plant-herbivore interactions (Waterman and McKey, 1990).

Animals of the tropical rain forests span a great range of size, from tiny insects to large predators or giant herbivores such as the elephant. The diversity of these many types of animals is regulated by many ecological, evolutionary, and historical factors. The two extremes of the range are: 1. groups of animals, particularly small insects, whose diversity is regulated by the structural and chemical heterogeneity of the plant community, and is therefore highly correlated with plant diversity and structure. Specialization and low populations densities of many of these species result in a low intensity of competitive interactions in some groups (cf., Strong, *et al.*, 1984); and 2. groups of animals at high trophic levels, including many of the vertebrates whose diversity is more strongly influenced by the primary productivity of the plants (and the secondary productivity of the prey species) than by the species diversity of the plants (e.g., Smythe, 1986).

Thus, the diversity of higher trophic levels, particularly mobile organisms, follows a different set of rules than do the sessile plants and the small animals at the base of the food web (see Chapters 4 and 5). Potentially intense competition among the highly mobile organisms at higher trophic levels can lead to competitive exclusion, and over long periods of time to the evolutionary avoidance of competition through character displacement and speciation. Two types of phenomena are fundamental to understanding animal diversity at higher trophic levels: 1. strong interspecific differentiation and resource partitioning; and 2. complete dependence on primary production of plants.

For these reasons, the relationship of animal diversity to site productivity is different from the pattern of plant diversity. Because of the inherent inefficiency of energy transfer from one trophic level to another (Lindeman, 1942; Odum, 1953), maximum diversity of the high trophic levels occurs at levels of productivity higher than those at which lower trophics reach their maximum diversity. The result of this 'trophic shift' in diversity is that the peak in animal diversity is at higher levels of primary productivity than the peak in plant diversity (see Chapter 5).

The Food Resource

Of even greater importance than the total amount of plant productivity available to animal consumers is the temporal variability of the food resource. The limitation on animal biomass that is imposed by the season with lowest food availability has been implicated in regulating the biomass (and size distribution) of primates in the tropics (Terborgh, 1986; Leighton and Leighton, 1982; Smythe, 1986). Seasonal variation in food availability has an even greater effect in the temperate zone and limits the total biomass and size range of animals in comparison with the tropics. Total mammal biomass is generally much higher in the tropics than in the temperate zone (Table 14.7). The long period of low food availability during the temperate winter (or a long tropical dry season) imposes severe limitations on the number, sizes, and food habits of animals that are able to survive the season of scarcity.

For herbivores in particular, a long leafless winter or dry season imposes a severe constraint. Thus, the humid tropics are able to support a larger biomass, and greater species diversity, of animals that use resources which are seasonally limited in the temperate zone and seasonal tropics. Examples include frugivores, insectivores, and nectivores, in addition to folivores. Much of the increase in mammal diversity in the tropics is the great diversity of bats adapted to use a wide variety of resources (Fig. 14.22)

In general, only animals large enough to store an adequate reserve of energy and able to digest low quality living and dead plant material are able to remain active during the non-productive season (Eisenberg, 1983). Thus, the proportion of small animals increases from the boreal zone to the tropics (Fig. 14.22). Small animals have access to a range of resources and environmental heterogeneity, such as small tree branches or bark crevices, that are not available to larger animals, which effectively increases the resource heterogeneity and spatial heterogeneity of their environment.

High plant productivity of resources such as flowers, nectar, and fruit is a prerequisite for high diversity of the many animals specialized for these resources in the tropics. Brown (1982) reports that high butterfly diversity (and low endemism) is associated with forests on fertile soils. Likewise, the high primate diversity studied by Terborgh (1983, 1985, 1986) in Manu National Park in Peru is in alluvial forests along a river draining the relatively fertile Andean foothills. Although the high degree of habitat heterogeneity produced by the meandering of the Manu River certainly contributes to the high animal diversity of this area, the high productivity of the forests, particularly in terms of fruit, makes possible the survival and coexistence of both the primates and the other mammals. Within the tropics variation in primate body size, total primate biomass, and primate diversity shows a pattern consistent with expectations based on productivity and the seasonality of productivity. Several facts support the interpretation that the higher species richness of Old World primate communities in comparison to New World primates is related to higher

Table 14.7. *Mammal biomass (kg/ha) in tropical and temperate ecosystems*

Temperate		
U.S. Tall Grass Prairie[a]	Cattle	49.0
Managed Chaparral, Ca.[b]	Black-tailed Deer	18.6
Mixed forest, Isle Royale, Michigan[c]	Moose	3.8
Mixed Forest, Upper Michigan[d]	White-tailed Deer	3.5
Beech-maple-fir Forest Czechoslovakia[e]	Herbivores and Omnivores	6.0
Coniferous Forest, Ontario[f]	Woodland Caribou	0.02
Tundra, Canadian Arctic[g]	Barrenground Caribou	0.79
Tropical		
Grass-Brush, Uganda[h]	Elephant, hippo, buffalo	175.0
Acacia Savannah, Kenya Masailand[i]	Wild ungulates	175.0
Acacia Brushland Kenya-Tanganyika[i]	Wild ungulates	52.5
Cocha Cashu Rainforest Manu, Peru[j]	All Mammals Carnivora	1,416 66
Barro Colorado Island, Panama	Sloths[k] Mammalian Frugivores[l] Total Nonvolant Mammals[l]	22.9 14.4 53.0

[a]Watts *et al.*, 1934
[b]Taber and Dasmann, 1958
[c]Mech, 1966
[d]McCullough, 1970
[e]Turček, 1969
[f]Simkin, 1965

[g]Banfield, 1954; Kelsall, 1957
[h]Petrides and Swank, 1966
[i]Talbot and Talbot, 1963
[j]Janson and Emmons, 1990
[k]Montgomery and Sunquist, 1975
[l]Eisenberg and Thorington, 1973

useable productivity and higher soil fertility (Table 14.3) in Old World rain forests. First, the biomass per unit area of Old World primate communities is significantly larger than that of New World primate communities (Table 14.8). Second, the average size of Old World primates is larger, and the total size range is much smaller than that of New World

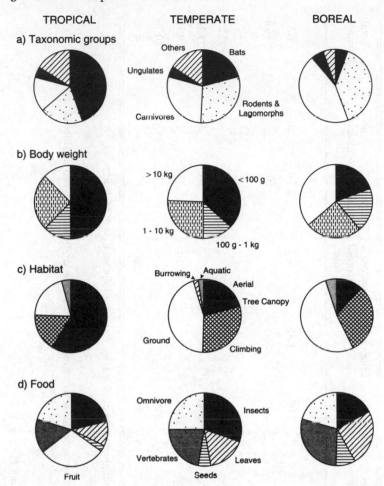

TROPICAL   TEMPERATE   BOREAL

a) Taxonomic groups

Others
Bats
Ungulates
Rodents &
Lagomorphs
Carnivores

b) Body weight

> 10 kg
< 100 g
1 - 10 kg
100 g - 1 kg

c) Habitat

Burrowing   Aquatic
Aerial
Tree Canopy
Ground
Climbing

d) Food

Omnivore
Insects
Vertebrates   Leaves
Fruit   Seeds

Fig. 14.22 Boreal to tropical changes in characteristics of mammal community. Sites are a Panamanian rain forest (9° N, 70 species), a deciduous temperate forest in Michigan (42° N, 35 species), and a boreal coniferous forest in Alaska (65° N, 15 species). (Based on data from Flemming, 1973, figure from Deshmukh, 1986.)

primates. Third, the home range area of primate groups is usually much larger in the Neotropics than in the Old World. The smaller average size, the presence of very small primates, and the lower total biomass of New World primates are consistent with lower productivity and perhaps greater seasonality of productivity in New World rain forests. This apparent positive relationship between productivity (both primary and

Table 14.8. *Biomass by trophic class in some New and Old World primate communities (from Terborgh, 1983)*

	Folivore-frugivore	Frugivore-folivore	Frugivore	Frugivore-folivore-insectivore	Frugivore-insectivore	Gum-insects	Total Biomass
New World Localities							
Cocha Cashu, Peru		2.00	1.75		2.75	.01	6.51
Barro Colorado Is., Panama[b]		3.65	.05		.50		4.20
Old World Localities							
Kibale, Uganda[c]	20.10		1.85		3.30		25.25
Morondava, Malagasy Rep.[d]	22.00	4.00			.80	.40	27.20
Polonnaruwa, Sri Lanka[e]	15.00	12.00		2.50	.25		29.75
Kuala Lompat, Malaysia[f]	2.20	6.00		2.50			10.70
Kutai Reserve, Kalimantan[g]	1.20	.80	.70	2.00			4.70

[a] Classes based on feeding times: Folivore-frugivore if > 50% leaves; frugivore-folivore if > 50% fruit; frugivore if < 10% leaves; insectivorous if > 10% insects.
[b] From Eisenberg and Thorington, 1973.
[c] From Struhsaker and Leland, 1979. The data represent five common species; data are lacking for two additional uncommon species, *Pan troglodytes* and *Cercopithecus l'hoesti*
[d] From Hladik, 1979
[e] From Hladik, 1975
[f] From Chivers, 1973, and Clutton-Brock and Harvey, 1977
[g] From Rodman, 1978

secondary) and primate diversity may be an example of the trophic shift of diversity along a gradient of primary productivity.

It is important to note that not all tropical forests have a high diversity of animals. Low productivity forests on poor or very poor soils often have a low biomass and diversity of animals, particularly vertebrates, even though the plant diversity may be relatively high (Emmons, 1984). Bates (1864) remarked on the silence and low density of animal life of the lowland primary forest in the vicinity of Pará, and Malcolm (1990) discusses the low density of mammals in the central Amazonian forests near Manaus, Brazil.

Physical Heterogeneity

The great structural and resource heterogeneity provided by plants is the principal reason for the high animal diversity in the tropics. This, coupled with the relatively smaller size of tropical animals, results in local environmental heterogeneity that is effectively much higher than in the temperate zone. Estimates of 10 to 30 species of animals that are specific to each species of tree in the humid tropics (Raven, 1976) exemplify how the high plant diversity magnifies animal diversity. Duellman and Pianka (1990) speculate that the high diversity of frogs in the Neotropics, in comparison with the Old World tropics, is an ecological and evolutionary consequence of the more aseasonal climate of the Neotropics, and the presence of bromeliads, which provide an abundant and diverse habitat.

High habitat diversity also contributes to high animal diversity, particularly among smaller animals. The high diversity of birds (Donahue *et al.*, 1989) and butterflies (Lamas, 1985) in the Tambopata Reserve in Peru has been attributed to the high habitat diversity, which has been documented by botanical surveys (Gentry, 1988). For insects and other small animals, the diversity of plant species is more important than the physical heterogeneity of the environment. At a species-poor subtropical site in south Brazil (Benson, 1978) 'niche separation' is based on habitat and not host species, so a reduced number of *Heliconius* butterflies can coexist.

The great variety of food resources for insects, birds, and mammals provided by plants is directly related to the high diversity of these animals found in the tropics (e.g., Pratt and Stiles, 1985). In the proximate sense, many of these animals can be considered interstitial species, totally dependent on the plants but having relatively minor direct effects on the plants. However, in the cases of obligate pollinators and seed dispersers, mutually beneficial adaptations that are the apparent result of coevolution demonstrate that animals can play a critical role in plant speciation (e.g., Ehrlich and Raven, 1963; Stiles, 1975; Regal, 1977; Temple, 1977; Wiebes, 1979; Bawa, 1980; Wheelwright and Orians; 1982; Crepet, 1983; Janson, 1983; Howe and Smallwood, 1982; Howe, 1986; Herrera, 1985).

Physical heterogeneity is less important for maintaining diversity of large mobile animals such as primates. For large, mobile animals, productivity of their food source is the critical factor that allows high diversity. Plant diversity is thus less important than plant productivity. In the Cocha Cashu forest on the Manu river in Peru, less than 1% of the plant species provide over 90% of the food for the primate community (Terborgh, 1983). During much of the year primates are concentrated in the low diversity, high productivity vegetation in fertile areas near the river.

Effects of Animals on Plant Diversity

While the emphasis in this chapter, and the entire book, has been on explaining the diversity of plants and using plant structure and diversity to explain animal diversity, it is obvious that animals do have many significant effects on plants. In general, these effects fall into two main classes. First are beneficial effects such as pollination, seed dispersal, and occasionally protection from the harmful effects of other animals. The coevolutionary adaptation of plants and animals to one another provides some of the most fascinating and convincing examples of natural selection, and are the topic of numerous books on the subject (Futuyma and Slatkin, 1983; Howe and Westley, 1988). Second are the negative effects of animals on plants, which generally take the form of herbivory, seed predation, or other actions that cause damage or mortality. These negative effects are of particular interest because of the influence they can have on plant growth and competition, and thus on the structure of plant communities and the animals that depend on them.

Over short time periods, herbivory can result in a significant reduction in the growth and reproductive output of rainforest plants (Marquis, 1984). Over longer time periods, herbivory can lead to an evolutionary response of increased energy expenditures for defensive chemicals (McKey, 1979; Coley *et al.*, 1985) with a corresponding reduction in growth and reproductive output. Such reductions in growth rate can reduce the rate of competitive displacement among competing plants and allow higher diversity within the plant community.

Seed predation, which can reduce the rate of population growth of plants, has been suggested as a mechanism that could contribute to the high diversity of tropical plants (Janzen, 1969, 1970; Connell, 1971). Populations of seed predators, primarily insects such as weevils, build up rapidly with an abundant food source, and consequently may affect the abundant species more than rarer species. Any density-dependent processes such as seed predation, mortality caused by pathogens (Augspurger, 1983a,b; Augspurger and Kelly, 1984), or intraspecific competition could clearly contribute to the maintenance of high species diversity. The potential importance of this type of process is illustrated by the change in spatial pattern from clumped or random for small size classes of some rain forest tree species to a more uniform pattern at larger sizes (Clark and Clark, 1984; Sterner *et al.*, 1986; Hubbell and Foster, 1990).

Four Neotropical Rain Forests: A Comparison

A recently published compendium of data from four of the most thoroughly studied neotropical rain forest sites (Gentry, ed., 1990a) offers the opportunity to examine the patterns of species richness among different groups of plants and animals. In spite of the research effort that has gone into these sites, comparisons are still difficult because of missing data, different methods, and different sample sizes. Nonetheless, the available information makes an attempt at some synthesis irresistible.

A summary of species richness and other ecosystem properties based on information in Gentry (ed., 1990) is presented in Table 14.9. The four sites are arranged in order of presumed soil fertility and productivity, with the two Amazonian sites representing the extremes. The 25% higher basal area of the upper Amazonian site (Cocha Cashu, Manu National Park) is presumably a consequence of the fertile alluvial soils deposited by the whitewater Manu river. The basal areas of the other three sites are indistinguishable, although other evidence supports the hypothesized fertility gradient.

Tree species richness differs little between the upper and central Amazonian sites, in spite of extreme differences in soil fertility. This appears to contradict the predictions of the dynamic equilibrium model, as well as the intermediate productivity hypothesis (Grime, 1973). However, the spatial heterogeneity and disturbance frequency of the Manu floodplain site (Cocha Cashu) are certainly much higher than at the central Amazonian sites near Manaus (Ducke, MCSE). These two differences would be

Table 14.9. Summary of community and ecosystem properties of four neotropical rainforest sites. The climate diagrams indicate the seasonal pattern of precipitation (upper line, right scale) and temperature (lower line, left scale). Axes are scaled so that precipitation exceeds evapotranspiration when precipitation line is above temperature line. Potential water deficits occur briefly at B.C.I. and Manaus (black area below horizontal temperature line).

	Cocha Cashu / Manu National Park	La Selva	Barro Colorado	Reserva Ducke / Minimum Critical Size of Ecosystems Project
	Upper Amazon, Peru	Costa Rica, Central America	Panama, Central America	Central Amazon, Manaus, Brazil
Climate Diagrams				
Annual Precipitation (mm)	2028	3994	2656	2186
Soil Characteristics	Rich Alluvial Deposits	Oxisols on Volcanic Sediments	Oxisols on Volcanic Sediments	Nutrient-poor Yellow Latosols
Tree Basal Area (m² ha⁻¹)	~37	~30	~30	~30

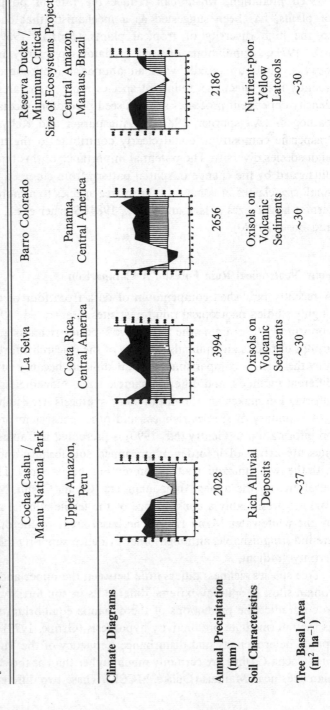

Tree Density (individuals ha^{-1})				
>10 cm dbh	~650	~450	~160	~215
>20 cm dbh				
Tree Species ha^{-1}				
>10 cm dbh	201 (155-283)	100 (88-118)	93 (76-116)	179 (>15 cm dbh)
>20 cm dbh			126	
>30 cm dbh	42		37	
Total Plant Species	1,307	1,668	1,320	825
Piper	25	44	21	<6
Psychotria	14	38	20	<6
Philodendron	12	31	13	<6
Anthurium	10	25	12	<6
Miconia	13	25	14	12
Inga	26	16	18	17
Ficus	34	15	16	<6
Pouteria	21	<6	<6	10
Plant Habitat Groups (percentage of total flora)				
Epiphytes	10	25	16	6
Trees	29	19	22	67
Herbs, Shrubs, Treelets	47	42	41	17
Climbers	9	11	20	7

Table 14.9. (*continued*)

	Cocha Cashu / Manu National Park Upper Amazon, Peru	La Selva Costa Rica, Central America	Barro Colorado Panama, Central America	Reserva Ducke / Minimum Critical Size of Ecosystems Project Central Amazon, Manaus, Brazil
Total Herp Species	131	134	133	133
Amphibians	77	48	52	48
Reptiles	54	86	81	89
Forest Bird Species	332	244	251	300
Insectivores	163	111	123	157
Omnivores	69	66	66	66
Frugivores	58	29	31	40
Insectivore-Nectarivores	14	18	12	12
Carnivores	26	15	18	21
Other	2	5	1	4
Nonflying Mammal Species	70	50	39	50
Carnivora	12	14	5	7

Based on data cited in Gentry, ed. (1990).

expected to allow maintenance of higher diversity among competing trees at the more fertile Manu site than would be expected without disturbance by the meandering river. Thus, these two sites may represent different dynamic equilibria along the diagonal of Fig. 5.5, with the increase in the rate of competitive displacement balanced by an increase in disturbance frequency.

Biogeographical considerations, based on the small total area of Central American rain forest and the distance of these sites from the center of diversity and radiation of most rainforest species in the Amazon basin, may explain the lower tree diversity of the Central American sites in comparison to the Amazonian sites (Gentry, 1990b). However, the relatively high soil fertility and lower frequency of major disturbances (neither La Selva nor Barro Colorado are affected by meandering rivers) are also consistent with the lower diversity of the Central American sites. Although tree diversity at La Selva, Costa Rica is lower than the Amazon sites, the large number of epiphyte and understory species results in a total species richness that is higher than many Amazonian sites (Gentry, 1990b).

The patterns of species richness show strong contrasts among different plant functional types (life forms, habitat groups). Epiphytes show the expected pattern of increasing species richness with increasing precipitation amount and decreasing seasonality. Likewise, the species richness of predominantly understory genera (e.g., *Piper, Psychotria, Philodendron, Anthurium, Peperomia, Thelypterus, Calathea, Heliconia*) also increases with increasing precipitation, possibly reflecting higher shade tolerance at the wetter sites.

Comparisons of vertebrate species richness are complicated by the relatively large human impact of hunting and habitat fragmentation at all sites except Manu National Park in Peru, where animal biomass and diversity would be predicted to be highest on the basis of the high primary productivity. Indeed, Manu is notable for its high density and diversity of mammals, including abundant predators that probably reduce the density of some of their prey species (Janson and Emmons, 1990). The low mammal density and diversity of the central Amazonian forests around Manaus (including Reserva Ducke and the MCSE) has been well documented, and is often attributed to the poor soils and low primary productivity (Emmons, 1984; Malcolm, 1990). The low number of carnivores on Barro Colorado can be attributed to poaching and the small size of the island (Glanz, 1990).

Few good comparative data are available for bird densities. The number of forest bird species is similar at both Amazonian sites, which are both higher than at the Central American sites (Karr et al., 1990). The total number of reptiles and amphibians is virtually identical at all four sites. However, it is reassuring that the relative proportion of reptiles and amphibians at the Amazonian sites conforms with expectations based on the lower seasonality and strong riparian influence at Manu National Park (Duellman, 1990).

Although much remains to be learned about the biota and ecology of these four sites, the data that are available are consistent with some of the concepts discussed in this book. The positive relationship of soil fertility to the density of mammals and the species richness of both birds and mammals at the two Amazonian sites contrasts with the absence of any difference in tree species richness. Although the higher fertility of the Manu site would be expected to lead to lower species richness of trees, the higher habitat heterogeneity and disturbance frequency caused by the river apparently compensate for a higher rate of competitive displacment. It is interesting to note that the floodplain forests along the Manu river are dominated by two species, a fig, *Ficus insipida* and a mahogany, *Cedrela odorata*, that achieve great size and dominate the forest for approximately 100 years. Foster (1990) estimates that the meandering river undercuts most forest before it gets older than 200 years. The highest diversity forest at Manu is found on upland areas away from the river, where soil fertility may be lower, and the forest regenerates primarily through treefalls and gap dynamics (Foster, 1990; Gentry and Terborgh, 1990).

Primary Productivity: A Tropical-Temperate Comparison

The importance ascribed to primary productivity in regulating plant species diversity within the humid tropics, as well as in determining patterns of plant endemism in North America, mediterranean climate regions, and the humid tropics, suggests that this mechanism could also be an explanation for the celebrated temperate-tropical diversity gradient. In fact, I believe that this explanation is the single most consistent difference between the temperate zone and the tropics, and is likely to explain much of the difference in both within-habitat and geographical diversity between these two regions (Huston, 1979).

The argument that plant productivity is lower in the tropics than in the temperate zone seems counterintuitive, given the popular images of lush, green rain forests with their 'vegetative frenzy' of plant growth. In

Table 14.10. *Estimates of average annual and monthly net primary productivity (NPP) for the world's main forest types.*

Vegetation Type	Annual NPP (t/ha/yr)	Growing Season (Months)	Monthly NPP (t/ha/mo)
Boreal Forests[a]			
Main and			
Southern Taiga	4.19	3	1.40
Other Conifer	6.00	3	2.00
Temperate Forests[a]			
Broad-leafed	6.04	6	1.01
Mixed	5.93	6	0.99
Tropical Forests[a]			
Broad-leafed	8.00	12	0.67
Dry Forest			
and Woodland	5.72	8	0.72
Savannah and			
Woodland	4.91	6	0.82
Montane Dwarf			
Woodland	6.67	12	0.56
Boreal Forest[b]	8(4-20)	3	2.7(1.3-6.7)
Temperate Forest[b]	13(6-30)	6	2.2(1.0-5.0)
Tropical Forest[b]	20(10-50)	12	1.7(0.8-4.2)

[a] Based on Olson *et al.* (1983).
[b] Based on Whittaker and Likens (1975).

fact, it is only recently that quantitative data have begun to accumulate that substantiate this difference, which has been suspected by some scientists for many years (Jordan 1971a,b; Willson, 1973; Jordan and Murphy, 1975; Kato *et al.*, 1974; Jordan, 1983). Dawkins (1959, 1964) observed that the natural basal areas of 'tropical high forest' were quite low, and the volume increments, except in the case of some plantation species grown at high elevations or higher latitudes, were not sufficient to support profitable forestry. Dawkins suggested that high respiration rates in the hot climate might contribute to the low productivity of most tropical forests.

It is certainly true that under favorable conditions of nutrient, moisture, and temperature, tropical crops such as sugarcane can be tremendously productive (8300 g dry matter m^{-2} year^{-1}) as can plantations of species such as pine and eucalyptus (Westlake, 1963). However, such high pro-

Table 14.11. *Latitudinal comparison of wood and litter production in broad-leaved mesic forest, subdivided into regions based on annual solar radiation.*

Annual Radiation (kcal cm^{-2} yr^{-1})	Forest type	Annual Production (g m^{-2} yr^{-1})	
		Wood (N)	Litter (N)
25-40	birch, beech, poplar	968±399 (14)	281±133 (23)
40-50	oak, maple, beech northern hardwoods	658±248 (11)	328±111 (24)
50-60	oak, tulip poplar poplar, basswood	610±537 (22)	367±116 (35)
60-70	subtropical, sclerophyll temperate rainforest	758±290 (3)	644±204 (20)
70	tropical rain forest, seasonal, moist forest	734±275 (10)	957±362 (33)
70	Plantations	1193±741 (15)	

Based on Jordan (1983).

ductivity is unfortunately the exception rather than the rule. Two primary factors contribute to the overall low productivity of tropical forests (and agricultural systems on cleared forest land): 1. the preponderance of highly weathered, acidic, nutrient-poor soils in the tropics (see Table 14.3); and 2. the high day-time and night-time temperatures of the lowlands which contribute to high respiration rates that use most of the carbon that is fixed by tropical plants.

Over the past thirty years, the estimates of tropical forest productivity used for calculating global carbon budgets have been steadily decreasing (Olson *et al.*, 1983) (Table 14.10). These values represent averages over vast areas and do not give any insight into the variation and extremes of productivity within any region. Nonetheless, these estimates clearly suggest that annual productivity increases from the poles toward the equator. However, this increase can be explained almost completely by length of the growing season, which is obviously much longer in the humid tropics, rather than on inherently faster growth and higher productivity during the period when plants are growing (Table 14.10). Expression of productivity in terms of growth per month of growing season demonstrates that the productivity of tropical forests is no higher, and may actually be lower than that of temperate forests.

Table 14.11. *(continued). Wood production per kcal is based on total annual radiation, not adjusted for growing season or sky conditions.*

Growing Season (Months)	Monthly Production (g m^{-2} mo^{-1}) Wood	Litter	Wood Production (g g^{-1} Leaf Litter)	Efficiency (mg kcal^{-1})
3	323	94	3.44	2.98
5	132	66	2.01	1.46
7	87	52	1.66	1.11
9	84	72	1.18	1.17
12	61	80	0.77	1.05

The ecological significance of this distinction between total annual productivity and productivity during the growing period is that plants do not interact when they are dormant. In particular, plants do not compete for light when they are leafless. The small relative differences in height that are critical for success or failure in competition for light are important only when the plants are growing and interacting. Thus, rapid growth during a short growing season potentially leads to a much higher rate of competitive displacement than slow growth over a long growing season, even though the total growth may be the same.

Compilations of available data from natural forests (Table 14.11, see also Table 14.1) show that the annual wood production of tropical rain forests is approximately equal to that of temperate forests (Jordan, 1971a,b, 1983). Wood production is actually more relevant than net primary productivity for evaluating growth rates and rates of competitive displacement since it is through investment in the physical structure of wood (and roots) that plants compete with one another. Thus, the fact that the monthly wood production of tropical rain forests is one-fifth to one-half that of northern hardwood or boreal forests is much more significant for evaluating the relative intensity of competition than is the fact that leaf (i.e., litter) production is approximately equal (Table 14.11). However, the higher annual leaf production, and its distribution throughout the year in tropical rain forests has important implications for the biomass and diversity of herbivores that can be supported.

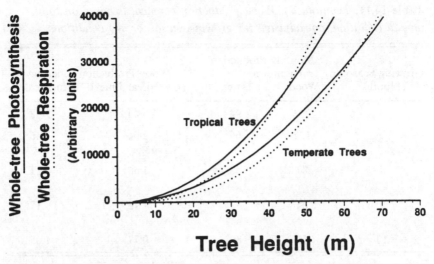

Fig. 14.23 Predicted effect of tree size on the balance between photosynthesis and respiration at temperatures characteristic of tropical (30 °C) and temperate (20 °C) forest ecosystems. This simple model assumes a Q_{10} (change in rate for every 10 °C change in temperature) for photosynthesis of 1.5 and a Q_{10} for respiration of 1.8. Photosynthetic area is assumed to increase as a squared function of height, while respiring tissue is assumed to increase as height to a power of 2.5.

Consideration of 'ecological' time rather than simply 'calendar' time is appropriate for evaluation of not only growth rates and productivity data but also of disturbance frequencies. A particular annual frequency of disturbance in a system with a 12 month growing season may represent a lower effective frequency of disturbance than the same rate in a system with a 6 month growth season. Since most studies show little difference between annual disturbance rates in tropical and temperate forests (e.g., Runkle, 1985), this interpretation suggests that tropical disturbance frequencies may actually be less than those of temperate forests.

In addition to the generally poor soils of the tropics, the unfavorable balance of respiration to photosynthesis that results from the high temperatures of the lowland tropics is a major contributor to the low growth and net primary productivity of natural forest vegetation, in spite of high gross primary productivity. Two estimates of the 'efficiency' of wood production, one based on wood production in terms of total annual radiation, and the other in terms of leaf weight (and presumably leaf area), document a dramatic **decrease** in growth efficiency as one moves toward the tropics (Jordan, 1971a; Table 14.11). These approximations are consistent with a recent study of solar energy conversion efficiencies of Amazonian rain forest stands that showed efficiencies more than an order of magnitude lower than those of some temperate forest stands (Saldarriaga and Luxmoore, 1991; Linder, 1985). Solar energy conversion efficiency inevitably decreases over the course of forest succession as the ratio of respiring biomass (e.g., trunks, branches) to photosynthetic biomass increases (e.g., Saldarriagga and Luxmoore, 1991).

The high temperatures under which tropical rain forest trees exist may contribute to the relatively short time that tropical trees survive once they have attained large size. Higher respiration rates for woody tissue in the tropics as compared to the temperate zone may push tropical trees past the break-even point for photosynthesis more quickly than temperate trees of the same size (Fig. 14.23). Although the difference would be less dramatic if tree age were expressed in 'ecological' time, the fact that respiration increases exponentially with temperature makes this phenomenon inevitable.

The high levels of precipitation that allow tropical rain forests to support large amounts of leaf area are another potentially important difference between tropical and temperate forests, because of the higher number of shade-tolerant functional types that can survive under wet conditions. However, many temperate forests have leaf area indices (LAI) as high as those of tropical rain forests (Woodward, 1987). The plant diversity of high LAI northern hardwood forests results primarily from the large number of subcanopy and understory species of shade-tolerant functional types, since the number of canopy tree species is relatively low. Although the plant diversity of some of these temperate broadleaf forests is high in relation to other temperate forests, it is still much lower than than of tropical forests. Thus, it is high diversity within plant functional types, as well as a large number of functional types that produces the extraordinary plant diversity of tropical forests.

Summary

The primary difference between high diversity tropical rain forests and other forests around the world is the low growth rate of trees in most tropical rain forests. Although contrary to common expectations, these low growth rates are the inevitable consequence of high respiration rates that result from high temperatures, and of the poor soils that result from high rates of decomposition, weathering, and leaching, all of which are caused by high temperatures and large amounts of precipitation (Fig. 14.24).

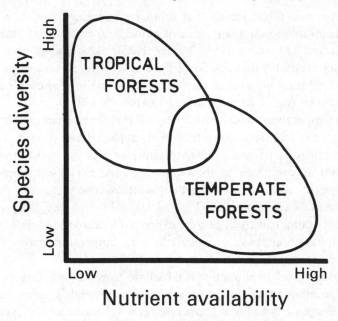

Fig. 14.24 Relationship between soil fertility and species diversity for tropical and temperate forests.

The low rates of competitive displacement coupled with the same types and frequencies of disturbance found in other forests around the world allow the nonequilibrium coexistence of many species of plants that are very similar morphologically and closely related genetically. Thus, high levels of species diversity and high rates of endemism in tropical forests are not merely the result of long periods of time for the evolution and accumulation of species, but result from the low rates of competitive displacement that allow both ancient and newly evolved species to avoid extinction and to coexist for long periods of time.

The high diversity of animals in the tropics can be attributed both to the high diversity of plants and to the low degree of seasonality which allows sufficient plant productivity over the course of the entire year to sustain large animal populations. In contrast to plant diversity, animal diversity usually increases monotonically with increasing primary productivity. High animal diversity is the consequence of processes such as character displacement, niche diversification, and ultimately speciation that occur in the context of the high diversity of physical structure and resources provided by the plants.

The same environmental conditions that result in a low rate of competitive displacement and prolonged coexistence of similar species, may also lead to high rates of speciation by reducing geneflow and dispersal. The physical environment of the tropics thus contributes to both the generation and the maintenance of high species diversity. These conditions are the direct and indirect results of those aspects of the tropical environment that are most strongly affected by the physical and climatic conditions that change along the latitudinal gradient, temperature and precipitation.

15

Concluding Comments: The Economics of Biological Diversity

An intellectual understanding of the natural regulation of biological diversity may seem like an irrelevant luxury in a world where 140 hectares of tropical rain forest are being cut with chainsaws or destroyed by bulldozers every minute, resulting in the extinction of an estimated 50 to 150 species per day (Reid and Miller, 1989). Can understanding the natural processes that influence diversity contribute to the preservation of species, or is it as useful as studying the reproductive endocrinology of the passenger pigeon? It is certainly my hope that the ideas developed in this book can make a useful contribution to the preservation of biological diversity and a wiser use of all of the world's natural resources.

There are many reasons for the destruction of natural habitats, including population pressure, urbanization, industrial development, conversion to agriculture, etc. Habitat destruction inevitably reduces biological diversity. The only defensible reason for the destruction of natural habitats is the improvement of the human condition, and in many cases the above factors can make significant contributions to improvements in the quality of life. However, there are many situations in which the conversion of natural habitats to other uses results in no improvement in the human condition, and may actually cause a deterioration in the quality of life of the people most directly affected.

By far the most extensive and biologically destructive use to which natural habitats are converted is agriculture, ranging from permanent cropping, to shifting cultivation, to conversion to pasture, to extraction of wood and other natural products. Obviously, the reason such conversions are made is that the economic productivity, and thus the contribution to quality of life, is presumed to be greater in the converted ecosystem than it is in the natural ecosystem. If natural ecosystems produced products of

558

sufficient agricultural and economic value, there would be no incentive to convert them to managed ecosystems (Plotkin, 1988).

However, there are numerous unfortunate and dramatic examples of situations in which the destruction of natural ecosystems and their conversion to agriculture did not result in an improved quality of life for those responsible, but has led to a cycle of continuing poverty and ecosystem destruction (Fearnside, 1990a). Virtually every effort to establish agricultural settlements in Brazil's Amazon basin has failed. These failures to establish settlements based on agricultural productivity almost always are the result of inherently low primary productivity of the natural ecosystem and thus of the managed agroecosystem as well.

The relationship between agricultural productivity and economics has driven patterns of human land use and population distribution throughout history, and this close relationship has only recently been distorted by politically motivated economic subsidies. The relationship between soils and the quality of life in agrarian societies is demonstrated by the spatial distribution of prehistoric and historic settlements, as well as by place names that have been preserved by geographers. For example, in the Ridge and Valley Geographical Province of southeastern North America, thrust faulting has produced a region of long, parallel ridges and valleys with geologic patterns repeated across the landscape. On maps of Grainger County, Tennessee, one can see 'Poor Valley' and the hills of the 'Poor Valley Knobs' running parallel to 'Richland Valley' and the 'Richland Valley Knobs'. Poor Valley is underlain by Devonian shales that are black, have high pyrite content, and produce an acidic nutrient-poor soil. In contrast, Richland valley is underlain by the Conosauga Group of shales and siltstones interbedded with limestone, which produce a much more fertile soil (D. Lietzke, personal communication). This pattern of alternating fertile and infertile valleys is found throughout the Ridge and Valley Province of Kentucky and Tennessee, and similar relationships between soils, local economic prosperity, and place names can undoubtedly be found throughout the world.

In the tropics, patterns of human populations have, until recently, reflected the influence of agricultural productivity. Tosi and Voertman (1964) determined the average human population densities in five different types of tropical forest ranging from very dry forest to rain forest (Fig. 15.1). Based on data from five Central American countries, the highest population densities were in Dry Forest and Moist Forest (*sensu* Holdridge, 1967), and the lowest densities in Wet Forest and Rain Forest. I believe that these patterns are not coincidental, but rather reflect an

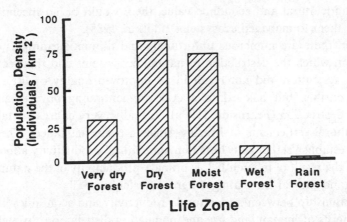

Fig. 15.1 Pattern of human population densities in the major forest types (Life Zones, *sensu* Holdridge, 1967) in five Central American countries. (From Tosi and Voertman, 1964.)

inherent and inevitable relationship between soil conditions and the economic strength of agrarian economies. While fertile soils can be found throughout the tropics, and many infertile soils can be productive if managed effectively (e.g., Sanchez and Buol, 1975; Sanchez, 1976), it is nonetheless true that most of the area of tropical rain forest is underlain by poor soils, particularly in the Neotropics.

Low agricultural productivity usually results from the same conditions that cause low primary productivity in natural ecosystems, that is, insufficient soil nutrients, insufficient or excessive water, or toxic concentrations of certain minerals. The expected relationship between **low** primary productivity and **high** species diversity of plants and lower trophic levels suggests that the success of conversion to agriculture is likely to be low in situations where species diversity is high. In general, land with high plant species diversity is not a good candidate for profitable agriculture or even for productive forestry. **Thus, there is no inherent conflict between the preservation of biological diversity and the economic improvement of the human condition.**

This suggests that the **positive** correlation between latitude and per capita income mentioned in Chapter 2 (Fig. 15.2) is not a spurious correlation, but in fact results from the same conditions that are responsible for the **negative** correlation between latitude and species diversity. If one examines the relationship between latitude and agricultural productivity, which is the ecologically relevant component of *per capita* income, one finds the same positive relationship, both for the economic value of agri-

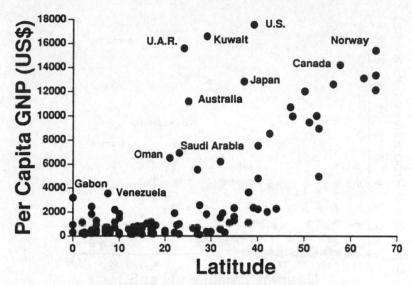

Fig. 15.2 The latitudinal gradient of per capita gross national product (GNP) in countries of the world in 1986, standardized as US dollars. (Based on data from the United Nations, summarized in World Resources 1990-1991.)

cultural productivity in terms of US dollars per hectare (Fig. 15.3) and for the weight of tuber and root crops produced per hectare (Fig. 15.4). The outliers in this distribution include both countries with soils that have naturally high fertility, such as El Salvador on the volcanic cordillera of Central America, and countries with highly subsidized agriculture, such as Israel, the Yemen Arab Republic, the Netherlands, and Belgium. The same pattern of lowest productivity in the tropics, with maximum productivity in the temperate zone (and a decline at the highest latitudes) is still apparent when the data are stratified by total precipitation (Figs 15.3 and 15.4).

Apparent exceptions to the general pattern of high biological diversity associated with low agricultural productivity include some of the most productive farmland in the world, that of the midwestern part of the United States. However, most of the apparent exceptions to this general pattern are still consistent with the dynamic equilibrium model of species diversity. The high diversity of the prairie grass and herb communities was maintained only because of the frequent fires that eliminated woody vegetation.

Thus, it should be possible to preserve most of the biological diversity contained in tropical rain forests (and other low productivity ecosystems)

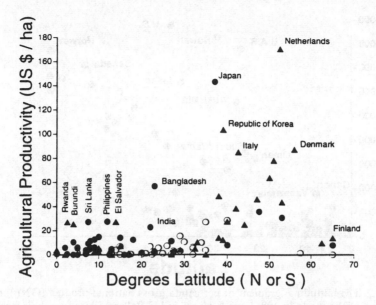

Fig. 15.3 The latitudinal gradient of agricultural productivity in terms of value in US dollars per hectare. Triangles indicate countries with annual precipitation between 55 and 130 cm/yr. Open circles indicate countries with annual precipitation less than 55 cm/yr. Closed circles indicate countries with annual precipitation greater than 130 cm/yr. (Based on data from the United Nations, summarized in World Resources 1990-1991.)

without imposing the economic burden that would result from sacrificing a significant amount of agricultural productivity. This trade-off of high productivity/low diversity land for low productivity/high diversity land can be achieved on a local scale and, indeed, is to some extent the inevitable result of the unimpeded functioning of an agriculturally based market economy.

However, in a truly global economy, it should be possible to sacrifice high productivity/low diversity temperate ecosystems to agricultural conversion in order to subsidize the preservation of a proportionally much larger area of low productivity/high diversity tropical ecosystems. To some extent this is already occurring as a result of international foreign aid and the activities of conservation organizations based in high latitude countries (Tangley, 1990). However, if this type of exchange can be seen as an economically profitable motivation for preserving biological diversity there should be a greater chance that such policies will be adopted. Understanding the relationship of agricultural and ecosystem productiv-

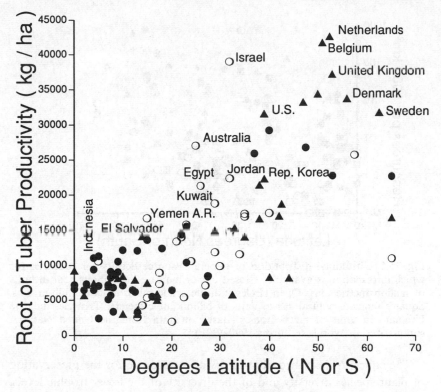

Fig. 15.4 The latitudinal gradient of agricultural productivity in terms of weight of root and tuber crops produced per hectare. Open circles indicate countries with annual precipitation less than 55 cm/yr. Triangles indicate countries with annual precipitation between 55 and 130 cm/yr. Closed circles indicate countries with annual precipitation greater than 130 cm/yr. (Based on data from the United Nations, summarized in World Resources 1990-1991.)

ity to biological diversity is a key step in establishing a successful global program for the preservation of biodiversity.

The argument that there is a deterministic relationship between ecosystem productivity and national economies should not be taken to mean that countries with high species diversity are doomed to have weak economies. Agricultural productivity is only one element of a national economy, and is generally a very small component in industrialized economies. In contrast to the latitudinal gradient of agricultural productivity, there is no latitudinal gradient of mineral resources (Fig. 15.5). Tropical countries can obviously have robust economies, and it is possible for a healthy industrial and service economy to subsidize high agricultural productivity even on relatively poor soils.

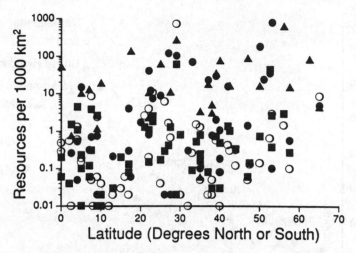

Fig. 15.5 Latitudinal distribution of mineral resource density for countries for which information is available. Closed circles indicate bituminous coal in units of million metric tons. Open circles indicate crude oil in units of million tons. Squares indicate natural gas in units of billion cubic meters. Triangles indicate Uranium in units of metric tons. (Based on data from the United Nations, summarized in World Resources 1990-1991.)

The concepts described above provide a rationale for the preservation of plant species diversity and of the diversity of the lower trophic levels, which are strongly affected by the spatial and resource heterogeneity created by the plants. However, this type of program will contribute much less to the preservation of higher trophic levels. Large predators are among the most endangered species on Earth, as are large animals in general. A special, heavily subsidized effort will have to be made to preserve these species, since their preservation almost inevitably involves a costly trade-off with agricultural productivity and human quality of life. Genetically viable populations of these species could potentially be maintained through *ex situ* conservation, possibly with the periodic release of animals from zoos or other captive breeding programs into the wild to maintain genetic variability (Reid and Miller, 1989). It will simply not be possible to save all species.

In a few cases, economic exploitation undoubtedly results in increases in biological diversity, through the reduction of very abundant species and a resulting increase in the evenness component of species diversity. This has happened in many of the world's commercial fisheries, and is reflected by an increased diversity of fish species available in markets and an increased number of fish recipes in cookbooks. Obviously,

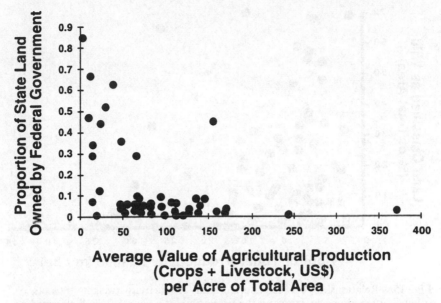

Fig. 15.6 Relationship of federal land ownership to the agricultural productivity of land in each state of the United States (total value of crop and livestock sales in 1987 divided by total land area of state). (Based on data from US Bureau of the Census, 1990.)

overexploitation of fisheries or any renewable resource is neither an economically sound policy nor a policy favorable to the long-term maintenance of biological diversity (Dasman *et al.*, 1973; Clark and Hollings, 1985; Clark, 1989).

The preservation of biological diversity through the set-aside of low productivity lands presently occurs in regions in which trade between different geographic units allows the subsidy of populations in low productivity regions by the excess economic productivity of high productivity regions. In low productivity regions it is possible to set aside large areas of land with relatively little effect on economic quality of life. This is clearly demonstrated by the pattern of land ownership in the United States (Fig. 15.6). The actual agricultural productivity of land in the US varies across three or perhaps four orders of magnitude. In regions of extremely high agricultural productivity, such as Iowa and Illinois, very little land is publically owned. In contrast, large proportions of western states, with extremely low agricultural productivity as a result of low precipitation, are owned by the federal government as National Parks, National Forests, and National Grasslands. While not all federal

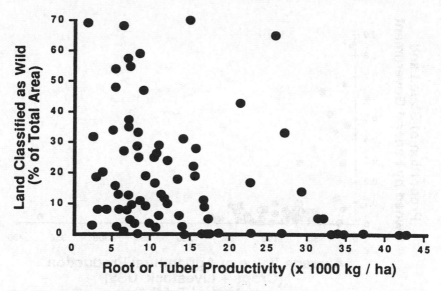

Fig. 15.7 Relationship between Agricultural Productivity (weight of roots or tubers harvested per hectare) and the proportion of a country's land classified as 'wild'. (Based on World Resources, 1990.)

lands receive sufficient protection to prevent degradation and the loss of biodiversity, in most cases public ownership reflects a much higher degree of protection for biodiversity than private ownership. Careful analysis of data such as that presented in Fig. 15.6 may reveal economic criteria that can be used to justify preservation of land for the maintenance of biodiversity. Figure 15.6 suggests that agricultural productivity of less than $40/acre is economically compatible with preservation (i.e., public ownership) of 10% to 60% of the total land area. Productivity of $50 to $150 can support set-aside of up to 10% of the land area. Land productivity in excess of $150/acre makes it difficult to justify preservation of more than a few per cent of the total land area. Figure 15.7 shows the same type of information on a global level, where subsidization of preserved areas in prosperous countries is evident.

One unfortunate consequence of the economic regulation of land use is that nearly all the productive ecosystems in the world have been destroyed and converted to agriculture. In case after case around the world, the productive lowland and alluvial areas have been almost completely converted to agriculture and the indigenous flora and fauna nearly eliminated, while the parks and nature preserves are located in the unproductive uplands, or on dry, infertile, hilly, or rocky sites that are

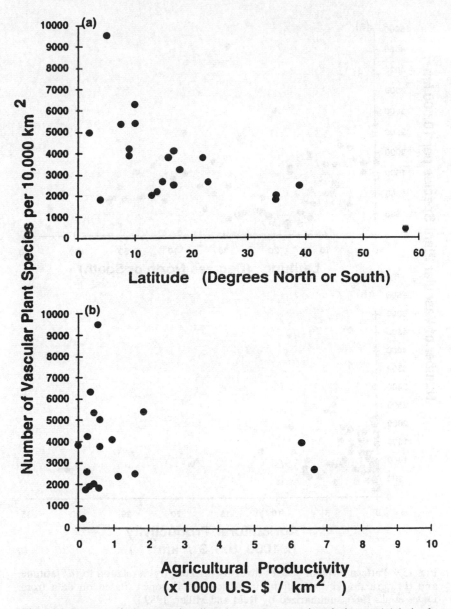

Fig. 15.8 Pattern of plant species diversity per country in relation to (*a*) latitude and (*b*) agricultural productivity in North, Central, and South America. (Based on data from Davis *et al.*, 1986, summarized by Reid and Miller, 1989.)

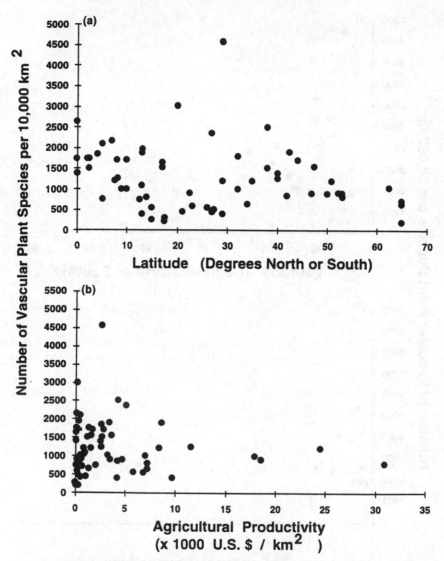

Fig. 15.9 Pattern of plant species diversity per country in relation to (*a*) latitude and (*b*) agricultural productivity in Africa and Europe. (Based on data from Davis et al., 1986, summarized by Reid and Miller, 1989.)

Asia and Australia

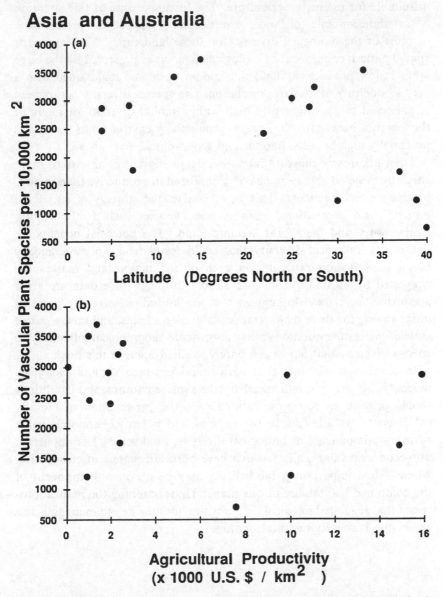

Fig. 15.10 Pattern of plant species diversity per country in relation to (a) latitude and (b) agricultural productivity in Australia, Southeast Asia and Asia. (Based on data from Davis *et al.*, 1986, summarized by Reid and Miller, 1989.)

unsuitable for intensive agriculture. The fortunate side of this situation is that the economics of land conversion have allowed the preservation of most of the biological diversity on these landscapes. The landscape spatial pattern component of biological diversity is preserved by the large areas of low productivity land that are set aside and maintained with a very low density of human population. The species diversity component is preserved by the inherently high within-habitat diversity of many of the low (not necessarily the lowest) productivity environments, as well as potentially high between-habitat and geographical diversity.

These arguments suggest that conservation efforts will have to be more carefully targeted and more heavily subsidized in productive than in non-productive environments. This is, of course, the strategy of successful national and international conservation agencies such as the Nature Conservancy and the World Wildlife Fund. The potential benefits for conserving biological diversity that could result from an economically based conservation program on each of the major land masses are suggested by Figs 15.8, 15.9, and 15.10. Although these data are gross approximations, they do suggest that the latitudinal diversity patterns differ among the three major transequatorial gradients, and raise a whole suite of interesting questions about large-scale geographical influences on species diversity that have been barely touched upon in this book.

In conclusion, the fact that agricultural productivity and biological diversity are strongly influenced by the same environmental conditions **should** provide an economic rationale for the preservation of biological diversity. In addition to the realized and potential economic value of many components of biological diversity, biodiversity has important non-economic values, all of which have been adequately discussed elsewhere. Most importantly, biodiversity may be a critical component of the continued habitability of our planet. Understanding the natural processes that regulate biological diversity provides the key to managing and preserving humanity's natural heritage.

References

Ab'Saber, A.N. (1982). The Paleoclimate and paleoecology of Brazilian Amazon. In *Biological Diversification in the Tropics*, ed. G.T. Prance, pp. 41-59. New York: Columbia University Press.

Abbott, R.T. (1968). *Seashells of North America* New York: Golden Press.

Abele, L.G. (1974). Species diversity of decapod crustaceans in marine habitats. *Ecology*, 55, 156-61.

Abele, L.G., and Walters, K. (1979). Marine benthic diversity: a critique and alternative explanation. *J. Biogeogr.*, 6, 115-26.

Aber, J.D., Botkin, D.B. and Melillo, J.M. (1979). Predicting the effects of different harvesting regimes on productivity and yield in northern hardwoods. *Canadian Journal of Forest Research*, 9, 10-4.

Aber, J.D., Melillo, J.M., and Federer, C.A. (1982). Predicting the effects of rotation length, harvest intensity, and fertilization on fiber yield from northern hardwood forests in New England. *Forest Science*, 28, 31-45.

Aber, J.D., and J.M. Melillo. (1980). Litter decomposition: Measuring relative contributions of organic matter and nitrogen to forest soils. *Canadian Journal of Botany*, 58, 416-21.

Aber. J.D., and J.M. Melillo. (1982a). Nitrogen immobilization in decaying hardwood leaf litter as a function of initial nitrogen and lignin content. *Canadian Journal of Botany*, 60, 2263-9.

Aber. J.D., and Melillo, J.M. (1982b). FORTNITE: a computer model of organic matter and nitrogen dynamics in forest ecosystems. University of Wisconsin Research Bulletin R3130.

Abrams, M.D. (1986). Historical development of gallery forests in northeast Kansas. *Vegetatio*, 65, 29-37.

Abrams, M.D. (1992). Fire and the development of oak forests. *BioScience*, 42, 346-53.

Abrams, M.D., and Hulbert, L.C. (1987). Effect of topographic position and fire on species composition in tallgrass prairie in northeast Kansas. *American Midland Naturalist*, 117, 442-5.

Abrams, M.D., and Nowacki, G.J. (1992). Historical variation in fire, oak recruitment and post-logging accelerated succession in central Pennsylvania. *Bulletin of the Torrey Botanical Club*, 119, 19-28.

Abrams, P. (1976). Limiting similarity and the form of the competition coefficient. *Theoretical Population Biology*, 8, 356-75.

571

Abrams, P.A. (1978a). The nonlinearity of competitive effects in models of competition for essential resources. *Theor. Pop. Biol.,* 32, 50-65.

Abrams, P.A. (1978b). The functional responses of adaptive consumers of two resources. *Theor. Pop. Biol.,* 32, 262-88.

Abrams, P.A. (1983). The theory of limiting similarity. *Annual Review of Ecology and Systematics,* 14, 359-76.

Abrams, P.A. (1986). Character displacement and niche shift analyzed using consumer-resource models of competition. *Theor. Pop. Biol.,* 29, 107-60.

Abrams, P.A. (1987a). Alternative models of character displacement and niche shift. 2. Displacement when there is competition for a single resource. *Am. Nat.,* 130, 271-82.

Abrams, P.A. (1988). Resource productivity - consumer species diversity: simple models of competition in spatially heterogeneous environments. *Ecology,* 69, 1418-33.

Abramsky, Z., and Rosenzweig, M.L. (1983). Tilman's predicted productivity-diversity relationship shown by desert rodents. *Nature,* 309, 150-1.

Abul-Atta, A.A. (1978). *Egypt and the Nile after the Construction of the Aswan High Dam.* Cairo: Ministry of Irrigation and Land Reclamation.

Acocks, J.P.H. (1975). Veld types of South Africa. *Memoirs of the Botanical Survey of South Africa, 40.* Pretoria: Government Printer.

Adams, C.C. (1902). Southeastern United States as a center of geographical distribution of fauna and flora. *Biological Bulletin,* 3, 115-31,

Adams, C.C. (1905). The post-glacial dispersal of the North American Biota. *Biological Bulletin,* 9, 53-71.

Adams, J.M., and Woodard, F.I. (1989). Patterns in tree species richness as a test of the glacial extinction hypothesis. *Nature,* 339, 699-701.

Adams, S.M., and DeAngelis, D.L. (1987). Indirect effects of early bass-shad interactions on predator population structure and food web dynamics. In *Predation in Aquatic Communities,* ed. W.C. Kerfoot and A. Sih, pp. 103-17. Hanover, New Hampshire: University Press of New England.

Aerts, R., and Berrendse, F. (1988). The effect of increased nutrient availability on vegetation dynamics in wet heathland. *Vegetatio,* 76, 63-9.

Ahti, T. (1977). Lichens of the boreal coniferous zone. In *Lichen Ecology,* ed. M.R.D. Seaward. pp. 147-81. London: Academic Press.

Aikman, D.P., and Watkinson, A.R. (1980). A model for growth and self-thinning in even-aged monocultures of plants. *Annals of Botany (London),* 45, 419-27.

Al-Mufti, M.M., Sydes, C.L., Furness, S.B., Grime, J.P., and Band, S.R. (1977). A quantitative analysis of shoot phenology and dominance in herbaceous vegetation. *Journal of Ecology.* 65, 759-91.

Albertson, F.W. (1937). Ecology of mixed prairie in west central Kansas. *Ecological Monographs,* 7, 481-547.

Aldous, A.E. (1934). Effect of burning on Kansas bluestem pasture. *Kansas Agricultural Experiment Station Technical Bulletin,* 38, 1-65.

Alexander, V. (1974). Primary productivity regimes of the nearshore Beaufort Sea, with reference to the potential roles of the ice biota. In *The Coast and Shelf of the Beaufort Sea.* ed. J.C. Reed and J.E. Sater. pp. 609-26. Calgary, Canada: Arctic Institute of North America.

Allan, H.H. (1936). Indigene versus alien in the New Zealand plant world. *Ecology,* 17, 187-93.

Allan, H.H. (1937). The origin and distribution of the naturalized plants of New Zealand. *Proceedings of the Linnean Society London,* 150, 25-46.

Allendorf, F.W., and Leary, R.F. (1986). Heterozygosity and fitness in natural populations of animals. In *Conservation Biology: the Science of Scarcity and Diversity,* ed. M.E. Soule', pp. 57-76. Sunderland, Massachusetts: Sinauer.

Alvarez, L.W., Alvarez, W., Asaro, F., and Michel, H.V. (1980). Extraterrestrial cause for the Cretaceous-Tertiary extinctions: experiment and theory. *Science,* 208, 1095-108.

Alvarez, W., Alvarez, L.W., Asaro, F., and Michel, H.V. (1982). Current status of the impact theory for the terminal Cretaceous extinction. In *Geological Implications of Impacts of Large Asteroids and Comets on the Earth. Geological Society of America Special Study,* 190, 305-15.

Alvim, P. de T., (1978). Perspectives de producaoagricola naregiao amazonica. *Intersciencia,* 3, 243-51.

Anderson, A.B. (1981). White sand vegetation of Brazilian Amazonia. *Biotropica,* 13, 199-210.

Anderson, A.B., and Benson, W.W. (1980). On the number of tree species in Amazonia. *Biotropica,* 12, 235-7.

Anderson, A.N., and Lonsdale, W.M. (1990). Herbivory by insects in Australian tropical savannas: a review. *J. Biogeogr.,* 17, 433-44.

Anderson, B.W., Ohmart, R.D., and Rice, J. (1983). Avian and vegetation community structure and their seasonal relationships in the lower Colorado River Valley. *Condor,* 85, 392-405.

Anderson, G.V.R., Ehrlich, A.H., Ehrlich, P.R., Roughgarden, J.D., Russell, B.C., and Talbot, F.H. (1981). The community structure of coral reef fishes. *American Naturalist,* 117, 476-95.

Anderson, H.E. (1970). Forest fuel ignitability. *Fire Technology,* 6, 312-9,322.

Anderson, R.C. (1972). The ecological relationships of meningeal worm and native cervids in North America. *Journal of Wildlife Disease,* 8, 304-10.

Andrewartha, H.A., and Birch, L.C. (1954). *The Distribution and Abundance of Animals.* Chicago: University of Chicago Press.

Archibald, E.E.A. (1949). The specific character of plant communities. II. A quantitative approach. *J. Ecol.,* 37,274-88.

Armstrong, R.A., and McGehee, R. (1976a). Coexistence of species competing for shared resources. *Theoretical Population Biology,* 9, 317-28.

Armstrong, R.A., and McGehee, R. (1976b). Coexistence of two competitors on one resource. *Journal of Theoretical Biology.* 56, 499-502.

Arnett, R.H. (1985). *American Insects: A Handbook of the Insects of America North of Mexico.* New York: Van Nostrand Reinhold.

Arnold, S.J. (1972). Species densities of predators and their prey. *American Naturalist,* 106, 220-36.

Arquit, A.M. (1990). Geological and hydrothermal controls on the distribution of megafauna in Ashes vent field, Juan de Fuca Ridge. *Journal of Geophysical Research,* 95, 12,947-60.

Aschmann, H. (1973). Distribution and peculiarity of Mediterranean ecosystems. In *Mediterranean Type Ecosystems, Origin and Structure,* ed. F. di Castri and H.A. Mooney, pp. 11-9. Berlin: Springer-Verlag.

Ashton, P.S. (1964). Ecological studies in the mixed Dipterocarp forests of Brunei state. *Oxford Forest Memoirs,* Number 25.

Ashton, P.S. (1971). The plants and vegetation of Bako National Park. *Malay Nature Journal,* 24, 151-62.

Ashton, P.S. (1977). A contribution of rain forest research to evolutionary theory. *Annals of the Missouri Botanical Garden,* 64, 694-705.

Ashton, P.S. (1988). Dipterocarp biology as a window to the understanding of tropical forest structure. *Annual Review of Ecology and Systematics,* 19, 347-70.

Ashton, P.S., et al. Comparative ecological studies in the mixed Dipterocarp forests of northern Borneo. III. Patterns of species richness. (in preparation).

Atjay, G.L., Ketner, P., and Duvigneaud, P. (1979). Terrestrial primary production and phytomass. In *The Global Carbon Cycle. SCOPE 13,* ed. B.Bolin, E.T. Degens, S. Kempe, and P. Ketner, pp. 129-81. Chichester: Wiley.

Auclair, A.N., and Goff, F.G. (1971). Diversity relations in upland forest in the western Great Lakes area. *American Naturalist,* 105, 499-528.

Augspurger, C.K. (1983a). Offspring recruitment around tropical trees: changes in cohort distance with time. *Oikos,* 20, 189-96.

Augspurger, C.K. (1983b). Seed dispersal of the tropical tree *Platypodium elegans,* and the escape of its seedlings from fungal pathogens. *Journal of Ecology,* 71, 759-71.

Augspurger, C.K., and Kelly, C.K. (1984). Pathogen mortality of tropical tree seedlings: expeimental studies of the effects of dispersal distance, seedling density, and light conditions. *Oecologia,* 61, 211-7.

Aung, L.H. (1974). Root-shoot relationships. In *The plant root and its environment,* ed. E.W. Carson, pp. 29-61. Charlottesville, Virginia: University Press.

Austin, M.P. (1980). Searching for a model for use in vegetation analysis. *Vegetatio,* 42, 11-21.

Austin, M.P. (1982). Use of a relative physiological performance value in the prediction of performance in multispecies mixtures from monoculture performance. *Journal of Ecology,* 70, 559-570.

Austin, M.P. (1985). Continuum concept, ordination methods, and niche theory. *Annual Review of Ecology and Systematics,* 16, 39-61.

Austin, M.P. (1987). Models for the analysis of species' response to environmental gradients. *Vegetatio,* 69, 35-45.

Austin, M.P., Groves, R.H., Fresco, L.F.M. and Kaye, P.E. (1985). Relative growth of six thistle species along a nutrient gradient with multispecies competition. *Journal of Ecology,* 73, 667-684.

Austin, M.P., and Austin, B.O. (1980). Behaviour of experimental plant communities along a nutrient gradient. *Journal of Ecology,* 68, 891-918.

Austin, M.P., and Cook, B.G. (1974). Ecosystem stability: a result from an abstract simulation. *Journal of Theoretical Biology,* 45, 435-58.

Austin, M.P., and Smith, T.M. (1989). A new model for the continuum concept. *Vegetatio,* 83, 35-47.

Axelrod, D.I. (1966). Origin of deciduous and evergreen habits in temperate forests. *Evolution,* 20, 1-15.

Axelrod, D.I. (1967). Drought, diastrophism, and quantum evolution. *Evolution,* 21, 201-9.

Axelrod, D.I. (1973). History of the mediterranean ecosystem in California. In *Mediterranean-type Ecosystems: Origin and Structure,* ed. F. di Castri and H.A. Mooney, pp. 225-77. Berlin:Springer-Verlag.

Axelrod, D.I. (1975). Evolution and biogeography of Madrean-Tethyan sclerophyll vegetation. *Annals of the Missouri Botanical Garden*, 62, 280-334.

Axelrod, D.I. (1985). Rise of the grassland biome, central North America. *Biological Review*, 51, 163-201.

Bailey, I.W., and Sinnott, E.W. (1915). A botanical index of Cretaceous and Tertiary climates. *Science*, 41, 831-4.

Bailey, I.W., and Sinnott, E.W. (1916). The climatic distribution of certain angiosperm leaves. *American Journal of Botany*, 3, 24-39.

Baird, A.M. (1977). Regeneration after fire in King's Park, Western Australia. *Journal of the Royal Society of Western Australia*, 60, 1-22.

Bak, R.P.M. (1975). Ecological aspects of the distribution of reef corals in the Netherlands Antilles. *Bijdragen Tot de Dierkunde*, 45, 181-90.

Bak, R.P.. (1976). The growth of coral colonies and the importance of crustose coralline algae and burrowing sponges in relation with carbonate accumulation. *Netherlands Journal of Sea Research*, 10, 285-337.

Bak, R.P.M., and Luckhurst, B.E. (1980). Constancy and change in coral reef habitats along depth gradients at Curacao. *Oecologia*, 47, 145-55.

Bakelaar, R.G., and Odum, E.P. (1978). Community and population level responses to fertilization in an old-field ecosystem. *Ecology*, 59, 660-5.

Baker, H.G. (1965). Characteristics and modes of origin of weeds. In *The Genetics of Colonizing Species*, ed. H.G. Baker and G.L. Stebbins, pp. 147-72. New York: Academic Press.

Baker, H.G. (1974). The evolution of weeds. *Annual Review of Ecology and Systematics*.5, 1-24.

Baker, H.G. (1986). Patterns of plant invasions in North America. In *Ecology of Biological Invasions of North America and Hawaii*, ed. H.A. Mooney and J.A. Drake, pp. 44-57. New York: Springer-Verlag.

Baker, H.G., Bawa, K.S., Frankie, G.W., and Opler, P.A. (1983). Reproductive biology of plants in tropical forests. In *Tropical Rain Forest Ecosystems*, Ecosystems of the World Vol. 14A, ed. F.B. Golley. pp. 183-215. Amsterdam: Elsevier.

Baker, P.A., and Weber, J.N. (1975). Coral growth rates: Variation with depth. *Earth and Planetary Science Letter*, 27, 57-61.

Bakker, J.P., de Leeuw, J., and van Wieren, S.E. (1983). Micro-patterns in grassland vegetation created and sustained by sheep-grazing. *Vegetatio*, 55, 153-61.

Banfield, A.W.F. (1954). Preliminary investigation of the barren ground caribou. *Canadian Wildlife Service Wildlife Management Bulletin Series 1*, 10A,a 1-79, 10B, 1-112.

Barbour, C.D., and Brown, J.H. (1974). Fish species diversity in lakes. *American Naturalist*, 108, 473-89.

Barbour, M.G., and Minnich, R.A. (1990). The myth of chaparral convergence. *Israel Journal of Botany*, 39, 453-63.

Bardach, J.E. (1958). On the movement of certain Bermuda reef fishes. *Ecology*, 39, 139-46.

Barden, L.S. (1977). Self-maintaining populations of *Pinus pungens* Lam. in the Southern Appalachian Mountains. *Castanea*, 42, 316-23.

Barden, L.S., and Woods, F.W. (1973). Characteristics of lightning fires in southern Appalachian Forests. *Proceedings of the Annual Tall Timbers Fire Ecology Conference*, 13, 345-61.

Barden, L.S., and Woods, F.W. (1976). Effects of fire on pine and pine-hardwood forests of the Southern Appalachians. *Forest Science*, 222, 339-403.

Barnes, D.J., and Taylor, D.L. (1973). In situ studies of calcification and photosynthetic carbon fixation in the coral *Montastrea annularis*. *Helgolander Wiss. Meeresunters*, 24, 284-91.

Barnes, P.W., Tiezen, L.L., and Ode, D.J. (1983). Distribution, production, and diversity of C_3- and C_4-dominated communities in a mixed prairie. *Canadian Journal of Botany*, 61, 741-51.

Barnett, M.A. (1983). Species structure and temporal stability of mesopelagic fish assemblages in the central gyres of the North and South Pacific Ocean. *Marine Biology (Berlin)*,74, 245-56.

Barrett, K.E.J. (1977). Provisional distribution maps of ants in the British Isles. In *Ants*, ed. M.V. Brian, pp. 203-16. London: Collins.

Bartholemew, W.V., Meyer, J., and Laudelot, H. (1953). Mineral nutrient immobilization under forest and grass fallow in the Yangambe (Belgian Congo) Region. I.N.E.A.C. Publ. No. 57. Serie Scientifique.

Bates, H.W. (1864). *The Naturalist on the River Amazons*. 1962 Edition. Berkeley: University of California Press.

Bawa, K.S. (1974). Breeding systems of tree species of a lowland tropical community. *Evolution*, 28, 85-92.

Bawa, K.S. (1980). The evolution of dioecy in flowering plants. *Annual Review of Ecology and Systematics*, 11, 15-39.

Bawa, K.S. (1990). Plant-pollinator interactions in tropical rain forests. *Annual Review of Ecology and Systematics*, 21, 399-422.

Baxter, R.M. (1977). Environmental effects of dams and impoundments. *Annual Review of Ecology and Systematics*, 8, 255-83.

Bazzaz, F.A. (1975). Plant species diversity in oldfield successional ecosystems in southern Illinois. *Ecology*, 56, 485-88.

Bazzaz, F.A. (1979). The physiological ecology of plant succession. *Annual Review of Ecology and Systematics*, 10, 351-71.

Bazzaz, F.A., Chiarello, N.R., Coley, P.D., and Pitelka, L.F. (1987). Allocating resources to reproduction and defense. *BioScience*, 37, 58-67.

Bazzaz, F.A., and Pickett, S.T.A. (1980). Physiological ecology of tropical succession: A comparative review. *Annual Review of Ecology and Systematics*, 11, 287-310.

Bazzaz, F.A., and Reekie, E.G. (1985) The meaning and measurement of reproductive effort in plants. In *Studies on Plant Demography: A Festschrift for John L. Harper*, ed. J. White, pp. 373-87. London: Academic.

Beadle, N.C.W. (1966). Soil phospate and its role in molding segments of the Australian flora and vegetation, with special reference to xeromorphy and sclerophylly. *Ecology*, 47, 992-1007.

Beard, J.S. (1946). The *Mora* forest of Trinidad, British West Indies. *Journal of Ecology*, 33, 173-92.

Beard, J.S. (1983). Ecological control of the vegetation of South-western Australia: moisture versus nutrients. In *Mediterranean-Type Ecosystems*, ed. F.J. Kruger, D.T. Mitchell, and J.U.M. Jarvis. pp. 66-73. Berlin: Springer-Verlag.

Beaver, R.A. (1979). Host specificity of temperate and tropical animals. *Nature (London)*, 281, 139-41.

Bell, R.H.V. (1970). The use of the herb layer by grazing ungulates in the

Serengeti. In *Animal Populations in Relation to Their Food Resources*, ed. A. Watson, pp. 111-23. Oxford: Blackwell.

Bell, R.H.V. (1971). A grazing system in the Serengeti. *Scientific American*, 224, 86-93.

Below, F.E., Christensen, L.E., Reed, A.J., and Hageman, R.H. (1981). Availability of reduced N and carbohydrates for ear development of maize. *Plant Physiology*, 68, 1186-90.

Ben-Eliahu, M.N., and Safriel, U.N. (1982). A comparison between species diversities of polychaetes from tropical and temperate structurally similar rocky intertidal habitats. *Journal of Biogeography*, 9, 371-90.

Benayahu, Y., and Loya, Y. (1977). Seasonal occurrence of benthic-algae communities and grazing regulation by sea urchins at the coral reefs of Eilat, Red Sea. *Proceedings of the Third International Coral Reef Symposium*, pp. 383-9.

Benson, W.W. (1978) Resource partitioning in passion vine butterflies *Evolution* 32, 493–518.

Benson, W.W. (1982). Alternative models for intrageneric diversification in the humid tropics: tests with passion vine butterflies. In *Biological Diversification in the Tropics*, ed. G.T. Prance, pp. 608-40. New York: Columbia University Press.

Benzing, D.H. (1990). *Vascular Epiphytes*. Cambridge: Cambridge University Press.

Berendse, F. (1985). The effect of grazing on the outcome of competition between plant populations with different nutrient requirements. *Oikos*, 44, 35-9.

Berger, A., Imbrie, J., Kukla, G., and Saltzman, B. (1984). *Milankovitch and Climate*. Dordrecht, The Netherlands: D. Reidel.

Bernhard-Reversat, F., Huttel, C., and Lemée, G. (1978). Structure and functioning of evergreen rain forest ecosystems of the Ivory Coast. In *Tropical Forest Ecosystems: A State-of-Knowledge Report*. pp. 557-74. Paris: UNESCO.

Berwick, S. (1974). *The Community of Wild Ruminants in the Gir Forest Ecosystem*. Thesis, Yale University, New Haven, Connecticutt.

Beven, S., Connor, E.F., and Beven, K. (1984). Avian biogeography in the Amazon basin and the biological model of diversification. *J. Biogeography*, 11, 383-99.

Bigarella, J.J., and Andrade-Lima, D. de. (1982). Paleoenvironmental changes in Brazil. In *Biological Diversification in the Tropics*, ed. G. Prance. pp. 27-40. New York: Columbia University Press.

Billings, D.W. (1938). The structure and development of old field short-leaf pine stands and certain associated physical properties of the soil. *Ecological Monographs*, 8, 437-99.

Billings, W.D. (1990). *Bromus tectorum*, a biotic cause of ecosystem impoverishment in the Great Basin. In *The Earth in Transition: Patterns and Processes of Biotic Impoverishment*,

Birkeland, C. (1977). The importance of rate of biomass accumulation in early succcessional stages of benthic communities to the survival of coral recruits. *Proceedings of the Third International Coral Reef Symposium*, pp. 15-21.

Birks, H.J.B. (1989). Holocene isochrone maps and patterns of tree-spreading in the British Isles. *J. Biogeography*, 16, 503-40.

Blackmore, A.C., Mentis, M.T., and Scholes, R.J. (1990). The origin and extent of nutrient-enriched patches within a nutrient-poor savanna in South Africa. *J. Biogeogr.,* 17, 463-70.

Blaisdell, J.P. (1953). Ecological effects of planned burning of sagebrush-grass range on the upper Snake River plains. Tech. Bull. 1075. Washington, D.C.: U.S. Department of Agriculture.

Blaisdell, R.S., Wooten, J., and Godfrey, R.K. (1973). The role of magnolia and beech in forest processes in the Tallahassee Florida, Thomasville, Georgia Area. *Proceedings of the Annual Tall Timbers Fire Ecology Conference,* 13, 363-97.

Bliss, L.C., and Cox, G.W. (1964). Plant community and soil variation within a northern Indiana prairie. *American Midland Naturalist,* 72, 113-28.

Bloom, A.J., Chapin, F.S., and Mooney, H.A. (1985). Resource limitation in plants - an economic analysis. *Annual Review of Ecology and Systematics,* 16, 363-92.

Boardman, N.K. (1977). Comparative photosynthesis of sun and shade plants. *Annual Review of Plant Physiology,* 28, 355-77.

Bonan, G. (1988). The size structure of theoretical plant populations: spatial patterns and neighborhood effects. *Ecology,* 69, 1721-30.

Bond, W. (1983). On alpha diversity and the richness of the Cape flora: a study in southern Cape Fynbos. in *Mediterranean Type Ecosystems: the Role of Nutrients,* ed. F.J. Kruger, D.T. Mitchell, and J.U.M. Jarvis. pp. 337-56. New York:Springer Verlag.

Bonem, R. M., and Stanley, G. D. (1977). Zonation of a lagoonal patch reef: analysis, comparison, and implications for fossil biohermal assemblages. *Proceedings of the Third International Coral Reef Symposium,* 2, 178-82.

Bonham, C.D., and Lerwick, A. (1976). Vegetation changes induced by prairie dogs on shortgrass prairie. *Journal of Range Management,* 29, 221-5.

Booth, W. (1989). Africa is becoming an elephant graveyard. *Science,* 243, 732.

Booth, W.L. (1989). Monitoring the fate of forests from space. *Science,* 243, 1428-9.

Boring, L.R., and Swank, W.T. (1984). The role of black locust (*Robinia pseudo-acacia*) in forest succession. *Journal of Ecology,* 72, 749-66.

Bormann, F.H., Likens, G.H., Fisher, D.W., and Pierce, R.S. (1968). Nutrient loss accelerated by clear-cutting of a forest ecosystem. *Science,* 159, 882-4.

Bormann, F.H., and Likens, G.H. (1979). *Pattern and process in a forested ecosystem.* New York: Springer-Verlag.

Botkin, D.B., Janak, J.F., and Wallis, J.R. (1972). Some ecological consequences of a computer model of forest growth. *Journal of Ecology,* 60, 849-72.

Boucher, D.H., James, S., and Keeler, K.H. (1982). The ecology of mutualism. *Ann. Rev. Ecol. Syst.,* 13, 315-47.

Bourlière, F. (1965). Densities and biomasses of some ungulate populations in Eastern Congo and Rwanda, with notes on population structure and lion/ungulate ratios. *Zool. Afr.,* 1, 199-207.

Bourlière, F. (1983a). The savanna mammals: introduction. In *Ecosystems of the World 13. Tropical Savannas,* ed. F. Bourlière, pp. 359-61. Amsterdam: Elsevier.

Bourlière, F. (1983b). The place of parasites in savanna ecosystems: introduction. In *Ecosystems of the World 13: Tropical Savannas,* ed. F. Bourlière, pp. 559-62. Amsterdam: Elsevier.

Bourlière, F. (ed.) (1983c). *Ecosystems of the World 13 Tropical Savannas.* Amsterdam: Elsevier.

Bourlière, F. (1983d). Mammals as secondary consumers in savanna ecosystems. In *Ecosystems of the World 13: Tropical Savannas*, ed. F. Bourlière, pp. 463-75. Amsterdam: Elsevier.

Bourlière, F., and Hadley, M. (1983). Present-day savannas: an overview. In *Ecosystems of the World 13: Tropical Savannas*. ed. F. Bourlière, pp. 1-17. Amsterdam: Elsevier.

Bowers, M.A., and Brown, J.H. (1982). Body size and coexistence in desert rodents: chance or community structure. *Ecology*, 63, 391-400.

Box, E.O. (1981). *Macroclimate and plant forms: an introduction to predictive modeling in phytogeography*. Boston: W. Junk.

Boyce, M.S. (1984). Restitution of *r*- and *K*-selection as a model of density-dependent natural selection. *Annual Review of Ecology and Systematics*, 15, 427-47.

Bradbury, I.K. (1981). Dynamics, structure and performance of shoot populations of the rhizomatous herb *Solidago canadensis* L., in abandoned pastures. *Oecologia* (Berlin), 48, 271-6.

Bradshaw, A.D., Chadwick, M.J., Jowett, D., and Snaydon, R.W. (1964). Experimental investigations into the mineral nutrition of several grass species. IV. Nitrogen level. *Journal of Ecology*, 52, 665-76.

Bragg, T.B., and Hulbert, L.C. (1976). Woody plant invasion of unburned Kansas bluestem prairie. *Journal of Range Management*, 29, 19-24.

Braudel, F. (1979a). *Civilisation Matérielle, Économie et Capitalisme, XVe-XVIIIe Siécle. Tome 1. Les Structures du Quotidien: le Possible et l'Impossible*. Paris: Armand Colin (published in English by Collins with the title *The Structures of Everyday Life.*).

Braudel, F. (1979b). *Civilisation Matérielle, Économie et Capitalisme, XVe-XVIIIe Siécle. Tome 2. Les Jeux de l'échange*. Paris: Armand Colin (published in English by Collins with the title *the Wheels of Commerce.*).

Braudel, F. (1979c). *Civilisation Matérielle, Économie et Capitalisme, XVe-XVIIIe Siécle. Tome 3. Le temps du monde*. Paris: Armand Colin (published in English by Collins with the title *The Perspective of the World.*).

Braun, E.L. (1950). *Deciduous Forests of Eastern North America*. Philadelphia: Blakiston.

Breitburg, D.L. (1985). Development of a subtidal epibenthic community: factors affecting species composition and the mechanisms of succession. *Oecologia*, 65, 173-84.

Brenchley, W.E. (1958). *The Park Grass Plots at Rothamsted*. Harpendon: Rothamsted Experimental Station.

Brian, A.D. (1957). Differences in the flowers visited by four species of bumblebees and their causes. *Journal of Animal Ecology*, 26, 71-98.

Bridgewater, P.B., and Backwhall, D.J. (1981). Dynamics of some Western Australian ligneous formations with special reference to the invasion of exotic species. *Vegetatio*, 46, 141-8.

Brock, R.E. (1979) . An experimental study on the effects of grazing by parrotfishes and role of refuges in benthic community structure. *Marine Biology*, 51, 381-8.

Brodskij, A.K. (1959). Leden in der Tiefe des Polarbeckens. *Naturwiss. Rundsch.*, 12, 52-6.

Brokaw, N. (1985a). Gap phase regeneration in a tropical forest. *Ecology*, 66, 682-687.

Brokaw, N.V.L. (1985a). Treefalls, regrowth, and community structure in tropical

forests. In *The Ecology of Natural Disturbance and Patch Dynamics,* ed.
S.T.A. Pickett and P.S. White, pp. 53-68. Orlando, Florida: Academic Press.

Brokaw, N.V.L. (1985b). Treefalls, regrowth, and community structure in
tropical forests. In *The Ecology of Natural Disturbance and Patch Dynamics.*
ed. S.T.A. Pickett and P.S. White. pp. 53-69. Orlando, FL: Academic Press.

Brokaw, N.V.L. (1982). Treefalls, frequency, timing, and consequences. In
Seasonal Rhythms in a Tropical Forest, ed. E.G. Leigh, A.S. Rand, and
D.M. Windsor, pp. 101-8. Washington, D.C.: Smithsonian Institution Press.

Brooke, R.K., Lloyd, P.H., and De Villiers, A.L. (1986). Alien and translocated
vertebrates in South Africa. In *The Ecology and Management of Biological
Invasions in Southern Africa,*

Brooks, J.L., and Dodson, S.I. (1965). Predation, body size, and the composition
of the plankton. *Science,* 150, 28-35.

Brown, A.H.D, and Marshall, D.R. (1981). Evolutionary changes accompanying
colonization in plants. In *Evolution Today, Proceedings of the Second
International Congress on Systematic and Evolutionary Biology, Vancouver,
B.C.* pp. 351-63. New York: Carnegie-Mellon University, Hunt Institute for
Botanical Documentation.

Brown, E.V., and Davis, K.P. (1973). *Forest Fire: Control and Use.* New York:
McGraw-Hill.

Brown, J. H., and Bourn, T.G. (1973). Patterns of soil moisture depletion in a
mixed oak stand. *Forest Science,* 19, 23-30.

Brown, J.H. (1971). Mammals on mountaintops: nonequilibrium insular
biogeography. *American Naturalist,* 105, 467-78.

Brown, J.H. (1973). Species diversity of seed-eating desert rodents in sand dune
habitats. *Ecology,* 54, 775-87.

Brown, J.H. (1981). Two decades of homage to Santa Rosalis: Toward a general
theory of diversity. *American Zoologist,* 21, 877-88.

Brown, J.H., and Davidson, D.W. (1977). Competition between seed-eating
rodents and ants in desert ecosystems. *Science,* 196, 880-2.

Brown, J.H., and Gibson, A.C. (1983). *Biogeography.* St. Louis: Mosby.

Brown, J.H., and Kodric-Brown, A. (1977). Turnover rates in insular
biogeography: effect of immigration on extinction. *Ecology,* 58, 445-9.

Brown, K.S. (1975). Geographical patterns of evolution in neotropical
Lepidoptera. Systematics and derivation of known and new *Heliconiini*
Nymphalidae: Nymphalinae). *Journal of Entomology (B),* 44, 201-41.

Brown, K.S. (1979). *Ecologia Geographica e Evolucao nas Florestas Neotropicais,*
Sao Paulo: Universidade Esataual de Campinas.

Brown, K.S. (1982). Palaeoecology and regional patterns of evolution in
neotropical forest butterflies.In *Biological Diversification in the Tropics,* ed.
G.T. Prance, pp. 255-308. New York: Columbia University Press.

Brown, R.G.B. (1980). Seabirds as marine animals. In *Behavior of Marine
Animals, Vol. 4, Marine Birds,* ed. J. Burger, B. Olla, and H.W. Winn, pp.
1-39. New York: Plenum.

Brown, S., Lugo, A.E., Silander, S., and Liegel, L. (1983). *Research History and
Opportunities in the Luquillo Experimental Forest.* USDA Southern Forest
Experiment Station General Technical Report SO-44. New Orleans:
USDA/SFES.

Brown, V.K. (1984). Secondary succession: insect-plant relationships. *BioScience,*
34, 710-6.

Brown, V.K. (1985). Insect herbivores and plant succession. *Oikos,* 44, 17-22.

Brown, V.K., Jepsen, M., and Gibson, C.W.D. (1988). Insect herbivory: effects on early old field succession. *Oikos,* 52, 293-302.

Brown, V.K., and Southwood, T.R.E. (1983). Trophic diversity, niche breadth and generation times of exopterygote insects in secondary succession. *Oecologia,* 56, 220-5.

Brown, W.L., and Wilson, E.O. (1956). Character displacement. *Systematic Zoolology,* 5, 49-64.

Browne, R.A. (1981). Lakes as islands: biogeographic distribution, turnover rates, and species composition in the lakes of central New York. *Journal of Biogeography,* 8, 75-83.

Brubaker, L.B. (1981). Long-term forest dynamics. In *Forest Succession Concepts and Applications,* ed. D.C. West, H.H. Shugart, and D.B. Botkin, pp. 95-106. New York: Springer-Verlag.

Brun, W.A., and Cooper, R.L. (1967). Effects of light intensity and carbon dioxide concentration on photosynthetic rate of soybean. *Crop Science,* 7, 451-4.

Brunig, E.F. (1972). Species richness and stand diversity in relation to site and succession of forests in Sarawak and Brunei (Borneo). *Amazoniana,* 4, 293-320.

Brunig, E.F. (1983). Vegetation structure and growth. In *Ecosystems of the World, Vol. 14A Tropical Rain Forest Ecosystems,* ed. F.B. Golley, pp. 49-75. Amsterdam: Elsevier.

Bruun, A.F. (1957). Deep sea and abyssal depths. In *Treatise on Marine Ecology and Paleoecology,* ed. J.W. Hedgpeth, pp. 641-72. Geological Society of America Memoir No. 67.

Bryan, F.G. (1975). Food habits, functional digestive morphology, and assimilation efficiency of the rabbit fish *Sianus spinus* (Pisces: Siganidae) on Guam. *Pacific Science,* 29, 269-77.

Bryant, J.P., Chapin, F.S., and Klein, D.R. (1983). Carbon/nutrient balance of boreal plants in relation to vertebrate herbivory. *Oikos,* 40, 357-68.

Buddemeier, R. W., and Kinzie, R. A. (1976). Coral growth. *Oceanography and Marine Biology Annual Review,* 14, 183-225.

Buddemeier, R.W. (1974). Environmental controls over annual and lunar monthly cycles in hermatypic coral calcification. *Proceedings of the Second International Coral Reef Symposium,* 2, 259-68.

Buddemeier, R.W., Maragos, J.E., and Knutson, D.W. (1974). Radiographic studies of reef coral exoskeletons: rates and patterns of coral growth. *Journal of Experimental Marine Biology,* 14, 179-200.

Budowski, G. (1965). Distribution of tropical American rainforest species in the light of successional processes. *Turrialba,* 15, 40-2.

Budowski, G. (1970). The distinction between old secondary and climax species in tropical Central American lowland forests. *Tropical Ecology,* 11, 44-8.

Bull, J.J., and Vogt, R.C. (1979). Temperature-dependent sex determination in turtles. *Science,* 206, 1186-8.

Bunt, J.S. (1975). Primary productivity of marine ecosystems. In *Primary Productivity of the Biosphere.* ed. H. Lieth and R.H. Whittaker. pp. 169-83. New York: Springer-Verlag.

Buol, S.W., and Eswaran, H. (1988). *International Committee on Oxisols: Final Report.* Technical Monograph No. 17. Washington, D.C.: Soil Management Support Services.

Burdon, J.J., Jarosz, A.M., and Kirby, G.C. (1989). Pattern and patchiness in plant-pathogen interactions - causes and consequences. *Annual Review of Ecology and Systematics*, 20, 119-36.

Burdon, J.J., and Chilvers, G.A. (1977). Preliminary studies on a native Australian eucalypt forest invaded by exotic pines. *Oecologia (Berlin)*, 31, 1-12.

Burgess, P.F. (1969). Preliminary observations on the autecology of *Shorea curtisii* Dyer ex King in the Malay Peninsula. *Malaysian Forester*, 32, 438.

Burns, C.W. (1968). The relationship between body size of filter feeding Cladocera and the maximum size of particle ingested. *Limnology and Oceanography*, 13, 675-8.

Bush, M.B., Colinvaux, P.A., Wiemann, M.C., Piperno, D.R., and Liu, K. (1990). Late Pleistocene temperature depression and vegetation change in Ecuadorian Amazonia. *Quaternary Research*, 34, 330-45.

Bush, M.B., and Colinvaux, P.A. (1990). A pollen record of a complete glacial cycle from lowland Panama. *Journal of Vegetation Science*, 1, 119-24.

Buss, L.W., Jackson, J.B.C. (1979). Competitive networks: nontransitive competitive relationships in cryptic coral reef environments. *American Naturalist*, 113, 223-34.

Buzas, M.A., and Givson, T.G. (1969). Species diversity: benthic Foraminifera in western North Atlantic. *Science*, 163, 72-5.

Byram, G.M. (1959). Combustion of forest fuels. In *Forest Fire: Control and Use*. ed. K.P. Davis. pp. 61-89. New York: McGraw-Hill.

Cain, S.A. (1934). The species-area curve. *American Midland Naturalist*, 19, 573-81.

Campbell, D.G., Daly, D.C., Prance, G.T., and Maciel, U.N. (1986). Quantitative ecological inventory of *terra firme* and *várzea* tropical forest on the Rio Xingu, Brazilian Amazon. *Brittonia*, 38, 369-93.

Campbell, D.G., Stone, J.L., and Rosas, A. (1992). A comparison of the phytosociology and dynamics of three floodplain (*Várzea*) forests of known ages, Rio Juruá, western Brazilian Amazon. *Biological Journal of the Linnean Society*, 108, 213-37.

Carey, A.G. (1985). Marine ice fauna: Arctic. In *Sea Ice Biota*, ed. R.A. Horner, pp. 174-90. Boca Raton, Florida: CRC Press.

Carlquist, S. (1965). *Island Life*. Garden City, New York: Natural History Press.

Carlquist, S. (1974). *Island Biology*. New York: Columbia University Press.

Carpenter, J.R. (1939). The biome. *American Midland Naturalist*, 21, 75-91.

Carpenter, R.C. (1985). Sea urchin mass mortality: effects on reef algal abundance, species composition, and metabolism and other coral reef herbivores. *Proceedings of the Fifth International Coral Reef Congress*, 4, 53-60.

Carpenter, R.C. (1986). Partitioning herbivory and its effects on coral reef algal communities. *Ecological Monographs*, 56, 345-63.

Carpenter, S., Kitchell, J.F., and Hodgson, J.R. (1985). Cascading trophic interactions and lake productivity. *Bioscience*, 35, 364-9.

Carpenter, S.R., Kitchell, J.F., Hodgson, J.R., Cochran, J.J., Elser, M.M. Lodge, D.M., Kretchmer, D., He, X., and von Endem C.N. (1987). Regulation of lake primary productivity by food web structure. *Ecology*, 68, 1863-76.

Carpenter, S.R., and Kitchell, J.F. (1984). Plankton community structure and limnetic primary production. *American Naturalist*, 124, 159-72.

Carvell, K.L, and Maxey, W.R. (1969). Wildfire adversely affects composition of cove hardwood stands. *West Virginia Agricultural Experiment Station Bulletin,* 2, 4-5.

Carvell, K.L., and Tyron, E.H. (1961). The effect of environmental factors on the abundance of oak regeneration beneath mature oak stands. *Forest Science,* 7, 98-105.

Caswell, H. (1978). Predator mediated coexistence: a non-equilibrium model. *American Naturalist,* 112, 127-54.

Caswell, H. (1982). Life history theory and the equilibrium status of populations. *American Naturalist,* 120, 317-39.

Caswell, H. (1988). Theories and models in ecology: a different perspective. *Bulletin of the Ecological Society of America,* 69, 102-9.

Caswell, H. (1989). *Matrix Population Models.* Sunderland, Massachusetts: Sinauer Associates.

Caswell, H., and Cohen, J.H. (1991). disturbance, interspecific interactions and diversity in metapopulations. *Biological Journal of the Linnean Society,* 42, 193-218.

Chaloner, W.G., and Lacey, W.S. (1973). The distribution of Late Paleozoic floras. *Special Papers in Palaeontology,* 12, 271-89.

Chamberlain, T.C. (1898). The ulterior basis of time divisions and the classification of geologic history. *Journal of Geology,* 6, 449-62.

Chapin, F.S. (1980). The mineral nutrition of wild plants. *Annual Review of Ecology and Systematics,* 11, 233-60.

Chapin, F.S., Bloom, A.J., Field, C.B., and Waring, R.H. (1987). Plant responses to multiple environmental factors. *BioScience,* 37, 49-57.

Chapin, F.S., Vitousek, P.M., and Van Cleve, K. (1986). The nature of nutrient limitation in plant communities. *American Naturalist,* 127, 48-58.

Chapman, V.J. (1976). *Mangrove Vegetation.* Vaduz, Liechtenstein: J. Cramer.

Chazdon, R.L. (1986). The costs of leaf support in understory palms: economy versus safety. *American Naturalist,* 127, 9-30.

Chesson, P. (1978). Predator-prey theory and variability. *Annual Review Ecolology Systematics,* 9, 323-47.

Chew, R.M., and Chew, A.E. (1965). The primary productivity of a desert-shrub (*Larrea tridentata*) community. *Ecological Monographs,* 35, 355-75.

Christensen, K., and Culver, D. (1968). Geographical variation and evolution in *Pseudosinella hirsuta. Evolution,* 22, 237-55.

Christensen, N.L. (1985). Shrubland fire regimes and their evolutionary consequences. In *The Ecology of Natural Disturbance and Patch Dynamics,* ed. S.T.A. Pickett and P.S. White, pp. 85-100. Orlando, Florida: Academic Press.

Christensen, N.L. Agee, J.K, Brussard, P.F., Huges, J., Knight, D.H., Minshall, G.W., Peek, J.M., Pyne, S.J., Swanson, F.J., Thomas, J.W., Wells, S., Williams, S.E., and Wright, H.A. (1989). Interpreting the Yellowstone fires of 1988. *BioScience,* 39, 678-85.

Christensen, N.L., and Peet, R.K. (1984). Convergence during secondary forest succession. *Journal of Ecology,* 72, 25-36.

Clapham, W.B. (1973). *Natural Ecosystems,* New York: Collier-Macmillan.

Clark, D.A., and Clark, D.B. (1984). Spacing dynamics of a tropical rain forest tree: evaluation of the Janzen-Connell model. *American Naturalist,* 124, 769-88.

Clark, J.S. (1988a). Charcoal-stratigraphic analysis on petrographic thin sections: recent fire history in northwest Minnesota. *Quaternary Research,* 30, 67-80.

Clark, J.S. (1988b). Effect of climate change on fire regimes in northwestern Minnesota. *Nature,* 334, 233-5.

Clark, J.S. (1989a). Ecological disturbance as a renewal process: theory and application to fire history. *Oikos,* 56, 17-30.

Clark, J.S. (1989b). Effects of long-term water balances on fire regime, north-western Minnesota. *Journal of Ecology,* 77, 989-1004.

Clark, J.S. (1989c). The forest is for burning. *Natural History,* 1/89, 51-3.

Clark, J.S. (1990). Fire and climate change during the last 750 yr in northwestern Minnesota. *Ecological Monographs,* 60, 135-59.

Clark, W.C. (1989). Managing planet earth. *Scientific American,* 261, 47-54.

Clark, W.C., and Hollings, C.S. (1985). Sustainable development of the biosphere: human activities and global change. In *Global Change: Proceedings of a Symposium,* ed. T.F. Malone and J.G. Roederer, pp. 474-90. New York: Cambridge University Press.

Clarke, T.A. (1978). Diel feeding patterns of 16 species of mesopelagic fishes from Hawaiian waters. United States National Marine Fisheries Service Fishery Bulletin 76, 495-513.

Clausen, C.P. (ed). (1978). *Introduced Parasites and Predators of Arthropod Pests and Weeds: A World Review.* Agriculture Handbook 480. Washington, D.C.: U.S. Department of Agriculture.

Clausen, J., Keck, D.D., and Hiesey, W.M. (1941). Regional differentiation in plant species. *American Naturalist,* 75, 231-50.

Clements, F.E. (1904). *The Development and Structure of Vegetation. Botanical Survey of Nebraska 7. Studies in the Vegetation of the State.* Lincoln, Nebraska.

Clements, F.E. (1905). *Research Methods in Ecology.* Lincoln: University Publishing Company. Reprinted New York: Arno Press (1977).

Clements, F.E. (1910). The life history of lodgepole burn forests. *United States Forest Service Bulletin* 79. Washington, D.C.

Clements, F.E. (1916). *Plant Succession: An Analysis of the Development of Vegetation.* Publication No. 242. Washington, D.C.: Carnegie Institution of Washington. 1-512

Clements, F.E. (1935). Experimental ecology in the public service. *Ecology,* 16, 142-63.

Clements, F.E., Weaver, J.E. and Hanson, H.C., (1929). *Plant Competition.* Carnegie Institute of Washington Publication 398, 1-340.

Clements, F.E., and Shelford, V.E. (1939). *Bio-Ecology.* New York: Wiley.

Cochran, G.R., and Rowe, J.S. (1969). Fire in the tundra at Rankin Inlet, N.W.T. *Proceedings of the Annual Tall Timbers Fire Ecology Conference,* 9, 61-74.

Cody, M.L. (1966). The consistency of intra- and inter-continental bird species counts. *American Naturalist,* 100, 371-6.

Cody, M.L. (1968). On the methods of resource division in grassland bird communities. *American Naturalist,* 102, 107-48.

Cody, M.L. (1974). *Competition and the Structure of Bird Communities.* Princeton: Princeton University Press.

Cody, M.L. (1975). Towards a theory of continential species diversities: bird distributions over Mediterranean habitat gradients. In *Ecology and Evolution of Communities.* ed. M.L. Cody and J.M. Diamond. pp. 214-57. Cambridge, Massachusetts: Belknap Press.

Cody, M.L. (1981). Continental diversity patterns and convergent evolution in bird communities. In *Mediterranean-type Ecosystems: The Role of Nutrients, Ecological Studies Volume 43*, ed. F.J. Kruger, D.T. Mitchell, and J.U.M. Jarvis, pp. 357-402. Berlin: Springer-Verlag.

Cody, M.L. (1986). Diversity, rarity, and conservation in Mediterranean-climate regions. In *Conservation Biology: the Science of Scarcity and Diversity*. ed. M.E. Soule, pp. 122-52. Sunderland, Massachusetts: Sinauer.

Cody, M.L., and Mooney, H.A. (1978). Convergence versus nonconvergence in Mediterranean-climate ecosystems. *Annual Review of Ecology and Systematics*, 9, 265-321.

Coe, M.J., Cumming, D.H., and Phillipson, J. (1976). Biomass and production of large African herbivores in relation to rainfall and primary production. *Oecologia*, 22, 341-54.

Coffin, D.P. and Lauenroth, W.K. (1990). A gap dynamics simulation model of succession in a semiarid grassland. *Ecological Modelling*, 49, 229-66.

Cohen, J.E. (1977). Ergodicity of age structure in populations with Markovian vital rates. II: General states, III: Finite-state moments and growth rate; an illustration. *Advances in Applied Probability*, 9, 18-37, 462-75.

Cohen, J.E. (1978). *Food Webs and Niche Space*. Princeton: Princeton University Press.

Cohen, J.E. (1979a). Comparative statics and stochastic dynamics of age-structured populations. *Theoretical Population Biology*, 16, 159-71.

Cohen, J.E. (1979b). Long-run growth rates of discrete multiplicative processes in Markovian environments. *Journal of Mathematical Analysis and Applications*, 69, 243-51.

Coley, P.D., Bryant, J.P., and Chapin, F.S. III. (1985). Resource availability and plant antiherbivore defense. *Science*, 230, 895-9.

Colinvaux, P.A. (1973). *Introduction to Ecology*. New York: Wiley.

Colinvaux, P.A. (1986). *Ecology*. New York: Wiley.

Colinvaux, P.A. (1987). Amazon diversity in light of the paleoecological record. *Quaternary Science Review*, 6, 93-114.

Collette, B.B., and Earle, S.A. (eds). (1972). Results of the Tektite program: ecology of coral reef fishes. *Science Bulletin Natural History Museum, Los Angeles County*, 14, 1-180.

Collins, D.D. (1969). Macroclimate and the grassland ecosystem. In *The Grassland Ecosystem, A Preliminary Synthesis*, R.L. Dix, ed. Range Science Department Science Series No. 2, pp. 29-39. Fort Collins: Colorado State University Press.

Collins, S.L. (1987). Interaction of disturbances in tallgrass prairie: a field experiment. *Ecology*, 68, 1243-50.

Collins, S.L., and Barber, S.C. (1985). Effects of disturbance on diversity in mixed-grass prairie. *Vegetatio*, 64, 87-94.

Collins, S.L., and Uno, G.E. (1983). The effects of early spring burning on vegetation in buffalo wallows. *Bulletin of the Torrey Botanical Club*, 110, 474-81.

Colwell, R.K. (1979). Toward a unified approach to the study of species diversity. In *Ecological Diversity in Theory and Practice*, ed. J.F. Grassle, G.P. Patil, and C. Taillie, pp. 75-91. Fairland, Maryland: International Co-operative Publishing House.

Colwell, R.K., and Fuentes, E.R. (1975). Experimental studies of the niche. *Annual Review of Ecology and Systematics*, 6, 281-310.

Connell, J.H. (1961). The influence of interspecific competition and other factors on the distribution of the barnacle *Chthamalus stellatus*. *Ecology*, 42, 710-23.

Connell, J.H. (1970). A predator-prey system in the marine intertidal region. I. *Balanus glandula* and several predatory species of *Thais*. *Ecological Monographs*, 40, 49-78.

Connell, J.H. (1971). On the role of natural enemies in preventing competitive exclusion in some marine animals and in rainforest trees. In *Dynamics of Populations*, ed. P.J. den Boer and G.R. Gradwell. Wageningen: Centre for Agricultural Publishing.

Connell, J.H. (1978). Diversity in tropical rain forests and coral reefs. *Science*, 199, 1302-9.

Connell, J.H. (1980). Diversity and the coevolution of competitors, or the ghost of competition past. *Oikos*, 35, 131-8.

Connell, J.H. (1983). On the prevalence and relative importance if interspecific competition: evidence from field experiments. *The American Naturalist*, 122, 661-96.

Connell, J.H. and R.O. Slatyer. (1977). Mechanisms of succession in natural communities and their role in community stability and organization. *American Naturalist*, 111, 1119-44.

Connell, J.H., and Keough, M.J. (1985). Disturbance and patch dynamics of subtidal animals on hard substrata. In *The Ecology of Natural Disturbance and Patch Dynamics*. ed. S.T.A. Pickett and P.S. White. pp. 125-51. Orlando, FL: Academic Press.

Connell, J.H., and Orias, E. (1964). The ecological regulation of species diversity. *American Naturalist*, 98, 399-414.

Connor, E.F., and McCoy, E.D. (1979). The statistics and biology of the species-area relationship. *American Naturalist*, 113, 791-833.

Connor, E.F., McCoy, E.D., and Cosby, B.J. (1983). Model discrimination and expected slope values in species-area studies. *American Naturalist*, 122, 789-96.

Cook, R.E. (1969). Variation in species density of North American birds. *Systematic Zoology*, 18, 63-84.

Cooper, A.W. (1981). Above-ground biomass accumulation and net primary production during the first 70 years of succession in *Populus grandidentata* stands on poor sites in northern lower Michigan. In *Forest succession concepts and applications*, D.C. West, H.H. Shugart, and D.B. Botkin, pp. 336-60. New York: Springer-Verlag.

Cooper, G.P. (1936). Food habits, rate of growth and cannibalism of young largemouth bass (*Aplites salmoides*) in state-operated rearing ponds in Michigan during 1935. *Transactions of the American Fisheries Society*, 66, 242-66.

Cooper, W.S. (1913). The climax forest of Isle Royale, Lake Superior and its development. *Botanical Gazette*, 15, 1-44, 115-40, 189-235.

Cooper, W.S. (1923). The recent ecological history of Glacier Bay, Alaska. *Ecology*, 4, 93-128, 223-46, 355-65.

Cooper, W.S. (1931). A third expedition to Glacier Bay, Alaska. *Ecology*, 12. 61-95.

Cooper, W.S. (1939). A fourth expedition to Glacier Bay, Alaska. *Ecology*, 20, 130-55.

Coppock, D.L., Detling, J.K., Ellis, J.E., and Dyer, M.I. (1983). Plant-herbivore interactions in North American mixed-grass prairie. I. Effects of black-tailed prairie dogs on intraseasonal aboveground biomass and nutrient dynamics and plant species diversity. *Oecologia,* 56, 1-9.

Corfield, T.F. (1973). Elephant mortality in Tsavo National Park, Kenya. *East African Wildlife Journal,* 11, 339-68.

Corliss, J.B., Dymond, J., Gordon, L.I., *et al.,* (1979). Submarine thermal springs on the Galapágos Rift. *Science,* 203, 1073-83.

Cottam, G., and McIntosh, R.P. (1966). Vegetation Continuum. *Science,* 152, 546-7.

Cottam, W.O., and Stewart, G. (1940). Plant succession as a result of grazing and of meadow desiccation by erosion since settlement in 1862. *Journal of Forestry,* 38, 613-26.

Covington, W.W., and J.D. Aber. (1980). Leaf production during secondary succession in northern hardwoods. *Ecology,* 61, 200-4.

Cowan, I.R. (1982). Regulation of water use in relation to carbon gain in higher plants. In *Physiological Plant Ecology II. Water Relations and Carbon Assimilation,* ed. O.L. Lange, P.S. Nobel, C.B. Osmond, and H. Ziegler, pp. 589-613. Berlin: Springer-Verlag.

Cowan, I.R. (1986). Economics of carbon fixation in higher plants. In *On the economy of plant form and function,* ed. T.J. Givnish, pp. 133-170. Cambridge: Cambridge University Press.

Cowles, H.C. (1899). The ecological relations of the vegetation of the sand dunes of Lake Michigan. *Botanical Gazette,* 27, 95-117, 167-202, 281-308, 361-91.

Cowles, H.C. (1901). The physiographic ecology of Chicago and vicinity: A study of the origin, development, and classification of plant societies. *Botanical Gazette,* 31, 73-108, 145-182.

Cowles, H.C. (1911). The causes of vegetative cycles. *Botanical Gazette,* 51, 161-83.

Cowling, R.M. (1983a). Phytochorology and vegetation history in the south-eastern Cape, South Africa. *Journal of Biogeography,* 10, 393-419.

Cowling, R.M. (1983b). Diversity relations in Cape shrublands and other vegetation in the southeastern Cape, South Africa. *Vegetatio,* 54, 103-27.

Cox, C.B., and Moore, P.D. (1985). *Biogeography: An Ecological and Evolutionary Approach.* Oxford: Blackwell.

Cox, D.R., and Smith, W.L. (1953). The superimposition of several strictly random periodic sequences of events. *Biometrika,* 40, 1-11.

Cox, D.R., and Smith, W.L. (1954). On the superimposition of renewal processes. *Biometrika,* 41, 91-9.

Cracraft, J. (1974). Continental drift and vertebrate distribution. *Annual Review of Ecology and Systematics,* 5, 215-61.

Craig, N., and Halais, P. (1934). The influence of maturity and rainfall on the properties of lateritic soils in Mauritius. *Empire Journal of Experimental Agriculture,* 2, 349-58.

Crawley, M.J. (1986). The population biology of invaders. *Philosophical Transactions of the Royal Society of London,* B314, 711-31.

Crawley, M.J. (1987). What makes a community invasible? In *Colonization, Succession, and Stability,* ed. M.J. Crawley, P.J. Edwards, and A.J. Gray, pp. 629-54. London: Blackwell.

Crepet, W.L. (1983). The role of insect pollination in the evolution of the angiosperms. In *Pollination Biology,* ed. L. Real, pp. 29-50. New York: Academic.

Crisp, D.J. (1975). Secondary productivity in the sea. In *Productivity of World Ecosystems.* pp. 71-89. Washington, D.C.: National Academy of Science.

Croat, T.B. (1985). A revision of *Anthurium* for Mexico and Central America. Part 2: Panama. *Monograph of the Missouri Botanical Garden.*

Crocker, R.L., and Major, J. (1955). Soil development in relation to vegetation and surface age at Glacier Bay, Alaska. *Journal of Ecology,* 43, 427-48.

Cronan, W. (1983). *Changes in the Land: Indians, Colonists, and the Ecology of New England.* New York: Hill and Wang.

Crosby, A.W. (1986). *Ecological Imperialism. The Biological Expansion of Europe, 900 - 1900.* Cambridge: Cambridge University Press.

Crowley, P.H. (1977). Spatially distributed stochasticity and the constancy of ecosystems. *Bulletin of Mathematical Biology,* 38, 157-62.

Crowley, P.H. (1981). Dispersal and the stability of predator-prey interactions. *American Naturalist,* 118, 673-701.

Culver, D., Holsinger, J.R., and Baroody, R. (1973). Toward a predictive cave biogeography: the Greenbriar Valley as a case study. *Evolution,* 27, 689-95.

Cummins, K.W. (1973). Trophic relations of aquatic insects. *Annual Review of Entomology,* 18, 183-206.

Cummins, K.W. (1974). Structure and function of stream ecosystems. *BioScience,* 24, 631-41.

Cummins, K.W., and Klug, M.J. (1979). Feeding ecology of stream invertebrates. *Annual Review of Ecology and Systematics,* 10, 147-72.

Currie, D.J. (1991). Energy and large-scale patterns of animal- and plant-species richness. *American Naturalist,* 137, 27-49.

Currie, D.J., and Paquin, V. (1987). Large-scale biogeographical patterns of species richness of trees. *Nature,* 329, 326-7.

Curtis, J.T. (1959). *The Vegetation of Wisconsin.* Madison: University of Wisconsin Press.

Curtis, J.T. (1962). The modification of mid-latitude grasslands and forests by man. In *Man's Role in Changing the Face of the Earth,* ed. W.L. Thomas, pp. 721-36. Chicago: University of Chicago Press.

Curtis, J.T., and Partch, M.L. (1948). Effects of fire on the competition between bluegrass and certain prairie plants. *American Midland Naturalist,* 39, 437-43.

Dahl, B.E. (1963). Soil moisture as a predictive index to forage yield for the sandhills range type. *Journal of Range Management,* 16, 128-32.

Dahlsten, D.L. (1986). Control of invaders. In *Ecology of Biological Invasions of North America and Hawaii,* ed. H.A. Mooney and J.A. Drake, pp. 275-302. New York: Springer-Verlag.

Dakin, W.J., and Colefax, A.N. (1940). The plankton of the Australian coastal waters off New South Wales. *University of Sidney Department of Zoology Publications Monograph,* 1, 1-215.

Dana, T.F. (1976). Reef-coral dispersion patterns and environmental variables on a Caribbean coral reef. *Bulletin Marine Science,* 26, 1-13.

Dansereau, P. (1957). *Biogeography: An Ecological Perspective.* New York: Ronald.

Dansereau, P., and Segadas-Vianna, F. (1952). Ecological study of the peat bogs of eastern North America. *Canadian Journal of Botany,* 30, 490-520.

Dart, J.K.G. (1972). Echinoids, algal lawn, and coral recolonization. *Nature,* 239, 50-1.

Darwin, C. (1859). *On The Origin of Species by Means of Natural Selection or the Preservation of Favored Races in the Struggle for Life.* London: Murray.

Darwin, C. (1872). *The Origin of Species.* Vol. 1, 6th Ed. London: John Murray.

Darwin, C. (1898). *Journal of Researches into the Natural History and Geology of the Countries visited during the Voyage of the H.M.S. Beagle Round the World, under the Command of Capt. Fitz Roy, R.N.* New York: D. Appleton.

Darwin, C. (1942). *The Voyage of the Beagle,* Natural History Library Edition 1962, ed. L. Engel. Garden City: Doubleday.

Darwin, C., and Wallace, A.R. (1858). On the tendency of species to form varieties: and on the perpetuation of varieties and species by natural means of selection. *J. Linnean Society (Zool.),* 3(1858), 45.

Dasman, R.F., Milton, J.P., and Freeman, P.F. (1973). *Ecological Principles for Economic Development.* New York: Wiley.

Daubenmire, R.F. (1947). *Plants and Environment: A Textbook of Plant Autecology.* New York: Wiley.

Daubenmire, R. (1966). Vegetation: identification of typical communities. *Science,* 151, 291-8.

Daubenmire, R. (1968). Ecology of fire in grasslands. *Advances in Ecological Research,* 5, 209-66.

Daubenmire, R., and Prusso, D.C. (1963). Studies of the decomposition rates of tree litter. *Ecology,* 44, 589-92.

Davies, W., and Jones, T.E. (1930). The yield and response to manures of contrasting pasture types. *Welsh Journal of Agriculture,* 8, 170-92.

Davis, M.B. (1969). Palynology and environmental history during the Quaternary Period. *American Scientist,* 57, 317-32.

Davis, M.B. (1976). Pleistocene biogeography of temperate deciduous forests. *Geosciences and Man,* 13, 13-26.

Davis, M.B. (1981). Quaternary history and the stability of forest communities. In *Forest Succession Concepts and Applications,* ed. D.C. West, H.H. Shugart, and D.B. Botkin, pp. 132-53. New York: Springer-Verlag.

Davis, M.B., Woods, K.D., Webb, S.L., and Futyma, R.B. (1986). Dispersal versus climate: expansion of *Fagus* and *Tsuga* into the upper Great Lakes region. *Vegetatio,* 67, 93-103.

Davis, S.D., Droop, S.J.M., Gregerson, P., Henson, L., Leon, C.J., Villa-Lobos, J.L., Synge, H., and Zantovska, J. (1986). *Plants in Danger: What Do We Know?* Gland, Switzerland: International Union for Conservation of Nature and Natural Resources.

Dawkins, H.C. (1959). The volume increment of natural tropical high-forest and limitations on its improvement. *Empire Forestry Review,* 38, 175-80.

Dawkins, H.C. (1964). The productivity of lowland tropical high forest and some comparisons with its competitors. *Oxford University Forestry Society Journal, Fifth Series,* 12, 1-8.

Day, G.M. (1953). The Indian as an ecological factor in the Northeastern forest. *Ecology,* 34, 329-46.

Day, R.T., Keddy, P.A., McNeill, J., and Carleton, T. (1988). Fertility and disturbance gradients: a summary model for riverine marsh vegetation. *Ecology,* 69, 1044-54.

Dayton, P.K. (1971). Competition, disturbance, and community organization: the provision and subsequent utilization of space in a rocky intertidal community. *Ecological Monographs*, 41, 351-89.

Dayton, P.K., and Hessler, R.R. (1972). The role of biological disturbance in maintaining diversity in the deep sea. *Deep-Sea Research*, 19, 199-208.

De Vos, A., and Dowsett, R.J. (1966). The behavior and population structure of three species of the genus *Kobus*. *Mammalia*, 30, 30-55.

DeAngelis, D.L. (1975). Stability and connectance in food web models. *Ecology*, 56, 238-43.

DeAngelis, D.L. (1980). Energy flow, nutrient cycling, and ecosystem resiliance. *Ecology*, 61, 764-71.

DeAngelis, D.L. (1991). *The Dynamics of Nutrient Cycling and Food Webs*. New York: Chapman and Hall.

DeAngelis, D.L., Adams, S.M., Breck, J.E., and Gross, L.J. (1984). A stochastic predation model: application to largemouth bass observations. *Ecological Modelling*, 24, 21-41.

DeAngelis, D.L., Cox, D.C., and Coutant, C.C. (1979). Cannibalism and size dispersal in young-or-the-year largemouth bass: experiments and model. *Ecological Modelling*, 8, 133-48.

DeAngelis, D.L., Elwood, J.W., Mulholland, P.J., Palumbo, A.V., and Steinman, A.D. (1990). Biogeochemical cycling constraints on recovery of disturbed stream ecosystems. *Environment Management*, 14, 685-97.

DeAngelis, D.L., Gardner, R.H., Mankin, J.B., Post, W.M., and Carney, J.H. (1978) Energy flow and the number of trophic levels in ecological communities. *Nature*, 273, 406-7.

DeAngelis, D.L., Godbout, L., and Shuter, B.J. (1991). An individual-based approach to predicting density-dependent dynamics in smallmouth bass populations. *Ecolology Modelling*, 57, 91-115.

DeAngelis, D.L., Post, W.M., and Travis, C. (1986). *Positive Feedback in Natural Systems*. Heidelberg: Springer-Verlag.

DeAngelis, D.L., and Gross, L.J. eds. (1992). Populations and Communities: An Individual-based Perspective. New York: Chapman and Hall.

DeAngelis, D.L., and Huston, M.A. (1987). Effects of growth rate in models of size distribution formation in plants and animals. *Ecolology Modelling*, 36, 119-37.

DeAngelis, D.L., and Waterhouse, J.C. (1987). Equilibrium and nonequilibrium concepts in ecological models. *Ecological Monographs*, 57, 1-21.

DeBach, P. (1966). The competitive displacement and coexistence principles. *Annual Review of Entomology*, 11, 183-212.

DeBach, P. (1974). *Biological Control by Natural Enemies*. Cambridge: Cambridge University Press.

DeBach, P., and Sundby, R.A. (1963). Competitive displacement between ecological homologues. *Hilgardia*, 34, 105-66.

den Boer, P.J. (1968). Spreading the risk and stabilization of animal numbers. *Acta Biotheoretica (Leiden)*, 18, 165-94.

den Boer, P.J. (1981). On the survival of populations in a heterogeneous and variable environment. *Oecologia (Berlin)*, 50, 39-53.

de Ruyter van Steveninck, E.D., and Bak, R.P.M. (1986). Changes in abundance of coral-reef bottom components related to mass mortality of the sea urchin *Diadema antillarum*. *Marine Ecology Progress Series*, 34, 87-94.

DeSelm, H.R., Clebsch, E.E.C., Nichols, G.M., and Thor, E. (1973). Response of herbs, shrubs, and tree sprouts in prescribed-burn hardwoods in Tennessee. *Proceedings of the Tall Timbers Fire Ecology Conference,* 12, 132-6.

DeSteven, D. (1991b). Experiments on mechanisms of tree establishment in old-field succession: seedling survival and growth. *Ecology,* 72, 1076-88.

Dcacon, H.J. (1983). The comparative evolution of mediterranean-type ecosystems: a southern perspective. In *Mediterranean-type Ecosystems: The Role of Nutrients, Ecological Studies Volume 43,* ed. F.J. Kruger, D.T. Mitchell, and J.U.M. Jarvis, pp. 3-40. Berlin: Springer-Verlag.

Delcourt, P.A. and Delcourt, H.R. (1987) *Long-term Forest Dynamics of the Temperate Zone. A Case Study of Late-Quaternary Forests in eastern North America.* New York: Springer-Verlag.

Dementeव́, G.P., Gladkov, N.A., Ptushenkov, E.S., Spangenberg, E.P., and Sudilovskaya, A.M. (1951-1954). *Birds of the Soviet Union (Ptitsy Sovetskogo Soyuza).* Vols. I-VI. ed. G.P. Dement'ev and N.A. Gladkov. Israel Program for Scientific Translations, Jerusalem, 1966. U.S. Department of Commerce, Clearinghouse for Federal Scientific and Tcchnical Information, Springfield, Va.

Denholm-Young, P.A. (1978). Studies on decomposing cattle dung and its associated fauna. D. Phil. thesis. Oxford University.

Denny, M.W., Daniel, T.L., and Koehl, M.A.R. (1985). Mechanical limits to size in wave-swept organisms. *Ecological Monographs,* 55, 69-102.

Denslow, J.S. (1987). Tropical rain forest gaps and tree species diversity. *Annual Review of Ecology Systematics,* 18, 431-51.

Denton, G.H., and Hughes, T.J. (1981). *The Last Great Ice Sheets.* New York: Wiley.

Deshmukh, I. (1986). *Ecology and Tropical Biology.* Palo Alto, California: Blackwell Scientific Publications.

Dexter, D. (1972). Comparison of the community structure in a Pacific and Atlantic Panamanian sandy beach. *Bulletin Marine Sciences,* 22, 449-62.

Diamond, J.M. (1970). Ecological consequences of island colonization by Southwest Pacific birds. I. Types of niche shifts. *Proceedings of the National Academy of Science,* 67, 529-36.

Diamond, J.M. (1973). Distributional ecology of New Guinea birds. *Science,* 179, 759-69.

Diamond, J.M. (1975). Assembly of species communities. In *Ecology and Evolution of Communities.* ed. M.L. Cody and J.M. Diamond. pp. 342-444. Cambridge, Massachusetts: Belknap Press.

Diamond, J.M. (1986). The design of a nature reserve system for Indonesian New Guinea. In *Conservation Biology: The Science of Scarcity and Diversity.* ed. M.E. Soulé. pp. 485-503. Sunderland, Massachusetts: Sinauer Associates, Inc.

Diamond, J.M. and Mayr, E. (1976). Island biogeography and the design of natural reserves. In *Theoretical Ecology: Principles and Applications.* ed. R.M.May. pp. 228-52. *Oxford: Blackwell Scientific Publications.*

di Castri, F. (1981). Mediterranean-type shrublands of the world. In *Ecosystems of the World, Vol. 11, Mediterranean Type Shrublands,* ed. F. di Castri, D.W. Goodall, and R.L. Specht, pp. 1-52. Amsterdam: Elsevier.

di Castri, F. (1989a). History of biological invasions with special emphasis on the Old World. In *Biological Invasions: A Global Perspective,* ed. J.A. Drake, H.A. Mooney, F. di Castri, R.H. Groves, F.J. Kruger, M. Rejmánek, and M. Williamson, pp. 1-30. Chichester: Wiley and Sons.

di Castri, F. (1989b). On invading species and invaded ecosystems: a play of historical chance and biological necessity. In *Biological Invasions in Europe and the Mediterranean Basin*, ed. F. di Castri, A.J. Hansen, and M. Debussche, pp. 3-15. Dordrecht: Kluwer.

di Castri, F. Goodall, D.W., and Specht, R.L. (1981). *Ecosystems of the World, Vol. 11, Mediterranean Type Shrublands*. Amsterdam: Elsevier.

di Castri, F., and Mooney, H.A. (eds.) (1973). *Mediterranean Type Ecosystems, Origin and Structure*. Berlin: Springer-Verlag.

Diggle, P.J. (1976). A spatial-stochastic model of inter-plant competition. *Journal of Applied Probability*, 13, 662-71.

Dinerstein, E. (1980). An ecological survey of the Royal Karnali-Bardia wildlife reserve, Nepal. Part III. Ungulate populations. *Biological Conservation*, 18, 5-38.

Ditlev, H. (1978). Zonation of corals (Scleractinia: Coelenterata) on intertidal reef flats at Ko Phuket, Eastern Indian Ocean. *Marine Biology*, 47, 29-39.

Dix, R.L. (1960). Effects of burning on the mulch structure and species composition of grasslands in western North Dakota. *Ecology*, 41, 49-56.

Dix, R.L., and Smeins, F. (1967). The prairie, meadow and marsh vegetation of Nelson County, North Dakota. *Canadian Journal of Botany*. 45, 21-58.

Dobzhansky, T. (1950). Evolution in the tropics. *American Scientist*, 38, 209-21.

Doctors van Leeuwen, W.M. (1929). Krakatau's new flora. *Proceedings of the Fourth Pacific Science Congress (Batavia) Part 2*, pp. 56-71.

Dodd, A.P. (1940). *The Biological Campaign against Prickly Pear*. Commonwealth Prickly Pear Board. Brisbane: Government Printer.

Dodd, A.P. (1959). The biological control of prickly pear in Australia. In *Biogeography and Ecology in Australia*, ed. A. Keast, R.L. Crocker, and C.S. Christian, *Monograph in Biology*, 8, 565-77.

Dodd, J.L., and Lauenroth, W.K. (1979). Analysis of the response of a grassland ecosystem to stress. In *Perspectives in Grassland Ecology*, ed. N.R. French, pp. 43-58. New York: Springer-Verlag.

Dodge, R.E., Allen, R.C., and Thompson, J. (1974). Coral growth related to resuspension of bottom sediments. *Nature*, 247, 574-7.

Dodge, R.E., and Thompson, J. (1974). The natural radiochemical and growth records in contemporary hermatypic corals from the Atlantic and Caribbean. *Earth and Planetary Science Letters*, 23, 313-22.

Dolph, G.E., and Dilcher, D.L. (1980). Variation in leaf size with respect to climate in the tropics of the Western Hemisphere. *Bulletin of the Torrey Botanical Club*, 107, 154-62.

Donahue, P.K., Parker, T., Sorrie, B., and Scott, D. (1989). *Birds of the Tambopata Reserve*, Peru: Department of Madre de Dios.

Donald, C.M. (1958). The interaction of competition for light and for nutrients. *Australian Journal of Agricultural Research*, 9, 421-35.

Dowsett, R.J. (1966). Wet season game populations and biomass in the Ngoma area of the Kafue National Park. *Puku*, 4, 135-46.

Doyle, T. (1981). The role of disturbance in the gap dynamics of a montane rain forest: an application of a tropical forest succession model. In *Forest succession concepts and applications*, ed. D.C. West, H.H. Shugart, and D.B. Botkin, pp. 56-73. New York: Springer-Verlag.

Drake, J. (1990). The mechanics of community assembly and succession. *Journal of Theoretical Biology*, 147, 213-33.

Drake, J. A. (1991). Community-assembly mechanisms and the structure of an experimental species assemblage. *American Naturalist,* 137, 1-26.

Drake, J.A., Mooney, H.A. diCastri, F., Groves, R.H., Kruger, F.J., Rejmánek, M., and Williamson, M. (1989). *Biological Invasions, A Global Perspective. SCOPE 37.* Chichester: Wiley.

Dressler, R.L. (1981). *The Orchids: Natural History and Classification.* Cambridge, Mass.: Harvard University Press.

Drury, W.H., and Nisbet, I.C.T. (1973). Succession. *Journal of the Arnold Arboretum,* 54, 331-68.

Duellman, W.E. (1982). Quaternary climatic-ecological fluctuations in the lowland tropics: frogs and forests. In *Biological Diversification in the Tropics,* ed. G. Prance. pp. 389-402. New York: Columbia University Press.

Duellman, W.E. (1990). Herpetofaunas in neotropical rainforests: comparative composition, history, and resource use. In *Four Neotropical Rainforests,* ed. A.H. Gentry, pp. 455-505. New Haven: Yale University Press.

Duellman, W.E., and Pianka, E.R. (1990). Biogeography of nocturnal insectivores: historical events and ecological filters. *Annual Review of Ecology and Systematics,* 21, 57-68.

Dunbar, M.J. (1968). *Ecological Development in Polar Regions.* Englewood Cliffs, New Jersey: Prentice-Hall.

Dyksterhuis, E.J. (1958). Ecological principles in range evaluation. *Botanical Reviews,* 24, 253-72.

Dyrness, C. T. (1973). Early stages of plant succession following logging and burning in the western Cascades of Oregon. *Ecol.,* 54, 57-69.

Edmond, J.M., and Von Damm, K.L. (1983). Hot springs on the ocean floor. *Scientific American,* 248, 78-93.

Edmond, J.M., and Von Damm, K.L. (1985). Chemistry of ridge crest hot springs. *Biological Society of Washington Bulletin,* 1985, 43-7.

Eggeling, W.J. (1947). Observations on the ecology of Budongo rain forest, Uganda. *Journal of Ecology,* 34, 20-87.

Egler, F.E. (1942). Indigene versus alien in the development of arid Hawaiian vegetation. *Ecology,* 23, 14-23.

Egler, F.E. (1954). Vegetation science concepts. I. Initial floristic composition: a factor in old field vegetation development. *Vegetatio,* 4, 412-417.

Ehler, L.E., and Hall, R.W. (1982). Evidence for competitive exclusion of introduced natural enemies in biological control. *Environmental Entomology,* 11, 1-4.

Ehleringer, J.R. (1984). Intraspecific competitive effects on water relations, growth, and reproduction in *Encelia farinosa. Oecologia,* 63, 153-8.

Ehrenreich, J.H., and Aikman, J.M. (1963). An ecological study of the effects of certain management practices on native prairie in Iowa. *Ecological Monographs,* 33, 113-30.

Ehrlich, P., and Raven, P.H. (1964). Butterflies and plants: a study in coevolution. *Evolution,* 18, 586-608.

Ehrlich, P.R. (1975). The population biology of coral reef fishes. *Annual Review of Ecology and Systematics,* 6, 211-47.

Ehrlich, P.R. (1986). Which animals will invade? In *Ecology of Biological Invasions of North America and Hawaii,* ed. H.A. Mooney and J.A. Drake, pp. 79-95. New York: Springer-Verlag.

Ehrlich, P.R., and Ehrlich, A.H. (1973). Coevolution: heterotypic schooling in Caribbean reef fishes. *American Naturalist,* 107, 157-60.

Eisenberg, J.F. (1980). The density and biomass of tropical mammals. In *Conservation Biology: An Evolutionary-Ecological Perspective*, ed. M.E. Soulé and B.A. Wilcox. pp. 35-55. Sunderland, Massachusetts: Sinauer Associates, Inc.

Eisenberg, J.F. (1983). Behavioral adaptations of higher vertebrates to tropical forests. In *Tropical Rain Forest Ecosystems. Biogeography and Ecology*, Ecosystems of the World. Vol. 14B, ed. H. Lieth and M.J.A. Werger, pp. 267-78. Amsterdam: Elsevier.

Eisenberg, J.F., O'Connell, M.A., and August, P.V. (1979). Density, productivity, and distribution of mammals in two Venezuelan habitats. In *Vertebrate Ecology in the Northern Neotropics*, ed. J.F. Eisenberg, pp. 187-207. Washington, D.C.: Smithsonian Institution Press.

Eisenberg, J.F., and Lockhart, M. (1972). An ecological reconnaissance of Wilpattu National Park, Ceylon. *Smithsononian Contributions Zoolology*, 101, 1-118.

Eisenberg, J.F., and Thorington, R.W. (1973). A preliminary analysis of a neotropical mammal fauna. *Biotropica*, 5, 150-61.

Eldredge, N., and S.J. Gould. (1972). Punctuated equilibria: An alternative to phyletic gradualism. In *Models in Paleobiology*, ed. T.J.M. Schopf. pp. 82-115. San Francisco: Freeman and Cooper.

Ellenberg, H. (1953). Physiologisches und okologisches Verhalten derselben Pflanzenarten. *Berichte derr Deutschen Botanischen Gesellschaft*,65, 351-62.

Ellenberg, H. (1954). Uber einige Fortschritte der Kausalen Vegetationskunde. *Vegetatio*, 5/6, 199-211.

Elliott, J.P., and Cowan, I.M. (1978). Territoriality, density, and prey of the lion in Ngorongoro Crater, Tanzania. *Canadian Journal of Zoology*, 56, 1726-34.

Elton, C.S. (1927). *Animal Ecology*. New York: Macmillan

Elton, C.S. (1958). *The Ecology of Invasions by Plants and Animals*. London: Methuen.

Eltringham, S.K. (1979). *The Ecology and Conservation of Large African Mammals*. London: Macmillan.

Emberger, L. (1930). La vegetation de la region mediterraneenne. Essai d'une classification des groupements vegetaux. *Rev. Gen. Bot.*, 42, 641-62, 705-21.

Emberger, L. (1955). Une classification biogeographique des climats. *Recl. Trav. Lab. Bot., Geol. Zool., Ser. Bot.*, 7, 3-43.

Emmons, L.H. (1984). Geographic variation in densities and diversities of non-flying mammals in Amazonia. *Biotropica*, 16, 210-22.

Endler, J.A. (1977). *Geographic Variation, Speciation, and Clines*. Monographs in Population Biology No. 10. Princeton: Princeton University Press.

Endler, J.A. (1982a). Pleistocene forest refuges: fact or fancy. In *Biological Diversification in the Tropics*, ed. G.T. Prance, pp. 641-57. New York: Columbia University Press.

Endler, J.A. (1982b). Problems in distinguishing historical from ecological factors in biogeography. *American Zoologist*, 22, 441-52.

Engle, D.M., and Bultsma, P.M. (1984). Burning of northern mixed priarie during drought. *Journal of Range Management*, 37, 398-401.

Enright, J.T., Newman, W.A., Hessler, R.R., and McGowan, J.A. (1981). Deep-ocean hydrothermal vent communities. *Nature*, 289, 219-21.

Eppley, R.W., Renger, E.H., Venrick, E.L., and Mullin, M.M. (1973). A study of plankton dynamics and nutrient cycling in the Central Gyre of the North Pacific Ocean. *Limnology and Oceanography*, 18, 534-51.

Eriksson, O., Hansen, A., and Sunding, P. (1974). *Flora of Macronesia: Checklist of Vascular Plants*, Sweden: University of Umeå.

Erwin, T.L. (1982). Tropical forests: their richness in Coleoptera and other Arthropod species. *Coleoptera Bulletin*, 36, 74-5.

Erwin, T.L. (1983). Beetles and other insects of tropical forest canopies at Manaus, Brazil, sampled by insecticidal fogging. In *Tropical Rain Forest: Ecology and Management*, ed. S.L. Sutton, T.C. Whitmore, and A.C. Chadwick. Edinburgh: Blackwell

Erwin, T.L. (1988). The tropical forest canopy: The heart of biotic diversity. In *Biodiversity*, ed. E.O. Wilson. Washington: National Academy Press.

Evans, F.C., and Dahl, E. (1955). The vegetational structure of an abandoned field in southeastern Michigan and its relation to environmental factors. *Ecology*, 36, 685-97.

Ewel, J. (1971). Biomass changes in early tropical succession. *Turrialba*, 21, 110-2.

Ewel, J.J. (1986). Invasibility: lessons from South Florida. In *Ecology of Biological Invasions of North America and Hawaii*, ed. H.A. Mooney and J.A. Drake, pp. 214-30. New York: Springer-Verlag.

Fagerstrom, J.A. (1987). *The Evolution of Reef Communities*. New York: Wiley.

Farquhar, G.D., and Sharkey, T.D. (1982). Stomatal conductance and photosynthesis. *Annual Review of Plant Physiology*, 33, 317-45.

Faure, G. (1974). Morphology and bionomy of the coral reef discontinuities in Rodriguez Island. *Proceedings of the Second International Coral Reef Symposium*, 2, 161-72.

Fearnside, P. (1979). Cattle yield prediction for the Transamazon Highway of Brazil. *Intersciencia*, 4, 220-5.

Fearnside, P. (1985). Agriculture in Amazonia. In *Amazonia*, ed. G.T. Prance and T.E. Lovejoy. pp. 393-418. Oxford: Pergamon.

Fearnside, P. (1987). Causes of deforestation in the Brazilian Amazon. In *The Geophysiology of Amazonia*, ed. R.E. Dickinson, pp. 37-61. New York: Wiley.

Fearnside, P.M. (1990a). Deforestation in Brazilian Amazonia. In *The Earth in Transition: Patterns and Processes of Biotic Impoverishment*, ed. G.M. Woodwell, pp. 211-38. Cambridge: Cambridge University Press.

Fearnside, P.M. (1990b). Fire in the tropical rainforest of the Amazon basin. In *Fire in the Tropical Biota*, ed. J.G. Goldammer, pp. 106-16. Berlin: Springer-Verlag.

Feinsinger, P., and Colwell, R.K. (1978). Community organization among neotropical nectar-feeding birds. *The American Zoologist*, 18, 779-95.

Fenchel, T. (1975). Character displacement and coexistence in mud snails (Hydrobiidae). *Oecologia*, 20, 19-32.

Fenchel, T. (1988). Marine plankton food chains. *Ann. Rev. Ecol. Syst.*, 19, 19-38.

Fenchel, T., and Kofoed, L. (1976). Evidence for exploitative interspecific competition in mud snails (Hydrobiidae). *Oikos*, 27, 367-76.

Ferguson, M.W.J., and Jaonen, T. (1982). Temperature of egg incubation determines sex in *Alligator mississippiensis*. *Nature*, 296, 850-3.

Ferrar, A.A., and Walker, B.H. (1974). An analysis of herbivore/habitat relationships in Kyle National Park, Rhodesia. *Journal of the South African Management Association*, 4, 137-47.

Fishelson, L. (1973). Ecological and biological phenomena influencing coral-species composition on the reef tables at Eilat (Gulf of Aquaba, Red Sea). *Marine Biology,* 19, 183-96.

Fisher, R.A., Corbet, A.S., and Williams, C.B. (1943). The relationship between the number of species and the number of individuals in a random sample of an animal population. *Journal of Animal Ecology,* 12, 42-58.

Fitter, A.H., and Hay, R.K.M. (1981). *Environmental Physiology of Plants.* London: Academic Press.

Fittkau, E.J. (1971). Ökologische Gliederung des Amazonas-Gebietes auf geochemischer Grundlage. *Münster. Forsch. Geol. Paläont.* 20/21, 35-50.

Fittkau, E.J., Irmler, J., Junk W.J., Feiss, F., and Schmidt, G.W. (1975). Productivity, biomass, and population dynamics in Amazonian water bodies. In *Tropical Ecological Systems,* ed. F.B Golley and E. Medina, pp. 281-311. New York: Springer-Verlag.

Fleming, T.H., Breitwisch, R., and Whitesides, G.H. (1987). Patterns of tropical vertebrate frugivore diversity. *Annual Review of Ecology and Systematics,* 18, 91-110.

Flemming, T.H. (1973). Numbers of mammal species in North and Central American forest communities. *Ecology,* 54, 555-63.

Flessa, K.W., and Thomas, R.H. (1985). Modeling the biogeographic regulation of evolutionary rates. In *Phanerozoic Diversity Patterns, Profiles in Macroevolution,* ed. J.W. Valentine, pp. 355-76. Princeton: Princeton University Press.

Flint, H.R. (1925). Fire resistance of northern Rocky Mountain conifers. *Idaho Forester,* 7, 7-10, 41-3.

Forbes, E. (1844). On the light thrown on geology by submarine researches. *New Philosophical Journal* (Edinburgh), 36, 318-27.

Forbes, S.A. (1880a). On some interactions of organisms. *Bulletin Illinois State Laboratory of Natural History,* 1, 3-17.

Forbes, S.A. (1880b). The food of fishes. *Bulletin Illinois State Laboratory of Natural History,* 1, 18-65.

Forbes, S.A. (1883). The first food of the common whitefish. *Bulletin Illinois State Laboratory of Natural History,* 1, 95-109.

Forbes, S.A. (1887). The lake as a microcosm. *Bulletin Science Association of Peoria,* 1887, 77-87.

Ford, E.D. (1975). Competition and stand structure in some even-aged plant monocultures. *Journal of Ecology,* 63, 311-33.

Ford, E.D., and Diggle, P.J. (1981). Competition for light in a plant monoculture as a spatial stochastic process. *Annals of Botany (London),* 48, 481-500.

Ford, E.D., and Newbould, P.J. (1970). Stand structure and dry weight production through the sweet chestnut (*Castanea sativa* Mill.) coppice cycle. *Journal of Ecology,* 58, 275-96.

Foster, J.B., and Coe, M.J. (1968). The biomass of game animals in Nairobi National Park. *Journal of Zoology, London,* 155, 413-25.

Foster, R.B. (1990). Long-term change in the successional forest community of the Rio Manu floodplain. In *Four Neotropical Rainforests,* ed. A.H. Gentry, pp. 565-72. New Haven: Yale University Press.

Foster, T. (1976) *Bushfire: History, Prevention, Control.* Sydney: Reed.

Fox, B.J. (1981). Mammal species diversity in Australian heathlands: the importance of pyric succession and habitat diversity. In *Mediterranean-type Ecosystems: The Role of Nutrients, Ecological Studies Volume 43,* ed. F.J.

Kruger, D.T. Mitchell, and J.U.M. Jarvis, pp. 473-89. Berlin: Springer-Verlag.

Fox, J.F. (1979). Intermediate disturbance hypothesis. *Science.* 204, 1344-5.

Fox, M.D.,and Fox, B.J. (1986). The susceptibility of natural communities to invasion. In *Ecology of Biological Invasions: an Australian Perspective,* ed. R.H. Groves and J.J. Burdon, pp. 57-66. Canberra: Australian Academy of Science.

Frank, P.W. (1952). A laboratory study of intraspecies and interspecies competition in *Daphnia pulicer* and *Simocephalus vetulus. Physiological Zoology,* 25, 178-204.

Frank, P.W. (1957).Coactions in laboratory populations of two species of *Daphnia. Ecology,* 38, 510-7.

Frank, P.W., Boll, C.D., and Kelly, R.W. (1957). Vital statistics of laboratory cultures of *Daphnia pulex* De Geer as related to density. *Physiological Zoology,* 30, 287-305.

Frankie, G.W., Baker, H.G., and Opler, P.A. (1974). Comparative phenological studies of trees in tropical wet and dry forests in the lowlands of Costa Rica. *Journal of Ecology,* 62, 881-919.

Franklin, J.F., and Hemstrom, M.A. (1981). Aspects of succession in the coniferous forests of the Pacific Northwest. In *Forest Succession: Concepts and Application,* ed. D.C. West, H.H. Shugart, and D.B. Botkin. pp. 212-29

Fransden, W.H. (1971). Fire spread through porous fuels from the conservation of energy. *Combustion and Flame,* 16, 9-16.

Fransden, W.H. (1973). Effective heating of fuel ahead of spreading fire. Research Paper INT-140. Ogden, Utah: U.S. Department of Agriculture Forest Service.

Frenkel, R.E. (1970). Ruderal vegetation along some California roadsides. *University of California Publications in Geography,* 20, 1-163.

Fridricksson, S. (1975). *Surtsey: Evolution of Life on a Volcanic Island.* London: Butterworths.

Fry, C.H. (1983). Birds in savanna ecosystems. In *Ecosystems of the World 13: Tropical Savannas,* ed. F. Bourlière, pp. 337-57. Amsterdam: Elsevier.

Fryer, G. (1959). Some aspects of evolution in Lake Nyasa. *Evolution,* 13, 440-51.

Fryer, G., and Iles, T.D. (1972). *The Cichlid Fishes of the Great Lakes of Africa: Their Biology and Evolution.* Edinburgh: Oliver and Boyd.

Futuyma, D.J., and Slatkin, M. eds. (1983). *Coevolution.* Sunderland, Massachusetts: Sinauer.

Gadgil, M, and Solbrib, O.T. (1972). The concept of *r*- and *K*-selection: evidence from wild flowers and some theorteical considerations. *American Naturalist,* 106, 14-31.

Gaines, S.D., and Lubchenco, J. (1982). A unified approach to marine plant-herbivore interactions. II. Biogeography. *Ann. Rev. Ecol. Syst.,* 13, 111-38.

Gambell, R. (1985). Birds and mammals - Antarctic whales. In *Key Environments - Antarctica,* ed. W.N. Bonner and and D.W.H. Walton, pp. 223-41. Oxford: Pergamon.

Gardner, M.R., and Ashby, W.R. (1970). Connectance of large dynamical (cybernetic) systems: critical values of stability. *Nature,* 228, 784.

Gardner, R.H., Cale, W.G., and O'Neill, R.V. (1982). Robust analysis of aggregation error. *Ecology,* 63, 1771-9.

Gardner, R.H., O'Neill, R.V., Mankin, J.B., and Carney, J.H. (1981). A comparison of sensitivity analysis and error analysis based on a stream ecosystem model. *Ecological Modelling*, 12, 173-90.

Garren, K.H. (1943). Effects of fire on vegetation of the Southeastern United States. *Botanical Review*, 9, 617-54.

Garwood, N.C., Janos, D.P., and Brokaw, N. (1979). Earthquake caused landslides: a major disturbance to tropical forests. *Science*, 205, 997-9.

Gates, C.T. (1968). Water deficits and growth of herbaceous plants. In *Water Deficits and Plant Growth II. Plant Water Consumption and Response*, ed. T.T. Kozlowski, pp. 135-190. New York: Academic Press.

Gates, D.J. (1978). Bimodality in even-aged plant monocultures. *Journal of Theoretical Biology*, 71, 525-40.

Gates, D.M. (1980). *Biophysical Ecology*. New York: Springer-Verlag.

Gauch, H.G., and Whittaker, R.H. (1972). Coenocline simulation. *Ecology*, 53, 446-51.

Gause, G.F. (1934). *The Struggle for Existence*. Baltimore: Williams and Williams.

Gause, G.F. (1935). Experimental demonstration of Volterra's periodid oscillation in the numbers of animals. *Journal of Experimental Biology*, 12, 44-8.

Geister, J. (1977). The influence of wave exposure on the ecological zonation of caribbean coral reefs. *Proceedings of the Third International Coral Reef Symposium*, pp. 23-9.

Gentilli, J. (1949). Foundation of Australian bird geography. *Emu*, 49, 85-130.

Gentry, A.H. (1982). Patterns of Neotropical plant species diversity. In *Evolutionary Biology*, ed. M.K. Hecht, B. Wallace, and G.T. Prance. 15, 1-84. New York: Plenum.

Gentry, A.H. (1986). Endemism in tropical versus temperate plant communities. In *Conservation Biology: The Science of Scarcity and Diversity*, ed. M. Soulé. pp, 153-81. Sunderland, Massachusetts: Sinauer Associates.

Gentry, A.H. (1986b). Species richness and floristic composition of Choco' region plant communities. *Caldasia*, 15, 71-91.

Gentry, A.H. (1988). Changes in plant community diversity and floristic composition on environmental and geographical gradients. *Annals of the Missouri Botanical Garden*, 75, 1-34.

Gentry, A.H. ed. (1990a). *Four Neotropical Rainforests*. New Haven: Yale University Press.

Gentry, A.H. (1990b). Floristic similarities and differences between southern Central America and upper and central Amazonia. In *Four Neotropical Rainforests*, ed. A.H. Gentry, pp. 141-57. New Haven: Yale University Press.

Gentry, A.H., and Dodson, C.H. (1987a). Contribution of nontrees to species richness of a tropical rain forest. *Biotropica*, 19, 149-56.

Gentry, A.H., and Dodson, C.H. (1987b). Diversity and biogeography of neotropical vascular epiphytes. *Annals of the Missouri Botanical Gardens*, 74, 205-33.

Gentry, A.H., and Emmons, L.H. (1987). Geographical variation in fertility, phenology, and composition of the understory of neotropical forests. *Biotropica*, 19, 216-27.

Gentry, A.W., and Terborgh, J. (1990). Composition and dynamics of the Cocha Cashu 'mature' floodplain forest. In *Four Neotropical Rainforests*, ed. A.W. Gentry, pp. 542-64. New Haven: Yale University Press.

George, A.S. (1984). *The Banksia Book*. Kenhurst, New South Wales: Kangaroo Press.

Getty, T. (1981). Poisson patterns in behavioural time series: the perception of randomness in complexity. *Animal Behaviour, 29*, 960.

Gibbs, R.J. (1967). The geochemistry of the Amazon river system. Part I: The factors that control the salinity and the composition and concentrations of the suspended solids. *Geolological Society of America Bulletin, 78*, 1203-32.

Gibbs, R.J. (1972). Water chemistry of the Amazon River. *Geochimica Cosmochimica Acta, 36*, 1061-6.

Gibson, C.W.D., Brown, V.K., and Jepsen, M. (1987a). Relationships between the effects of insect herbivory and sheep grazing on seasonal changes in an early successional plant community. *Oecologia, 71*, 245-53.

Gibson, C.W.D., Dawkins, H.C., Brown, V.K., and Jepsen, M. (1987b). Spring grazing by sheep: effects on seasonal changes during early old field succession. *Vegetatio, 70*, 33-43.

Gibson, D.J. (1988). Regeneration and fluctuation of tallgrass prairie vegetation in response to burning frequency. *Bulletin of the Torrey Botanical Club, 115*, 1-12.

Gibson, D.J., and Hulbert, L.C. (1987). Effects of fire, topography and year-to-year climatic variation on species composition in tallgrass prairie. *Vegetatio, 72*, 175-85.

Gifford, R.M., and Evans, L.T. (1981). Photosynthesis, carbon partitioning, and yield. *Annual Review of Plant Physiol.,32*, 485-509.

Gilbert, F.S. (1980). The equilibrium theory of island biogeography: fact or fiction? *J. Biogeography, 7*, 209-35.

Gilbert, J.M. (1960). Regeneration of *Eucalyptus regnans* in the Florentine Valley. *Appita, 13*, 132-5.

Gilbert, J.M. (1963). Fire as a factor in the development of vegetational types. *Australian Forestry, 27*, 67-70.

Gilbert, L.E. (1975). Ecological consequences of coevolved mutualism between butterflies and plants. In *Coevolution of Animals and Plants*, ed. L.E. Gilbert and P.H. Raven, pp. 210-40. Austin: University of Texas Press.

Gilbert, L.E. (1980). Food Web Organisation and the conversation of neotropical biodiversity. In *Conservation Biology: An Evolutionary–Ecological Perspective*, ed. M.E. Soulé and B.A. Wilcox, pp11-33. Sunderland, Mass: Sinauer Associates.

Gilbert, L.E., and Smiley, J.T. (1978). Determinants of local diversity in phytophagous insects: host specialists in tropical environments. In *Diversity of Insect Faunas*, ed. L.A. Mound and L. Waloff. London: Blackwell.

Gill, A.M. (1975). Fire and the Australian flora: a review. *Australian Forestry, 38*, 4-25.

Gill, A.M. (1981). Fire adaptive traits of vascular plants. In *Fire Regimes and Ecosystem Properties*, ed. H.A. Mooney, T.M. Bonnicksen, N.L. Christensen, J.E. Lotan, and W.A. Reiners, pp. 208-230. GTR-WO-26. Washington, D.C.: U.S. Forest Service - Washington Office.

Gill, A.M., Hoare, J.R.L., and Cheney, N.P. (1990). Fires and their effects in the wet-dry tropics of Australia. In *Fire and the Tropical Biota*. ed. J.G. Goldammer. pp. 159-78. Berlin: Springer-Verlag.

Gill, A.M., and Groves, R.H. (1981). Fire regimes in heathlands and their plant ecologic effects. In *Ecosystems of the World. Vol. 9B: Heathlands and Related Shrublands*, ed. R.L. Specht, pp. 61-84. Amsterdam: Elsevier.

Gill, D.E. (1986). Individual plants as genetic mosaics: ecological organisms versus evolutionary individuals. In *Plant Ecology*. ed. M.J. Crawley. pp. 321-44. Oxford: Blackwell.

Gill, D.S., and Marks, P.L. (1991). Tree and shrub seedling colonization of old fields in central New York. *Ecological Monographs*, 61, 183-205.

Gillon, Y. (1983). The invertebrates of the grass layer. In *Ecosystems of the World 13: Tropical Savannas*, ed. F. Bourlière, pp. 289-311. Amsterdam: Elsevier.

Gilpin, M.E., and Soulé, M.E. (1986). Minimum viable populations: processes of species extinction. In *Conservation Biology: The Science of Scarcity and Diversity*. ed. M.E. Soulé. pp. 19-34. Sunderland, Massachusetts: Sinauer Associates, Inc.

Gilpin, M.E.. (1972). Enriched predator-prey systems: theoretical stability. *Science*, 177, 902-4.

Gilpin, M.E.. (1975). Limit cycles in competition communities. *American Naturalist*, 109, 51-60.

Gilpin, M.E.. (1979). Spiral chaos in a predator-prey model. *American Naturalist*, 113, 306-8.

Givnish, T.J. (1978). On the adaptive significance of compound leaves, with particular reference to tropical trees. In *Tropical Trees as Living Systems*, ed. P.B. Tomlinson, and M.H. Zimmermann, pp. 351-80. Cambridge: Cambridge University Press.

Givnish, T.J. (1979). On the adaptive significance of leaf form. In *Topics in Plant Population Biology*, ed. O.T. Solbrig, S. Jain, G.B. Johnson and P.H. Raven, pp. 375-407. New York: Columbia University Press.

Givnish, T.J. (1982). On the adaptive significance of leaf height in forest herbs. *American Naturalist*, 120, 353-81.

Givnish, T.J. ed. (1986). *On the economy of plant form and function*. Cambridge: Cambridge Univ. Press.

Gladfelter, W.B., and Gladfelter, E.H. (1978). Fish community structure as a function of habitat structure on West Indian Patch reefs. *Revista de Biologia Tropical*, 26 (supplement 1), 65-84.

Glanz, W.E. (1990). Neotropical mammal densities: how unusual is the community on Barro Colorado Island, Panama. In *Four Neotropical Rainforests*, ed. A.H. Gentry, pp. 287-311. New Haven: Yale University Press.

Gleason, H.A. (1913). The relation of forest distribution and prairie fires in the middle west. *Torreya*, 13, 173-81.

Gleason, H.A. (1917). The structure and development of the plant association. *Bulletin of the Torrey Botanical Club*, 43, 463-81.

Gleason, H.A. (1922). On the relation of species and area. *Ecology*, 3, 158-62.

Gleason, H.A. (1923). The vegetational history of the Middle West. *Annals of the Association of American Geographers*, 12, 39-85.

Gleason, H.A. (1925). Species and area. *Ecology*, 6, 66-74.

Gleason, H.A. (1926). The individualistic concept of the plant association. *Bulletin of the Torrey Botanical Club*, 53, 1-20.

Gleason, H.A. (1939). The individualistic concept of the plant association. *American Midland Naturalist*, 21, 92-110.

Gleick, J. (1987). Chaos. New York: Viking.

Glenn-Lewin, D.C. (1977). Species diversity in North American temperate forests. *Vegetatio*, 33, 153-62.

Glitzenstein, J.S., Harcombe, P.A., and Strong, D.R. (1986). Disturbance, succession, and maintenance of species diversity in an east Texas forest. *Ecological Monographs*, 56, 243-58.

Glynn, P.W. (1973). Ecology of a Caribbean coral reef. The Porites reef-flat biotype: Part II. Plankton community with evidence for depletion. *Marine Biology*, 22, 1-21.

Glynn, P.W. (1974). Rolling stones among the Scleractinea: Mobile coralliths in the Gulf of Panama. *Proceedings of the Second International Coral Reef Symposium*, 2, 183-98.

Glynn, P.W. (1976). Some physical and biological determinants of coral community structure in the eastern pacific. *Ecological Monographs*, 46, 431-56.

Glynn, P.W. (1977). Coral growth in upwelling and nonupwelling areas off the Pacific coast of Panama. *Journal of Marine Research*, 35, 567-85.

Glynn, P.W., Stewart, R.H., McCosker, J.E. (1972). Pacific coral reefs of Panama: structure, distribution, and predators. *Geologische Rundschau*, 61, 483-519.

Glynn, P.W., and Stewart, R.H. (1973). Distribution of coral reefs in the Pearl Islands (Gulf of Panama) in relation to thermal conditions. *Limnology and Oceanography*, 18, 367-79.

Goh, B.S. (1979). Robust stability concepts for ecosystem models. In *Theoretical Systems in Ecology*. ed. E. Halfon. pp 467-87. New York: Academic Press.

Goldammer, J.B., and Siebert, B. (1990). The impact of droughts and forest fires on tropical lowland rain forest of East Kalimantan. In *Fire in the Tropical Biota*, ed. J.G. Goldammer, pp. 11-31. Berlin: Springer-Verlag.

Goldammer, J.G. (1987). Wildfires and forest development in tropical and subtropical Asia: outlook for the year 2000. In *Proceedings of the Symposium on Wildland Fire 2000, U.S. Forest Service General Technical Report PSW-101*, ed. J.B. Davis and R.E. Martin, pp. 164-76. Berkeley, California: Pacific Southwest Forest and Range Experiment Station.

Goldammer, J.G. ed. (1990). *Fire in the Tropical Biota: Ecosystem Processes and Global Challenges*. Berlin: Springer-Verlag.

Goldammer, J.G., and Pénafiel, S.R. (1990). Fire in the pine-grassland biome of tropical and subtropical Asia. In *Fire in the Tropical Biota*, ed. J.G. Goldammer, pp. 45-62. Berlin: Springer-Verlag.

Goldberg, D.E., and Miller, T.E. (1990). Effects of different resource additions on species diversity in an annual plant community. *Ecology*, 71, 213-25.

Goldberg, W.M. (1973). The ecology of the coral-octocoral communities off the southeast Florida coast: geomorphology, species composition, and zonation. *Bulletin of Marine Science*, 23, 465-88.

Goldblatt, P. (1978). An analysis of the flora of Southern Africa: its characteristics, relationships, and origins. *Annals of the Missouri Botanical Garden*, 65, 369-436.

Goldman, B., and Talbot, F.H. (1976). Aspects of the ecology of coral reef fishes. In *Biology and Geology of Coral Reefs. Vol. III. Biology 2*. ed. O.A. Jones and R. Endean. pp. 125-54. New York: Academic Press.

Golley, F.B., McGinnis, J.T., Clements, R.G., Child, G.I., and Duever, M.J. (1975). *Mineral Cycling in a Tropical Moist Forest Ecosystem*. Athens, Georgia: University of Georgia Press.

Goodland, R.J.A. (1980). Environmental ranking of Amazonian development projects in Brazil. *Environmental Conservation*, 7, 9-26.

Goreau, T.F. l959a. The ecology of Jamaican reefs. I. Species composition and zonation. *Ecology*, 40, 67-90.

Goreau, T.F. l959b. The physiology of skeleton formation in corals. I. A method for measuring the rate of calcium deposition by corals under different conditions. *Biological Bulletin of the Marine Biological Laboratory, Woods Hole*, 116, 59-75

Goreau, T.F., and Goreau, N.I. (1959). The physiology of skeleton formation in corals II. Calcium deposition by hermatypic corals under various conditions in the reef. *Biological Bulletin of the Marine Biological Laboratory, Woods Hole*, 117, 239-50.

Goreau, T.F., and Goreau, N.I. (1973). The ecology of Jamaican coral reefs. II. Geomorphology, zonation, and sedimentary phases. *Bulletin of Marine Science*, 23, 399-464.

Goreau, T.F., and Wells, J.W. (1967). The shallow-water Scleractinea of Jamaica: revised list of species and their vertical ranges. *Bulletin of Marine Science*, 17, 442-53.

Gorman, M.L. (1979). *Island Ecology*. London: Chapman and Hall.

Goudie, A. (1990). *The Human Impact*. Oxford: Blackwell.

Gould, F.W. (1968). *Grass Systematics*. New York: McGraw-Hill.

Gould, S.J. (1989). *Wonderful Life: The Burgess Shale and the Nature of History*. New York: W.W. Norton and Company.

Grace, J.B., and Tilman, D. eds. (1990). *Perspectives on Plant Competition*. San Diego: Academic Press.

Graham, A. (1982). Diversification beyond the Amazon basin. In *Biological Diversification in the Tropics*, ed. G. Prance. pp. 78-90. New York: Columbia University Press.

Grant, P.R. (1972). Convergent and divergent character displacement. *Biological Journal of the Linnean Society*, 4, 39-68.

Grant, P.R. (1975). The classical case of character displacement. In *Evolutionary Biology*. ed. T. Dobzhansky, M.K. Hecht, and W.C. Steere. Vol. 8, pp. 237-337. New York: Plenum.

Grant, P.R. (1986). *The Ecology and Evolution of Darwin's Finches*. Princeton: Princeton University Press.

Grant, P.R., and Abbott, I. (1980). Interspecific competition, island biogeography and null hypotheses. *Evolution*, 34, 322-41.

Grant, P.R., and Boag, P.T. (1980). Rainfall on the Galapagos and the demography of Darwin's finches. *Auk*, 97, 227-44.

Grant, P.R., and Schluter, D. (1984). Interspecific competition inferred from patterns of guild structure. In *Ecological Communities, Conceptual Issues and the Evidence*, ed. D.R. Strong, D. Simberloff, L.G. Abele, A.B. Thistle, pp. 201-31. Princeton: Princeton University Press.

Granville, J. de. (1982). Rain forest and xeric flora refuges in French Guiana. In *Biological Diversification in the Tropics*, ed. G. Prance. pp. 159-81. New York: Columbia University Press.

Grassle, J.F. (1985). Hydrothermal vent animals: distribution and biology. *Science*, 229, 713-7.

Grassle, J.F. (1991). Deep-sea benthic biodiversity. *BioScience*, 41, 464-9.

Gray, W.M. (1990). Strong association between West African rainfall and U.S. landfall of intense hurricanes. *Science*, 249, 1251-6.

Greathead, D.J. (1971). *A Review of Biological Control in the Ethiopian Region.* Technical Communication No. 5, Commonwealth Agricultural Bureau, Farnham Royal.

Greathead, D.J. (1976). *A Review of Biological Control in Western and Southern Europe.* Technical Communication No. 7, Commonwealth Agricultural Bureau, Farnham Royal.

Grebmeier, J.M., Feder, H.W., and McRoy, C.P. (1989). Pelagic-benthic coupling on the shelf of the northern Bering and Chukchi Seas. II. Benthic community structure. *Marine Ecology Progress Series,* 51, 253-68.

Grebmeier, J.M., McRoy, C.P., and Feder, H.M. (1988). Pelagic-benthic coupling on the shelf of the northern Bering and Chukchi Seas. I. Food supply and benthic biomass. *Marine Ecology Progress Series,* 48, 57-67.

Greenslade, P.J.M. (1983). Adversity selection and the habitat templet. *American Naturalist,* 122, 352-65.

Greenwood, P.H. (1965). The cichlid fishes of Lake Nabugabo, Uganda. *Brit. Mus. Nat. Hist. Bull. (Zool.),* 12, 315-57.

Greenwood, P.H. (1974). The cichlid fishes of Lake Victoria, East Africa: The biology and evolution of a species flock. *British Museum (Natural History) Bull.* Suppl. 6. 134 pp.

Grenney, W.J., Bella, D.A. and Curl, H.C. (1973). A theoretical approach to interspecific competition in phytoplankton communities. *American Naturalist,* 107, 405-25.

Grice, G.D., and Hart, A.D. (1962). The abundance, seasonal occurrence and distribution of the epizooplankton between New York and Bermuda. *Ecological Monographs,* 32, 287-309.

Grieg-Smith, P. (1957). *Quantitative Plant Ecology.* London: Butterworths Scientific.

Grigg, R.W. (1983). Community structure, succession, and development of coral reefs in Hawaii. *Marine Ecology, Progress Series,* 11, 1-14.

Grigg, R.W., Maragos, J.E. (1974). Recolonization of hermatypic corals on submerged lava flows in Hawaii. *Ecology,* 55, 387-95.

Grime, J.P. (1973). Control of species density in herbaceous vegetation. *Journal of Environmental Management,* 1, 151-67.

Grime, J.P. (1974). Vegetation classification by reference to strategies. *Nature,* 250, 26-31.

Grime, J.P. (1977). Evidence for the existence of three primary strategies in plants and its relevance to ecological and evolutionary theory. *American Naturalist,* 111, 1169-94.

Grime, J.P. (1979). *Plant Strategies and Vegetation Processes.* New York: John Wiley.

Grime, J.P., and Hunt, R. (1975). Relative growth rate: Its range and adaptive significance in a local flora. *Journal of Ecology,* 63, 393-422.

Grime, J.P., and Jeffrey, D.W. (1965). Seedling establishment in vertical gradients of sunlight. *Journal of Ecology,* 53, 621-42.

Grime, J.P., and Lloyd, P.S. (1973). *An Ecological Atlas of Grassland Plants.* London: Edward Arnold.

Grinnell, J. (1904). The origin and distribution of the Chestnut-backed Chickadee. *Auk,* 21, 364-82.

Groves, R.H. (1989). Ecological control of invasive terrestrial plants. In *Biological Invasions: a Global Perspective. SCOPE 37,* ed. J.A. Drake *et al.,* pp. 437-61. Chichester: Wiley.

Groves, R.H., and Burdon, J.J. eds. (1986). *Ecology of Biological Invasions: An Australian Perspective*. Canberra: Australian Academy of Science.

Grubb, P.J. (1977). The maintenance of species-richness in plant communities: the importance of the regeneration niche. *Biological Review of the Cambridge Philosophical Society*, 52, 107-45.

Grubb, P.J. (1982). Refuges and dispersal in the speciation of African forest mammals. In *Biological Diversification in the Tropics*, ed. G. Prance. pp. 537-53. New York: Columbia University Press.

Grubb, P.J. 1986. Some generalizing ideas about colonization and succession in green plants and fungi. In *Colonization, Succession, and Stability*, ed. A.J. Gray, M.J. Crawley, and P.J. Edwards,

Gulmon, S.L. (1977). A comparative study of the grassland of California and Chile. *Flora*, 166, 261-78.

Gurney, W.S.C., and Nisbet, R.M. (1978). Single-species population fluctuations in patchy environments. *American Naturalist*, 112, 1075-90.

Guth, A.R., and Cohen, S.B. (1989). *Red Skies of '88*. Missoula, Montana, USA: Pictorial Histories Publishing Company.

Gwynne, M.D., and Bell, R.H.V. (1968). Selection of grazing components by grazing ungulates in the Serengeti National Park. *Nature, London*, 220, 390-3.

Gérard, P. (1960). Etude écologique de la forêt dense á *Gilbertiodendron dewevrei* dans la région de l'Uele. INEAC (Inst. Natl. Etude Agron. Congo) Ser. Sci. no 87. Brussels.

Haedrich, R.L, and Rowe, G.T. (1977), Megafaunal biomass in the deep sea. *Nature*, 269, 141-2.

Haedrich, R.L., Rowe, G.T., and Polloni, P.T. (1975). Zonation and faunal composition of epibenthic populations on the continental slope south of New England. *Journal of Marine Research*, 33, 191-212.

Haedrich, R.L., Rowe, G.T., and Polloni, P.T. (1980). The megabenthic fauna in the deep sea south of New England. *Marine Biology*, 57, 165-79.

Haffer, J. (1969). Speciation in Amazonian forest birds. *Science*, 165, 131-7.

Haffer, J. (1974). *Avian Speciation in Tropical South America*. Cambridge, Massachussetts: Nuttall Ornithological Club. 390 pp.

Haffer, J. (1981). Aspects of Neotropical bird speciation during the Cenozoic. In *Vicariance Biogeography: A Critique*, ed. G. Nelson and D.E. Rosen. pp. 371-94. New York: Columbia University Press.

Haffer, J. (1982). General aspects of the refuge theory. In *Biological Diversification in the Tropics*, ed. G.T. Prance. New York: Columbia University Press.

Hairston, N.G., Smith, F.E., and Slobodkin, L.B. (1960). Community structure, population control, and comeptition. *Am. Nat.*, 94, 421-5.

Halkka, O. (1978). Influence of spatial and host-plant population on polymorphism in *Philaenus spumarius*. In *Diversity of Insect faunas*, ed. L.A. Mound and N Waloff, pp. 41-55. Oxford: Blackwell.

Halkka, O., Halkka, L. Hovinen, R., Raatikaninen, M., and Vasarainen, A. (1975). Genetics of *Philaenus* colour polymorphism: the 28 genotypes. *Hereditas*, 79, 308-10.

Hall, D.J. (1971). Predator-prey relationships between yellow perch and Daphnia in a large temperate lake. *Transactions of the American Microscopic Society*, 90, 106-7.

Hall, D.J., Cooper, W.J., and Werner, E.E. (1970). An experimental approach to the production dynamics and structure of freshwater *Limnology and Oceanography*, 15, 839-928.

Hall, D.J., Threlkeld, S.K., Burns, C.W., and Crowley, P.H. (1976). The size-efficiency hypothesis and the size structure of zoolplankton communities. *Annual Review of Ecology and Systematics*, 7, 177-208.

Hall, J.B., and Swaine, M.D. (1976). Classification and ecology of closed-canopy forest in Ghana. *Journal of Ecology*, 64, 913-51.

Hall, R.W., and Ehler, L.E. (1979). Rate of establishment of natural enemies in classical biological control. *Bulletin of the Entomological Society of America*, 25, 280-2.

Hallé, F. (1974). Architecture of trees in the rain forest of Morobe District, New Guinea. *Biotropica*, 6, 43-50.

Hallé, F., and Oldemann, R.A.A. (1975). *Essay on the Architecture and Dynamics of Growth of Tropical Trees*. Penerbit University, Kuala Lumpur, Malaya.

Hamilton, A. (1989). African Forests. In *Ecosystems of the World. 14B. Tropical Rain Forest Ecosystems: Biogeographical and Ecological Studies*. ed. H. Lieth and M.J. A. Werger. pp. 155-82. Amsterdam: Elsevier.

Hamilton, W.D. (1971). Geometry for the selfish herd. *Journal of Theoretical Biology*, 31, 295-311.

Hamrick, J.L., Linhart, Y.B., and Mitton, J.B. (1979). Relationships between life history characteristics and electrophoretically detectable genetic variation in plants. *Annual Review of Ecology and Systematics*, 10, 175-200.

Hanby, J.P., and Bygott, J.D. (1979). Population changes in lions and other predators. In *Serengeti: Dynamics of an Ecosystem*, ed. A.R.E. Sinclair and M. Norton-Griffiths, pp. 249-65. Chicago, Ill: University of Chicago Press.

Hanes, T.L. (1971). Succession after fire in the chaparral of southern California. *Ecological Monographs*, 41, 27-52.

Happold, D.C.D. (1983). Rodents and lagomorphs. In *Ecosystems of the World 13: Tropical Savannas*, ed. F. Bourlière, pp. 363-400. Amsterdam: Elsevier.

Harcombe, P.A. (1977a). Nutrient accumulation by vegetation during the first year of recovery of a tropical forest ecosystem. In *Recovery and Restoration of Damaged Ecosystems*, ed. J. Cairns, K.L. Dickson, and E.E. Herricks, pp. 347-78. Charlottesville, Virginia: University Press of Virginia.

Harcombe, P.A. (1977b). The influence of fertilization on some aspects of succession in a humid tropical forest. *Ecology*, 58, 1375-1383.

Hardin, G. (1960). The competitive exclusion principle. *Science*, 131, 1292-7.

Hare, R.C. (1965). Contribution of bark to fire resistance of southern trees. *Journal of Forestry*. 63, 248-51.

Harman, W.N. (1972). Benthic substrates: their effect on fresh-water mollusca. *Ecology*, 53, 271-7.

Harmon, M.E., Franklin, J.F., Swanson, F.J., *et al*. (1986). Ecology of coarse woody debris in temperate ecosystems. *Advances in Ecological Research*, 15, 133-302.

Harper, J.L. (1969). The role of predation in vegetational diversity. *Brookhaven Symposium on Biology No. 22, Diversity and Stability in Ecological Systems.*, pp. 48-62.

Harper, J.L. (1977). The contribution of terrestrial plant studies to the development of the theory of ecology. In *Changing Scences in the Life Sciences, 1776-1976*. C.E. Goulden, ed. pp. 139-57. Special Publication. 12. Philadelphia: Academy of Natural Science.

Harper, J.L., Lovell, P.H., and Moore, K.G. (1970). The sizes and shapes of seeds. *Annual Review of Ecology and Systematics*, 1, 327-56.

Harris, L.D. (1984). *The Fragmented Forest*. Chicago, Illinois: University of Chicago Press.

Harris, W., and Thomas, V.J. (1972). Competition among pasture plants. II. Effects of frequency and height of cutting on competition between *Agrostis tenuis* and two ryegrass cultivars. *New Zealand Journal Agricultural Research*, 15, 19-32.

Hart, R. (1980). The coexistence of weeds and restricted native plants on serpentine barrens in southeastern Pennsylvania. *Ecology*, 61, 688-701.

Hart, T.B., Hart, J.A., and Murphy, P.G. (1989) Monodominant and species-rich forests of the humid tropics: causes for their co-occurrence. *American Naturalist* 133, 613-33.

Hartman, O. (1965). Deep-water benthic polychaetous annelids off New England to Bermuda and other North Atlantic areas. *Occassional Papers of the Allan Hancock Foundation*, 28, 1-378.

Hartmeyer, R. (1911). Tunicata. Chap. XVII. Die geographicshe Verbreitung. In *Dr. H.G. Bronn's Klassen und Ordungen des Tier-Reiches: wissenschlaftlich dargestellt in Wort und Bild. III, Suppl.* ed. H.G. Bronn. Leipzig: Winter.

Hartshorn, G.S. (1972). The ecological life history and population dynamics of *Pentaclethra macroloba*, a tropical wet forest dominant, and *Stryphnodendron excelsum*, an occassional associate. Ph.D. Thesis, University of Washington.

Hartshorn, G.S. (1978). Tree falls and tropical forest dynamics. In *Tropical trees as living systems*, ed. Tomlinson, P.B. and M.H. Zimmermann, pp. 617-638. Cambridge: Cambridge University Press.

Harvey, H.W. (1955). *The Chemistry and Fertility of Sea Waters*. Cambridge: Cambridge University Press.

Harvey, T.H., Shellhammer, H.S., and Stecker, R.E. (1980). *Giant Sequoia Ecology*. Science Monograph Series 12. Washington, D.C.:U.S. Department of the Interior, National Park Service.

Hatcher, B.G., and Larkum, A.W.D. (1983). An experimental analysis of factors controlling the standing crop of the epilithic algal community on a coral reef. *Journal of Experimental Marine Biology and Ecology*, 69, 61-84.

Hay, M. E. (1984). Patterns of fish and urchin grazing on caribbean coral reefs: are previous results typical? *Ecology*, 65, 446-54.

Hay, M.E., and Fenicel, W. (1988) Marine plant-herbivore interactions: the ecology of chemical defense. *Annual Review of Ecology and Systematics*, 19, 111-45.

Hays, J.D., Imbrie, J., and Shackleton, N. (1976). Variations in the earth's orbit: pacemaker of the ice volume cycle. *Science*, 194, 1121-32.

Hayward, T.L, Venrick, E.L., and McGowan, J.A. (1983). Environmental heterogeneity and plankton community structure in the central North Pacific. *Journal of Marine Research*, 41, 711-29.

Hayward, T.L., and McGowan, J.A. (1979). Pattern and structure in an oceanic zooplankton community. *American Zoologist*, 19, 1045-55.

Heady, H.F. (1977). Valley grassland. In *Terrestrial Vegetation of California*, ed. M.G. Barbour and J. Major, pp. 491-514. New York: Wiley.

Heck, K., van Belle, G., and Simberloff, D. (1975). Explicit calculation of the rarefaction diversity measurement and the determination of sufficient sample size. *Ecology*, 56, 1459-61.

Heck, K.L. (1979). Some determinants of the composition and abundance of motile macroinvertebrate species in tropical and temperate turtlegrass (*Thalassia testudinum*) meadows. *Journal of Biogeography,* 6, 183-200.

Heddle, E.M., and Specht, R.L. (1975). The Dark Island Heath (Ninety-Mile Plain, South Australia) VII: the effects of fertilizers on composition and growth, 1950-72. *Australian Journal of Botany,* 23, 151-64.

Hedrick, P.W., Ginevan, M.E., and Ewing E.P. (1976). Genetic polymorphism in heterogeneous environments. *Annual Review of Ecology and Systematics,* 7, 1-32.

Heinemann, D., Hunt, G., and Everson, I. (1989). Relationships between the distributions of marine avian predators and their prey, *Euphausia superba,* in Bransfield Strait and southern Drake Passage, Antarctica. *Marine Ecology Progress Series,* 58, 3-16.

Heinselman, M.L. (1973). Fire in the virgin forests of the Boundary Waters Canoe Area. *Quaternary Research,* 3, 329-82.

Heinselman, M.L. (1981a). Fire and succession in the conifer forests of northern North America. In *Forest Succession Concepts and Application,* ed. D.C. West, H.H. Shugart, and D.B. Botkin, pp. 374-405. New York: Springer-Verlag.

Heinselmann. (1981b). Fire intensity and frequency as factors in the distributon and structure of northern ecosystems. In *Fire Regimes and Ecosystem Properties,* ed. H.A. Mooney, T.M. Bonnicksen, N.L. Christensen, J.E. Lotan, and W.A. Reiners, pp. 7-57. GTR-WO-26. Washington, D.C.: U.S. Forest Service - Washington Office.

Heinselman, M.L., and Wright, H.E. (1973). The ecological role of fire in natural conifer forests of western and northern North America. *Quaternary Research,* 3, 317-482.

Hemstrom, M.A., and Franklin, J.F. (1982). Fire and other disturbances of the forests in Mount Ranier National Park. *Quaternary Research,* 18, 32-51.

Herrera, C.M. (1985). Determinants of plant-animal coevolution: the case of mutualistic vertebrate seed dispersal systems. *Oikos,* 44, 132-44.

Hershkovitz, P. (1978). *Living New World Monkeys (Platyrrhini). Introduction to Primates.* Chicago: University of Chicago Press.

Herwitz, S.R. (1986). Episodic stemflow inputs of magnesium and potassium to a tropical forest floor during heavy rainfall events. *Oecologia. (Berlin),* 70, 423-5.

Hessler, R.R., and Jumars, P.A. (1974). Abyssal community analysis from replicate box cores in the central North Pacific. *Deep-Sea Research,* 21, 185-209.

Hessler, R.R., and Sanders, H.L. (1967). Faunal diversity in the deep-sea. *Deep-Sea Research,* 14, 65-78.

Hessler, R.R., and Smithey, W.M. (1983). The distribution and community structure of megafauna at the Galapagos Rift hydrothermal vents. In *Hydrothermal Processes at Seafloor Spreading Centers,* ed. P.A. Rona, K. Böstrom, L. Laubier, and K.L. Smith, pp. 735-70. New York: Plenum.

Hewatt, W.G. (1935). Ecological succession in the *Mytilus californianus* habitat as observed in Monterey Bay, California. *Ecology,* 16, 244-51.

Hewetson, C.E. (1956). A discussion on the climax concept in relation to the tropical rain and deciduous forest. *Empire Forestry Review,* 35, 274-91.

Heywood, V.H. (1989). Patterns, extents, and modes of invasions by terrestrial plants. In *Biological Invasions: a Global Perspective. SCOPE 37*, ed. J.A. Drake et al, pp. 31-60. Chichester: Wiley.

Hiatt, R.W., and Strasburg, D.W. (1960). Ecological relationships of the fish fauna on coral reefs of the Marshall Islands. *Ecological Monographs*, 30, 65-127.

Highsmith, R.C. (1979). Coral growth rates and environmental control of density banding. *Journal of Experimental Marine Biology*, 37, 105-25.

Hilborn, R. (1975). The effect of spatial heterogeneity on the persistence of predator-prey interactions. *Theoretical Population Biology*, 8, 346-55.

Hildrew, A.G., and Townsend, C.R. (1982). Predators and prey in a patchy environment: a freshwater study. *Journal of Animal Ecology*, 51, 797-816.

Hill, M.O. (1973). Diversity and evenness: a unifying notion and its consequences. *Ecology*, 54, 427-31.

Hilton, T.E., and Kown-tsei, J.Y. (1972). The impact of the Volta scheme on the lower Volta flood plains. *Journal of Tropical Geography*, 1972, 29-37.

Hladik, A. (1986). Données comparatives sur la richesse spécifique et les structures des peuplements des forêts tropicales d'Afrique et d'Amérique. *Memoires du Museum National D'Histoire Naturelle*, 132, 9-18.

Hobbs, N.T., and Schimel, D.S. (1984). Fire effects on nitrogen mineralization and fixation in mountain shrub and grassland communities. *Journal of Range Management*, 37, 402-5.

Hobbs, R.J. (1989). The nature and effects of disturbance relative to invasions. In *Biological Invasions: a Global Perspective. SCOPE 37*, ed. J.A. Drake *et al.*, pp. 389-405. Chichester: Wiley.

Hobbs, R.J., and Gimingham, C.H. (1984a). Studies on fire in Scottish heathland communities. I. Fire characteristics. *Journal of Ecology*, 72, 223-40.

Hodgkins, E.J. (1958). Effects of fire on undergrowth vegetation in upland southern pine forests. *Ecology*, 39, 36-46.

Hogeweg, P, and Hesper, B. (1981). Two predators and one prey in a patchy environment: an application of MICMAC modelling. *Journal of Theoretical Biology*, 93, 411-32.

Holdgate, M.W. (1960). The fauna of the mid-Atlantic Islands. *Proceedings of the Royal Society (London)*, B152, 550-67.

Holdridge, L.R. (1947). Determination of world plant formations from simple climatic data. *Science*, 105, 367-8.

Holdridge, L.R., (1967). *Life Zone Ecology*. Tropical Science Center, San Jose, Costa Rica.

Holdridge, L.R., Grenke, W.C., Hatheway, W.H., Liang, T., and Tosi, J.A. (1971). *Forest Environments in Tropical Life Zones: A Pilot Study*. New York: Pergamon.

Holloway, J.D. (1977). *The Lepidoptera of Norfolk Island, their Biogeography and Ecology. Series Entomologica 13*. The Hague: Junk.

Holloway, J.D. (1979). *A Survey of the Lepidoptera, Biogeography and Ecology of New Caledonia. Series Entomologica 15*. The Hague: Junk.

Holmes, J.C. (1961). Effects of concurrent infections of *Hymenolepis diminuta* (Destoda) and *Moniliformis dubius* (Acanthocephala). I. General effects and comparison with crowding. *Journal of Parasitology*, 47, 209-16.

Holmes, R.T., Sherry, T.W., Marra, P.P., and Petit K.E. (**In press**). Multiple brooding, nesting success, and annual productivity of a neotropical migrant, the Black-throated Blue Warbler

Holmes. R.T., Sherry, T.W., and Sturges, F.W. (1986). Bird community dynamics in a temperate deciduous forest: Long-term trends at Hubbard Brook. *Ecology,* 56, 201-20.

Holmes, R.T., Sherry, T.W., and Sturges, F.W. (1992). Numerical and demographic responses of temperate forest birds to fluctuations in their food resources. *Proceedings International Ornithological Congress,* 20, in press.

Holt, B.R. (1972). Effect of arrival time on recruitment, mortality and reproduction in successional plant populations. *Ecology,* 53, 668-73.

Holt, J.A., and Conventry, R.J. (1990). Nutrient cycling in Australian savannas. *Journal of Biogeography,* 17, 427-32.

Holt, R.D. (1977). Predation, apparent competition, and the structure of prey communities. *Theoretical Population Biology,* 12, 197-229.

Holt, R.D., and Pickering, J. (1985). Infectious disease and species coexistence: a model of Lotka-Volterra form. *American Naturalist,* 126, 196-211.

Hopkins, B. (1955). The species-area relations of plant communities. *Journal of Ecology,* 43, 409-26.

Hopper, S.D. (1979). Biogeographical aspects of speciation in the southwest Australian flora. *Annual Review of Ecology and Systematics,* 10, 399-422.

Horak, I.G. (1983). Helminth, arthropod, and protozoan parasites of mammals in African savannas. In *Ecosystems of the World 13: Tropical Savannas,* ed. F. Bourlière, pp. 563-81. Amsterdam: Elsevier.

Horn, H.S. (1974). The ecology of secondary succession. *Annual Review of Ecology and Systematics,* 5, 25-37.

Horn, H. (1975). Markovian properties of forest succession. In *Ecology and Evolution of Communities,* ed. M.L. Cody and J.M. Diamond, pp. 196-211. Cambridge, Massachusetts: Harvard University Press.

Horn, H.S., and MacArthur, R.H. (1972). Competition among fugitive species in a harlequin environment. *Ecology,* 53, 749-52.

Horn, H.S., and May, R.M. (1977). Limits to similarity among coexisting competitors. *Nature* (London), 270, 660-1.

Horn, M.H., and Allen, L.G. (1978). A distributional analysis of California coastal marine fishes. *Journal of Biogeography,* 5, 23-42.

Howe, H.F. (1986). Seed dispersal by fruit-eating birds and mammals. In *Seed Dispersal,* ed. D.R. Murray, pp. 123-90. Sydney: Academic Press.

Howe, H.F., and Smallwood, J. (1982). Ecology of seed dispersal. *Annual Review of Ecology and Systematics,* 13, 201-28.

Howe, H.F., and Westley, L.C. (1988). *Ecological Relationships of Animals and Plants.* New York: Oxford University Press.

Hubbard, J.A.E.B. (1974). Scleractinian coral behavior in calibrated current experiment: an index to their distribution patterns. *Proceedings of the Second International Coral Reef Symposium,* 2, 107-26.

Hubbard, J.A.E.B., Pocock, Y.P. (1972). Sediment rejection by recent Scleractinian corals: a key to paleo-environmental reconstruction. *Geologische Rundschau,* 61, 598-626.

Hubbell, S.G., and Foster, R.B. (1986).Canopy gaps and the dynamics of a tropical rain forest. In *Plant Ecology,* ed. M.J. Crawley, pp. 75-95. Oxford: Blackwell.

Hubbell, S.P., and Foster, R.B. (1987). Biology, chance, and history and the structure of tropical rain forest tree communities. In *Community Ecology,* ed. J. Diamond and T.J. Case. pp. 314-29. New York: Harper and Row.

Hubbell, S.P., and Foster, R.B. (1990). Structure, dynamics and equilibrium status of old-growth forest on Barro Colorado Island. In *Four Neotropical Rainforests*. ed. A.H. Gentry. pp. 522-41. New Haven: Yale University Press.

Hubendick. B. (1962). Aspects on the diversity of the fresh-water fauna. *Oikos,* 35, 214-29.

Huenneke, L.F., Hamburg, S.P., Koide, R., Mooney, H.A., and Vitousek, P.M. (1990). Effects of soil resources on plant invasion and community structure in Californian serpentine grassland. *Ecology,* 71, 478-91.

Huffaker, C.B. (1958). Experimental studies on predation: dispersion factors and predator-prey oscillations. *Hilgardia,* 27, 343-83.

Huffaker, C.B., Shea, K.P., and Herman, S.G. (1963). Experimental studies on predation. *Hilgardia,* 34, 305-30.

Huges, R.G., and Thomas, M.L. (1971). The classification and ordination of shallow-water benthic samples from Prince Edward Island, Canada. *Journal of Experimental Marine Biology and Ecology,* 7, 1-39.

Hughes, T.P. (1990). Recruitment limitation, mortality, and population regulation in open systems: a case study. *Ecology,* 71, 12-20.

Hughes, T.P., Reed, D.C, and Boyle, M-J. (1987). Herbivory on coral reefs: community structure following mass mortalities of sea urchins. *Journal of Experimental Marine Biology and Ecology,* 113, 39-59.

Hulbert, L.C. (1969). Fire and litter effects in undisturbed bluestem prairie in Kansas. *Ecology,* 50, 874-7.

Hulbert, L.C. (1973). Management of Konze Prairie to approximate pre-white-man fire influences. *Proceedings of the Third Midwest Prairie Conference,* ed. L.C. Hulbert. Manhattan, Kansas: Kansas State University Press.

Humboldt, A. von, and Bonpland, A. (1807). *Essai sur la Geographie des Plantes.* Paris: Librarie Lebrault Schoell.

Humphrey, R.R. (1962). *Range Ecology.* New York: Ronald Press.

Humphrey, R.R. (1974). Fire in the deserts and desert grassland of North America. In *Fire and Ecosystems,* ed. T.T. Kozlowski and C.E. Ahlegren, pp. 365-400. New York: Academic Press.

Hunt, C.B. (1975). *Death Valley. Geology, Ecology, Archaeology.* Berkeley, California: University of California Press.

Hunt, G.L., and Schneider, D.C. (1987). Scale-dependent processes in the physical and biological environment of marine birds. In *Seabirds, Feeding Ecology and Role in Marine Ecosystems,* ed. J.P. Croxall, pp. 7-41. Cambridge: Cambridge University Press.

Hunt, R., and Nicholls, A.O. (1986). Stress and the coarse control of growth and root-shoot partitioning in herbaceous plants. *Oikos,* 47, 149-58.

Huntley, B., and Birks, H.J.B. (1983). *An Atlas of Past and Present Pollen Maps for Europe: 0-13000 years ago.* Cambridge: Cambridge University Press.

Huntley, B., and Webb, T. (1989). Migration: species' response to climatic variations caused by changes in the earth's orbit. *Journal of Biogeography,* 16, 5-19.

Hurd, L.E., Mellinger, M.V., Wolf, L.L., and McNaughton, S.J. (1971). Stability and diversity at three trophic levels in terrestrial successional ecosystems. *Science,* 173, 1134-6.

Hurlbert, S.H. (1971). The nonconcept of species diversity: a critique and alternative parameters. *Ecology,* 52, 577-86.

Huston, M. (1985b). Changes in coral growth rates with depth at Discovery Bay, Jamaica. *Coral Reefs,* 4, 19-25.

Huston, M.A. (1979). A general hypothesis of species diversity. *American Naturalist*. 113, 81-101.

Huston, M.A. (1980a). Soil nutrients and tree species richness in Costa Rican forests.

Huston, M.A. (1980b). Patterns of species diversity in an oldfield ecosystem. *Bulletin of the Ecological Society of America*, 61, 110.

Huston, M.A. (1982). *The Effect of Soil Nutrients and Light on Tree Growth and Interactions during Tropical Forest Succession: Experiments in Costa Rica.* Ph.D. Thesis, University of Michigan.

Huston, M.A. (1985a). Patterns of species diversity on coral reefs. *Annual Review of Ecology and Systematics*, 16, 149-77.

Huston, M.A. (1985b). Changes in coral growth rates with depth at Discovery Bay, Jamaica. *Coral Reefs*, 4, 19-25.

Huston, M.A. (1985c). Patterns of species diversity in relation to depth at Discovery Bay, Jamaica. *Bulletin of Marine Science*, 37, 928-35.

Huston, M.A. (1986). Size bimodality in plant populations: an alternative hypothesis. *Ecology*, 67, 265-9.

Huston, M.A. (1991). Use of individual-based forest succession models to link physiological whole-tree models to landscape-scale ecosystem models. *Tree Physiology*, 9, 293-306.

Huston, M.A. (1992). Individual-based forest succession models and the theory of plant competition. In *Populations and Communities: An Individual-based Perspective*, ed. D.L. DeAngelis and L.J. Gross. pp. 408-20. New York: Chapman and Hall.

Huston, M.A., DeAngelis, D.L., and Post, W.M. (1988). New computer models unify ecological theory. *BioScience*, 38, 682-91.

Huston, M.A., and DeAngelis, D.L. (1987). Size bimodality in monospecific populations: a critical review of potential mechanisms. *American Naturalist*, 129, 678-707.

Huston, M.A., and Smith, T.M. (1987). Plant Succession: Life history and competition. *American Naturalist*, 130, 168-98.

Hutchinson, G.E. (1941). Ecological aspects of succession in natural populations. *American Naturalist*. 75, 406-18.

Hutchinson, G.E. (1948). Circular causal systems in ecology. *Annals of the New York Academy of Sciences*, 50, 221-46.

Hutchinson, G.E. (1953). The concept of pattern in ecology. *Proceedings of the Academy of Natural Sciences, Philadelphia*. 105, 1-12.

Hutchinson, G.E. (1957). Concluding remarks. *Cold Spring Harbor Symposium on Quantitative Biology*, 22, 415-27.

Hutchinson, G.E. (1959). Homage to Santa Rosalia; or, why are there so many kinds of animals. *American Naturalist*, 93, 145-59.

Hutchinson, G.E. (1961). The paradox of the plankton. *American Naturalist*. 95, 137-45.

Hutchinson, G.E. (1965). *The Ecological Theater and the Evolutionary Play*. New Haven: Yale University Press.

Hutchinson, G.E. (1967). *A Treatise on Limnology: Vol. 2, Introduction to Lake Biology and the Limnoplankton*. New York: Wiley.

Huttel, C. (1975). Recherches sur l'ecosytéme de la forêt subéquatoriale de basse Côte d'Ivoire. III. Inventaire et structure de la végétation ligneuse. Programme ORS-TOM - Forêt dense (P.B.I.). *Terre Vie, Rev. Ecol. Appl.*, 29, 178-91.

Huttel, C., and Bernhard-Reversat, F. (1975). Recherches sur l'ecosysteme de la foret subequatoriale de basse Cote d'Ivore. 5. Biomasse vegetale et productivite' primaire. *Terre Vie*, 29, 203-28.

Huxley, C.R. (1980). Symbioses between ants and epiphytes. *Biological Review*, 55, 231-40.

Hyman, J.B., McAninch, J.B., and DeAngelis, D.L. (1991). An individual-based simulation model of herbivory in a heterogeneous landscape. In *Quantitative Methods in Landscape Ecology*, ed. M. Turner and R. Gardner, pp. 443-75. New York: Springer-Verlag.

Imbrie, J. (1985). A theoretical framework for the Pleistocene ice ages. *Journal of the Geological Society of London*, 142, 417-32.

Istock, C.A., and Scheiner, S.M. (1987). Affinities and higher-order diversity within landscape mosaics. *Evolutionary Ecology*, 1, 11-29.

Jacard, P. (1912). The distribution of the flora in the alpine zone. *New Phytologist*, 11, 37-50.

Jackson, A.S. (1965). Wildfires in the Great Plains grasslands. *Proceedings of the Annual Tall Timbers Fire Ecology Conference*, 4, 241-59.

Jacobs, M.R. (1954). Silvicultural problems in the mixed eucalypt forests of the east coast of Australia. *Empire Forestry Review*, 33, 30-8.

James, F.C., and Wamer, N.O. (1982). Relationships between temperate forest bird communities and vegetation structure. *Ecology*, 63, 159-71.

Jannasch, H.W. (1985). The chemosynthetic support of life and the microbial diversity at deep-sea hydrothermal vents. *Proceedings of the Royal Society of London, Series B*, 225, 277-97.

Jannasch, H.W., and Wirsen, C.O. (1979). Chemosynthetic primary production at East Pacific sea floor spreading centers. *BioScience*, 29, 592-8.

Janson, C.H. (1983). Adaptation of fruit morphology to dispersal agents in a neotropical forest. *Science*, 219, 187-9.

Janson, C.H., and Emmons, L.H. (1990). Ecological structure of the nonflying mammal community at Cocha Cashu biological Station, Manu National Park, Peru. In *Four Neotropical Rainforests*, ed. A.H. Gentry, pp. 314-38. New Haven: Yale University Press.

Janzen, D.H. (1967). Why mountain passes are higher in the tropics. *American Naturalist*, 101, 233-49.

Janzen, D.H. (1969). Seed eaters versus seed size, number, toxicity, and dispersal. *Evolution*, 23, 1-27.

Janzen, D.H. (1970). Herbivores and the number of tree species in tropical forests. *American Naturalist*, 104, 501-28.

Janzen, D.H. (1974). Tropical blackwater rivers, animals, and mast fruiting by Dipterocarpaceae. *Biotropica* 6, 69-103.

Janzen, D.H. (1981). The peak in North American ichneumonid species richness lies between 38° and 42°. *Ecology*, 62, 532-7.

Janzen, D.H. (1983). Insects: Introduction. In *Costa Rican Natural History*. ed. D.H. Janzen, pp. 619-45. Chicago, Illinois: University of Chicago Press.

Jarman, P.J. (1974). The social organization of antelope in relation to their ecology. *Behaviour*, 48, 215-66.

Jarvis, P.G. (1989). Atmospheric carbon dioxide and forests. *Philosophical Transactions of the Royal Society (London) B*, 324, 369-92.

Jarvis, P.G., James, G.B., and Landsberg, J.J. (1976). Coniferous forest. In *Vegetation and the Atmosphere, Volume 2*, ed. J.L. Monteith, pp. 246-72. London: Academic Press.

Jaubert, J.M., and Vasseur, P. (1974). Light measurements: duration, aspect, and the distribution of benthic organisms in an Indian Ocean coral reef (Tulear, Madagascar). *Proceedings of the Second International Coral Reef Symposium, 2,* 127-42

Jenny, H. (1941). *Factors of Soil Formation.* New York: McGraw-Hill.

Jenny, H. (1980). *The Soil Resource (Ecological Studies, Vol. 37),* New York: Springer-Verlag.

Jenny, H., Gessel, S.P., and Bingham, F.T. (1949) Comparative study of decomposition rates of organic matter in temperate and tropical regions. *Soil Science,* 68, 419-32.

Jepson, W.L. (1931). The role of fire in relation to the differentiation of species in the chaparral. *Proceedings of the Fifth International Botanical Congress,* 193, 114-6.

Johannes, R.E., Cole, S.L., and Kuenzel, N.T. (1970). The role of zooplankton in the nutrition of some scleractinean corals. *Limnology and Oceanography,* 15, 579-86.

Johnson, D.W., Cole, D.W., and Gessel, S.P. (1975). Processes of nutrient transfer in a tropical rain forest. *Biotropica,* 7, 208-15.

Johnson, D.W., Cole, D.W., Gessel, S.P., Singer, M.J., and Minden, R.V. (1977). Carbonic acid leaching in a tropical temperate, subalpine, and northern forest soil. *Arctic and Alpine Research,* 9, 329-43.

Johnson, M.P., and Raven, P.H. (1970). Natural regulation of plant species diversity. In *Evolutionary Biology,* Vol. 4, ed. T. Dobzhansky, M.K. Hecht, W.C. Steer, pp. 127-62. New York: Appleton-Century-Crofts.

Johnson, M.P., and Simberloff, D.S. (1974). Environmental determinants of island species numbers in the British Isles. *J. Biogeography,* 1, 149-54.

Johnson, N.K. (1975). Controls of number of bird species on montane islands in the Great Basin. *Evolution,* 29, 545-67.

Johnson, W.C., Sharpe, D.M., DeAngelis, D.L., Fields, D.E., and Olson, R.J. (1981). Modeling seed dispersal and forest island dynamics. In *Forest Island Dynamics in Man-dominated Landscapes,* ed. R.L. Burgess and D.M.Sharpe, pp. 215-39. New York: Springer-Verlag.

Johnston, D.W., and Odum, E.P. (1956). Breeding bird populations in relation to plant succession on the piedmont of Georgia. *Ecology,* 37, 50-62.

Jokiel, P.L. (1978). Effects of water motion on reef corals. *Journal of Experimental Marine Biology,* 35, 87-97.

Jones, N.S., and Sanders, H.L. (1972). Distribution of Cumacea in the deep Atlantic. *Deep-Sea Research,* 19, 737-45.

Jordan, C.F. (1971a). A world pattern in plant energetics. *American Scientist,* 59, 425-33.

Jordan, C.F. (1971b). Productivity of a tropical forest and its relation to a world pattern of energy storage. *Journal of Ecology,* 59, 127-42.

Jordan, C.F. (1983). Productivity of tropical rain forest ecosystems and the implications for their use as future wood and energy sources. In *Tropical Rain Forest Ecosystems. Structure and Function, Ecosystems of the World.* Vol. 14A, ed. F.B. Golley, pp. 117-36. Amsterdam: Elsevier.

Jordan, C.F. (1985a). Soils of the Amazon rainforest. In *Amazonia,* ed. G.T. Prance and T.E. Lovejoy, pp. 83-94. Oxford: Pergamon.

Jordan, C.F. (1985b). *Nutrient Cycling in Tropical Forest Ecosystems,* Chichester, Great Britain: Wiley.

Jordan, C.F., and Kline, J.R. (1972). Mineral cycling: some basic concepts and their application in a tropical rain forest. *Annual Review of Ecology and Systematics*, 3, 33-50.

Jumars, P.A. (1975a). Methods for measurment of community structure in deep-sea macrobenthos. *Marine Biology*, 30, 245-52.

Jumars, P.A. (1975b). Environmental grain and polychaete species diversity in a bathyal benthic community. *Marine Biology*, 30, 253-66.

Jumars, P.A. (1976). Deep-sea species diversity: does it have a characteristic scale? *Journal of Marine Research*, 34, 217-46.

Junk, W.J., and Furch, K. (1985). The physical and chemical properties of Amazonian waters and their relationships with the biota. In *Amazonia*, ed. G.T. Prance and T.E. Lovejoy, pp. 3-17. Oxford: Pergamon.

Järvinen, O. (1979). Geographical gradients of stability in European land bird communities. *Oecologia*, 38, 51-69.

Kadambi, K. (1942). The evergreen Ghat rain forest of the Tunga and Bhadra river sources, Kadur District, Mysore State. Parts, 1,2. *Indian Forestry*, 68, 233-40, 305-12.

Karl, D.M. (1987). Bacterial production at deep-sea hydrothermal vents and cold seeps: Evidence for chemosynthetic primary production. In *Ecology of Microbial Communities. SGM 41*, pp. 319-60. New York: Cambridge University Press.

Karl, D.M., Wirsen, C.O., and Jannasch, H.W. (1980). Deep-sea primary production at the Galapagos hydrothermal vents. *Science*, 207, 1345-7.

Karl, T.R. (1988). Multi-year fluctuations of temperature and precipitation: the gray area of climate change. *Climate Change*, 12, 179-97.

Karr, J.R. (1968). Habitat and avian diversity on strip-mined land in east-central Illinois. *Condor*, 70, 348-57.

Karr, J.R. (1971). Structure of avian communities in selected Panama and Illinois habitats. *Ecological Monographs*, 41, 207-233.

Karr, J.R. (1982). Avian extinction on Barro Colorado Island: a reassessment. *Am. Nat.*, 119, 220-39.

Karr, J.R., Robinson, S.K., Blake, J.G., and Bierregaard, R.O. (1990). Birds of four neotropical forests. In *Four Neotropical Rainforests*, ed. A.H. Gentry, pp. 237-69. New Haven: Yale University Press.

Karr, J.R., and Roth, R.R. (1971). Vegetation structure and avian diversity in several New World areas. *American Naturalist*, 105, 423-35.

Kato, R., Tadaki, Y., and Ogawa, H. (1978). Plant biomass and growth increment studies in Pasoh Forest. *Malay Naturalist Journal*, 30, 211-24.

Kauffman, E.G., and Fagerstrom, J.A. (1993). The Phanerozoic evolution of reef diversity. In *Biological Diversity*, ed. R.E. Ricklefs and D. Schluter, in press. Chicago: The University of Chicago Press.

Kauffman, J.B., Uhl, C., and Cummings, D.L. (1988). Fire in the Venezuelan Amazon 1: Fuel biomass and fire chemistry in the evergreen rainforest of Venezuela. *Oikos*, 53, 167-75.

Kaufman, L. (1977). The three spot damselfish: effects on benthic biota of Caribbean coral reefs. *Proceedings Third International Coral Reef Symposium*, 1, 559-64.

Kaufmann, J.B., and Uhl, C. (1990). Interactions of anthropogenic activities, fire, and rain forests in the Amazon Basin. In *Fire in the Tropical Biota*, ed. J.G. Goldammer, pp. 117-34. Berlin: Springer-Verlag.

Kawaguti, S., and Sakumoto, D. (1948). The effect of light on the calcium deposition of corals. *Bulletin of the Oceanographic Institute, Taiwan*, 4, 65-70.

Keast, A, and Eadie, J.M. (1985). Growth depensation in year-0 largemouth bass: the influence of diet. *Transactions of the American Fisheries Society*, 114, 204-13.

Keast, A. (1961). Bird speciation on the Australian continent. *Bulletin of the Museum of Comparative Zoology*, 123, 305-495.

Keast, A. (1969). A comparison of the contemporary mammalian faunas of the southern continents. *Quarterly Review of Biology*, 44, 121-67.

Keast, A. (1973). Comparisons of contemporary mammal faunas on southern continents. In *Evolution, Mammals, and Southern Continents*, ed. A. Keast, F.C. Erk, and B. Glass, pp. 19-87. Albany, New York: State University of New York Press.

Keast, A., Erk, F.C., and Glass, B. eds. (1972). *Evolution, Mammals, and Southern Continents*. Albany: State University of New York Press.

Keddy, P.A., and MacLellan, P. (1990). Centrifugal organization in forests. *Oikos*, 59, 75-84.

Keeley, J.E., and Keeley, S.C. (1988). Chaparral. In *North American Terrestrial Vegetation*, ed. M.G. Barbour and D.W. Billings, pp. 166-207. Cambridge: Cambridge University Press.

Keeley, S.C., Hutchinson, J.E., and Johnson, A.W. (1981). Postfire-succession of the herbaceous flora in the southern California chaparral. *Ecology*, 62, 1608-21.

Keeley, S.C., and Johnson, A.W. (1977). A comparison of the pattern of herb and shrub growth in comparable sites in Chile and California. *American Midland Naturalist*, 97, 120-32.

Keetch, J.J. (1944). Sprout development on once-burned and repeatedly-burned areas in the Southern Appalachians. U.S. Department of Agriculture, Forest Service, Southeastern Forest Experiment Station Technical Note 59. Washington, D.C.: U.S. Government Printing Office.

Keetch, J.J., and Byram, G.M. (1968). Drought index for forest fire control. U.S. Department of Agriculture, Forest Service Research Paper, SE-38.

Keever, C. (1950). Causes of succession on old fields of the Piedmont, North Carolina. *Ecological Monographs*, 20, 229-50.

Keller, M.A. (1984). Reassessing evidence for competitive exclusion of introduced natural enemies. *Environmental Entomology*, 13, 192-5.

Kelsall, J. (1957). Continued barren-ground caribou studies. *Canadian Wildlife Service Wildlife Management Bulletin Series 1*, 12, 1-148.

Kemp, W.M., and Mitsch, W.J. (1979). Turbulence and phytoplankton diversity: a general model of the "paradox of plankton." *Ecological Modelling*. 7, 201-22.

Kempton, R.A. (1979). The structure of species abundance and measurement of diversity. *Biometrics*, 35, 307-21.

Kerfoot, W.C. ed. (1980). *Evolution and Ecology of Zooplankton Communities*. Hanover, New Hampshire: University Press of New England.

Kerfoot, W.C., and Sih, A. eds. (1987). *Predation: Direct and Indirect Impacts on Aquatic Communities*. Hanover, New Hampshire: University Press of New England.

Kiester, A.R. (1971). Species density of North American amphibians and reptiles. *Systematic Zoology*, 20, 127-37.

Kikkawa, J., and Williams, W.T. (1971). Altitudinal distributions of land birds in New Guinea. *Search (Sydney)*, 2, 64-5.

Kilgore, B.M. (1973). The ecological role of fire in Sierran conifer forests. Its application to National Park Management. *Quaternary Research*, 3, 496-513.

King, L.J. (1966). *Weeds of the World: Biology and Control*. London: L. Hill.

Kinzey, W.G. (1982). Distribution of primates and forest refuges. In *Biological Diversification in the Tropics*, ed. G. Prance. pp. 455-82. New York: Columbia University Press.

Kinzie, R.F. III. (1973). The zonation of West Indian gorgonians. *Bulletin of Marine Science*, 23, 93-155.

Kira, T., and Shidei, T. (1967). Primary production and turnover of organic matter in different forest ecosystems of the western Pacific. *Japanese Journal of Ecology*, 17, 70-87.

Kitchell, J.A., and Carr, T.R. (1985). Nonequilibrium model of diversification: faunal turnover dynamics. In *Phanerozoic Diversity Patterns, Profiles in Macroevolution*, ed. J.W. Valentine, pp. 277-310. Princeton: Princeton University Press.

Kitchell, J.F., and Carpenter, S.R. (1987). Piscivores, planktivores, fossils, phorbins. In *Predation: Direct and Indirect Impacts on Aquatic Communities*. ed. W.C. Kerfoot and A. Sih pp. 132-146 Hanover: University Press of New England.

Klienfeldt, S.E. (1978). Ant-gardens: the interactions of *Codonanthe crassifolia* (Gesneriaceae) and *Crematogaster longispina* (Formicidae). *Ecology*, 59, 449-56.

Klinge, H, and Herrera, R. (1978). Biomass studies in Amazon Caatinga forest in southern Venezuela. I. Standing crop of composite root mass in selected stands. *Tropical Ecology*, 19, 93-110.

Klinge, H., and Medina, E. (1978). Rio Negro caatingas and campinas, Amazonas states of Venezuela and Brazil. In *Ecosystems of the World*, Vol. 9, ed. R.L. Specht, pp. 483-8. Amsterdam: Elsevier.

Klock, G.O., and Helvey, J.D. (1974). Soil-water trends following wildfire on the Entiat Experimental Forest.*Proceedings of the Annual Tall Timbers Fire Ecology Conference*, 15, 193-200.

Knapp, A.K., and Seastedt, T.R. (1986). Detritus accumulation limits productivity of tallgrass prairie. *BioScience*, 36, 662-8.

Kneidel, K.A. (1984). Competition and disturbance in communities of carrion-breeding diptera. *Journal of Animal Ecology*, 53, 849-65.

Knight, D.H. (1987). Parasites, lightning, and the vegetation mosaic in wilderness landscapes. In *Landscape Heterogeneity and Disturbance*, ed. M.G. Turner, pp. 59-83. New York: Springer-Verlag.

Koblentz-Mishke, I.J., Volkovinsky, V.V., and Kabanova, J.B. (1970). Plankton primary production of the world ocean. In *Scientific Exploration of the South Pacific*. ed. W.S. Wooster. Washington: National Academy of Science Press.

Koch, A.L. (1974a). Coexistence resulting from an alteration of density-independent and density-dependent growth. *Journal of Theoretical Biology*, 44, 373-86.

Koch, A.L. (1974b). Competitive coexistence of two predators utilizing the same prey under constant environmental conditions. *Journal of Theoretical Biology*. 44, 387-95.

Kohn, A.J. (1967). Environmental complexity and species diversity in the gastropod genus *Conus* on Indo-West Pacific reef platforms. *American Naturalist*, 101, 251-60.

Kohn, A.J. (1968). Microhabitats, abundance, and food of *Conus* on atoll reefs in the Maldive and Chagos Islands. *Ecology*, 49, 1046-62.

Komarek, E.V. (1964). The natural history of lightning. *Proceedings of the Third Annual Tall Timbers Fire Ecology Conference*, 3, 139-83.

Komarek, E.V. (1969). Fire and animal behavior. *Proceedings of the Annual Tall Timbers Fire Ecology Conference*, 9, 161-207.

Komarek, E.V. (1974). Effects of fire on temperate forests and related ecosystems: Southeastern United States. In *Fire and Ecosystems*, ed. T.T. Kozlowski and C.E. Ahlgren, pp. 251-77. New York: Academic Press.

Komarek, E.V. (1981). History of prescribed fire and controlled burning in wildlife management in the south. In *Prescribed Fire and Wildlife in Southern Forests. Proceedings of a Symposium*, ed. G.W. Wood, pp. 1-15. Georgetown, South Carolina: Belle W. Baruch Forest Science Institute of Clemson University.

Konikoff, M., and Leis, W.M. (1974). Variation in weight of cage-reared channel catfish. *Progress in Fish-Culture*, 36, 138-44.

Koonce, A.L., and González-Cabán, A. (1990). Social and ecological aspects of fire in Central America. In *Fire in the Tropical Biota*, ed. J.G. Goldammer, pp. 135-58. Berlin: Springer-Verlag.

Koopman, M.J.F., and Verhoef, L. (1938). *Eusideroxylon zwageri* T. & B., het ijzerhout van Borneo en Sumatra. *Tectona*, 31, 381-99.

Korstian, C.F., and Coile, T.S. (1938). Plant competition in forest stands. *Duke University School of Forestry, Bulletin 3*. 125 pp.

Kotler, B.P., and Brown, J.S. (1988). Environmental heterogeneity and the coexistence of desert rodents. *Annual Review of Ecology and Systematics*, 19, 281-307.

Kozhov, M. (1963). *Lake Baikal and its Life*. The Hague: W. Junk.

Kozlowski, T.T. (1976). Water supply and leaf shedding. In *Water Deficits and Plant Growth IV. Soil Water Measurement, Plant Responses, and Breeding for Drought Resistance*. Ed. T.T. Kozlowski, pp. 191-231. New York: Academic Press.

Kozlowski, T.T. (1982). Water supply and tree growth. Part I. Water Deficits. *Commonwealth Forestry Abstracts*, 43, 57-95.

Kramer, P.J. (1969). *Plant and Soil Water Relationships: A Modern Synthesis*. New York: McGraw-Hill.

Kramer, P.J. (1983). *Water Relations of Plants*. New York: Academic Press.

Kramer, P.J., and Kozlowski, T.T. (1979). *Physiology of Woody Plants*. New York: Academic Press.

Kruger, F.J. (1977). Invasive woody plants in the Cape fynbos with special reference to the biology and control of *Pinus pinaster*. *Proceedings of the Second National Weeds Conference of South Africa*. Cape Town: A.A. Balkema.

Kruger, F.J., Breytenbach, G.J., MacDonald, I.A.W., and Richardson, D.M. (1989). The characteristics of invaded Mediterranean-climate regions. In *Biological Invasions: a Global Perspective. SCOPE 37*, ed. J.A. Drake *et al.*, pp. 181-213. Chichester: Wiley.

Kruger, F.J., Richardson, D.M., and van Wilgen, B.W. (1986). Processes of invasion by plants. In *The Ecology and Management of Biological Invasions of South Africa,* ed. I.A.W Macdonald, F.J. Kruger, and A.A. Ferrar, pp. 145-55. Cape Town: Oxford University Press.

Kruger, F.J., and Taylor, H.C. (1979). Plant species diversity in Cape Fynbos: gamma and delta diversity. *Vegetatio,* 41-2, 85-93.

Kubitzki, K. (1985). The dispersal of forest plants. In *Amazonia,* ed. G.T. Prance and T.E. Lovejoy, pp. 192-206. Oxford: Pergamon.

Kucera, C.L., and Koelling, M. (1964). The influence of fire on composition of Central Missouri Prairie. *American Midland Naturalist,* 72, 142-7.

Kusenov, N. (1957). Numbers of species of ants in faunae of different latitudes. *Evolution,* 11, 298-9.

Kutiel, P., and Danin, A. (1987). Annual-species diversity and aboveground phytomass in relation to some soil properties in the sand dunes of the northern Sharon Plains, Israel. *Vegetatio,* 70, 45-9.

Kwan, W.Y., and Whitmore, T.C. (1970). On the influence of soil properties on species distribution in a Malayan lowland Dipterocarp rain forest. *Malayan Forester*, 33, 42-54.

L'Hertier, P. and Tessier, G. (1935). Étude d'une population de Drosophiles en équilibre. *Comptes Rendus de Academie des Seances (Paris),* 197, 1765-7.

LaGory, K.E., LaGory, M.K., and Perino, J.V. (1982). Response of big and little bluestem (*Andropogon*) seedlings to soils and moisture conditions. *Ohio Journal of Science,* 82, 19-23.

Lacey, W.S. (1975). Some problems in the 'mixed' floras in the Permian of Gondwanaland. In *Gondwana Geology.* ed. K.S.W. Campbell. pp. 125-34. Canberra: Australian National University Press.

Lack, D. (1944). Ecological aspects of species formation in passerine birds. *Ibis,* 86, 260-86.

Lack, D. (1945). The ecology of closely related species with special reference to Cormorant (*Phalacrocorax carbo*) and Shag (*P. aristotelis*). *Journal of Animal Ecology,* 14, 12-6.

Lack, D. (1947). *Darwin's Finches.* Cambridge: Cambridge University Press.

Lack, D. (1954). *The Natural Regulation of Animal Numbers.* Oxford: Clarendon Press.

Lack, D. (1969). Subspecies and sympatry in Darwin's finches. *Evolution,* 23, 252-63.

Lack, D. (1976). *Island Birds.* Oxford: Blackwell Scientific Publications.

Laing, J.E., and Hamai, J. (1976). Biological control of insect pests and weeds by imported parasites, predators, and pathogens. In *Theory and Practice of Biological Control,* ed. C.B. Huffaker and P.S. Messenger, pp. 685-743. New York: Academic Press.

Lamas, G. (1985). Los Papilionoidae (Lepidoptera) de la Zona Reservada de Tambopata, Madre de Dios, Peru. I. Papilionidae, Pieridae, y Nymphalidae (En Parte). *Revista Peruana Entomologia,* 27, 59-73.

Lamotte, M. (1983). Amphibians in savanna ecosystems. In *Ecosystems of the World 13: Tropical Savannas,* ed. F. Bourlière, pp. 313-23. Amsterdam: Elsevier.

Lamprey, H.F. (1963). Ecological separation of the large mammal species in the Tarangire Game Reserve, Tanganyika. *East African Wildlife Journal,* 1, 63-92.

Lamprey, H.F. (1964). Estimation of the large mammal densities, biomass and energy exchange in the Tarangire Game Reserve and the Masai steppe in Tanganyika. *East African Wildlife Journal,* 2, 1-46.

Lang, J. (1973). Interspecific aggression by Scleractinian corals. 2. Why the race is not only to the swift. *Bulletin of Marine Science*, 23, 260-79.

Langer, P. (1974). Stomach evolution in the Artiodactyla. *Mammalia*, 38, 295-314.

Larcher, W., (1980). *Physiological Plant Ecology*. Berlin: Springer-Verlag.

Lassen, H.H. (1975). The diversity of freshwater snails in view of the equilibrium theory of island biogeography. *Oecologia*, 19, 1-18.

Lathwell, D.J., and Grove, T.L. (1986). Soil-plant relationships in the tropics. *Annual Review of Ecology and Systematics*, 17, 1-16.

Lattin, J.D. (1990). Arthropod diversity in Northwest old-growth forests. *Wings*, 15, 7-10.

Lauenroth, W.K., Dodd, J.L, and Sims, P.L. (1978). The effects of water- and nitrogen-induced stresses on plant community structure in a semiarid grassland. *Oecologia*, 36, 211-22.

Lawes, J., and Gilbert, J. (1880). Agricultural, botanical, and chemical results of experiments on the mixed herbage of permanent grassland, conducted for many years in succession on the same land. I. *Philosophical Transactions of the Royal Society*, 171, 189-416.

Lawes, J.B., Gilbert, J.H., and Masters, M.T. (1882). Agricultural, botanical, and chemical results of experiments on the mixed herbage of permanent grassland, conducted for more than twenty years in succession on the same land. Part II, The botanical results. *Philosophical Transactions of the Royal Society (London)*, A and B, 173, 1181-1413.

Lawrence, D.B. (1958). Glaciers and vegetation in southeastern Alaska. *American Scientist*, 46, 89-122.

Laws, R.M. (1970). Elephants as agents of habitat and landscape change in East Africa. *Oikos*, 21, 1-15.

Laws, R.M. (1985). The ecology of the Southern Ocean. *American Scientist*, 73, 26-40.

Lawton, J.H. (1984). Non-competitive populations, non-convergent communities, and vacant niches: The herbivores of bracken. In*Ecological Communities: Conceptual Issues and the Evidence*, ed. D.E. Strong, D. Simberloff, L.G. Abele, and A.B. Thistle. pp. 67-100. Princeton: Princeton University Press.

Lawton, R.M. (1978). A study of the dynamic ecology of Zambian vegetation. *Journal of Ecology*, 66, 175-98.

Le Houerou, H.N. (1974). Fire and vegetation in the Mediterranean Basin. *Proceedings of the Annual Tall Timbers Fire Ecology Conference*, 13, 237-77.

LeBarron, R.K., and Eyre, F.H. (1939). The release of seeds from jack pine cones. *Journal of Forestry*, 37, 305-9.

Ledig, F.T. (1986). Heterozygosity, heterosis, and fitness in outbreeding plants. In *Conservation Biology: the Science of Scarcity and Diversity*, ed. M.E. Soulé, pp. 77-104. Sunderland, Massachusetts: Sinauer.

Lee, P.C. (1967). Ecological studies on *Dryobalanops aromatica* Gaertn. Ph.D. Diss. University of Malaysia, Kuala Lumpur.

Lehman, J.T. (1980). Release and cycling of nutrients between planktonic algae and herbivores. *Limnology and Oceanography*, 25, 620-32.

Lehman, J.T., Botkin, D.B., and Likens, G.E. (1975). The assumptions and rationales of a computer model of phytoplankton population dynamics. *Limnology and Oceanography*. 20, 343-64.

Leighton, M. and Leighton, D.R. (1982). The relationship of size of feeding aggregate to size of food patch: howler monkeys (*Alouatta pallida*) feeding

620

in *Trichilia cipo* fruit trees on Barro Colorado Island. *Biotropica*, 14, 81-90.

Leighton, M., and Wirawan, N. (1986). Catastrophic drought and fire in Borneo tropical rain forest associated with the 1982-1983 El Niño Southern Oscillation Event. In *Tropical Rain Forests and the World Atmosphere*, ed. G. Prance, pp. 75-102. Boulder, Colorado: Westview Press.

Lessios, H. A., Robertson, D. R., and Cubit, J. D. (1984). Spread of *Diadema* mass mortality through the Caribbean. *Science*, 226, 335-7.

Lessios, H.A. (1988). Mass mortality of *Diadema antillarum* in the Caribbean: What have we learned? *Annual Review of Ecology and Systematics*, 19, 371-93.

Leuthold, W. (1977). *African Ungulates*. Berlin: Springer-Verlag.

Leuthold, W., and Leuthold, B.M. (1975). Temporal patterns of reproduction in ungulates of Tsavo East National Park, Kenya. *East African Wildlife Journal*, 13, 159-69.

Levey, D.J. (1988a). Tropical wet forest treefall gaps and distributions of understory birds and plants. *Ecology*, 69, 1076-89.

Levey, D.J. (1988b). Spatial and temporal variation in Costa Rican fruit and fruit-eating bird abundance. *Ecological Monographs*, 58, 251-69.

Levin, S.A. (1970). Community equilibrium and stability, and an extension of the competitive exclusion principle. *American Naturalist*, 104, 413-23.

Levin, S.A. (1974). Dispersion and population interactions. *American Naturalist*, 108, 207-28.

Levin, S.A. (1976). Population dynamic models in heterogeneous environments. *Annual Review of Ecology and Systematics*, 7, 287-310.

Levin, S.A., and Paine, R.T. (1974). Disturbance, patch formation, and community structure. *Proceedings of the National Academy of Sciences*, 71, 2744-7.

Levin, S.A., and Paine, R.T. (1975). The role of disturbance in models of community structure. In *Ecosystem Analysis and Prediction*, ed. S.A. Levin, pp. 56-67. Philadelphia: Society for Industrial and Applied Mathematics.

Levine, S.H. (1976). Competitive interactions in ecosystems. *American Naturalist*, 110, 903-10.

Levins, R. (1968). *Evolution in Changing Environments*. Princeton, New Jersey: Princeton University Press.

Levins, R. (1969). Some demographic consequences of environmental heterogeneity for biological control. *Bulletin of the Entomological Society of America*, 15, 237-240.

Levins, R. (1979). Coexistence in a variable environment. *American Naturalist*, 114, 765-83.

Levins, R., and Culver, D. (1971). Regional coexistence of species and competition between rare species. *Proceedings of the National Academy of Sciences*, 68, 1246-8.

Lewis, C.E., and Harshbarger, T.J. (1976). Shrub and herbaceous vegetation after 20 years of prescribed burning in the South Carolina coastal plain. *J. Range Management*, 29, 13-8.

Lewis, J.R. (1976). *The Ecology of Rocky Shores*, London: Hodder and Stoughton.

Lewis, S.M. (1986). The role of herbivorous fishes in the organization of a Caribbean reef community. *Ecological Monographs*, 56, 183-200.

Lewis, W.M. (1978). Dynamics and succession of the phytoplankton in a tropical lake: Lake Lanao, Philippines. *Journal of Ecology*, 66, 849-80.

Lewis, W.M. (1987). Tropical limnology. *Annual Review Ecology Systematics*, 18, 159-84.

Liddell, W. D. and Ohlhorst, S. L. (1981). Geomorphology and community composition of two adjacent reef areas. Discovery Bay, Jamaica. *Journal of Marine Research*, 39, 791-804.

Liddell, W. D., Ohlhorst, S. L., and Boss, S. K. (1984). Community patterns on the Jamaican fore reef (15-56 M). *Paleontographic Americana*, 54, 385-9.

Liddell, W.D., Ohlhorst, S.L., and Coates, A.G. (1984). *Modern and Ancient Carbaonate Environments of Jamaica. Sedimenta X*. Miami Beach, Florida: Rosensteil School of Marine and Atmospheric Science.

Liddell, W.D., and Ohlhorst, S.L. (1986). Changes in benthic community composition following the mass mortality of *Diadema* at Jamaica. *Journal of Experimental Marine Biology and Ecology*, 95, 271-8.

Lieth, H., and Whittaker, R.H. (1975). *Primary Productivity of the Biosphere*. Berlin: Springer-Verlag.

Lillegraven, J.A. (1974). Biogeographical considerations of the marsupial-placental dichotomy. *Annual Review of Ecology and Systematics*, 5, 263-83.

Lindemann, R.L. (1942). The trophic-dynamic aspect of ecology. *Ecology*, 23, 399-418.

Linder, S. (1985). Potential and actual production of Australian forest stands. In *Research for Forest Management*. ed. J.J. Landsberg and W. Parsons. pp. 11-35. East Melbourne: Commonwealth Scientific and Industrial Research Organization.

Linder, S., McDonald, J., and Lohammar, T. (1981). Effect of nitrogen status and irradiance during cultivation on photosynthesis and respiration in birch seedlings. Energy Forest Project (EFP). Swedish University Agricultural Science, Upsala.

Lindsay, D.C. (1977) Lichens of cold deserts. In *Lichen Ecology*, ed. M.R.D. Seaward. pp. 183-209. London: Academic Press.

Linhart, Y.B. (1974). Intra-population differentiation in annual plants. I. *Veronica peregrina* L. raised under non-competitive conditions. *Evolution*, 28, 232-43.

Linhart, Y.B. (1976). Evolutionary studies of plants in vernal pools. In *Vernal Pools- Their Ecology and Conservation, Publication No. 9*, ed. S. Jain, pp. 40-6. Davis, California: Institute of Ecology, University of California.

Liogier, A.H. (1981). Flora of Hispaniola. Part 1, *Phytologia Memoirs*, 3, 1-218.

Little, S. (1946). The effects of forest fires on the stand history of New Jersey's pine region. *U.S. Forest Service, Northeast Forest Experiment Station, Forest Management Paper*, No. 2, 1-43.

Little, S. (1979). Fire and plant succession in the New Jersey Pine Barrens. In *Pine Barrens: Ecosystems and Landscapes*, ed. R.T.T. Forman, pp. 297-314. New York: Academic Press.

Livingstone, D.A. (1975). Late Quaternary climatic change in Africa. *Annual Review of Ecology and Systematics*, 6, 249-78.

Livingstone, D.A., and van der Hammen, T. (1978). Palaeogeography and palaeoclimatology. In *Tropical Forest Ecosystems: a State-of-Knowledge Report*, pp. 61-90. Paris: UNESCO.

Loach, K. (1967). Shade tolerance in tree seedlings. I. Leaf photosynthesis and respiration in plants raised under artificial shade. *New Phytologist*, 66, 607-21.

Loehle, C. (1988a). Forest decline: endogenous dynamics, tree defenses, and the elimination of spurious correlation. *Vegetatio*, 77, 65-78.

Loehle, C. (1988b). Tree life history strategies: the role of defense. *Canadian Journal of Forest Research*, 18, 209-22.

Lomnicki, A. (1988). *Population Ecology of Individuals*. Princeton, N.J.: Princeton University Press.

Lomolino, M.V. (1989). Interpretations and comparisons of constants in the species-area relationship: an additional caution. *American Naturalist*, 133, 277-80.

Lonsdale, P. (1977). Clustering of suspension-feeding macrobenthos near abyssal hydrothermal vents at oceanic spreading centers. *Deep Sea Research*, 24, 857-63.

Loomis, R.M. (1977). Wildfire effects on an oak-hickory forest in southeast Missouri. U.S. Department of Agriculture, Forest Service, North Central Forest Experiment Station Research Note NC-219.

Loope, L.L., and Gruell, G.E. (1973). The ecological role of fire in the Jackson Hole Area, Northwestern Wyoming. *Quaternary Research*, 3, 425-43.

Lorimer, C.G. (1977). The presettlement forest and natural disturbance cycle of northeastern Maine. *Ecology*, 58, 139-48.

Lorimer, C.G. (1980). Age structure and disturbance history of a southern Appalachian virgin forest. *Ecology*, 61, 1169-84.

Lorimer, C.G., and Gough, W.R. (1988). Frequency of drought and severe fire weather in north-eastern Wisconsin. *Journal of Environmental Management*, 26, 203-19.

Lorimer, C.G., and Krug, A.G. (1983). Diameter distributions in even-aged stands of shade-tolerant and midtolerant tree species. *American Midland Naturalist*, 109, 331-345.

Lotan, J.E. (1974). Cone serotiny-fire relationships in lodgepole pine. In *Proceedings of the Annual Tall Timbers Fire Ecology Conference*, 14, 267-78.

Lotan, J.E., Alexander, M.E., Arno, S.F., French, R.E., Langdon, O.G., Loomis, R.M., Norum, R.A., Rothermel, R.C., Schmidt, W.C., and Van Wagtendonk, J. (1981). *Effects of Fire on Flora: A State-of-Knowledge Review*. U.S. Department of Agriculture, Forest Service General Technical Report WO-16. Washington, D.C.:U.S. Government Printing Office.

Lotka, A.J. (1925). *Elements of Physical Biology*. Baltimore: Williams and Wilkins. (Reprinted 1956 by Dover Publ., New York, as *Elements of Mathematical Biology*.)

Loucks, O.L. (1970). Evolution of diversity, efficiency, and community stability. *American Zoologist*, 10, 17-25.

Louda, S.M. (1982a). Limitation of the recruitment of the shrub *Haplopappus squarrosus* (Asteraceae) by flower- and seed-feeding insects. *Journal of Ecology*, 70, 43-53.

Louda, S.M. (1982b). Distribution ecology: variation in plant recruitment over a gradient in relation to insect seed predation. *Ecological Monographs*, 52, 25-41.

Louda, S.M. (1983). Seed predation and seedling mortality in the recruitment of a shrub, *Haplopappus venetus* (Asteraceae), along a climatic gradient. *Ecology*, 64, 511-21.

Louda, S.M., Keeler, K.H., and Holt, R.D. (1990). Herbivore influences on plant performance and competitive interactions. In *Perspectives on Plant Competition.* ed. J.B. Grace and D. Tilman. pp. 413-44. San Diego: Academic.

Louis, J. (1947). Contribution á l' étude des foréts equatoriales congolaises. In *Comptes rendes de la Semaine agricole de Yangambi,* pp. 902-15. INEAC (Inst. Etude Agron. Congo), Brussels.

Loveless, M.D., and Hamrick, J.L. (1984). Ecological determinants of genetic structure in plant populations. *Annual Review of Ecology and Systematics,* 15, 65-95.

Lovelock, J. (1988). *The Ages of Gaia.* New York: W.W. Norton and Company.

Lowe, V.P.W. (1969). Population dynamics of the red deer (*Cervus elaphus* L.) on Rhum. *Journal of Animal Ecology,* 38, 425-57.

Lowe-McConnell, R.H. (1987). *Ecological Studies in Tropical Fish Communities.* Cambridge: Cambridge University Press.

Loya, Y. (1972). Community structure and species diversity of hermatypic corals at Eilat, Red Sea. *Marine Biology,* 13, 100-23.

Loya, Y. (1976). Effects of water turbidity and sedimentation on the community structure of Puerto Rican corals. *Bulletin of Marine Science,* 26, 450-66.

Loya, Y., Slobodkin, L.B. (1971) . The coral reefs of Eilat (Gulf of Eilat, Red Sea). In *Regional Variation in Indian Ocean Coral Reefs,* ed. D.R. Stoddart and C.M. Yonge, pp. 117-39. London: Academic Press

Lubchenco, J. (1978). Plant species diversity in a marine intertidal community: importance of herbivore food preference and algal comeptitive abilities. *American Naturalist,* 112, 23-39.

Lubchenco, J. (1980). Algal zonation in a New England rocky intertidal community: an experimental analysis. *Ecology,* 61, 333-44.

Lubchenco, J., and Cubit, J. (1980). Heteromorphic life histories of certain marine algae as adaptations to variation in herbivory. *Ecology,* 61, 676-87.

Lubchenco, J., and Gaines, S.D. (1981). A unified approach to marine plant-herbivore interactions. I. Populations and communities. *Annual Review of Ecology and Systematics,* 12, 405-37.

Lubchenco, J., and Menge, B.A. (1978). Community development and persistence in a low rocky intertidal zone. *Ecological Monographs,* 48, 67-94.

Ludwig, J.A., Whitford, W.G., and Cornelius, J.M. (1989). Effects of water, nitrogen and sulfur amendments on cover, density and size of Chichuahuan Desert ephemerals. *Journal of Arid Environments,* 16, 35-42.

Lugo, A.E. Biotropica palm forest reference.

Lugo, A.E., and Snedaker, S.C. (1974). The ecology of mangroves. *Annual Review of Ecology and Systematics,* 5, 39-64.

Lutz, H.J. (1956). *Ecological Effects of Forest Fires in the Interior of Alaska.* U.S. Forest Service Technical Bulletin No. 1133.

Lyon, L.J., and Stickney, P.F. (1976). Early vegetal succession following large northern Rocky Mountain wildfires. *Proceedings of the Tall Timbers Fire Ecology Conference,* 14, 355-75.

MacArthur, R.H. (1958). Population ecology of some warblers of northeastern coniferous forests. *Ecology,* 39, 599-619.

MacArthur, R. (1959). On the breeding distribution patterns of North American migrant birds. *Auk,* 318-25.

MacArthur, R.H. (1962). Some generalized theorems of natural selection. *Proceedings of the National Academy of Science (USA),* 48, 1893-7.

MacArthur, R.H. (1964). Environmental factors affecting bird species diversity. *American Naturalist*, 98, 387-97.

MacArthur, R.H. (1965). Patterns of species diversity. *Biological Review*, 40, 510-33.

MacArthur, R.H. (1968). The theory of the niche. In *Population Biology and Evolution*, ed. R.C. Lewontin, pp. 159-76. Syracuse: Syracuse University Press.

MacArthur, R.H. (1969). Patterns of communities in the tropics. *Biological Journal of the Linnean Society*, 1, 19-30.

MacArthur, R.H., Recher, H., and Cody, M. (1966). On the relation between habitat selection and species diversity. *American Naturalist*, 100, 319-32.

MacArthur, R.H., and Levins, R. (1967). The limiting similarity, convergence, and divergence of coexisting species. *American Naturalist*, 101, 377-85.

MacArthur, R.H., and MacArthur, J. (1961). On bird species diversity. *Ecology*, 42, 594-8.

MacArthur, R.H., and Wilson, E.O. (1963). An equilibrium theory of insular zoogeography. *Evolution*, 17, 373-87.

MacArthur, R.H., and Wilson, E.O. (1967). *The Theory of Island Biogeography*. Princeton, N.J.: Princeton University Press.

MacDonald, I.A.W., Loope, L.L., Usher, M.B., and Hamann, O. (1989). Wildlife conservation and the invasion of nature reserves by introduced species: a global perspective. In *Biological Invasions: A Global Perspective*, ed. J.A. Drake, *et al.*, pp. 215-55. Chichester: Wiley.

MacDonald, K.B. (1969). Quantitative studies of salt marsh mollusc faunas from the North American Pacific Coast. *Ecological Monographs*, 39, 33-60.

MacMahon, J.A. (1980). Ecosystems over time: succession and other types of change. In *Forests: Fresh Perspectives from Ecosystem Analysis*. ed. R.H. Waring. pp. 27-58. Corvallis: Oregon State University Press.

MacMahon, J.A. (1981). Successional processes: comparisons among biomes with special reference to probable roles of and influences on animals. In *Forest succession concepts and applications*. ed. D.C. West, H.H. Shugart, and D.B. Botkin. pp. 277-304. New York: Springer-Verlag.

MacNae, W. (1968). A general account of the fauna and flora of mangrove swamps and forests of the Indo-West-Pacific region. *Advances in Marine Biology*, 6, 73-270.

Macdonald, I.A.W., and Jarman, M.L. (eds.) (1984). Invasive alien organisms in the terrestrial ecosystems of the fynbos biome, South Africa. *South African National Scientific Programmes Report 85*, Pretoria: CSIR.

Macdonald, I.A.W. (1984). Is the fynbos biome especially susceptible to invasion by alien plants? A reanalysis of available data. *South African Journal of Science*, 80, 369-77.

Macdonald, I.A.W., Loope, L.L, Usher, M.B., and Hamann, O. (1989). Wildlife conservation and the invasion of nature reserves by introduced species: a global perspective. In ed. J.A. Drake *et al.*

Macdonald, I.A.W., Powrie, F.J., and Siegfried, W.R. (1986). The differential invasion of Soith Africa's biomes and ecosystems by alien plants and animals. In *The Ecology and Management of Biological Invasions of South Africa*, ed. I.A.W Macdonald, F.J. Kruger, and A.A. Ferrar.

Macdonald, I.A.W., and Richardson, D.M. (1986). Alien species in terrestrial ecosystems of the fynbos biome. In *The Ecology and Management of Biological Invasions in Southern Africa*, ed. I.A.W. Macdonald,

F.J. Kruger, and A.A. Ferrar, pp. 77-91. Cape Town: Oxford University Press.

Macintyre, I .G., and Smith, S.V. (1974). X-radiographic studies of skeletal development in coral colonies. *Proceedings of the Second International Coral Reef Symposium, 2*, 277-87.

Mack, R.N. (1981). Invasion of *Bromus tectorum* into western North America: an ecological chronicle. *Agro-Ecosystems, 7*, 145-65.

Mack, R.N. (1985). Invading plants: their potential contribution to population biology. In *Studies in Plant Demography: A Festschrift for John Harper*, ed. J. White, pp. 127-41. London: Academic Press.

Mack, R.N. (1986). Alien plant invasion into the Intermountain West: a case history. In *Ecology of Biological Invasions of North America and Hawaii*, ed. H.A. Mooney and J.A. Drake, pp. 191-213. New York: Springer-Verlag.

Mack, R.N. (1989). Temperate grasslands vulnerable to plant invasions: characteristics and consequences. In *Biological Invasions: a Global Perspective. SCOPE 37*, ed. J.A. Drake *et al.*, pp. 155-79. Chichester: Wiley.

Maddock, L. (1979). The 'migration' and grazing succession. In *Serengeti: Dynamics of an Ecosystem*, ed. A.R.E. Sinclair and M. Norton-Griffiths, pp. 104-29. Chicago, Illinois University of Chicago Press.

Madenjian, C.P., Johnson, B.M., and Carpenter, S.R. (1991). Stocking strategies for fingerling walleyes: an individual-based model approach. *Ecological Applications, 1*, 280-8.

Madenjian, C.P., and Carpenter, S.R. (1991). Individual-based model for growth of young-of-the-year walleye: a piece of the recruitment puzzle. *Ecological Applications, 1*, 268-79.

Madison, M. (1977). Vascular epiphytes: Their systematic occurrence and salient featurs. *Selbyana, 2*, 1-13.

Magurran, A.E. (1988). *Ecological Diversity and its Measurement*. Princeton: Princeton University Press.

Mahdi, A., Law, R., and Willis, A.J. (1989). Large niche overlaps among coexisting plant species in a limestone grassland community. *Journal of Ecology, 77*, 386-400.

Mahmoud, A., and Grime, J.P. (1976). Analysis of competitive ability in three perennial grasses. *New Phytologist, 77*, 431-5.

Malcolm, J.R. (1990). Estimation of mammalian densities in continuous forest north of Manaus. In *Four Neotropical Rainforests*, ed. A.H. Gentry, pp. 339-57. New Haven: Yale University Press.

Malingreau, J.P., Stephens, G., and Fellows, L. (1985). Remote sensing of forest fires: Kalimantan and North Borneo in 1982-1983. *Ambio, 14*, 314-21.

Mann, K.H. (1982). *Ecology of Coastal Waters: A Systems Approach*. Berkeley, California: University of California Press.

Marcuzzi, G. (1989). Migratory phenomena in European animal species. In *Biological Invasions in Europe and the Mediterranean Basin*, ed. F. di Castri, A.J. Hansen, and M. Debussche, pp. 217-28. Dordrecht: Kluwer.

Margules, C.R., Nichols, A.O., and Austin, M.P. (1987). Diversity of *Eucalyptus* species predicted by a multi-variable environmental gradient. *Oecologia, 71*, 229-32.

Markham, R.H., and Babbedge, A.J. (1979). Soil and vegetation catenas on the forest-savannah boundary in Ghana. *Biotropica, 11*, 224-34.

Marks, P.L. (1974). The role of pin cherry (*Prunus pensylvanica* L.) in the maintenance of stability in northern hardwood ecosystems. *Ecological Monographs*, 44, 73-88.

Marks, P.L., and Bormann, F.H. (1972). Revegetation following forest cutting: mechanisms for return to steady state nutrient cycling. *Science*, 176, 914-5.

Marquis, R.J. (1984). Leaf herbivores decrease fitness of a tropical plant. *Science*, 226, 537-9.

Marshall, L.G., Webb, S.D., Sepkowski, J.J., and Raup, D.M. (1982). Mammalian evolution and the great American interchange. *Science*, 215, 1351-7.

Martin, R.E. (1963). A basic approach to fire injury of tree stems. In *Proceedings of the Second Annual Tall Timbers Fire Ecology Conference*. 2, 186-90. Tallahassee, Florida: Tall Timbers Research Station.

Martin, S.C. (1975). Ecology and management of southwestern semidesert grass-shrub ranges: the status of our knowledge. Res. Paper RM-156. Fort Collins, Colorado: U.S. Department of Agriculture Forest Service.

Martin, W.H. (1989). *The Role and History of Fire in the Daniel Boone National Forest*. Winchester, Kentucky: Daniel Boone National Forest.

Maslin, B.R., and Pedley, L. (1982). The distribution of *Acacia* (Leguminosae: Mimosoideae) in Australia. Part 1. Species Distribution Maps. *Western Australia Herbarium Research Notes*, 6, 1-128.

Matthew, W.D. (1915). Climate and Evolution. *Annals of the New York Academy of Science*, 24, 171-318.

Matthews, J.A. (1978). Plant colonisation patterns on a gletschervorfeld, southern Norway: a meso-scale geographical approach to vegetation change and phytometric dating. *Boreas*, 7, 155-78.

Matthews, J.A. (1979). A study of the variability of some successional and climax plant assemblage types using multiple discriminant analysis. *Journal of Ecology*, 67, 255-71.

Maxwell, H. (1910). The use and abuse of the forests by the Virginia Indians. *William and Mary College Quarterly Historical Magazine*, 19, 73-103.

May, R.M. (1972). Will a large complex system be stable? *Nature*, 238, 413-4.

May, R.M. (1973). *Stability and Complexity in Model Ecosystems*. Princeton, Princeton University Press.

May, R.M. (1975). Patterns of species abundance and diversity. In *Ecology and Evolution of Communities*. ed. M.L. Cody, and J.M. Diamond. pp. 81-120. Cambridge, Massachusetts: Belknap Press.

May, R.M. (1985). Population dynamics: communities. In *The Study of Populations*. ed. H. Messell. pp. 31-44. Rushcutters Bay, Australia: Pergamon Press.

May, R.M., and Oster, G. (1976). Bifurcations and dynamic complexity in simple ecological models. *American Naturalist*, 110, 573-99.

Mayr, E. (1969). *Principles of Systematic Zoology*. New York: McGraw-Hill.

McArthur, J.V., Kovacic, D.A., and Smith, M.H. (1988). Genetic diversity in natural populations of a soil bacterium across a landscape gradient. *Proceedings of the National Academy of Science, USA*, 85, 9621-4.

McCarthy, E.F. (1933). *Yellow poplar characteristics, growth, and management*. U.S. Department of Agriculture Technical Bulletin 356. Washington, D.C.: U.S. Government Printing Office.

McClaugherty, C.A., Pastor, J., Aber, J.D., and Muratore, J.F. (1985). Forest litter decomposition in relation to soil nitrogen dynamics and litter quality. *Ecology*, 66, 266-75.

McCormick, P.V., and Stevenson, R.J. (1989). Effects of snail grazing on benthic algal community structure in different nutrient environments. *Journal of the North American Benthological Society*, 8, 162-72.

McCoy, E.D., and Connor, E.F. (1980). Latitudinal gradients in the species diversity of North American mammals. *Evolution*, 34, 193-203.

McCullough, D.R. (1970). Secondary production of birds and mammals. In *Analysis of Temperate Forest Ecosystems*, ed. D.E. Reichle, pp. 107-30. New York: Springer-Verlag.

McCune, B., and Allen, T.F.H. (1985). Will similar forests develop on similar sites? *Canadian Journal of Botany*, 63, 367-76.

McGee, C.E. (1979). Fire and other factors related to oak regeneration. In *Regenerating Oaks in Upland Hardwood Forests*, ed. H.A. Holt and B.C. Fischer, pp. 75-81. West Lafayette, Indiana: Purdue University Department of Forest and Natural Resources.

McGowan, J.A., and Walker, P.W. (1979). Structure in the copepod community of the North Pacific Central Gyre. *Ecological Monographs*, 49, 195-226.

McGowan, J.A., and Walker, P.W. (1985). Dominance and diversity maintenance in an oceanic ecosystem. *Ecological Monographs*, 55, 103-18.

McIntosh, R.P. (1967). The continuum concept of vegetation. *Biological Review*, 33, 130-87.

McIntosh, R.P. (1981). Succession and ecological theory. In *Forest succession concepts and applications*, ed. D.C. West, H.H. Shugart, and D.B. Botkin, pp. 10-23. New York: Springer-Verlag.

McIntosh, R.P. (1985). *The Background of Ecology, Concept and Theory*. Cambridge: Cambridge University Press.

McIntosh, R.P. (1987). Pluralism in ecology. *Annual Review of Ecology and Systematics*, 18, 321-41.

McIntosh, R.P. ed. (1978). *Phytosociology. Benchmark Papers in Ecology/6*. Stroudsburg, Pennsylvania: Dowden, Hutchinson, and Ross.

McKee, W.H. (1982). Changes in soil fertility following prescribed burning on Coastal Plain pine sites. U.S. Department of Agriculture Forest Service, Research Paper SE-234.

McKey, D. (1979) The distribution of secondary compounds within plants. In *Herbivores. Their Interactions with Secondary Plant Constituents*, ed. G.A. Rosenthal and D.H. Janzen, pp. 56-133. New York: Academic Press.

McKey, D.B., Waterman, P.G., Mbi, C.N., Gartlan, J.S., and Strusaker, T.T. (1978). Phenolic content of vegetation in two African rain-forests: ecological implications. *Science*, 202, 61-4.

McLaughlin, S.B., and Shriner, D.S. (1980). Allocation of resources to defense and repair. In *Plant Diseases. Vol. IV*, ed. J.G. Horsfall and E.B. Cowling, pp. 407-31. New York: Academic.

McNab, B.K. (1963). Bioenergetics and the determination of home range size. *American Naturalist*, 97, 130-40.

McNaughton, S.J. (1968). Structure and function in California grasslands. *Ecology*, 49, 962-72.

McNaughton, S.J. (1984). Grazing lawns: animals in herds, plant form, and coevolution. *American Naturalist*, 124, 863-86.

McPherson, J.K., and Muller, C.H. (1969). Allelopathic effects of *Adenostoma fasiculatum* 'chamise,' in the Californian chaparral. *Ecological Monographs*, 39, 177-98.

Mech, L.D. (1966). *The Wolves of Isle Royale*. National Parks Fauna Series, No. 7.

Medina, E., and Silva, J.F. (1990). Savannas of northern South America: a steady state regulated by water-fire interactions on a background of low nutrient availability. *Journal of Biogeography*, 17, 403-13.

Meentemeyer, V. (1978). Macroclimate and lignin control of litter decomposition rates. *Ecology*, 59, 266-75.

Melillo, J.M., Aber, J.D., and Muratore, J.F. (1982). Nitrogen and lignin control of hardwood leaf litter decomposition dynamics. *Ecology*, 63, 621-6.

Mellinger, M.V., and McNaughton, S.J. (1975). Structure and function of successional vascular plant communities in central New York. *Ecological Monographs*, 45, 161-82.

Menaut, J.-C. (1983). The vegetation of African savannas. In *Ecosystems of the World 13: Tropical Savannas*, ed. F. Bourlière, pp. 109-49. Amsterdam: Elsevier.

Menge, B.A. (1976). Organization of the New England rocky intertidal community: role of predation, competition, and environmental heterogeneity. *Ecological Monographs*, 46, 355-93.

Menge, B.A., and Sutherland, J.P. (1976). Species diversity gradients: synthesis of the roles of predation competition, and temporal heterogeneity. *American Naturalist*, 110, 351-69.

Menzies, R.J., George, R.Y., and Rowe, G.T. (1973). *Abyssal Environment and Ecology of the World Oceans*. New York: Wiley and Sons.

Mergner, H., and Scheer, G. (1974). The physiographic zonation and ecological conditions of some South Indian and Ceylon coral reefs. *Proceedings of the Second International Coral Reef Symposium*, 2, 3-31.

Merrell, D.J. (1951). Interspecific competition between *Drosophila funebris* and *Drosophila melanogaster*. *American Naturalist*, 85, 159-69.

Mertz, D.B. (1972). The *Tribolium* model and the mathematics of population growth. *Annual Review of Ecology and Systematics*, 3, 51-78.

Meyer, F.H., and Göttsche, D. (1971). Distribution of root tips and tender roots of beech. In *Integrated Experimental Ecology: Methods and Results of Ecosystem Research in the German Solling Project (Ecological Studies, Vol. 2)*, ed. H. Ellenberg, pp. 48-52. Heidelberg: Springer-Verlag.

Michaux, J., Cheylan, G., and Croset, H. (1989). Of mice and men. In *Biological Invasions in Europe and the Mediterranean Basin*, ed. F. di Castri, A.J. Hansen, and M. Debussche, pp. 263-84. Dordrecht: Kluwer.

Michener, C.D. (1979). Biogeography of the bees. *Annals of the Missouri Botanical Garden*, 66, 277-347.

Mielke, D.L., Shugart, H.H., and West, D.C. (1977). User's manual for FORAR, a stand model for upland forests of southern Arkansas. ORNL/TM-5767. Oak Ridge, Tennessee: Oak Ridge National Laboratory.

Mielke, D.L., Shugart, H.H., and West, D.C. (1978). A stand model for upland forests of southern Arkansas. ORNL/TM-6225. Oak Ridge, Tennessee: Oak Ridge National Laboratory.

Mikhailov, V.M. (1964). Hydrology and formation of river-mouth bass. *Humid Tropical Research UNESCO*, 24, 59-64.

Milchunas, D.G., Sala, O.E., and Lauenroth, W.K. (1988). A generalized model of the effects of grazing by large herbivores on grassland community structure. *American Naturalist*, 132, 87-106.

Milchunas, D.G., and Lauenroth, W.K. (1991). A quantitative global assessment of the effects of grazing by large herbivores on vegetation and soils. *Bulletin of the Ecological Society of America,* 72:195.

Milewski, A.V. (1981). A comparison of reptile communities in relation to soil fertility in the mediterrean and adjacent arid parts of Australia and southern Africa. *Journal of Biogeography,* 8, 493-503.

Milewski, A.V. (1982). The occurrence of seeds and fruits taken by ants versus birds in mediterranean Australia and southern Africa, in relation to the availability of soil potassium. *Journal of Biogeography,* 9, 505-16.

Milewski, A.V. (1983). A comparison of ecosystems in mediterranean Australia and Southern Africa: nutrient-poor sites at the Barrens and the Caledon Coast. *Annual Review of Ecology and Systematics,* 14, 57-76.

Miller, A.C. (1974). A comparison of gastropod species diversity and trophic · structure in the rocky intertidal zone of the temperate and tropical West Americas. Ph.D. Thesis, University of Oregon.

Miller, R.R. (1958). Origin and affinities of the freshwater fish fauna of western North America. In *Zoogeography.* ed. C.L. Hubbs. Publ. 51. pp. 187-222. Washington, D.C.. American Association for the Advancement of Science.

Miller, R.S. (1964). Ecology and distribution of pocket gophers (Geomyidae) in Colorado. *Ecology,* 45, 256-72.

Miller, R.S. (1967). Pattern and process in competition. *Advances in Ecological Research,* 4, 1-74.

Miller, R.S. (1969). Competition and species diversity. *Brookhaven Symposium in Biology,* 22, 63-70.

Miller, T.E. (1982) Community diversity and interactions between the size and frequency of distrubance. *American Naturalist,* 120, 533-6.

Milliman, J.D., Qin, Y.S., Ren, M.E., and Saita, Y. (1987). Man's influence on erosion and transport of sediment by Asian rivers: the Yellow River (Huanghe) example. *Journal of Geology,* 95, 751-62.

Milton, W.E.J. (1940). The effect of manuring, grazing and cutting on the yield, botanical and chemical composition of natural hill pastures. *Journal of Ecology,* 28, 326-56.

Milton, W.E.J. (1947). The yield, botanical, and chemical composition of natural hill herbage under manuring, controlled grazing, and hay conditions. I. Yield and botanical. *Journal of Ecology,* 35, 65-89.

Mitchell, H.L., and Chandler, R.F. (1939). The nitrogen nutrition and growth of certain deciduous trees of northeastern United States. *Black Rock Forest Bulletin,* 11, 1-91.

Mohler, C.L., Marks, P.L., and Sprugel, D.G. (1978). Stand structure and allometry of trees during self-thinning of pure stands. *Journal of Ecology,* 66, 599-614.

Molles, M.C. (1978). Fish species diversity on model and natural reef patches: experimental insular biogeography. *Ecological Monographs,* 48, 289-305.

Monk, C.D. (1966). Ecological importance of root/shoot ratios. *Bulletin of the Torrey Botanical Club,* 93, 402-6.

Monk, C.D. (1967). Tree species diversity in the eastern deciduous forest with particular reference to north central Florida. *American Naturalist,* 101, 173-87.

Monro, J. (1967). The exploitation and conservation of resources by populations of insects. *Journal of Animal Ecology* 36, 531-47.

Monsi, N. (1968). Mathematical models of plant communities. *Functioning of*

Terrestrial Ecosystems at the Primary Production Level, ed. F. Eckardt, pp. 131-49 Paris: UNESCO.

Monsi, N., and Murata, Y. (1970). Development of photosynthetic systems as influenced by distribution of matter. In *Prediction and Measurement of Photosynthetic Productivity,* pp. 115-29. Wageningen, Netherlands: Cent. Agr. Publ. Doc.

Montgomery, G.G., and Sundquist, M.E. (1975). Impact of sloths on neotropical forest energy flow and nutrient cycling. In *Tropical Ecological Systems,* ed. F.B. Golley and E.Medina, pp. 69-98. Berlin: Springer-Verlag.

Montgomery, K.R., and Strid, T.W. (1976). Regeneration of introduced species of *Cistus* (Cistaceae) after fire in southern California. *Madrono, 23,* 417-27.

Mooney, H.A. (1972). The carbon balance of plants. *Annual Review of of Ecology and Systematics, 3,* 315-46.

Mooney, H.A. (ed.) (1977). *Convergent Evolution in Chile and California, Mediterranean Climate Ecosystems.* Stroudsburg, Pennsylvania: Dowden, Hutchinson and Ross.

Mooney, H.A., Hamburg, S.P., and Drake, J.A. (1986). The invasion of plants and animals into California. In *Ecology of Biological Invasions of North America and Hawaii,* ed. H.A. Mooney and J.A. Drake, pp. 250-72. New York: Springer-Verlag.

Mooney, H.A., and Bartholomew, B. (1974). Comparative carbon balance and reproductive models of two Californian *Aesculus* species. *Botanical Gazette, 135,* 306-13.

Mooney, H.A., and Dunn, E.L. (1970). Convergent evolution of mediterranean-climate evergreen scherlphyll shrubs. *Evolution, 24,* 292-303.

Mooney, H.A., and Gulmon, S.L. (1979). Environmental and evolutionary constraints on the photosynthetic characteristics of higher plants. In *Topics in Plant Population Biology,* In *Topics in Plant Population Biology.* ed. O.T. Solbrig, S. Jain, G.B. Johnson, and P.H. Raven, pp. 316-37. New York: Columbia University Press.

Mooney, H.A., and Gulmon, S.L. (1982). Constraints on leaf structure and function in reference to herbivory. *Bioscience, 32,* 198-206.

Mooney, H.A., and Parsons, D.J. (1973). Structure and function of the California chaparral - an example from San Dimas. In *Mediterranean-Type Ecosystems Origens and Structure,* ed. F. Di Castri and H.A. Mooney, pp. 83-113. New York: Springer-Verlag.

Moore, D.R.J.,and Keddy, P.A. (1989). The relationship between species richness and standing crop in wetlands: the importance of scale. *Vegetatio, 79,* 99-106.

Moore, J.A. (1952a). Competition between *Drosophila melanogaster* and *Drosophila simulans.* I. Population cage experiments. *Evolution, 6,* 407-20.

Moore, J.A. (1952b). Competition between *Drosophila melanogaster* and *Drosophila simulans,* II. The improvement of competitive ability through selection. *Proceedings of the National Academy of Science, 38,* 813-17.

Moore, R.C. (1954). Evolution of Late Paleozoic invertebrates in response to major oscillations of shallow seas. *Bulletin of the Museum of Comparative Zoology, 112,* 259-86.

Moore, R.M. (1975). *Australian Grasslands.* Canberra: Australian National University Press.

Moran, V.C. (1980). Interactions between phytophagous insects and their *Opuntia* hosts. *Ecological Monographs, 50,* 153-64.

Morat, P., Veillon, J.-M., and MacKee, H.S. (1984). Floristic relationships of New Caledonian rain forest panerogams. In *Biogeography of the Tropical Pacific,* ed. R Radovsky, R. Raven, and S. Sohmer, pp. 70-128. *Bishop Museum Special Publication,*

Morell, V. (1990). Running for their lives. *International Wildlife,* 20, 4-13.

Morris, S.C. (1979). The Burgess shale (Middle Cambrian) fauna. *Annual Review of Ecology and Systematics,* 10, 327-49.

Morrison, D. (1988). Comparing the effects of fish and sea urchin grazing in shallow and deeper coral reef algal communities. *Ecology,* 69, 1367-82.

Morrow, P.A. (1981). The role of sclerophyllous leaves in determining insect grazing damage. In *Mediterranean-type Ecosystems: The Role of Nutrients, Ecological Studies Volume 43,* ed. F.J. Kruger, D.T. Mitchell, and J.U.M. Jarvis, pp. 509-24. Berlin: Springer-Verlag.

Morton, J.K. (1972). Phytogeography of the West African mountains. In *Taxonomy, Phytogeography, and Evolution.* ed. D.H. Valentine. pp. 221-36. New York: Academic Press.

Motta, P.J. (1988). Functional morphology of the feeding apparatus of ten species of Pacific butterflyfishes (Perciformes, Chaetodontidae): an ecomorphological approach. *Environmental Biology of Fishes,* 22, 39-67.

Mueller, I.M., and Weaver, J.E. (1942). Relative drought resistance of seedlings of dominant prairie grasses. *Ecology,* 23, 387-98.

Mueller-Dombois, D. and Ellenberg, H. (1974). Aims and Methods of Vegetation Ecology. New York: Wiley.

Mueller-Dombois, D., Bridges, K.W., and Carson, H.L. (eds.) (1981). *Island Ecosystems.* Stroudsburg, Pennsylvania: Hutchinson Ross.

Mueller-Dombois, D., and 21 coauthors. (1981). Altitudinal distribution of organisms along an island mountain transect. In *Island Ecosystems: Biological Organization in Selected Hawaiian Communities.* ed. D. Mueller-Dombois, K.W. Bridges, and H.L. Carson. pp. 77-180. Stroudsburg, Pennsylvania: Hutchinson Ross.

Mueller-Dombois, D., and Sims, H.P. (1966). Response of three grasses to two soils and water-table depth gradient. *Ecology,* 47, 644-8.

Muir, P.S., and Lotan, J.E. (1985a). Serotiny and life history of *Pinus contorta* var. *latifolia. Canadian Journal of Botany,* 63, 938-45.

Muir, P.S., and Lotan, J.E. (1985b). Disturbance history and serotiny of *Pinus contorta* in western Montana. *Ecology,* 66, 1658-68.

Muller, C.H., and del Moral, R. (1966). Soil toxicity induced by terpenes from *Salvia leucophylla. Bulletin of the Torrey Botanical Garden,* 93, 130-7.

Murdoch, W.W., Chesson, J., and Chesson, P.L. (1985). Biological control in theory and practice. *American Naturalist,* 125, 344-66.

Murdoch, W.W., Evans, F.C., and Peterson, C.H. (1972). Diversity and pattern in plants and insects. *Ecology,* 53, 819-29.

Murphy, W.E. (1960). Ecological changes induced in moorland pastured by different fertilizer treatments. *Proceedings of the International Grasslands Congress,* 8, 86-9.

Murton, R.K., Isaacson, A.J., and Westwood, N.J. (1966). The relationships between wood pigeons and their clover food supply and the mechanism of population control. *Journal of Applied Ecology,* 3, 55-93.

Muscatine, L. (1973). Nutrition of Corals. pp. 77-115 in *Biology and Geology of Coral Reefs. Biology 1. Vol. 2,* ed. O.A. Jones and R. Endean. New York: Academic Press.

Mutch, R.W. (1970). Wildland fires and ecosystems - a hypothesis. *Ecology*, 51, 1046-51.

Myers, J.H. (1976). Distribution and dispersal in populations capable of resource depletion: a simulation model. *Oecologia*, 23, 255-69.

Myers, K. (1986). Introduced vertebrates in Australia, with emphasis on the mammals. In *Ecology of Biological Invasions: An Australian Perspective*, ed. R.H. Groves and J.J. Burdon, pp. 120-36. Canberra: Australian Academy of Science.

Myers, R.L. (1985). Fire and the dynamic relationship between Florida sandhill and sand pine scrub vegetation. *Bulletin Torrey Botanical Club*, 112, 241-52.

Müller, M.J. (1982). *Selected Climatic Data for a Global Set of Standard Stations for Vegetation Science*. The Hague: Junk.

NORPAC Committee. (1960). *Oceanic Observations of the Pacific: 1955, the NORPAC Atlas*. Berkeley, California and Tokyo, Japan: The University of California Press and the University of Tokyo Press.

Nanson, G.C., and Beach, H.F. (1977). Forest succession and sedimentation on a meandering-river floodplain, northeast British Columbia, Canada. *Journal of Biogeography*, 4, 229-51.

Naveh, Z. (1973). The ecology of fire in Israel. *Proceedings of the Annual Tall Timbers Fire Ecology Conference*, 13, 131-70.

Naveh, Z. (1974). Effects of fire in the Mediterranean region. In *Fire and Ecosystems*, ed. T.T. Kozlowski and C.E. Ahlgren, pp. 401-34. New York: Academic Press.

Neite, H., and Wittig, R. (1985). Correlation of chemical soil pattern with the floristic pattern in the trunk base area of the beech. *Acta Oecologia Plantarum*, 6, 375-86.

Nelson, C.H., and Johnson, K.R. (1987). Whales and walruses as tillers of the sea floor. *Scientific American*, [Feb. 1987], 112-117.

Nelson, E.W. (1934). The influence of precipitation and grazing upon black grama grass range. Tech. Bull. 409. Washington, D.C.: U.S. Department of Agriculture.

Nelson, G. and Platnick, N. (1981). *Systematics and Biogeography: Cladistics and Vicariance*. New York: Columbia University Press.

Nelson, G. and Rosen, D.E. eds. (1981). *Vicariance Biogeography: A Critique*, New York: Columbia University Press.

Nelson, J.R. (1974). Forest fire and big game in the Pacific Northwest. *Proceedings of the Annual Tall Timbers Fire Ecology Conference*, 15, 85-102.

Nelson, R.M. (1952). Observations of heat tolerance of southern pine needles. Southeastern Forest Experiment Station Paper Number 14. Asheville, North Carolina: U.S. Department of Agriculture, Forest Service.

Nerini, M.K., and Oliver, J.S. (1983). Gray whales and the structure of the Bering Sea benthos. *Oecologia*, 59, 224-5.

Neudecker, S. (1977). Transplant experiments to test the effect of fish grazing on coral distribution. *Proceedings of the Third International Coral Reef Symposium*, 1, 317-23.

Neumann, C.J., Coy, C.W., Caso, E.L., and Jarvinen, B.R. (1978). *Tropical cyclones of the North Atlantic Ocean*. Washington, D.C.: U.S. Dept. of Commerce, NOAA.

Nevo, E. (1979). Adaptive convergence and divergence of subterranean mammals. *Annual Review of Ecology and Systematics*, 10, 269-308.

Niering, W.A. (1981). The role of fire management in altering ecosystems. In

Fire Regimes and Ecosystems Properties, pp. 489-510. U.S. Department of Agriculture Forest Service General Technical Report WO-26.

Niering, W.A., and Lowe, C.H. (1984). Vegetation of the Santa Catalina Mountains: community types and dynamics. *Vegatatio*, 58, 3-28.

Niklas, K.J., Tiffney, B.H., and Knoll, A.H. (1983). Patterns in vascular land plant diversification. *Nature*, 303, 293-9.

Nowacki, G.J., Abrams, M.D., and Lorimer, C.G. (1990). Composition, structure, and historical development of northern red oak stands along an edaphic gradient in north-central Wisconsin. *Forest Science*, 36, 276-92.

Noy-Meir, I. (1973). Desert ecosystems: environment and producers. *Annual Review of Ecology and Systematics*, 4, 25-51.

Nudds, T.D. (1983). Niche dynamics and organization of waterfowl guilds in variable environments. *Ecology*, 64, 319-30.

Nutman, F.J. (1937). Studies in the physiology of *Coffea arabica* L. Photosynthesis of coffee leaves under natural conditions. *Annals of Botany N.S.*, 1, 353-67.

O'Neill, R.V. (1976). Paradigms of ecosystem analysis. In *Ecological Theory and Ecosystem Models*, ed. S.A. Levin, pp. 16-9. Indianapolis, Indianana: Institute of Ecology.

O'Neill, R.V., Harris, W.F., Ausmus, B.S., and Reichle, D.E. (1975). Theoretical basis for ecosystem analysis with particular reference to element cycling. In *Mineral Cycling in Southeastern Ecosystems*, ed. F.G. Howell, J.B. Gentry, and M.H. Smith, pp. 28-40. ERDS (Energy Research and Development Administration) Symposium Series CONF-740513.

O'Neill, R.V., and Giddings, J.M. (1979). Population interactions and ecosystem function: phytoplankton competition and community production. In *Systems Analysis of Ecosystems*, ed. G.S. Innis and R.V. O'Neill,

O'Neill, R.V., and Rust, B.W. (1979). Aggregation error in ecological models. *Ecological Modelling*, 7, 91-105.

Odum, E. P. (1960). Organic production and turnover in old field succession. *Ecology*, 41, 34-48.

Odum, E.P, Pomeroy, S.E., Dickinson, J.C., and Hutcheson, K. (1974). The effects of late winter litter burn on the composition, productivity and diversity of a 4-year old fallow-field in Georgia. *Proceedings of the Annual Tall Timbers Fire Ecology Conference*, 13, 399-427.

Odum, E.P. (1953). *Fundamentals of Ecology*. Philadelphia: Saunders.

Odum, H.T. (1970). Rain forest structure and nutrient cycling homeostasis. In *A Tropical Rain Forest. Book 3*, ed. H.T. Odum and R.F. Pigeon, pp. H3-H52. Washington, D.C.: U.S. Atomic Energy Commission.

Odum, H.T., Abbott, W. (1970)... *A study of Irradiation and Ecology at El Verde, Puerto Rico. Book 3*, ed. H.T. Odum and R.F. Pigeon, pp. I3-I19. Washington, D.C: U.S. Atomic Energy Commission.

Odum, H.T., Abbott, W., Sealander, R.K., Golley, F.B., and Wilson, R.F. (1970). Estimates of chlorophyll and biomasss of the Tabunuco forest of Puerto Rico. In *A Tropical Rainforest*.

Ogden, J.C., and Lobel, P.S. (1978). The role of herbivorous fishes and urchins in coral reef communities. *Environmental Biology of Fishes*, 3, 49-63.

Ohlhorst, S.L. (1980). Jamaican coral reefs: important biological and physical parameters. Ph.D. thesis, Yale University

Oksanen, L., Fretwell, S.D., Arruda, J., and Niemela, P. (1981). Exploitation ecosystems in gradients of primary productivity. *Am. Nat.*, 118, 240-61.

Oliver, E.G.H., Linder, H.P, and Rourke, J.P. (1983). Geographical distribution of present-day Cape taxa and their phytogeographical significance. *Bothalia*, 14, 427-40.

Olson, J.S. (1958). Rates of succession and soil changes on southern Lake Michigan sand dunes. *Botanical Gazette*, 119, 125-69.

Olson, J.S. (1963). Energy storage and the balance of producers and decomposers in scological systems. *Ecology*, 44, 322–31.

Olson, J.S., Watts, J.A., and Allison, L.J. (1983). *Carbon in Live Vegetation of Major World Ecosystems*. Oak Ridge National Laboratory Publication 5862.

Olsvig, L.S., Cryan, J.F., and Whittaker, R.H. (1979). Vegetational gradients of the Pine Plains and Barrens of Long Island, New York. In *Pine Barrens Ecosystems and Landscapes*. ed. R.T.T. Forman. pp. 265-82. New York: Academic Press.

Oosting, H.J. (1942). An ecological analysis of the plant communities of Piedmont, North Carolina. *American Midland Naturalist*, 28, 1-126.

Oosting, H.J., and Kramer, P.J. (1946). Water and light in relation to pine reproduction. *Ecology*, 27, 47-53.

Opler, P.A., Baker, H.G., and Frankie, G.W. (1977). Recovery of tropical lowland forest ecosystems. In *Recovery and restoration of damaged ecosystems*. ed. J. Cairns, K. L. Dickson, and E. E. Herricks. pp. 379-421. Charlottesville, Virginia: Univ. Press of Virginia.

Orians, G.H. (1982). The influence of tree-falls in tropical forests on tree species richness. *Tropical Ecology*, 23, 255-77.

Orians, G.H. (1986). Site characteristics favoring invasion. In *Ecology of Biological Invasions of North America and Hawaii*, ed. H.A. Mooney and J.A. Drake, pp. 133-48. New York: Springer-Verlag.

Orians, G.H., and Paine, R.T. (1983). Convergent evolution at the community level. In *Coevolution*, ed. D.J. Futuyma and M. Slatkin, pp. 431-58. Sunderland, Massachusetts: Sinauer.

Orians, G.H., and Solbrig, O.T. (1977). A cost-income model of leaves and roots with special reference to arid and semiarid areas. *American Naturalist*, 111, 677-90.

Osmond, C.B. (1978). Crassulacean acid metabolism: a curiosity in context. *Annual Review of Plant Physiology*, 29, 379-414.

Osmond, C.B., Austin, M.P., Berry, J.A., Billings, W.D., Boyer, J.S., Dacey, J.W.H., Nobel, P.S., Smith, S.D., and Winner, W.E. (1987). Stress physiology and the distribution of plants. *Bioscience*, 37, 38-48.

Osonubi, O., and Davies, J.W. (1980). The influence of water stress on the photosynthetic performance and stomatal behavior of tree seedlings subjected to variation in temperature and irradiance. *Oecologia*, 45, 3-10.

Ott, B. (1975). Community patterns on a submerged barrier reef at Barbados, West Indies. *Int. Revue. Ges. Hydrobiol.*, 60, 719-36.

Overpeck, J.T., Rind, D., and Goldberg, R. (1990). Climate-induced changes in forest disturbance and vegetation. *Nature*, 343, 51-3.

Owen, D.F., and Owen, J. (1974). Species diversity in temperate and tropical Ichneumonidae. *Nature (Lond.)*, 249, 583-4.

Padian, K. and Clemens, W.A. (1985). Terrestrial vertebrate diversity: episodes and insights. In *Phanerozoic Diversity Patterns: Profiles in Macroevolution*. ed. J.W. Valentine, pp. 41-96. Princeton, N.J.: Princeton University Press.

Paine, R.T. (1966). Food web complexity and species diversity. *American Naturalist*, 100, 65-75.

Paine, R.T. (1969).The *Pisaster-Tegula* interaction: prey patches, predator preference, and intertidal community structure. *Ecology, 50,* 950-61.

Paine, R.T., and Levin, S.A. (1981). Intertidal landscapes: Disturbance and the dynamics of pattern. *Ecological Monographs, 51,* 145-78.

Paine, R.T., and Vadas, R.L. (1969). The effects of grazing by sea urchins *Strongylocentrotus* spp. on benthic algal populations. *Limnology and Oceanography, 14,* 710-9.

Park, T. (1948). Experimental studies of interspecific competition. I. Competition between populations of the flour beetles *Tribolium confusum* Dival and *Tribolium castaneum* Herbst. *Ecological Monographs, 18,* 267-307.

Park, T. (1954). Experimental studies of interspecific competition. II. Temperature, humidity and competition in two species of *Tribolium. Physiological Zoology, 27,* 177-238.

Park, T. (1962). Beetles, competition and populations. *Science, 138,* 1369-75.

Park, T., Mertz, D.B., Grodzinski, W., and Prus, T. (1965). Cannibalistic predation in populations of flour beetles. *Physiological Zoology, 38,* 289-321.

Parker, S.P. (ed). (1982). *Synopsis and Classification of Living Organisms.* New York: McGraw-Hill.

Parkhurst, D.G., and Loucks, O.L. (1972). Optimal leaf size in relation to environment. *Journal of Ecology, 60,* 505-37.

Parminter, J. (1991). Fire history and effects on vegetation in three biogeoclimatic zones of British Columbia. In *Fire and the Environment: Ecological and Cultural Perspectives,* ed. S.C. Nodvin and T.A. Waldrop, pp. 263-72. Asheville, NC: Southeastern Forest Experiment Station.

Parsons, R.F. (1968a). The significance of growth rate comparisons for plant ecology. *American Naturalist, 102,* 295-7.

Parsons, R.F. (1968b). Ecological aspects of growth and mineral nutrition of three mallee species of Eucalyptus. *Oecologia Plantarum, 3,* 121-36.

Pastor, J. and Post, W.M. (1985). *Development of linked forest productivity-soil process model.* ORNL/TM-9519. Oak Ridge National Lab, Oak Ridge, TN.

Pastor, J. and Post, W.M. (1986). Influence of climate, soil moisture, and succession on forest carbon and nitrogen cycles. *Biogeochemistry, 2,* 3-17.

Pastor, J. and Post, W.M. (1988). Response of northern forests to CO_2-induced climatic change: Dependence on soil water and nitrogen availabilities. *Nature, 334,* 55-8.

Pastor, J., Naiman, R.J., Dewey, B., and McInnes, P. (1988). Moose, microbes, and the boreal forest. *BioScience, 38,* 770-7.

Pastor, J., and Huston, M. (1986). Predicting ecosystem properties from physical data: a case study of nested soil moisture-climatic gradients along the Appalachian chain. In *Coupling of Ecological Studies with Remote Sensing: Potentials at Four Biosphere Reserves in the United States,* ed. M.I. Dyer and D.A. Crossley, pp. 82-95. U.S. Department of State Publication 9504. Washington, D.C.: U.S. Government Printing Office.

Patil, G.P., and Taillie, C. (1982). Diversity as a concept and its measurement. *Journal of the American Statistical Association, 77,* 548-67.

Patrick, R. (1963). The structures of diatom communities under varying ecological conditions. *Annals of the New York Academy of Science, 108,* 353-8.

Paul, B.G. (1930). The applications of silviculture in controlling the specific gravity of wood. USDA Tech. Bull. 168.

Pavlik, B.M. (1989). Phytogeography of sand dunes in the Great Basin and Mojave Deserts. *Journal of Biogeography*, 16, 227-38.

Pearson, T.H., and Rosenberg, R. (1978). Macrobenthic succession in relation to organic enrichment and pollution of the marine environment. *Oceanography and Marine Biology Annual Review*, 16, 229-311.

Peet, R.K. (1974). The measurement of species diversity. *Annual Review of Ecology and Systematics*, 5, 285-307.

Peet, R.K. (1978). Forest vegetation of the Colorado Front Range: patterns of species diversity. *Vegetatio*, 37, 65-78.

Peet, R.K. (1981). Changes in biomass and succession during secondary forest succession. In *Forest Succession Concepts and Application*, ed. D.C. West, H.H. Shugart, and D.B. Botkin, pp. 324-38. New York: Springer-Verlag.

Peet, R.K., and Christensen, N.L. (1980). Succession: a population process. *Vegetatio*, 43, 131-40.

Peet, R.K., and Loucks, O.L. (1977). A gradient analysis of southern Wisconsin forests. *Ecology*, 58, 485-99.

Pennington, W.A. (1986). Lags in adjustment of vegetation to climate caused by the pace of soil development. *Vegetatio*, 67, 105-18.

Pennycuick, L. (1975). Movements of the migratory wildebeest population in the Serengeti area between 1960 and 1973. *East African Wildlife Journal*, 13, 65-87.

Pereira, M. da Costa, and Setzer, A.W. (1986). *Detecao de queimadas de plumas de funaca na Amazonia atraves de imagens de satelites NOAA*. Publication number INPE-3924-PRE/958. Sao Jose dos Campos, Sao Paulo, Brazil: Instituto de Pesquisas Espaciais.

Periera, H.C. (1973). *Land Use and Water Resources in Temperate and Tropical Climates*. Cambridge: Cambridge University Press.

Perlez, J. (1990). Can he save the elephants? *The New York Times Magazine*, January 7, 28-33.

Peters, R. (1980). From natural history to ecology. *Perspectives in Biology and Medicine*, Winter 1980, 191-203.

Peterson, D.L., and Bazzaz, F.A. (1978). Life cycle characteristics of *Aster pilosus* in early successional habitats. *Ecology*. 59, 1005-13.

Peterson, G.L., and Abbott, P.L. (1979). Mid-Eocene climatic change, south-western California and north-western Baja California. *Palaeogeography, Palaeoclimatology, Palaeoecology*, 26, 73-87.

Petrides, G.A., and Swank, W.G. (1965). (1965). Population densities and the range-carryying capacity for large mammals in Queen Elizabeth National Park, Uganda. *Zool. Afr.*, 1, 209-25.

Petrides, G.A., and Swank, W.G. (1966). Estimating the productivity and energy relations of an African elephant population. *Proceedings of the International Grassland Congress, Sao Paulo, Brazil*, 9, 831-942.

Petts, G.E. (1979). *Impounded Rivers: Perspectives for Ecological Management*, Chichester: Wiley.

Phillipson, J. (1975). Rainfall, primary production and 'carrying capacity' of Tsavo National Park (East), Kenya. *East African Wildlife Journal*, 13, 171-202.

Pianka, E.R. (1967). On lizard species diversity: North American flatland deserts. *Ecology*, 333-51.

Pianka, E.R. (1970). On *r*- and *K*-selection. *American Naturalist*, 104, 592-7.

Pianka, E.R. (1973). The structure of lizard communities. *Annual Review of Ecology and Systematics*, 4, 53-74.

Pianka, E.R. (1976). Competition and niche theory. In *Theoretical Ecology*, ed. R.M. May. pp. 167-96. Sunderland, Massachusetts: Sinauer.

Pianka, E.R. (1983). *Evolutionary Ecology*, 3rd edition. New York: Harper and Row.

Pickard, J., and Seppelt, R.D. (1984). Phytogeography of Antarctica. *Journal of Biogeography*, 11, 83-102.

Pickett, S.T.A. (1976). Succession: An evolutionary interpretation. *American Naturalist*, 110, 107-19.

Pickett, S.T.A., and White, P. eds. (1985). *The Ecology of Natural Disturbance and Patch Dynamics*. Orlando, Florida: Academic Press.

Pickford, G.D. (1932). The influence of continued heavy grazing and of promiscuous burning on spring-fall ranges in Utah. *Ecology*, 13, 159-71.

Pielou, E.C. (1975). *Ecological Diversity*. New York: Wiley and Sons.

Pielou, E.C. (1977). *Mathematical Ecology*. New York: Wiley and Sons.

Piemeisel, R.L. (1951). Causes affecting change and rate of change in a vegetation of annuals in Idaho. *Ecology*, 32, 53-72.

Pike, D.G., and Welch, H.E. (1990). Spatial and temporal distribution of sub-ice macrofauna in the Barrow Strait area, Northwest Territories. *Canadian Journal of Fisheries and Aquatic Science*, 47, 81-91.

Pimentel, D. (1986). Biological invasions of plants and animals in agriculture and forestry. In *Ecology of Biological Invasions of North America and Hawaii*, ed. H.A. Mooney and J.A. Drake, pp. 149-62. New York: Springer-Verlag.

Pimm, S.L. (1982). *Food Webs*. London: Chapman and Hall.

Pimm, S.L., and Lawton, J.H. (1977). Number of trophic levels in ecological communities. *Nature*, 268, 329-31.

Pinder, J.E. (1975). Effects of species removal on an old-field plant community. *Ecology*, 56, 747-51.

Pineda, F.D., Nicolas, J.P., Ruiz, M., Peco, B., and Bernaldez, F.G. 1981a. Succession, diversite' et amplitude de niche dans les paturages du centre de la penineule iberique. *Vegetatio*, 47, 267-77.

Pincda, F.D., Nicolas, J.P., Ruiz, M., Peco, B., and Bernaldez, F.G. 1981b. Ecological succession in oligotrophic pastures of central Spain. *Vegetatio*, 44, 165-76.

Pires, J.M., and Prance, G.T. (1985). The vegetation types of the Brazilian Amazon. In *Key Environments: Amazonia*. ed. G.T. Prance and T.J. Lovejoy. pp. 109-45. Oxford: Pergamon.

Platt, J.R., (1964). Strong inference. *Science*, 146, 347-53.

Platt, W.J. (1975). The colonization and formation of equilibrium plant species associations on badger disturbances in a tall-grass prairie. *Ecological Monographs*, 45, 285-305.

Plotkin, M.J. (1988). The outlook for new agricultural and industrial products from the tropics. In *Biodiversity*, ed. E.O. Wilson and F.M. Peter, pp. 106-16. Washington, D.C.: National Academy Press.

Polis, G.A. (1988). Exploitation competition and the evolution of interference, cannibalism and intraguild predation in age/size structured populations. In *Size Structured Populations: Ecology and Evolution*, ed. L. Perrson, B. Ebenmann, pp. 185-202. New York: Springer-Verlag.

Polis, G.A., Myers, C.A., and Holt, R.D. (1989). The ecology and evolution of intraguild predation: Potential competitors that eat each other. *Annual Review of Ecology and Systematics*, 20, 297-330.

Polis, G.A., and McCormick, S. (1986). Scorpions, spiders and solpugids: predation and competition among distantly related taxa. *Oecologia*, 71, 111-6.

Popenoe, H. (1957). The influence of shifting cultivation cycles on soil properties in Central America. *Ninth Pacific Science Congress Proceedings, Bangkok*, 1, 72-7.

Porter, J.W. (1974a). Community structure of coral reefs on opposite sides of the Isthmus of Panama. *Science*, 186, 543-5. Page-306, 307

Porter, J.W. (1974b). Zooplankton feeding by the Caribbean reef-building coral *Montastrea cavernosa*. *Proceedings of the Second International Coral Reef Symposium*, 1, 111-25.

Porter, J.W. (1976). Autotrophy, heterotrophy, and resource partitioning in Caribbean reef-building corals. *American Naturalist*, 110, 731-42.

Porter, K.G. (1973). Selective grazing and differential digestion of algae by zooplankton. *Nature*, 244, 179-80.

Porter, K.G. (1976). Enhancement of algal growth and productivity by grazing zooplankton. *Science*, 192, 1332-4.

Post, W.M., Emanuel, W.R., Zinke, P.J., and Stangeberger, A.G. (1982). Soil carbon pools and world life zones. *Nature*, 298, 156-9.

Post, W.M., and Pastor, J. (1990). An individual-based forest ecosystem model for projecting forest response to nutrient cycling and climate change. In *Forest Simulation Systems*, ed. L.C. Wensel and G.S. Biging, pp. 61-74. University of California, Division of Agricultural and Natural Resources, Bulletin 1927.

Potts, D.C. (1977). Suppression of coral populations by filamentous algae within damselfish territories. *Journal of Experimental Marine Biology*, 28, 207-16.

Pound, R., and Clements, F.E. (1898). *The phytogeography of Nebraska*, 1st ed. Lincoln, Nebr. (2nd ed., 1900) Reprinted New York: Arno Press (1977).

Powell, T. and Richerson, P.J. (1985). Temporal variation, spatial heterogeneity, and competition for resources in plankton systems: a theoretical model. *American Naturalist*, 125, 431-64.

Power, D.M. (1972). Numbers of bird species on the Californian Islands. *Evolution*, 26, 451-63.

Prance, G.T. (1973). Phytogeographic support for the theory of Pleistocene forest refugees in the Amazon Basin, based on evidence from distribution patterns in Caryocaraceae, Chrysobalanaceae, Dichapetalaceae, and Lecythidaceae. *Acta Amazonica*. 3, 5-28.

Prance, G.T. (1981). Discussion of aspects of Neotropical bird speciation during the Cenozoic. In *Vicariance Biogeography: A Critique*, ed. G. Nelson and D.E. Rosen. pp. 395-405. New York: Columbia University Press.

Prance, G.T. (1982). Forest refuges: Evidence from woody angiosperms. In *Biological Diversification in the Tropics*, ed. G. Prance. pp. 137-58. New York: Columbia University Press.

Prance, G.T. ed. (1982). *Biological Diversification in the Tropics*. New York: Columbia University Press.

Prance, G.T. (1985). The changing forests. In *Amazonia*, ed. G.T. Prance and T.E. Lovejoy. pp. 146-165. Oxford: Pergamon.

Pratt, D.J., Greenway, P.J., and Gwynne, M.D. (1966). A classification of East

African rangeland, with an appendix on terminology. *Journal of Applied Ecology*, 3, 369-82.

Preston, F.W. (1948). The commonness, and rarity, of species. *Ecology*, 29, 254-83.

Preston, F.W. (1960). Time and space and the variation of species. *Ecology*, 41, 611-27.

Preston, F.W. (1962). The canonical distribution of commonness and rarity. *Ecology*, 43, 185-215, 410-32.

Preston, F.W. (1969). Diversity and stability in the biological world. *Brookhaven Symposium in Biology*, 22, 1-12.

Pringle, C.M. (1990). Nutrient spatial heterogeneity: effects on community structure, physiognomy, and diversity of stream algae. *Ecology*, 71, 905-20.

Procter, D.L.C. (1984). Towards a biogeography of free-living soil nematodes. I. Changing species richness, diversity and densities with changing latitude. *Journal of Biogeography*, 11, 103-17.

Pyke, G.II., Pulliam, II.R., and Charnov, E.L. (1977). Optimal foraging: a selective review of theory and tests. *Quarterly Review of Biology*, 52, 137-54.

Pyne, S.J. (1982). *Fire in America*. Princeton, N.J.: Princeton University Press.

Pyne, S.J. (1984). *Introduction to Wildland Fire: Fire Management in the United States*. New York: Wiley.

Quetzal, P., Barbero, M., Bonin, G., and Loisel, R. (1989). Recent plant invasions in the Circum-Mediterranean region. In *Biological Invasions in Europe and the Mediterranean Basin*, ed. F. di Castri, A.J. Hansen, and M. Debussche, pp. 51-60. Dordrecht: Kluwer.

Rabinovitch-Vin, A. (1983). Influence of nutrients on the composition and distribution of plant communities in Mediterranean-type Ecosystems of Israel. In *Mediterranean-type Ecosystems: The Role of Nutrients, Ecological Studies Volume 43*, ed. F.J. Kruger, D.T. Mitchell, and J.U.M. Jarvis, pp. 74-85. Berlin: Springer-Verlag.

Rabinowitz, D. (1978a). Dispersal properties of mangrove propagules. *Biotropica*, 10, 47-57.

Rabinowitz, D. (1978b). Early growth of mangrove seedlings in Panama, and an hypothesis concerning the relationship of dispersal and zonation. *Journal of Biogeography*, 5, 113-33.

Rabinowitz, D. (1979). Bimodal distributions of seedling weight in relation to density of *Festuca paradoxa* Desv. *Nature* (London). 277, 297-8.

Radke, K., Arndt, W.-H., and Wakimoto, R.H. (1982). Fire history at the Santa Monica Mountains. In *Dynamics and Management of Mediterranean-Type Ecosystems*, ed. C.E. Conrad and W.C. Oechel, USDA Forest Service Pacific Southwest Forest and Range Experiment Station General Technical Report PSW-58. pp. 438-443.

Raison, R.J. (1979). Modification of the soil environment by vegetation fires with particular reference to nitrogen transformation: a review. *Plant and Soil*, 51, 73-108.

Ralph, C.J. (1985). Habitat association patterns of forest and steppe birds of northern Patagonia, Argentina. *The Condor*, 87, 471-83.

Ramakrishnan, P.S., and Vitousek, P.M. (1989). Ecosystem-level processes and the consequences of biological invasions. In *Biological Invasions: a Global Perspective. SCOPE 37*, ed. J.A. Drake *et al.*, pp. 281-300. Chichester: Wiley.

Rand, A.L. (1948). Glaciation, an isolating factor in speciation. *Evolution*, 2, 314-21.

Randall, J.E. (1974). The effect of fishes on coral reefs. *Proceedings of the Second International Coral Reef Symposium*, 1, 159-66.

Randolph, A.D., and Larson, M.A. (1971). *Theory of Particulate Processes*. New York: Academic.

Rapoport, E.H. (1975). *Areografia: estrategias de las especies.* Mexico City: Fondo de Cultura Economica.

Rapoport, E.H. (1982). *Areography: geographical strategies of species.* transl. E. Drausal. New York: Pergamon.

Rasanen, M.E., Salo, J.S., and Kalliola, R.J. (1987). Fluvial perturbance in the western Amazon basin - regulation by long-term sub-Andean tectonics. *Science*, 238, 1398-1401.

Rashit, E., and Bazin, M. (1987). Environmental fluctuations, productivity, and species diversity: An experimental study. *Microbial Ecology*, 14, 101-12.

Rathcke, B. (1983). Competition and facilitation among plants for pollination. In *Pollination Biology*, ed. L. Real. pp. 305-29. New York: Academic Press.

Rathcke, B.J. (1976a). Competition and coexistence within a guild of herbivorous insects. *Ecology*, 57, 76-87.

Rathcke, B.J. (1976b). Insect-plant patterns and relationships in the stem-boring guild. *American Midland Naturalist*, 99, 98-117.

Raunkiaer, C. (1934). *The Life Forms of Plants and Statistical Plant Geography*. Oxford: Oxford University Press.

Raup, D.M. (1979). Size of the Permo-Triassic bottleneck and its evolutionary implications. *Science*, 206, 217-8.

Raup, D.M., and Sepkoski, J.J. (1982). Mass extinctions in the marine fossil record. *Science*, 215, 1501-3.

Rauscher, M.D. (1978). Search image for leaf shape in a butterfly. *Science*, 200, 1071-3.

Raven, P.H. (1976). Ethics and attitudes. In *Conservation of Threatened Plants*, ed. J.B. Simmons, R.I. Beyer, P.E. Brandham, G.L. Lucas, and V.T.H. Parry, pp. 155-79. New York: Plenum.

Raven, P.H. (1977). The California flora. In *Terrestrial Vegetation of California*, ed. M.G. Barbour and J. Major, pp. 109-37. New York: Wiley.

Raven, P.H. (1988). Our diminishing tropical forests. In *Biodiversity*. ed. E.O. Wilson. pp. 119-22. Washinton, D.C.: National Academy Press.

Raven, P.H., and Axelrod, D., (1978). Origin and relationships of the California flora. *University of California Publications in Botany*, 72, 1-135.

Ray, G.C. (1988). Ecological diversity in coastal zones and oceans. In *Biodiversity*, ed. E.O. Wilson and F.M. Peter, pp. 36-50. Washington, D.C.: National Academy Press.

Reader, R.J., and Best, B.J. (1989). Variation in competition along an environmental gradient: *Hieracium floribundum* in an abandoned pasture. *Journal of Ecology*, 77, 673-84.

Recher, H.F. (1969). Bird species diversity and habitat diversity in Australia and North America. *American Naturalist*, 103, 75-80.

Reddingius, J., and den Boer, P.J. (1970). Simulation experiments illustrating stabilization of animal numbers by spreading of risk. *Oecologia (Berlin)*, 5, 240-84.

Redmann, R.E. (1972). Plant communities and soils of an eastern North Dakota prairie. *Bulletin of the Torrey Botanical Club*, 99, 65-76.

Reed, D.C., and Foster, M.S. (1984). The effects of canopy shading on algal recruitment and growth in a giant kelp forest. *Ecology,* 65, 937-48.

Reekie, E.G., and Bazzaz, F.A. (1987a). Reproductive effort in plants. 1. Carbon allocation to reproduction. *American Naturalist,* 129, 876-96.

Reekie, E.G., and Bazzaz, F.A. (1987b). Reproductive effort in plants. 3. Effect of reproduction on vegetative activity. *American Naturalist,* 129, 907-19.

Regal, P.J. (1977). Ecology and evolution of flowering plant dominance. *Science,* 196, 622-9.

Regal, P.J. (1982). Pollination by wind and animal: ecology of geographic patterns. *Annual Review of Ecology and Systematics,* 13, 497-524.

Regehr, D.L., and Bazzaz, F.A. (1976). Low temperature photosynthesis in successional winter annuals. *Ecology,* 57, 1297-1303.

Reichle, D.E. ed. (1970). *Analysis of Temperate Forest Ecosystems.* New York: Springer-Verlag.

Reichle, D.E., and Auerbach, S.I. (1972). Analysis of ecosystems. In *Challenging Biological Problems: Directions toward their Solution,* ed. J. Behnke, pp. 260-80. New York: Academic Press.

Reid, J.L. (1962). On circulation, phosphate-phosphorus content, and zooplankton volumes in the upper part of the Pacific Ocean. *Limnology and Oceanography,* 7, 287-306.

Reid, W.V., and Miller, K.R. (1989). *Keeping Options Alive: The Scientific Basis for Conserving Biodiversity.* Washington, D.C.: World Resources Institute.

Renshaw, J.F., and Doolittle, W.T. (1965). Yellow poplar (*Liriodendron tulipifera* L.) In *Silvics of Forest Trees of the United States, Agriculture Handbook No. 271,* ed. H.A. Fowler, pp. 256-65. Washington, D.C.: USDA Forest Service.

Rex, M.A. (1973). Deep-sea species diversity: decreased gastropod diversity at abyssal depths. *Science,* 181, 1051-3.

Rex, M.A. (1976). Biological accomodation in the deep-sea benthos: comparative evidence on the importance of predation and productivity. *Deep-sea Research,* 23, 975-87.

Rex, M.A. (1981). Community structure in the deep sea benthos. *Annual Review of Ecology and Systematics,* 12, 331-53.

Rex, M.A. (1983). Geographical patterns of species diversity in the deep-sea benthos. In *The Sea, Vol. 8,* ed. G.T. Rowe, pp. 453-72. New York: Wiley.

Rex, M.A., Etter, R.J., and Nimeskern, P.W. (1990). Density estimates for deep-sea gastropod assemblages. *Deep-sea Research, Part A - Oceanographic Research Papers,* 37, 555-69.

Rex, M.A., and Ware'n, A. (1981). Evolution in the deep sea: taxonomic diversity of gastropod assemblages. In *Biology of Pacific Ocean Depths,* ed. N.G. Vinogradova. Vladivostok: Nauka.

Rey, J.R. (1981). Ecological biogeography of arthropods on *Spartina* islands in northwest Florida. *Ecological Monographs,* 51, 237-65.

Reynolds, C.S. (1984). *The Ecology of Freshwater Phytoplankton.* Cambridge: Cambridge University Press.

Ribbink, A.J., Marsh, B.A., Marsh, A.C., and Sharp, B.J. (1983). A preliminary survey of the cichlid fishes of rocky habitats in L. Malawi. *South African Journal of Zoology,* 18, 160 pp.

Rice, K., and Jain, S. (1985). Plant population genetics and evolution in disturbed environments. In *The Ecology of Natural Disturbance and Patch Dynamics* ed. S.T.A. Pickett and P.S. White, pp. 287-303. Orlando: Academic Press.

Richards, P.W. (1952). *The Tropical Rain Forest*. Cambridge: Cambridge University Press.

Richards, P.W. (1969). Speciation in the tropical rainforest and the concept of the niche. In *Speciation in Tropical Environments*, ed. R.H. Lowe-McConnell. New York: Academic Press.

Richardson, C.A., Dustan, P., Lang, J.C. (1979). Maintenance of living space by sweeper tentacles of *Montastrea cavernosa*, a Caribbean Reef Coral. *Marine Biology*, 55, 181-6.

Richerson, P.R., Armstrong, R., and Goldman, C.R. (1970). Contemporaneous disequilibrium, a new hypothesis to explain the "paradox of the plankton." *Proceedings of the National Academy of Science (USA)*, 67, 1710-4.

Richter, D.D., and Babbar, L.I. (1991). Soil diversity in the tropics. *Advances in Ecological Research*, 21, 316-89.

Ricklefs, R.E. (1973). *Ecology*. Newton, Massachusetts: Chiron.

Ricklefs, R.E. (1987). Community diversity: relative roles of local and regional processes. *Science*, 235, 167-71.

Ricklefs, R.E., and O'Rourke, K. (1975). Aspect diversity in moths: a temperate-tropical comparison. *Evolution*, 29, 313-24.

Ridley, H.N. (1930). *The Dispersal of Plants Throughout the World*. Ashford: L. Reeve and Company.

Riebesell, J.F. (1974). Paradox of enrichment in competitive systems. *Ecology*, 55, 183-7.

Robertson, G.P. (1982). Factors regulating nitrification in primary and secondary succession. *Ecology*, 63, 1561-73.

Robertson, G.P., Huston, M.A., Evans, F.C., and Tiedje, J.M. (1988). Spatial variability in a successional plant community: patterns of nitrogen mineralization, nitrification, and denitrification. *Ecology*, 69, 1517-24.

Robinson, J.V., and Dickerson, J.E. (1987). Does invasion sequence affect community structure. *Ecology*, 68, 587-95.

Robinson, J.V., and Edgemon, M.A. (1988). An experimental evaluation of the effect of invasion history on community structure. *Ecology*, 69, 1410-7.

Rodin, L.E., Bazilevich, N.I., and Rozov, N.N. (1975). Productivity of the world's main ecosystems. In *Productivity of World Ecosystems*, pp. 13-26. Washington, D.C.: National Academy of Sciences.

Roff, D.A. (1974a). Spatial heterogeneity and the persistence of populations. *Oecologia (Berlin)*, 15, 245-58.

Roff, D.A. (1974b). The analysis of a population model demonstrating the importance of dispersal in a heterogeneous environment. *Oecologia (Berlin)*, 15, 259-75.

Rogers, D.H. (1961). Measuring the efficiency of fire control in California chaparral. *Journal of Forestry*, 49, 697-703.

Rogers, R.W. (1977). Lichens of hot arid and semi-arid lands. In *Lichen Ecology*, ed. M.R.D. Seaward. pp. 211-52. London: Academic Press.

Rogers, W.A., Owens, C.F., and Homewood, K.M. (1982). Biogeography of East African forest mammals. *Journal of Biogeography*, 9, 41-54.

Rogler, G.A., and Haas, H.J. (1947). Range production as related to soil moisture and precipitation on the Northern Great Plains. *Journal of the American Society of Agronomy*, 39, 378-89.

Romer, A.S. (1966). *Vertebrate Paleontology*. Chicago: University of Chicago Press.

Romme, W.H. (1982). Fire and landscape diversity in subalpine forests of Yellowstone National Park. *Ecological Monographs, 52,* 199-221.

Romme, W.H., and Despain, D.G. (1989). Historical perspective on the Yellowstone fires of 1988. *BioScience, 39,* 695-9.

Romme, W.H., and Knight, D.H. (1981) Fire frequency and subalpine forest succession along a topographic gradient in Wyoming. *Ecology, 62,* 319-26.

Root, J.B. (1967). The niche exploitation pattern of the blue-grey gnat catcher. *Ecological Monographs, 37,* 317-50.

Root, R. (1973). Organization of a plant-arthropod association in simple and diverse habitats: the fauna of collards (*Brassica oleracea*). *Ecological Monographs, 43,* 95-124.

Rorison, I. (1968). The response to phosphorus of some ecologically distinct plant species: I. Growth rates and phosphorus absorption. *New Phytologist, 67,* 913-23.

Rosatti, T.J. (1982). Trichome variation and the ecology of *Arctostaphylus* in Michigan. *The Michigan Botanist, 21,* 171-80.

Rosatti, T.J. (1987). Field and garden studies of *Arctostaphylus uva-ursi* (Ericaceae) in North America. *Systematic Botany, 12,* 61-77.

Rosenzweig, M.L. (1971). Paradox of enrichment: destabilization of exploitation ecosystems in ecological time. *Science, 171,* 385-7.

Rosenzweig, M.L. (1973). Exploitation in three trophic levels. *American Naturalist, 107,* 275-94.

Rosenzweig, M.L. (1975). On continental steady states of species diversity. In *Ecology and Evolution of Communities,* ed. M.L. Cody and J.M. Diamond, pp. 121-40. Cambridge, Mass.:Belknap Press.

Rosenzweig, M.L., and MacArthur, R.H. (1963). Graphical representation and stability condition of prey-predator interactions. *American Naturalist, 97,* 209-23.

Rosenzweig, M.L., and Winakur, J. (1966). Population ecology of desert rodent communities: habitats and environmental complexity. *Ecology, 50,* 558-72.

Rotenberry, J.T. (1978). Components of avian diversity along a multifactorial climatic gradient. *Ecology, 59,* 693-9.

Roth, I. (1984). *Stratification of Tropical Forests as Seen in Leaf Structure.* The Hague, Junk.

Rothermel, R.C. (1972). A mathematical model for predicting fire spread in wildland fuels. Research Paper INT-115. Ogden, Utah: U.S. Department of Agriculture Forest Service.

Rothstein, S.I. (1973). The niche-variation model – is it valid? *American Naturalist, 107,* 598-620.

Rourke, J.P. (1980). *The Proteas of Southern Africa.* Cape Town: Purnell.

Rowe, G.T. (1971). Benthic biomass and surface productivity. In *Fertility in the Sea, Vol. 2,* ed. J.D. Costlow, pp. 441-54. New York: Gordon and Breach.

Rowe, G.T., Polloni, P.T., and Horner, S.G. (1974). Benthic biomass estimates from the northwest Atlantic Ocean and the northern Gulf of Mexico. *Deep-sea Research, 21,* 641-50.

Rowe, J.S., Bersteinsson, J.L., Padbury, G.A., and Hermesh, G.A. (1974). Fire studies in the Mackenzie Valley. Canadian Department of Indian Northern Affairs, Publication No QS-1567-000-EE-A1.

Rowe, J.S., and Scotter, G.W. (1973). Fire in the boreal forest. *Quaternary Research, 3,* 444-64.

644   *References*

Roy, K.J., and Smith, S.V. (1971). Sedimentation and coral reef development in turbid water: Fanning Lagoon. *Pacific Sciience*, 25, 234-48.

Rundel, P.W. (1981). The matorral zone of central Chile. In *Ecosystems of the World, Vol. 11, Mediterranean Type Shrublands*, ed. F. di Castri, D.W. Goodall, and R.L. Specht, pp. 175-201. Amsterdam: Elsevier.

Rundel, P.W. (1983). Impact of fire on nutrient cycles in Mediterranean-type ecosystems with reference to chaparral. In *Mediterranean-type Ecosystems. The Role of Nutrients*, ed. F.J. Kruger, D.T. Mitchell, and J.U.M. Jarvis, pp. 192-207. Berlin: Springer-Verlag.

Runkle, J.R. (1981). Gap regeneration in some old-growth forests of the eastern United States. *Ecology*, 62, 1041-51.

Runkle, J.R. (1985). Disturbance regimes in temperate forests. In *The Ecology of Natural Disturbance and Patch Dynamics*, ed. S.T.A. Pickett and P.S. White, pp. 17-33. Orlando, Florida: Academic Press.

Runkle, J.R., and Yetter, T.L. (1987). Treefalls revisited: gap dynamics in the southern Appalachians. *Ecology*, 68, 417-24.

Russell, E.W.B. (1983). Indian-set fires in forests of the northeastern United States. *Ecology*, 64, 78-88.

Russell, F.S. (1934). The zooplankton. III. A comparison of the abundance of zooplankton in the Barrier Reef Lagoon with that of some regions in Northern European waters. *Great Barrier Reef Expedition Scientific Reports*, 2, 176-201.

Russell-Hunter, W.D. (1970). *Aquatic Productivity*. New York: Macmillan.

Ryan, M.G. (1991). Effects of climate change on plant respiration. *Ecological Applications*, 11, 157-67.

Ryther, J.H. (1969). Photosynthesis and fish production in the sea. *Science*, 166, 72-6.

Sailer, R.I. (1983). History of insect introductions. In *Exotic Plant Pests and North American Agriculture*, ed. C. Graham and C. Wilson, pp. 15-38. New York: Academic Press.

Sailer, R.I. (1978). Our immigrant insect fauna. *Bulletin of the Entomological Society of America*, 24, 3-11.

Saldarriaga, J.G., and Luxmoore, R.J. (1991). Solar energy conversion efficiencies during succession of a tropical rain forest in Amazonia. *Journal of Tropical Ecology*, 7, 233-42.

Sale, P.F. (1977). Maintenance of high diversity in coral reef fish communities. *American Naturalist*, 111, 337-59.

Sale, P.F. (1978). Coexistence of coral reef fishes - a lottery for living space. *Environmental Biology of Fishes*, 3, 85-102.

Sale, P.F. (1979). Recruitment, loss and coexistence in a guild of territorial coral reef fishes. *Oecologia (Berlin)*, 42, 159-77.

Sale, P.F. (1980a). Assemblages of fish on patch reefs - predictable or unpredictable? *Environmental Biology of Fishes*, 5, 243-9.

Sale, P.F. (1980b). The ecology of fishes on coral reefs. *Oceanography and Marine Biology*, 18, 367-421.

Sale, P.F., and Dybdahl, R. (1975). Determinants of community structure for coral reef fishes in an experimental habitat. *Ecology*, 56, 1345-55.

Salisbury, E.J. (1929). The biological equipment of species in relation to competition. *Journal of Ecology*, 17, 197-222.

Salisbury, E.J. (1942). *The Reproductive Capacity of Plants*. London: G. Bell and Sons.

Salisbury, H.E. (1989). *The Great Black Dragon Fire: A Chinese Inferno*. Boston: Little, Brown.

Salo, J. (1987). Pleistocene forest refuges in the Amazon - evaluation of the biostratigraphical, lithostratigraphical, and geomorphological data. *Annales Zoologici Fennici*, 24, 203-11.

Salo, J., Kallioloa, R., Häkkinen, I., Mäkinen, Y., Niemelä, P., Puhakka, M., and Coley, P.D. (1986). River dynamics and the diversity of Amazon lowland forest. *Nature*, 322, 254-8.

Salt, G.W. (1979). A comment on the use of the term emergent properties. *American Naturalist*, 113, 145-8.

Sambrotto, R.N., Goering, J.J., and McRoy, C.P. (1984). Large yearly production of phytoplankton in the western Bering Strait. *Science*, 225, 1147-50.

Sammarco, P.W. (1980). *Diadema* and its relationship to coral spat mortality: grazing, competition, and biological disturbance. *Journal of Experimental Marine Biology*, 45, 245-72.

Sammarco, P.W. (1982a). Echinoid grazing as a structuring force in coral communities: whole reef manipulations. *Journal of Experimental Marine Biolology and Ecology*, 61, 31-55.

Sammarco, P.W. (1982b). Effects of grazing by *Diadema antillarum* Philippi (Echinodermata: Echinoidea) on algal diversity and community structure. *Journal of Experimental Marine Biology and Ecology*, 65, 83-105.

Sammarco, P.W., Levinton, J.S., and Ogden, J.C. (1974). Grazing and control of coral reef community structure by *Diadema antillarum* Philippi (Echinodermata: Echinoidea): a preliminary study. *Journal of Marine Research*, 32, 47-53.

Sanchez, P.A. (1976). *Properties and Management of Soils in the Tropics*. New York: Wiley and Sons.

Sanchez, P.A., and Buol, S.W. (1975). Soils of the tropics and the world food crisis. *Science*, 188, 598-603.

Sanders, H.L. (1968). Marine benthic diversity: a comparative study. *American Naturalist*, 102, 243-82.

Sanders, H.L. (1969). Benthic marine diversity and the stability-time hypothesis. *Brookhaven Symposium on Biology*, 22, 17-81.

Sanders, H.L., Hessler, R.R., and Hampson, G.R. (1965). An introduction to the study of deep-sea benthic faunal assemblages along the Gay Head-Bermuda transect. *Deep-sea Research*, 12, 845-67.

Sanders, H.L., and Hessler, R.R. (1969). Ecology of the deep sea benthos. *Science*, 163, 1419-24.

Sanders, T.L, and Clark, F.B. (1971). *Reproduction of Upland Hardwood Forests in the Central United States*. U.S. Department of Agriculture, Agricultural Handbook 405. Washington, D.C.: U.S. Government Printing Office.

Sanford, R.L., Saldarriaga, J., Clark, K.E., Uhl, C., and Herrera, R. (1985). Amazon rain forest fires. *Science*, 216, 821-7.

Sanford, W.W. (1969). The distribution of epiphytic orchids in Nigeria in relation to each other and to geographic location and climate, type of vegetation and tree species. *Biological Journal of the Linnean Society*, 1, 247-85.

Sarmiento, G. (1983). The savannas of Tropical America. In *Ecosystems of the World 13: Tropical Savannas*, ed. F. Bourlière, pp. 245-88. Amsterdam: Elsevier.

Sarmiento, G., and Monasterio, M. (1975). A critical consideration of the environmental conditions associated with the occurrence of savanna ecosystems in tropical America. In *Tropical Ecological Systems. Trends in Terrestrial and Aquatic Research. Ecological Studies 11,* ed. F.B. Golley and E. Medina, pp. 223-50. New York: Springer-Verlag.

Sauer, C.O. (1950). Grassland, climax, fire, and man. *Journal of Range Management,* 3, 16-22.

Saunders, B.P. (1973). Meningeal worm in white-tailed deer in northwest Ontario and moose populations densities. *Journal of Wildlife Management,* 37, 327-30.

Schaffer, W.M. (1974). Optimal reproductive effort in fluctuating environments. *American Naturalist,* 108, 783-90.

Schaffer, W.M. (1984). Stretching and folding in lynx fur returns: evidence for a strange attractor in nature? *American Naturalist,* 124, 798-820.

Schaffer, W.M., and Kot, M. (1985). Do strange attractors govern ecological systems. *BioScience,* 35, 324-50.

Schall, J.J., and Pianka, E.R. (1978). Geographical trends in numbers of species. *Science (Washington, D.C.),* 201, 679-86.

Schaller, G.B. (1967). *The Deer and the Tiger. A Study of Wildlife in India.* Chicago, Illinois: University of Chicago Press.

Schaller, G.B. (1972). *The Serengeti Lion. A Study of Predator-Prey Relations.* Chicago, Illinois: University of Chicago Press.

Scheiner, S.M. (1990). Affinity analysis: effects of sampling. *Vegetatio,* 86, 175-81.

Scheiner, S.M., and Istock, C.A. (1987). Affinity analysis: methodologies and statistical inference. *Vegetatio,* 72, 89-93.

Schimper, A.F.W. (1888). *Die epiphytische Vegetation Amerikas. Bot. Mitt. Tropen II.* Jena: Fischer.

Schimper, A.F.W. (1898). *Pflanzengeographie auf physiologischer Grundlage* (ed. 2). Jena.

Schimper, A.F.W. (1903). *Plant-geography upon a physiological basis.* Transl. by W.R. Fischer. Ed. P. Groom and I.B Balfour. Oxford.

Schimper, A.F.W., and Faber, F.C. von. (1935). *Pflanzengeographie auf physiologischer Grundlage.* 3rd ed. Jena: Fischer.

Schluter, D. (1988). The evolution of finch communities on islands and continents: Kenya va. Galápagos. *Ecological Monographs,* 58, 229-49.

Schluter, D., and Grant, P.R. (1984). Determinants of morphological patterns in communities of Darwin's Finches. *American Naturalist,* 123, 175-96.

Schoener, T.W., (1965). The evolution of bill size differences among sympatric congeneric species of birds. *Evolution,* 19, 189-213.

Schoener, T.W. (1969). Models of optimal size for solitary predators. *American Naturalist,* 103, 277-313.

Schoener, T.W. (1974). Resource partitioning in ecological communities. *Science,* 185, 27-39.

Schoener, T.W. (1974). Temporal resource partitioning and the compression hypothesis. *Proceedings of the National Academy of Science, USA,* 71, 4169-72.

Schoener, T.W. (1983). Field experiments on interspecific competition. *The American Naturalist,* 122, 240-85.

Schoener, T.W. (1986). Mechanistic approaches to community ecology: a new reductionism. *American Zoologist,* 26, 81-106.

Schoener, T.W., and Schoener, A. (1978). Inverse relation of survival of lizards with island size and avifaunal richness. *Nature,* 274, 685-7.

Scholes, R.J. (1990). The influence of soil fertility on the ecology of southern African dry savannas. *Journal of Biogeography*, 17, 415-19.

Schrieber, H. (1978). Dispersal centres of Sphingidae (Lepidoptera) in the Neotropical region. *Biogeographica 10*. The Hague: Junk.

Schuhmacher, H. (1974). On the conditions accompanying the first settlement of corals on artificial reefs with special reference to the influence of grazing sea urchins (Eilat, Red Sea). *Proceedings of the Second International Coral Reef Symposium*, 1, 257-67.

Schullery, P. (1989). The fires and fire policy. *BioScience*, 39, 868-94.

Schulz, J.P. (1960). *Ecological Studies on Rain Forest in Northern Suriname.* Amsterdam: N.V. Noord.

Schulze, E.D. (1982). Plant life forms and their carbon, water, and nutrient relations. In *Water Relations and Carbon Assimilation*, ed. O.L. Lange, P.S. Nobel, C.B. and Osmond, pp. 616-76. Berlin: Springer-Verlag.

Schulze, E.D. (1986). Whole-plant responses to drought. *Australian Journal of Plant Physiology*, 13, 127-41.

Schulze, E.D., Robichaux, R.H., Grace, J., Rundel, P.W., and Ehleringer, J.R. (1987). Plant water balance. *BioScience*, 37, 30-7.

Schwartz, C.C., and Ellis, J.E. (1981). Feeding ecology and niche separation in some native and domestic ungulates on the shortgrass prairie. *Journal of Applied Ecology*, 18, 343-53.

Scott, P. ed. (1974). *The World Atlas of Birds.* New York: Crescent.

Scriber, J.M. (1973). Latitudinal gradients in larval feeding specialization of the world Papilionidae (Lepidoptera). *Psyche (Cambridge, Massachusetts)*, 80, 355-73.

Scriber, J.M. (1984). Larval foodplant utilization by the world Papilionidae (Lep.): latitudinal gradients reappraised. *Tokurana (Acta Rhopalocerologica) Mishima-shi, Japan*, 6/7, 1-50.

Seidensticker, J. (1976). Ungulate populations in Chitawan valley, Nepal. *Biological Conservation*, 10, 183-210.

Semeniuk, V. (1983). Mangrove distribution in Northwestern Australia in relationship to regional and local freshwater seepage. *Vegetatio*, 53,11-31.

Sen, A.K. (1973) *On Economic Inequality.* Oxford: Clarendon Press.

Senft, R.L., Rittenhouse, L.R., and Woodmansee, R.G. (1985). Factors influencing patterns of cattle grazing behavior on shortgrass steppe. *Journal of Range Management*, 38, 82-7.

Sepkoski, J.J., and Hulver, M.L. (1985) An atlas of Phanerozoic clade diversity diagrams. In *Phanerozoic Diversity Patterns: Profiles in Macroevolution*, ed. J.W. Valentine, pp. 11-39. Princeton, New Jersey: Princeton University Press.

Shannon, C.E., and Weaver, W. (1949). *The Mathematical Theory of Communication.* Urbana, Ill.: University of Illinois Press.

Shantz, H.L. (1947). *The Use of Fire as a Tool in the Management of the Brush Ranges in California.* California State Board of Forestry.

Shantz, H.L. (1954). The place of grasslands in the earth's cover of vegetation. *Ecology*, 35, 143-5.

Sharpe, P.J.H., Walker, J., Penridge, L.K., Wu, H., and Rykiel, E.J. (1986). Spatial considerations in physiological models of tree growth. *Tree Physiology*, 2, 403-21.

Sharples, F.E. (1983). Spread of organisms with novel genotypes: thoughts from an ecological perspective. *Recombinant DNA Technical Bulletin*, 6, 43-56.

Shearin, A.T., Bruner, M.H., and Goebel, N.B. (1972). Prescribed burning stimulates natural regeneration of yellow-poplar. *Journal of Forestry*, 70, 482-4.

Shelford, V.E. (1913). *Animal Communities in Temperate North America as Illustrated in the Chicago Region*. No. 5. Chicago: Bulletin of the Geographical Society of Chicago. Reprinted New York: Arno Press (1977).

Shelford, V.E. (1963). *The Ecology of North America*. Urbana, Illinois: University of Illinois Press.

Shelton, W.L, Davies, W.D., King, T.A., and Timmons, T.J. (1979). Variation in the growth of the initial year class of largemouth bass in West Point Reservoir, Alabama and Georgia. *Transactions of the American Fisheries Society*, 108, 142-9.

Sheppard, C.R.C. (1980). Coral cover, zonation and diversity on reef slopes of Chagos Atolls, and population structures of the major species. *Marine Ecology Progress Series*, 2, 193-205.

Sheppard, C. R. C. (1981). Illumination and the coral reef community beneath tabular *Acropora* species. *Marine Biology*, 64, 53-8.

Sherry, T.W., and Holmes, R.T. (1991). Population fluctuations in a long-distance neotropical migrant: demographic evidence for the importance of breeding season events in the American Redstart. In *Ecology and Conservation of Neotropical Migrant Landbirds*. ed. J.M. Hangan and D.W. Johnston. pp. 431-42. Washington, D.C.: Smithsonian Institution Press.

Shiflet, T.N., and Dietz, H.E. (1974). Relationship between precipitation and annual rangeland herbage production in Southeastern Kansas. *Journal of Range Management*, 27, 272-6.

Shinn, E.A. (1972). Coral reef recovery in Florida and the Persian Gulf. Environmental Conservation Department, Shell Oil Company, Houston, Texas. 9 pp.

Shipley, B., Keddy, P.A., Gaudet, C., and Moore, D.R.J. (1991). A model of species density in shoreline vegetation. *Ecology*, 72, 1658-67.

Shmida, A., and Wilson, M.V. (1985). Biological determinants of species diversity. *Journal of Biogeography*, 12, 1-20.

Shugart, H.H. (1984). *A theory of forest dynamics*. New York: Springer-Verlag.

Shugart, H.H., Hopkins, M.S., Burgess, I.P., Mortlock, A.T. (1981). The development of a succession model for subtropical rain forest and its application to assess the effects of timber harvest at Wiangarree State Forest, New South Wales. *Journal of Environmental Management*, 11, 243-65.

Shugart, H.H., West, D.C., and Emanuel, W.R. (1981). Patterns and dynamics of forests: An application of simulation models. In *Forest Succession: Concepts and Applications*, ed. D.C. West, H.H. Shugart, and D.B. Botkin, pp. 74-94. New York: Springer-Verlag.

Shugart, H.H., and Hett, J.M. (1973). Succession: similarities of turnover rates. *Science*, 180, 1379-81.

Shugart, H.H., and Noble, I.R. (1980). A computer model of succession and fire response of the high-altitude *Eucalyptus* forest of the Brindabella Range, Australian Capital Territory. *Australian Journal of Ecology*, 6, 149-64.

Shugart, H.H., and O'Neill, R.V. eds. (1979). *Systems Ecology*. Stroudsburg, Pennsylvania: Dowden, Hutchinson, and Ross.

Shugart, H.H., and West, D.C. (1977). Development of an Appalachain deciduous forest succession model and its application to assessment of the impact of the Chestnut blight. *Journal of Environmental Management*, 5, 161-79.

Shulman, M.J., and Ogden, J.C. (1987). What controls tropical reef fish populations: recruitment or benthic mortality: an example in the Caribbean reef fish *Haemulon flavolineatum*. *Marine Ecology Progress Series*, 39, 233-42.

Siegfried, W.R., and Crowe, T.M. (1981). Distribution and species diversity of birds and plants in fynbos vegetation of mediterranean-climate zone, South Africa. In *Mediterranean-type Ecosystems: The Role of Nutrients, Ecological Sudies Volume 43*, ed. F.J. Kruger, D.T. Mitchell, and J.U.M. Jarvis, pp. 3-40. Berlin: Springer-Verlag.

Signor, P.W. (1990). The geological history of diversity. *Annual Review of Ecology and Systematics*, 21, 509-39.

Silvertown, J.W. (1980). The dynamics of a grassland ecosystem: botanical equilibrium in the park grass experiment. *Journal of Applied Ecology*, 17, 491-504.

Silvertown, J.W. (1983). Plants in limestone pavements: tests of species interaction and niche separation against null hypotheses. *Journal of Ecology*, 71, 819-28.

Silvertown, J.W. (1984). History of a latitudinal diversity gradient: woody plants in Europe 13,000-1000 years B.P. *Journal of Biogeography*, 12, 519-25.

Simberloff, D.S. (1974). Equilibrium theory of island biogeography and ecology. *Annual Review of Ecology and Systematics*. 5, 161-82.

Simberloff, D.S. (1976). Experimental zoogeography of islands: effects of island size. *Ecology*, 57, 629-48.

Simberloff, D.S. (1981). The sick science of ecology: symptoms, diagnosis, and prescription. *Eidema*, 1, 49-54.

Simberloff, D.S. (1982). The status of competition theory in ecology. *Annales Zoologici Fennici*, 19, 241-53.

Simberloff, D. (1986). Introduced insects: a biogeographic and systematic perspective. In *Ecology of Biological Invasions of North America and Hawaii*, ed. H.A. Mooney and J.A. Drake, pp. 3-26. New York: Springer-Verlag.

Simberloff, D. (1989). Which insect introductions succeed and which fail? In *Biological Invasions: a Global Perspective. SCOPE 37*, ed. J.A. Drake et al, pp. 61-75. Chichester: Wiley.

Simberloff, D.S., and Wilson, E.O. (1969). Experimental zoogeography of islands: the colonization of empty islands. *Ecology*, 50, 278-96.

Simkin, D.W. (1965). A preliminary report of the woodland caribou study in Ontario. *Ontario Department of Lands and Forests Section Report (Wildlife)*, 59, 1-76.

Simpson, B.B., and Haffer, J. (1978). Speciation patterns in the Amazonian forest biota. *Annual Review of Ecology and Systematics*, 9, 497-518.

Simpson, E.H. (1949). Measurement of diversity. *Nature*, 163, 688.

Simpson, G.G. (1947). Holarctic mammalian faunas and continental relationships during the Cenozoic. *Bulletin of the Geological Society of America*, 58, 613-88.

Simpson, G.G. (1950). History of the fauna of Latin America. *American Scientist*, 38, 361-89.

Simpson, G.G. (1964). Species density of recent North American mammals. *Systematic Zoology* 13, 57-73.

Simpson, G.G. (1965). *The Geography of Evolution. Collected Essays*. Philadelphia: Chilton.

Sims, I.H. (1932). Establishment and survival of yellow poplar following a clearcutting in the Southern Appalachians. *Journal of Forestry*, 30, 409-14.

Sims, P.L. (1988). Grasslands. In *North American Terrestrial Vegetation*, ed. M.G. Barbour and W.D. Billings, pp. 265-86. Cambridge: Cambridge University Press.

Sinclair, A.R.E. (1977). *The African Buffalo*. Chicago: University of Chicago Press.

Sinclair, A.R.E. (1983). The adaptations of African ungulates and their effects on community function. In *Ecosystems of the World 13: Tropical Savannas* ed. F. Bourlière, pp. 401-26. Amsterdam: Elsevier.

Sinclair, A.R.E., and Norton-Griffiths, M. (1979). *Serengeti. Dynamics of an Ecosystem*. Chicago, Illinois: University of Chicago Press.

Sioli, H. (1950). Des Wasser im Amazonasgebiet. *Forsch. Fortschr.* 26(21/22), 274-80.

Sioli, H. (1964). General features of the limnology of Amazonia. *Verh. Int. Verein. Limnol.*, 15, 1053-8.

Sioli, H. (1975a). Tropical river: the Amazon. In *River Ecology*, ed. B.A. Whitton, pp. 461-88. Los Angeles: University of California Press.

Sioli, H. (1975b). Tropical rivers as expressions of their terrestrial environments. In *Tropical Ecological Systems. Trends in Terrestrial and Aquatic Research. Ecological Studies 11*, ed. F.B. Golley and E. Medina, pp. 275-88. New York: Springer-Verlag.

Sipman, H.J.M., and Harris, R.C. (1989). Lichens. In *Ecosystems of the World 14B. Tropical Rain Forests: Biogeographical and Ecological Studies*. ed. H. Lieth and M.J.A. Werger. pp. 303-9. Amsterdam: Elsevier.

Sirén, G. (1973) Some remarks on fire ecology in Finnish forestry. In *Proceedings of the Annual Tall Timbers Fire Ecology Conference*. 13, 191-209.

Slatkin, M. (1974). Competition and regional coexistence. *Ecology*, 55, 128-34.

Slobodkin, L.B. (1964). Experimental populations of Hydrida. *Journal of Animal Ecology*, 33 (Suppl.), 131-48.

Slud, P. (1976). Geographic and climatic relationships of avifaunas with special reference to comparative distribution in the Neotropics. *Smithsonian Contributions in Zoology*, 212.

Smith, A.P., and Young, T.P. (1987). Tropical alpine plant ecology. *Annual Review of Ecology and Systematics*, 18, 137-58.

Smith, G.B. (1979). Relationship of eastern Gulf of Mexico reef-fish communities to the species equilibrium theory of insular biogeography. *Journal of Biogeography*, 6, 49-61.

Smith, J. (1950). *Distribution of tree species in the Sudan in relation to rainfall and soil type*. Khartoum: Agricultural Publishing Commission. 8

Smith, K.L. (1978). Benthic community respiration in the N.W. Atlantic Ocean: *in situ* measurements from 40-5200m. *Marine Biology*, 47, 337-47.

Smith, T.M. and Goodman, P.S. (1986). The role of competition on the structure and dynamics of Acacia savannas in southern Africa. *Journal of Ecology*, 74, 1031-44.

Smith, T.M. and Goodman, P.S. (1987). Successional dynamics in a semi-arid savanna: Spatial and temporal relationship between *Acacia nilotica* and *Euclea divinorum*. *Journal of Ecology*, 75, 603-10.

Smith, T.M., and Huston, M.A. (1989). A theory of the spatial and temporal dynamics of plant communities. *Vegetatio*, 83, 49-69.

Smith, T.M., and Urban, D.L. (1988). Scale and resolution of forest structural pattern. *Vegetatio,* 74, 143-50.

Smoliak, S. (1956). Influence of climatic conditions on forage production of shortgrass rangeland. *Journal of Range Management,* 9, 89-91.

Smythe, N. (1986). Competition and resource partitioning in the guild of neogropical terrestrial frugivorous mammals. *Annual Review of Ecology and Systematics,* 17, 169-88.

Soares, R.V. (1990). Fire in some tropical and subtropical South American vegetation types: an overview. In *Fire in the Tropical Biota,* ed. J.G. Goldammer, pp. 63-81. Berlin: Springer-Verlag.

Solomon, A.M. (1986). Transient responses of forests to CO^2-induced climate change: simulation modeling experiments in eastern North America. *Oecologia,* 68, 567-79.

Solomon, A.M., Delcourt, H.R., West, D.C., and Blasing, T.J. (1980). Testing a simulation for reconstruction of prehistoric forest stand dynamics. *Quaternary Research,* 14, 275-93.

Sombroek, W.G. (1984). Soils of the Amazon region. In *The Amazon: Limnology and Landscape Ecology of a Mighty Tropical River and its Basin,* ed. H. Sioli, pp. 521-35. Dordrecht: W. Junk.

Sommer, U. (1984). The paradox of the plankton: Fluctuations of phosphorus availability maintain diversity of phytoplankton in flow-through cultures. *Limnology and Oceanography,* 29, 633-6.

Sousa, W.P. (1979a). Experimental investigation of disturbance and ecological succession in a rocky intertidal algal community. *Ecological Monographs,* 49, 227-54.

Sousa, W.P. (1979b). Disturbance in marine intertidal boulder fields: the nonequilibrium maintenance of species diversity. *Ecology,* 60, 1225-39.

Sousa, W.P. (1980). The responses of a community to disturbance: the importance of successional age and species life histories. *Oecologia,* 45, 72-81.

Sousa, W.P. (1985). Disturbance and patch dynamics on rocky intertidal shores. In *The Ecology of Natural Disturbance and Patch Dynamics,* ed. S.T.A. Pickett and P.S. White, pp. 101-24. Orlando, Florida: Academic Press.

Southwood, T.R.E. (1977). Habitat, the templet for ecological strategies. *Journal of Animal Ecology,* 46, 337-65.

Southwood, T.R.E., May, R.M., Hassel, M.P., and Conway, G.R. (1974). Ecological strategies and population parameters. *American Naturalist,* 108, 791-804.

Sparks, S.R., West, N.E., and Allen, E.B. (1990). Changes in vegetation and land use at two townships in Skull Valley, western Utah. In *Proceedings - Symposium on Cheatgrass Invasion, Shrub Die-off, and Other Aspects of Shrub Biology and Management. Gen. Tech. Report INT-276,* pp. 4-10. Provo, Utah: U.S. Department of Agriculture Forest Service.

Specht, R.L. (1963). The Dark Island Heath (Ninety-Mile Plain, South Australia) VII: the effects of fertilizers on composition and growth, 1950-72. *Australian Journal of Botany,* 11, 67-94.

Specht, R.L. (1972) *The Vegetation of South Australia.* Adelaide: Government Printer.

Specht, R.L. (1981). Primary production of mediterranean-climate ecosystems regenerating after fire. In *Ecosystems of the World, Vol. 11: Mediterranean-Type Shrublands,* ed. F. DiCastri, D.W. Goodall, and R.L. Specht,

Specht, R.L., and Moll, E.J. (1983). Mediterranean-type heathlands and sclerophyllous shrublands of the world: An overview. In *Mediterranean-type Ecosystems: The Role of Nutrients, Ecological Studies Volume 43*, ed. F.J. Kruger, D.T. Mitchell, and J.U.M. Jarvis, pp. 41-65. Berlin: Springer-Verlag.

Spence, D.H.N. (1959). Studies on the vegetation of Shetland. II. Reasons for the restriction of the exclusive pioneers to serpentine debris. *Journal of Ecology*, 47, 641-9.

Spence, D.H.N. (1982). The zonation of plants in freshwater lakes. *Advances in Ecological Research*, 12, 37-125.

Spight, T.M. (1976). Censuses of rocky shore prosobranchs from Washington and Costa Rica. *Veliger*, 18, 309-17.

Spight, T.M. (1977). Diversity of shallow-water gastropod communities on temperate and tropical beaches. *American Naturalist*, 111, 1077-97.

Springer, A.M., McRoy, C.P., and Turco, K.R. (1989). The paradox of pelagic food webs in the northern Bering Sea - II. Zooplankton Communities. *Continental Shelf Research*, 9, 359-86.

Springer, A.M., Murphy, E.C., Roseneau, D.G., McRoy, C.P., and Cooper, B.A. (1987). The paradox of pelagic food webs in the northern Bering Sea during summer - I. Seabird food habits. *Continental Shelf Research*, 7, 895-911.

Springer, A.M., Roseneau, D.G., Lloyd, D.S., McRoy, C.P., and Murphy, E.C. (1986). Seabird responses to fluctuating prey availability in the eastern Bering Sea. *Marine Ecology Progress Series*, 32, 1-12.

Springer, V.G., and McErlean, A.J. (1962). A study of the behavior of some tagged South Florida coral reef fishes. *American Midland Naturalist*, 67, 386-97.

Spurr, S.H. (1954). The forests of Itasca in the nineteenth century as related to fire. *Ecology*, 35, 21-5.

Spurr, S.H., and Barnes, B.V. (1973). *Forest Ecology*, New York: Ronald Press.

Stanton, N.L. (1979). Patterns of species diversity in temperate and tropical litter mites. *Ecology*, 60, 295-304.

Stark, N.M. (1974). Fuel reduction-nutrient status and cycling relationships associated with understory burning in larch/Douglas-fir stands. *Proceedings of the Annual Tall Timbers Fire Ecology Conference*, 14, 573-96.

Stark, N.M., and Jordan, C.F. (1978). Nutrient retention by the root mat of an Amazonian rain forest. *Ecology*, 59, 434-7.

Starker, T.J. (1934). Fire resistance in the forest. *Journal of Forestry*, 32, 462-7.

Stearns, S.C. (1976). Life-history tactics: a review of the ideas. *Quarterly Review of Biology*, 51, 3-47.

Stearns, S.C. (1977). The evolution of life history traits: a critique of the theory and a review of the data. *Annual Review of Ecology and Systematics*, 8, 145-71.

Stebbins, G.L. (1974). *Flowering Plants: Evolution above the species level*. Cambridge, Massachusetts: Belknap Press.

Stebbins, G.L., and Majors, J. (1965). Endemism and speciation in the California flora. *Ecological Monographs*, 35, 1-35.

Steemann-Neilson, E. (1954). On organic production in the oceans. *Journal of the International Council for the Study of the Sea*, 19, 309-28.

Steenis, C.J. van. (1958). Ecology of mangroves. *Flora Malesiana Series 1, Spermatophyta*, 5, 429-48.

Stehli, F.G. (1968). Taxonomic diversity gradients in pole location: the recent model. In *Evolution and Environment*, ed. E.T. Drake, pp. 163-227. New Haven: Yale University Press.

Stehli, F.G., Douglas, R.G., and Newell, N.D. (1969). Generation and maintenance of gradients in taxonomic diversity. *Science, (Washington D.C.)*, 164, 947-9.

Stehli, F.G., McAlester, A.L., and Helsley, C.E. (1967). Taxonomic diversity of recent bivalves and some implications for geology. *Geological Society of America Bulletin* 78, 455-66.

Stehli, F.G., and Wells, J.W. (1971). Diversity and age patterns in hermatypic corals. *Systematic Zoology*, 20, 115-26.

Steinman, A.D., and McIntire, C.D. (1990). Recovery of lotic periphyton communities following disturbance. *Environmental Management*, 14, 589-604.

Stephenson, N.L. (1990). Climatic control of vegetation distribution: the role of water balance. *American Naturalist*, 135, 649-70.

Stephenson, T.A., and Stephenson, A. (1949). The universal features of zonation between tide-marks on Rocky Coasts. *Journal of Ecology*, 37, 289-305.

Stephenson, T.A., and Stephenson, A. (1950). Life between the tide-marks in North America: I, The Florida Keys. *Journal of Ecology*, 38, 354-402.

Stephenson, T.A., and Stephenson, A. (1952). Life between the tide-marks in North America: II, North Florida and the Carolinas. *Journal of Ecology*, 40, 1-49.

Stephenson, T.A., and Stephenson, A. (1954a). Life between the tide-marks in North America: IIIA, Nova Scotia and Prince Edward Island: Description of the region. *Journal of Ecology*, 42, 14-45.

Stephenson, T.A., and Stephenson, A. (1954b). Life between the tide-marks in North America: IIIB, Nova Scotia and Prince Edward Island: The geographical features of the region. *Journal of Ecology*, 42, 46-70.

Stephenson, T.A., and Stephenson, A. (1961). Life between the tide-marks in North America: IVA, IVB, Vancouver Island. *Journal of Ecology*, 49. 1-29, 229-43.

Sterner, R.W., (1989). Resource competition during seasonal succession toward dominance by cyanobacteria. *Ecology*, 70, 229-45.

Sterner, R.W., Ribic, C.A., and Schat, G.E. (1986). Testing for life historical changes in spatial patterns of four tropical tree species. *Journal of Ecology*, 74, 621-33.

Stevens, G.C. (1989). The latitudinal gradient in geographical range: how so many species coexist in the tropics. *American Naturalist*, 133, 240-56.

Stewart, G, and Hull, A.C. (1949). Cheatgrass (*Bromus tectorum L.*) - an ecological intruder in southern Idaho. *Ecology*, 30, 58-74.

Stewart, G, and Young, A.E. (1939). The hazard of basing permanent grazing capacity on *Bromus tectorum*. *American Society of Agronomy Journal*, 31, 1002-15.

Stiles, F.G. (1975). Ecology, flowering phenology and hummingbird pollination of some Costa Rican *Heliconia* species. *Ecology*, 56, 285-310.

Stocker, G.C., and Unwin, G.L. (1990). The rain forests of northeastern Australia - their environment, evolutionary history and dynamics. In *Ecosystems of the World. 14B. Tropical Rain Forest Ecosystems. Biogeography and Ecology*, ed. H. Lieth and M.J.A. Werger, pp. 241-59. Amsterdam: Elsevier.

Stoddart, D.R. (1969). Ecology and morphology of recent coral reefs. *Biological Reviews*, 44, 433-98.

Stone, E.C., and Juren, G. (1951). The effect of fire on the germination of the seed of *Rhus ovata*. *American Journal of Botany*. 38, 368-72.

Stott, P.A., Goldammer, J.G., and Werner, W.L. (1990). The role of fire in the tropical lowland deciduous forest of Asia. In *Fire in the Tropical Biota*, ed. J.G. Goldammer, pp. 32-44. Berlin: Springer-Verlag.

Stout, J., and Vandermeer, J.H. (1975). Comparison of species richness for stream-inhabiting insects in tropical and mid-latitude streams. *American Naturalist*, 109, 263-80.

Strong, D.R. (1977). Epiphyte loads, treefalls, and perennial forest disruption: a mechanism for maintaining higher tree species richness in the tropics without animals. *Journal of Biogeography*, 4, 215-8.

Strong, D.R. (1982a). Harmonious coexistence of hispine beetles on *Heliconia* in experimental and natural communities. *Ecology*, 63, 1039-49.

Strong, D.R. (1982b). Potential interspecific competition and host specificity: hispine beetles on *Heliconia*. *Ecological Entomology*, 7, 217-20.

Strong, D.R. (1984). Exorcising the ghost of competition past: phytophagous insects. In *Ecological Communities: Conceptual Issues and the Evidence*, ed. D.R. Strong, D. Simberloff, L.G. Abele, and A.B. Thistle, pp. 28-41. Princeton: Princeton University Press.

Strong, D.R., Lawton, J.H., and Southwood, T.R.E. (1984). *Insects on Plants: Community Patterns and Mechanisms*. Oxford: Blackwell Scientific

Strong, D.R., Syska, L.A., and Simberloff, D.S. (1979). Tests of community-wide character displacement against null hypotheses. *Evolution*, 33, 897-913.

Strong, D.R., and Levin, D.A. (1979). Species richness of plant parasites and growth form of their hosts. *American Naturalist*, 114, 1-22.

Strong, D.R., and Simberloff, D.S. (1981). Straining at gnats and swallowing ratios: character displacement. *Evolution*, 35, 810-2.

Struik, G.J., and Bray, J.R. (1970). Root-shoot ratios of native forest herbs and *Zea mays* at different soil moisture levels. *Ecology*, 51, 892-3.

Stuart, C.T., and Rex, M.A. (1989). A latitudinal gradient in deep-sea gastropod diversity. *American Zoologist*, 29, 26.

Sugihara, G. (1980). Minimal community structure: an explanation of species abundance patterns. *American Naturalist*, 116, 770-87.

Sumich, J.L. (1976). *An Introduction to the Biology of Marine Life*. Dubuque, Iowa: Wm. C. Brown.

Suter, G. W. (1981). Ecosystem theory and NEPA assessment. *Bulletin of the Ecological Society of America*, 62, 186-92.

Sverdrup, H.U., Johnson, M.W., and Fleming, R.H. (1942). *The Oceans, Their Physics, Chemistry, and General Biology*. Englewood Cliffs, New Jersey: Prentice-Hall.

Swain, A.M. (1973). History of fire and vegetation in northeastern Minnesota as recorded in lake sediments. *Quaternary Research*, 3, 383-96.

Swaine, M.D., and Hall, J.B. (1981). The monospecific tropical forest of the Ghanaian endemic tree, *Talbotiella gentii*. In *The Biological Aspects of Rare Plant Conservation: Proceedings of an International Conference*, ed. H. Synge, pp. 355-63. Botanical Society of the British Isles Conference Report 17. Chichester: Wiley.

Swan, F.R. (1970). Postfire responses of four plant communities in south-central New York state. *Ecology*, 51, 1074-82.

Swetnam, T.W., and Betancourt, J.J. (1990). Fire-Southern Oscillation relations in the southwestern United States. *Science*, 249, 1017-9.

Swift, M.J., Heal, O.W., and Anderson, J.M. (1979). *Decomposition in Terrestrial Systems*. Oxford: Blackwell.

Swingle, H.S. (1946). Experiments with combinations of largemouth black bass, Bluegills, and minnows in ponds. *Transactions of the American Fisheries Society*, 76, 46-62.

Sykora, K.V. (1989). History of the impact of man on the distribution of plant species, and the behaviour of exotic plant species in the country of introduction. In *Biological Invasions in Europe and the Mediterranean Basin*, ed. F. di Castri, A.J. Hansen, and M. Debussche, pp. 37-50. Dordrecht: Kluwer.

Taber, R.D., and Dasmann, R.F. (1958). The black-tailed deer of the chaparral. *California Fish and Game Department, Game Bulletin No. 8*.

Talbot, F.H., Russell, B.C., and Anderson, G.V.R. (1978). Coral reef fish communities: unstable, high diversity systems? *Ecological Monographs*, 48, 425-40.

Talbot, L.M., and Talbot, M.H. (1963). The high biomass of wild ungulates on East African savanna. *Transactions of the North Amererican Wildlife Conference*, 28, 465-76.

Tallaway, D.W. (1983). Equilibrium biogeography and its application to insect host-parasite systems. *American Naturalist*, 121, 244-54.

Tangley, L. (1990). Cataloging Costa Rica's diversity. *BioScience*, 40, 633-6.

Tansley, A.G., and Adamson, R.S. (1925). Studies of the vegetation of the English chalk. III. The chalk grasslands of the Hampshire-Sussex border. *Journal of Ecology*, 13, 177-223.

Taylor, A.D. (1990). Metapopulations, dispersal, and predator-prey dynamics: an overview. *Ecology*, 71, 429-33.

Taylor, D.L. (1969). Biotic succession of lodgepole pine forests of fire origin in Yellowstone National Park. Ph.D. Dissertation, University of Wyoming, Laramie.

Taylor, W.D. (1978). Maximum growth rates, size and commonness in a community of bactivorous ciliates. *Oecologia (Berlin)*, 36, 263-72.

Temple, S.A. (1977). Plant-animal mutualism: coevolution with dodo leads to near extinction of plant. *Science*, 197, 885-6.

Teraguchi, S., Stenzel, J., Sedlacek, J., and Deininger, R. (1981). Arthropod-grass communities: comparisons of communities in Ohio and Alaska. *Journal of Biogeography*, 8, 53-65.

Terborgh, J. (1971). Distribution on environmental gradients: theory and a preliminary interpretation of distributional patterns in the Avifauna of the Cordillera Vilcabamba, Peru. *Ecology*, 52, 23-40.

Terborgh, J. (1980). Causes of tropical species diversity. In *Acta XVII Congress International Ornithologica (Berlin)*, ed. R. Nohring. pp. 955-61. Berlin: Deutsche Ornithologen-Gesellschaft.

Terborgh, J. (1983). *Five New World Primates: A Study in Comparative Ecology*. Princeton, New Jersey: Princeton University Press.

Terborgh, J. (1985). The vertical component of plant species divesity in temperate and tropical forests. *American Naturalist*, 126, 760-66.

Terborgh, J. (1986). Keystone plant resources in tropical forests. In *Conservation Biology: the Science of Scarcity and Diversity*, ed. M.E. Soulé, pp. 330-44. Sunderland, Massachusetts: Sinauer.

Terborgh, J., and Robinson, S. (1986). Guilds and their utility in ecology. In

Community Ecology: Pattern and Process, ed. J. Kikkawa and D.J. Anderson. pp. 65-90. Melbourne: Blackwell Scientific.

Terborgh, J., and Winter, B. (1980). Some causes of extinction. In *Conservation Biology: An Evolutionary-Ecological Perspective,* ed. M.E. Soulé and B.A. Wilcox, pp. 119-33. Sunderland, Massachusetts: Sinauer Associates.

Terborgh, J., and Winter, B. (1982). Evolutionary circumstances of species with small ranges. In *Biological Diversification in the Tropics*, ed. G. Prance. pp. 587-600. New York: Columbia University Press.

ter Steege, H., and Cornelissen, J.H.C. (1990). Distribution and ecology of vascular epiphytes in lowland rain forest of Guyana. *Biotropica, 21,* 331-9.

Tester, J.R., and Marshall, W.H. (1962). Minnesota prairie management techniques and their wildlife implications. *Transactions of the North American Wildlife Conference, 27,* 267-87.

Tharp, M.L. (1978). Modeling major perturbations on a forest ecosystem. M.S. Thesis. Knoxville: University of Tennessee.

Thiel, H. (1975). The size structure of the deep-sea benthos. *Int. Rev. Ges. Hydrobiol. 60,* 575-606.

Thiel, H. (1979). Structural aspects of the deep-sea benthos. *Ambio Special Report 6, 25-31.*

Thirgood, J.V. (1981). *Man and the Mediterranean Forest - A History of Resource Depletion.* London: Academic Press.

Thistle, D. (1978). Harpacticoid dispersal patterns: implications for deep-sea diversity. *Journal of Marine Science, 36,* 377-97.

Thompson, J.N. (1978). Within-patch structure and dynamics in *Pastinaca sativa* and resource availability to a specialized herbivore. *Ecology, 59,* 443-8.

Thompson, J.N. (1985). Within-patch dynamics of life histories, populations, and interactions: Selection over time in small spaces. In *The Ecology of Natural Disturbance and Patch Dynamics,* ed. S.T.A. Pickett and P.S. White, pp. 253-64. Orlando, FL: Academic Press.

Thoreau, H.D. (1860). Succession of forest trees, An address to the Middlesex Agricultural Society, September, 1860. In *Excursions.* (1891). Boston: Houghton Mifflin.

Thorson, G. (1951). Sur jetzigen Lage der marinen *Bodetier-Ökologie. Verh. Dtsch. Zool. Ges. 1951. Zool. Anz. Suppl.,* 16, 276-327.

Thorson, G. (1957). Bottom communities (sublittoral and shallow shelf). In *Treatise of Marine Ecology and Paleoecology,* ed. H. Ladd. *Geological Society of America Memoir* 67.

Thurston, J.M. (1951). Some experiments and field observations on the germination of wild oat (*Avena fatua* and *A. ludoviciana*) seeds in soil and the emergence of seedlings. *Annals of Applied Biology,* 38, 812-32.

Thurston, J.M. (1969). The effect of liming and fertilizers on the botanical composition of permanent grassland, and on the yield of hay. In *Ecological Aspects of the Mineral Nutrition of Plants.* ed. I.H. Rorison. pp. 3-10. Oxford: Blackwell.

Tillyard, R.J. (1917). *The Biology of Dragonflies.* Cambridge: Cambridge University Press.

Tilman, D. (1980). Resources: a graphical-mechanistic approach to competition and predation. *American Naturalist,* 116, 362-93.

Tilman, D. (1982). *Resource Competition and Community Structure.* Princeton: Princeton University Press.

Tilman, D. (1985). The resource-ratio hypothesis of plant succession. *American Naturalist,* 125, 827-52.

Tilman, D. (1987a). The importance of the mechanisms of interspecific interaction. *American Naturalist,* 129, 769-74.

Tilman, D. (1987b). Secondary succession and patterns of plant dominance along experimental nitrogen gradients. *Ecological Monographs,* 57, 189-214.

Tilman, D. (1988). *Plant Strategies and the Dynamics and Structure of Plant Communities.* Princeton, New Jersey: Princeton University Press.

Tilman, D., Mattson, M., and Langer, S. (1981). Competition and nutrient kinetics along a temperature gradient: an experimental test of a mechanistic approach to niche theory. *Limnology and Oceanography,* 26, 1020-33.

Timmons, T.J., Shelton, W.L., and Davies, W.D. (1980). Differential growth of largemouth bass in West Point Reservoir, Alabama-Georgia. *Transactions of the American Fisheries Society,* 109, 176-86.

Timms, R.M., and Moss, B. (1984). Prevention of potentially dense phytoplankton populations by zooplankton grazing, in the presence of zooplanktivorous fish, in a shallow wetland ecosystem. *Limnology and Oceanography,* 29, 472-86.

Tinley, K.L. (1982). The influence of soil moisture balance on ecosystem patterns in Southern Africa. In *Ecology of Tropical Savannahs, Ecological Studies Volume 42,* ed. B.J. Hunley, and B.H. Walker, pp. 175-92. Berlin: Springer.

Tison, L.J. (1964). Problems de sedimentation dans les deltas. *Humid Tropical Research UNESCO,* 24, 57.

Titman, D. (1976). Ecological competition between algae: experimental confirmation of resource-based competition theory. *Science,* 192, 463-5.

Tomkins, D.J., and Grant, W.F. (1977). Effects of herbicides on species diversities of two plant communities. *Ecology,* 58, 398-406.

Tomoff, C.S. (1974). Avian species diversity in desert scrub. *Ecology,* 55, 396-403.

Tonn, W.M., and Magnuson, J.J. (1982). Patterns in the species composition and richness of fish assemblages in northern Wisconsin lakes. *Ecology,* 63, 1149-66.

Tosi, J., and Voertman, R.F. (1964). Some environmental factors in the economic development of the tropics. *Economic Geography,* 40, 189-205.

Toumey, J.W., and R. Kienholz. (1931). Trenched plots under forest canopies. *Yale University School of Forestry, Bulletin,* 30. 31 pp.

Towne, G., and Owensby, C. (1984). Long-term effects of annual burning at different dates in ungrazed Kansas tallgrass prairie. *Journal of Range Management,* 37, 392-7.

Townsend, C.R., and Hildrew, A.G. (1978). Predation strategy and resource utilization by *Plectrocnemia conspersa* (Curtis) (Trichoptera:Polycentropodidae). *Proceedings of the Second International Symposium on Trichoptera,* pp. 299-307. Junk, The Hague.

Tramer, E.J. (1969). Bird species diversity: components of Shannon's formula. *Ecology,* 50, 927-9.

Tramer, E.J. (1974). On latitudinal gradients in avian diversity. *Condor,* 76, 123-30.

Tremblay, C., Runge, J.A., and Legendre, L. (1989). Grazing and sedimentation of ice algae during and immediately after a bloom at the ice-water interface. *Marine Ecoloy Progress Series,* 56, 291-300.

Tsuda, R.T., and Bryan, P.G. (1973). Food preferences of juvenile *Siganus rostratus* and *S. spinus* in Guam. *Copeia,* 1973, 604-6.

Tsuda, R.T., and Kami, H.T. (1973). Algal succession on artificial reefs in a marine lagoon environment on Guam. *Journal of Phycology,* 9, 260-4.

Tunnicliffe, V.J. (1983). Caribbean staghorn coral populations: pre-hurricane Allen conditions in Discovery Bay, Jamaica. *Bulletin Marine Science,* 33, 132-51.

Turner, J.R.G. (1982). How do refuges produce biological diversity? In *Biological Diversification in the Tropics*, ed. G. Prance. pp. 309-35. New York: Columbia University Press.

Turner, J.T. (1981). Latitudinal patterns of calanoid and cyclopoid copepod diversity in estuarine waters of eastern North America. *Journal of Biogeography,* 8, 369-82.

Turner, R.D. (1973). Wood-boring bivalves, opportunistic species in the deep sea. *Science,* 180, 1377-9.

Túrcek, F.J. (1969). Large mammal secondary production in European broad leaved and mixed forests: some results and methods of recent research. *Biologia (Bratislava),* 24, 173-81.

U.S. Bureau of the Census. (1990). *Statistical Abstract of the United States: 1990.* Washington, D.C.: U.S. Department of Commerce.

UNESCO. (1978). *Tropical Forest Ecosystems: A State-of-knowledge Report.* Paris: United Nations Educational, Scientific, and Cultural Organization.

Uhl, C., and Buschbacher, R. (1985). A disturbing synergism between cattle ranch burning practices and selective tree harvesting in the eastern Amazon. *Biotropica,* 17, 265-8.

Uhl, C., and Kauffman, J.B. (1990). Deforestation, fire susceptibility, and potential tree responses to fire in the eastern Amazon. *Ecology,* 71, 437-49.

Uhl, C., Kauffman, J.B., and Cummings, D.L. (1988). Fire in the Venezuelan Amazon 2: Environmental conditions necessary for forest fires in the evergreen rainforest of Venezuela. *Oikos,* 53, 176-84.

Uhl, C., and Murphy, P.G. (1981). Composition, structure, and regeneration of a terra firme forest in the Amazon Basin of Venezuela. *Tropical Ecology,* 22, 219-37.

Underwood, A.J. (1978a). The detection of non-random patterns of distribution of species along a gradient. *Oecologia,* 36, 317-26.

Underwood, A.J. (1978b). A refutation of critical tidal levels as determinants of the structure of intertidal communities on British shores. *Journal of Experimental Marine Biology and Ecology,* 33, 261-76.

Underwood, A.J., and Denley, E.J. (1984). Paradigms, explanations, and generalizations in models for the structure of intertidal communities on rocky shores. In *Ecological Communities Conceptual Issues and the Evidence.* ed. D.R. Strong, D. Simberloff, L.G. Abele, and A.B. Thistle. pp. 151-80. Princeton: Princeton University Press.

Upchurch, G.R., and Wolfe, J.A. (1987). Mid-Cretaceous to Early Tertiary vegetation and climate: evidence from fossil leaves and woods. In *The Origin of the Angiosperms and their Biological Consequences,* ed. E.M. Friis, W.G. Chaloner, and P.R. Crane, pp. 75-105. Cambridge: Cambridge University Press.

Usinger, R.L. (1962). Foreword to *The Naturalist on the River Amazons.* (Bates, 1964). 1st Paperback Edition. Berkeley, California: University of California Press.

Utida, S. (1957). Cyclic fluctuations of population density intrinsic to the host-parasite system. *Ecology,* 38, 442-9.

Valentine, J.W. (1970). How many marine invertebrate fossil species? a new approximation. *Journal of Paleontology*, 44, 410-5.

Valentine, J.W. (1985). Biotic diversity and clade diversity. In *Phanerozoic Diversity Patterns: Profiles in Macroevolution*. ed. J.W. Valentine. pp. 419-24. Princeton, New Jersey: Princeton University Press.

Valentine, J.W., and Campbell, C.A. (1975). Genetic regulation and the fossil record. *American Scientist*, 63, 673-80.

Valentine, J.W., Foin, T.C., and Peart, D. (1978). A provincial model of Phanerozoic marine diversity. *Paleobiology*, 4, 55-66.

Van Auken, O.W., and Bush, J.K. (1985). Secondary succession on terraces of the San Antonio River. *Bulletin of the Torrey Botanical Club*, 112, 158-66.

Van Lear, D.H., and Waldrop, T.A. (1989). *History, Uses, and Effects of Fire in the Appalachians*. Southeastern Forest Experiment Station General Technical Report SE-54. Washington, D.C.: U.S. Department of Agriculture.

Vance, R.R. (1984). The effect of dispersal on population stability of one-species, discrete-space population growth models. *American Naturalist*, 123, 230-54.

van der Hammen, T. (1982). Paleoecology of tropical South America. In *Biological Diversification in the Tropics*, ed. G.T. Prance, pp. 60-6. New York: Columbia University Press.

van der Pijl, L. (1969). Evolutionary action of tropical animals on the reproduction of plants. *Biological Journal of the Linnean Society*, 1, 85-92.

van der Pijl, L. (1972) *Principles of Dispersal in Higher Plants*. Berlin: Springer-Verlag.

van der Werff, H. (1983). Species number, area and habitat diversity in the Galapagos Islands. *Vegetatio*, 54, 167-75.

Vandermeer, J.H. (1969). The competitive structure of communities: an experimental approach with protozoa. *Ecology*, 50, 362-71.

Vandermeer, J.H. (1973). On the regional stabilization of locally unstable predator-prey relationships. *Journal of Theoretical Biology*, 41, 161-70.

Vandermeer, J.H. (1980). Indirect mutualism: variations on a theme by Stephen Levine. *American Naturalist*, 116, 441-8.

Vanzolini, P.E. (1970). Zoologia sistematica geografia e a origem das cspccics. *Instituto Geografico Sao Paulo. Serie teses e monografias*, 3, 1-56.

Vasseur, P. (1974). The overhangs, tunnels, and dark reef galleries of Tulear (Madagascar) and their sessile invertebrate communities. *Proceedings of the Second International Coral Reef Symposium*, 2, 143-59.

Veblen, T.T., and Ashton, P.S. (1978). Catastrophic influence on the vegetation of the Valdevian Andes, Chile. *Vegetatio*, 36, 149-67.

Venrick, E.L. (1982). Phytoplankton in an oligotrophic ocean: observations and questions. *Ecological Monographs*, 52, 129-54.

Vermeer, J.G., and Berendse, F. (1983). The relationship between nutrient availability, shoot biomass and species richness in grassland and wetland communities. *Vegetatio*, 53, 121-6.

Vestal, A. (1949). Minimum areas for different vegetations. Their determination from species-area curves. *Illinois Biological Monographs*, 20(3).

Vesy-Fitzgerald, D.F. (1960). Grazing succession amongst East African game animals. *Journal of Mammalogy*, 41, 161-70.

Vesy-Fitzgerald, D.F. (1965). The utilization of natural pastures by wild animals in the Rukwa Valley, Tanganyika. *East African Wildlife Journal*, 3, 38-48.

Viereck, L.A. (1966). Plant succession and soil development on gravel outwash of the Muldrow Glacier, Alaska. *Ecological Monographs,* 36, 181-91.

Viereck, L.A. (1970). Forest succession and soil development adjacent to the Chena River in interior Alaska. *Arctic and Alpine Research,* 2, 1-26.

Viereck, L.A. (1973). Wildfire in the taiga of Alaska. *Quaternary Research,*3, 465-95.

Vigne, J.D. (1983). Le remplacement des faunes de petits mammifères en Corse lors de l'arrivée de l'homme. *Corsican Royal Society of Biogéography,* 59, 41-51.

Vincent, V., and Thomas, R.G. (1961). *An agricultural survey of southern Rhodesia. Part I. Agro-ecological survey.* Salisbury: Government Printer.

Vine, P.F. (1974). Effects of algal grazing and aggressive behavior of the fished *Pomacentrus lividus* and *Acanthurus sohal* on coral reef ecology. *Marine Biology,* 24, 131-6.

Vine, P.F. (1974). Effects of algal grazing and aggressive behavior of the fishes *Pomacentrus lividus* and *Acanthurus sohal* on coral reef ecology. *Marine Biology,* 24, 131-6.

Vinogradov, M.E. (1983). Open-ocean ecosystems. In *Marine Ecology, Volume V, Part 2,* ed. O. Kinne. Chichester: Wiley.

Vitousek, P.M. (1986). Biological invasions and ecosystem properties: can species make a difference. In *Ecology of Biological Invasions of North America and Hawaii,* ed. H.A. Mooney and J.A. Drake, pp. 163-76. New York: Springer-Verlag.

Vitousek, P.M., and Denslow, J.S. (1987). Differences in extractable phosphorus among soils of the La Selva Biological Station, Costa Rica. *Biotropica,* 19, 167-70.

Vitousek, P.M., and Matson, P.A. (1985). Disturbance, nitrogen availability, and nitrogen losses in an intensively managed loblolly pine plantation. *Ecological Monographs,* 66, 1360-76.

Vitousek, P.M., and Sanford, R.L. (1986). Nutrient cycling in moist tropical forest. *Annual Review of Ecology and Systematics,* 17, 137-67.

Vogl, R.J. (1964). The effects of fire on muskeg in northern Wisconsin. *Journal of Wildlife Management,* 28, 317-29.

Vogl, R.J. (1969). Fire adaptations of some southern California plants. In *Proceedings of the Annual Tall Timbers Fire Ecology Conference,* 7, 79-109.

Volterra, V. (1926). Variations and fluctuations in the number of individuals of animal species living together. In *Animal Ecology,* ed R.N. Chapman, pp. 409-48. New York: McGraw-Hill.

von Broembsen, S.L. (1989). Invasions of natural ecosystems by plant pathogens. In *Biological Invasions: a Global Perspective. SCOPE 37,* ed. J.A. Drake *et al.,* pp. 31-60. Chichester: Wiley.

Vrijenhoek, R.C. (1985). Animal population genetics and disturbance: the effects of local extinctions and recolonizations on heterozygosity and fitness. In *The Ecology of Natural Disturbance and Patch Dynamics* ed. S.T.A. Pickett and P.S. White, pp. 266-86. Orlando: Academic Press.

Wagner, W., Herbst, D.R., and Yee, R.S. (1985). Status of the flowering plants of the Hawaiian Islands. In *Hawaii's Terrestrial Ecosystems: Preservation and Management,* ed. C.P. Stone and J.M. Scott, pp. 23-104. Honolulu: Cooperative National Park Resources Study Unit and the University of Hawaii.

Walbran, P.D., Henderson, R.A., Jull, A.J.T., and Head, M.J. (1989). Evidence from sediments of long-term *Acanthaster planci* predation on corals of the Great Barrier Reef. *Science,* 245, 847-50.

Walde, S.J. (1986). Effect of an abiotic disturbance on a lotic predator-prey interaction. *Oecologia (Berlin),* 69, 243-7.

Waldrop, T.A., van Lear, D.H., Lloyd, F.T., and Harms, W.R. (1987). *Long-term Studies of Prescribed Burning in Loblolly Pine Forests of the Southeastern Coastal Plain.* U.S. Department of Agriculture, Forest Service General Technical Report SE 45,

Walker, B.H. (1981). Is succession a viable concept in African savanna ecosystems. In *Forest Succession Concepts and Applications,* ed. D.C. West, H.H. Shugart, and D.B. Botkin, pp. 431-47. New York: Springer-Verlag.

Walker, B.H., Ludwig, D., Holling, C.S., and Peterman, R.M. (1981). Stability of semi-arid savanna grazing systems. *Journal of Ecology,* 69, 473-98.

Walker, B.H., and Noy-Meir, I. (1982). Aspects of the stability and resilience of savanna ecosystems.In *Ecology of Tropical Savannahs,* ed. B.J. Huntley and B.H. Walker, pp. 556-90. Heidelburg: Springer-Verlag.

Walker, D. (1970). Direction and rate in some British post-glacial hydroseres. In *Studies in the Vegetational History of the British Isles,* ed. D. Walker and R. West. pp. 117-37. Cambridge: Cambridge University Press.

Walker, J., Sharpe, P.J.H., Penridge, L.K., and Wu, H. (1989). Ecological field theory: the concept and field tests. *Vegetatio,* 83, 81-95.

Walker, J., Thompson, C.H., Fergus, I.F., and Tunstall, B.R. (1981). Plant succession and soil development in coastal sand dunes of subtropical eastern Australia. In *Forest Succession: Concepts and Application,* ed. D.C. West, H.H. Shugart, and D.B. Botkin, pp. 107-31. New York: Springer-Verlag.

Walker, J., and Peet, R.K. (1983). Composition and species diversity of pine-wiregrass savannas of the Green Swamp, North Carolina. *Vegetatio,* 55, 163-79.

Walker, L.R., Zasada. J.C., and Chapin, F.S. (1986). The role of life history processes in primary succession on an Alaskan floodplain. *Ecology,* 67, 1243-53.

Walker, T.D., and J.W. Valentine. (1984). Equilibrium models of evolutionary species diversity and the number of empty niches. *American Naturalist,* 124, 887-99.

Wallace, A.R. (1878). Tropical nature and other essays. London: Macmillan.

Walsh, J.J., McRoy, C.P., Blackburn, T.H., *et al.* (1989). The role of Bering Strait in the carbon/nitrogen fluxes of polar marine ecosystems. In *Proceedings of the Sixth Conference of the Comite Artique International,* ed. L. Rey and V. Alexander. Leiden, The Netherlands: E.J. Brill.

Walter, H. (1939). Grasland, Savanne und Busch der ariden Teile Afrikas in ihrer ökologischen Bedingtheit. *Jb. Wiss. Bot.* 87, 750-860.

Walter, H. (1960). *Standortslehre. Phytologie, Vol 3. part 1, 2nd ed.* Stuttgart: Ulmer.

Walter, H. (1964). *Die Vegetation der Erde, I. Die tropischen und subtropischen Zonen.* Jena: Fischer-Verlag.

Walter, H. (1968). *Die Vegetation der Erde in Okophysiologischer Betrachtung, Vol. 2. Die gemassigten und arktischen Zonen.* Jena: Fischer.

Walter, H. (1971). *Ecology of Tropical and Sub-tropical Vegetation.* Trans. D. Mueller-Dombois. Edinburgh: Oliver and Boyd.

Walter, H. (1973). *Vegetation of the earth in relation to the eco-physiological conditions.* New York: Springer-Verlag.

Walter, H. (1985). *Vegetation of the Earth and Ecological Systems of the Geo-biosphere.* Third edition, translated from the Fifth German edition by Owen Muise. Berlin: Springer-Verlag.

Walter, H., and Steiner, M. (1936). Die Oekologie der Ost-Afrikanischen Mangroven. *Z. Botanische*, 30, 65-193.

Walton, S. (1981). U.S.-Egypt Nile project studies high dam's effects. *BioScience,* 31, 9-13.

Wangersky, P.J. (1978). Lotka-Volterra population models. *Annual Review of Ecology and Systematics,* 9, 189-218.

Waring, R.H. (1987). Characteristics of trees predisposed to die. *BioScience,* 37, 569-74.

Waring, R.H., and Franklin, J.F. (1979). Evergreen coniferous forests of the Pacific Northwest. *Science,* 1380-6.

Waring, R.H., and Schlesinger, W.H. (1985). *Forest Ecosystems: Concepts and Management.* Orlando, Florida: Academic Press.

Washburn, J.O. (1984). The gypsy moth and its parasites in North America: a community in equilibrium? *American Naturalist,* 124, 288-92.

Waterman, P.G., and McKey, D. (1990). Herbivory and secondary compounds in rain-forest plants. In *Tropical Rain Forest Ecosystems. Biogeography and Ecology,* Ecosystems of the World Vol 14B, ed. H. Lieth and M.J.A. Werger, pp. 513-33. Amsterdam: Elsevier.

Watson, G.E. (1964). Ecology and evolution of passerine birds on the islands of the Aegean Sea. Ph.D. thesis, Yale University (Dissertation microfilm 65-1956).

Watson, J.G. (1928). The mangrove swamps of the Malay Peninsula. *Malay Forest Record,* 6.

Watt, A.S. (1925). On the ecology of British beechwoods with special reference to their regeneration. Part II. The development and structure of beech communities on the Sussex Downs. *Journal of Ecology,* 13, 27-73.

Watt, A.S. (1934). The vegetation of the Chiltern Hills, with special reference to the beechwoods and their seral relationships. *Journal of Ecology,* 22, 230-70, 445-507.

Watt, A.S. (1947). Pattern and process in the plant community. *Journal of Ecology,* 35, 1-12.

Watts, L.R., Stewart, G., Connaughton, C., Palmer, L.J., and Talbot, M.W. (1936). The management of range lands. *U.S. Senate Document,* 199, 501-22.

Watts, W.A. (1970). The full-glacial vegetation of northwestern Georgia. *Ecology,* 51, 19-33.

Watts, W.A. (1973). Rates of change and stability in vegetation in the perspective of long periods of time. In *Quaternary Plant Ecology.* ed. H.J.B. Birks and R.G. West. pp. 195-206. Oxford: Blackwell.

Watts, W.A. (1979). Late-Quaternary vegetation of central Appalachia and the New Jersey coastal plain. *Ecological Monographs,* 49, 427-69.

Watts, W.A. (1983). Vegetational history of the eastern United States 25000 to 10000 years ago. In *Late-Quaternary Environments of the United States. Vol. 1. The Late Pleistocene,* ed. S.C. Porter, pp. 294-310. Minneapolis: University of Minnesota Press.

Weaver, H. (1967). Fire and its relationship to Ponderosa pine. *Proceedings of the Annual Tall Timbers Fire Ecology Conference,* 7, 127-49.

Weaver, J.E. (1960). Flood plain vegetation of the central Missouri Valley and contacts of woodland with prairie. *Ecological Monographs,* 30, 37-64.

Weaver, J.E., and Albertson, F.W. (1936). Effects of the great drought on the prairie of Iowa, Nebraska, and Kansas. *Ecology,* 17, 567-639.

Weaver, J.E., and Albertson, F.W. (1939). Major changes in grassland as a result of continued drought. *Botanical Gazette,* 100, 576-91.

Weaver, J.E., and Albertson, F.W. (1940). Deterioration of midwestern ranges. *Ecology,* 21, 216-36.

Weaver, J.E., and Fitzpatrick, T.J. (1934). The prairie. *Ecological Monographs,* 4, 109-295.

Weaver, J.E., and Roland, H.W. (1952). Effects of excessive natural mulch on development, yield, and structure of native grassland. *Botanical Gazette,* 114, 1-19.

Webb, L.J. (1959). A physiognomic classification of Australian rain forests. *Journal of Ecology,* 47, 551-70.

Webb, L.J., Tracey, J.G., Williams, W.T., and Lance, G.N. (1970). Studies in the numerical analysis of complex rain-forest communities. V. A comparison of the properties of floristic and physiognomic-structural data. *Journal of Ecology,* 58, 203-32.

Webb, L.J., Tracy, J.G., and Williams, W.T. (1972). Regeneration and pattern in the subtropical rain forest. *Journal of Ecology,* 60, 675-95.

Webb, S.D. (1977). A history of savanna vertebrates in the New World. Part I: North America. *Annual Review of Ecology and Systematics,* 8, 355-80.

Webb, S.D. (1978). A history of savanna vertebrates in the New World. Part II: South America and the great interchange. *Annual Review of Ecology and Systematics,* 9, 393-426.

Webb, T. (1986). Is vegetation in equilibrium with climate? How to interpret late-Quaternary pollen data. *Vegetatio,* 67, 75-91.

Webb, T. (1986). Vegetational change in eastern North America from 18,000 to 500 Yr B.P. In *Climate-Vegetation Interactions, OIES-2* ed. C. Rosenzweig and R. Dickenson. pp. 63-9. Boulder, Colorado: University Corporation for Atmospheric Research.

Webb, T. (1987). The appearance and disappearance of major vegetational assemblages: long-term vegetational dynamics in eastern North America. *Vegetatio,* 69, 177-87.

Webb, T., Cushing, E.J., and Wright, H.E. (1983). Holocene changes in the vegetation of the Midwest. In *Late-Quaternary Environments of the United States, Volume 2, The Holocene.* ed. H.E. Wright. pp. 142-65. Minneapolis: University of Minnesota Press.

Webb, T., and Bernabo, J.C. (1977). The contemporary distribution and Holocene stratigraphy of pollen in eastern North America. In *Contributions of Stratigraphic Palynology. Vol. 1, Cenozoic Palynology, Contr. Ser. No. 5A.* ed. W.C. Elsik. pp. 130-46. Dallas, Texas: American Association of Stratigraphic Palynologists.

Webb, T., and Bryson, R.A. (1972). Late and post-glacial climatic change in the northern midwest, USA: Quantitative estimates derived from fossil pollen spectra by multivariate statistical analysis. *Quaternary Research,* 2, 70-115.

Weber, J.N., Deines, P., White, E.W., and Weber, P.H. (1975). Seasonal high and low density bands in reef coral skeletons. *Nature,* 255, 697.

Weber, J.N., and White, E.W. (1974). Activation energy for skeletal aragonite deposition by the hermatypic coral *Platyyra* spp. *Marine Biology,* 26, 353-9.

Webster, J.R. (1979). Hierarchical organization of ecosystems. In *Theoretical Systems Ecology*, ed. E. Halfon. pp. 119-31. New York: Academic Press.

Weiner, J. (1985). Size hierarchies in experimental populations of annual plants. *Ecology*, 66, 743-52.

Weiner, J., and Solbrig, O.T. (1984). The meaning and measurement of size hierarchies in plant populations. *Oecologia (Berlin)*. 61, 334-6.

Weins, J.A. (1977). On competition and variable environments. *American Scientist*, 65, 590-7.

Weins, J.A. (1984). On understanding a non-equilibrium world: myth and reality in community patterns and processes. *Ecological communities: conceptual issues and the evidence.* eds. D.R. Strong, D. Simberloff, L.G. Abele, and A.B. Thistle. pp. 439-57. Princeton, New Jersey:Princeton University Press

Weins, J.A. (1989). *The Ecology of Bird Communities. Vol. 1. Foundations and Patterns.* Cambridge: Cambridge University Press.

Weisburd, S. (1988). African elephants: a dying way of life. *Science News*, 133, 333.

Weisser, P.J., and Marques, F. (1979). Gross vegetation changes in the dune areas between Richards Bay and Mfolozi River. *Bothalia*, 12, 711-21.

Weller, D.E. (1987). A re-evaluation of the -3/2 power rule of plant self-thinning. *Ecological Monographs*, 57, 23-43.

Wellington, G.M. (1980). Reversal of digestive interactions between Pacific reef corals: mediation by sweeper tentacles. *Oecologia*, 47, 340-3.

Wellington, G.M. (1982a). An experimental analysis of the effects of light and zooplankton on coral zonation. *Oecologia*, 52, 311-20.

Wellington, G.M. (1982b). Depth zonation of corals in the Gulf of Panama: Control and facilitation by resident reef fishes. *Ecological Monographs*, 52, 223-41.

Wells, C.G. (1971). Effects of prescribed burning on soil chemical properties and nutrient availability. In *Prescribed Burning Symposium Proceedings, Charleston, South Carolina* pp. 86-97. Asheville, North Carolina: U.S. Department of Agriculture, Forest Service, Southeastern Forest Experiment Station.

Wells, C.G., Campbell, R.E., DeBano, L.F., *et al.* (1979). *Effects of Fire on Soil: A State-of-Knowledge Review.* General Technical Report WO-7. Washington,D.C.: U.S. Department of Agriculture, Forest Service.

Wells, J.W. (1955). A survey of the distribution of reef coral genera in the Great Barrier Reef region. In *Reports of the Great Barrier Reef Committee. IV. Pt.2.* pp. 21-9. Brisbane, Queensland, Australia.

Wells, J.W. (1957). Coral reefs. *Memoirs of the Geological Society of America*, 67, 609-31.

Wells, M.J., Balsinhas, A.A., Joffe, H., Engelbrecht, V.M., Harding, G., and Stirton, C.H. (1986). A catalogue of problem plants in southern Africa incorporating the national weed list of South Africa. *Memoirs of the Botanical Survey of South Africa*, 53, 1-549.

Went, F.W. (1940). Soziologie der Epiphyten eines tropischen Urwaldes. *Ann. Jard. Bot. Buitenz.*, 50, 1-98.

Went, F.W. *et al.* (1952). Fire and biotic factors affecting germination. *Ecology*, 33, 351-64.

Werner, P.A., and Platt, W.J. (1976). Ecological relationships of co-occuring goldenrods (*Solidago*, Compositae). *American Naturalist*, 110, 959-71.

Wesson, G., and Wareing, P.F. (1969). The role of light in germination of naturally occurring populations of buried weed seeds. *Journal of Experimental Botany*, 20, 402-13.

West, N.E. (1982). Dynamics of plant communities dominated by chenopod shrubs. *International Journal of Ecology and Environmental Science*, 8, 73-84.

West, P.W., and Borough, C.J. (1983). Tree suppression and the self-thinning rule in a monoculture of *Pinus radiata* D. Don. *Annals of Botany* (London), 52, 149-58.

Western, D, and Ssemakula, J. (1981). The future of savannah ecosystems: ecological islands or faunal enclaves? *African Journal of Ecology*, 19, 7-19.

Westlake, D.F. (1963). Comparisons of plant productivity. *Biological Reviews*, 38, 385-425.

Westman, W.E., O'Leary, J.F., and Malanson, G.P. (1981). The effects of fire intensity, aspect, and substrate on post-fire growth of California coastal sage shrub. In *Components of Productivity of Mediterranean-Climate Regions - Basic and Applied*, ed. N.S. Margaris and H.A Mooney, pp. 151-79. The Hague: Junk.

Westoby, M. (1984). The self-thinning rule. *Advances in Ecological Research*, 14, 167-225.

Wethey, D. (1985). Catastrophe, extinction, and species diversity: a rocky intertidal example. *Ecology*, 66, 445-56.

Wetzel, R.G. (1975). *Limnology*. Philadelphia: Saunders.

Wetzel, R.G. (1983). *Limnology*, 2nd ed. New York: Saunders College Publishing.

Wheelwright, N.T., and Orians, G.H. (1982). Seed dispersal by animals: contrasts with pollen dispersal, problems of terminology, and constraints on coevolution. *American Naturalist*, 119, 402-13.

Whisenant, S.G. (1990). Changing fire frequencies on Idaho's Snake River plains: ecological and management implications. In *Proceedings - Symposium on Cheatgrass Invasion, Shrub Die-off, and Other Aspects of Shrub Biology and Management. Gen. Tech. Report INT-276*, pp. 4-10. Provo, Utah: U.S. Department of Agriculture Forest Service.

White, D.L., Waldrop, T.A., and Jones, S.M. (1991). Forty years of prescribed burning on the Santee fire plots: effects on understory vegetation. In *Fire and the Environment: Ecological and Cultural Perspectives*, ed. S.C. Nodvin and T.A. Waldrop, pp. 45-59. Asheville, NC: Southeastern Forest Experiment Station.

White, F. (1968). Zambia. *Acta Phytogeographica Seucica*, 54, 208-15.

White, H.H. (1963). Variation of stand structure correlated with altitude in the Luquillo Mountains. *Caribbean Forester*, 24, 46-52.

White, P.S. (1979). Pattern, process, and natural disturbance in vegetation. *Botanical Review*, 45, 229-99.

Whiteside, M.C., and Harmsworth, R.V. (1967). Species diversity in Chydorid (Cladocera) communities. *Ecology*, 48, 664-67.

Whitford, W.G., Steinberger, Y., and Ettershank, G. (1982). Contributions of subterranean termites to the 'economy' of Chihuahuan Desert ecosystems. *Oecologia*, 55, 289-302.

Whitham, T.G. (1981). Individual trees as heterogeneous environments: adaptation to herbivory or epigenetic noise? In *Insect Life History Patterns: Habitat and Geographic Variation*. ed. R.F. Denno and H. Dingle. pp. 9-27. New York: Springer-Verlag.

Whitham, T.G., Williams, A.G., and Robinson, A.M. (1984). The variation principle: individual plants as temporal and spatial mosaics of resistance to rapidly evolving pests. In *Novel Approaches to Interactive Systems*, ed. P.W. Price, C.N. Slobodchikoff, and W.S. Gaud, pp. 15-52. New York: Wiley.

Whitham, T.G., and Slobodchikoff, C.N. (1981). Evolution by individuals, plant-herbivore interactions, and mosaics of genetic variability: the adaptive significance of somatic mutations in plants. *Oecologia*, 49, 287-92.

Whitmore, T.C. (1984). *Tropical Rain Forests of the Far East*. Oxford: Clarendon Press.

Whittaker, R. H. (1975). *Communities and ecosystems*. New York: MacMillan.

Whittaker, R.H. (1956). Vegetation of the Great Smoky Mountains. *Ecological Monographs*, 26, 1-80.

Whittaker, R.H. (1960). Vegetation of the Siskiyou Mountains, Oregon and California. *Ecological Monographs*, 30, 279-338.

Whittaker, R.H. (1965). Dominance and diversity in land plant communities. *Science*, 147, 250-60.

Whittaker, R.H. (1966). Forest dimensions and production in the Great Smoky Mountains. *Ecology*, 47, 103-21.

Whittaker, R.H. (1967). Gradient analysis of vegetation. *Biological Reviews*, 42, 207-64.

Whittaker, R.H. (1970). The population structure of vegetation. In *Gesellschaftsmorphologie (Strukturforschung)*, ed. R. Tüxen, pp. 39-59. The Hague: Junk.

Whittaker, R.H. (1972). Evolution and measurement of species diversity. *Taxon*, 21, 213-51.

Whittaker, R.H. (1975). *Communities and Ecosystems, Second Edition*. New York: MacMillan.

Whittaker, R.H. (1977). Evolution of species diversity in land communities. *Evolutionary Biology*, 10, 1-67.

Whittaker, R.H. ed. (1973). *Ordination and Classification of Plant Communities*, The Hague: Junk.

Whittaker, R.H., Levin, S.A., and Root, R.P. (1973). Niche, habitat, and ecotope. *American Naturalist*, 107, 321-38.

Whittaker, R.H., and Levin, S.A. (1977). The role of moasic phenomena in natural communities. *Theoretical Population Biology*, 12, 117-39.

Whittaker, R.H., and Likens, G.E. (1975). The biosphere and man. In *Primary Productivity of the Biosphere*, ed. H. Lieth and R.H. Whittaker, pp. 305-28. New York: Springer-Verlag.

Whittaker, R.H., and Niering, W.A. (1965). Vegetation of the Santa Catalina Mountains, Arizona. (II) A gradient analysis of the south slope. *Ecology*, 46, 429-52.

Whittaker, R.H., and Niering, W.A. (1975). Vegetation of the Santa Catalina Mountains, Arizona. V. Biomass, production, and diversity along the elevation gradient. *Ecology*, 56, 771-90.

Whittaker, R.J., Bush, M.B., and Richards, K. (1990). Plant recolonization and vegetation succession on the Krakatau Islands, Indonesia. *Ecological Monographs*, 59, 59-123.

Whittle, K.J. (1977). Marine organisms and their contribution to organic matter in the oceans. *Marine Chemistry*, 5, 381-411.

Wiebes, J.T. (1979). Coevolution of figs and their pollinators. *Annual Review of Ecology and Systematics*, 10, 1-12.

Wieland, N.K., and Bazzaz, F.A. (1975). Physiological ecology of three codominant successional annuals. *Ecology*, 56, 681-8.

Wigley, T.M.L, Briffa, K.R., and Jones, P.D. (1984). Predicting plant productivity and water relations. *Nature*, 312, 102-3.

Wiley, E.O. (1988). Vicariance biogeography. *Annual Review of Ecology and Systematics*, 19, 513-42.

Willems, J.H. (1980). Observations on north-west European limestone grassland communities. An experimental approach to the study of species diversity and above-ground biomass in chalk grassland. *Proc. Koninklijke Nederlandse Akademie van Wetenschappen, Series C*, 83, 279-306.

Williams, C.B. (1947). The logarithmic series and the comparison of island floras. *Proceedings of the Linnean Society of London*, 158, 104-8.

Williams, C.B. (1964). *Patterns in the Balance of Nature and Related Problems in Quantitative Ecology. New York: Academic Press.*

Williams, D. McB. (1982). Patterns in the distribution of fish communities across the central Great Barrier Reef. *Coral Reefs*, 1, xx-xxx.

Williams, D.McB., and Sale, P.F. (1981). Spatial and temporal patterns of recruitment of juvenile coral reef fishes to coral habitats within One Tree Lagoon, Great Barrier Reef. *Marine Biology*, 65, 245-53.

Williams, I.J.M. (1972). A revision of the genus *Leucadendron* (Proteaceae). *Contribution of the Bolus Herbarium*, 3, 1-145. Rondebosch, C.P., South Africa.

Williams, L.G. (1964). Possible relations between plankton-diatom species numbers and water-quality estimates. *Ecology*, 45, 809-23.

Williamson, M. (1981). *Island Populations.* Oxford: Oxford University Press.

Williamson, M. (1989). The MacArthur and Wilson theory today: true but trivial. *Journal of Biogeography*, 16, 3-4.

Williamson, M.H. (1981). *Island Populations.* Oxford: Oxford University Press.

Willis, A.J. (1963). Braunton Burrows: the effect on the vegetation of the addition of mineral nutrients to the dune soils. *Journal of Ecology*, 51, 353-74.

Willis, E.O. (1974). Populations and local extinctions of birds on Barro Colorado Island, Panama. *Ecological Monographs*, 44, 153-69.

Willson, M.F. (1973). Tropical plant production and animal species diversity. *Tropical Ecology*, 14, 62-5.

Willson, M.F. (1974). Avian community organization and habitat structure. *Ecology*, 55, 1017-29.

Wilson, E.O. (1961). The nature of the taxon cycle in the Melanesian ant fauna. *American Naturalist*, 95, 169-93.

Wilson, E.O. (1965). The challenge from related species. In *The Genetics of Colonizing Species*, ed. H.G. Baker and G.L. Stebbins, pp. 7-27. New York: Academic Press.

Wilson, E.O. (1987). The arboreal ant fauna of Peruvian Amazon forests: a first assessment. *Biotropica*, 19, 245-51.

Wilson, E.O. (1988). The current state of biological diversity. In *Biodiversity*, ed. E.O. Wilson and F.M. Peter, pp 3-18. Washington: National Academy Press.

Wilson, G.D.F., and Hessler, R.R. (1987). Speciation in the deep sea. *Annual Review of Ecology and Systematics*, 18, 185-207.

Wilson, J.W. III. (1974). Analytical zoogeography of North American mammals. *Evolution*, 28, 124-40.

Wilson, M.V., and Mohler, C.L. (1983). Measuring compositional change along gradients. *Vegetatio,* 54, 129-41.

Wilson, S.D., and Keddy, P.A. (1986). Measuring diffuse competition along an environmental gradient: results from a shoreline plant community. *American Naturalist,* 127, 862-9.

Wilson, W.H. (1990). Competition and predation in marine soft-sediment communities. *Annual Review of Ecology and Systematics,* 21, 221-41.

Winsor, J. (1983). Persistence by habitat dominance in the annual *Impatiens capensis* (Balsaminaceae). *Journal of Ecology,* 71, 451-66.

Wishner, K.F. (1980). The biomass of the deep-sea benthopelagic plankton. *Deep-Sea Research,* 27, 203-16.

Withers, J.R. (1979). Studies on the status of unburnt Eucalyptus woodland at Ocean-Grove Victoria. IV. Effects of shading on seedling establishment. *Australian Journal of Botany,* 27, 47-66.

Witherspoon, C.P., Iwamoto, Y.R., and Piirto, D.D. (1985). *Proceedings of Workshop on Management of Giant Sequoia.* Gen. Tech. Report PSW-95. Berkeley,California: U.S. Department of Agriculture Forest Service.

Wittig, R., and Neite, H. (1985). Acid indicators around the trunk base of *Fagus sylvatica* in limestone and loess beechwoods: distribution pattern and phytosociological problems. *Vegetatio,* 64, 113-9.

Wolda, H. (1983). Spatial and temporal variation in abundance in tropical animals. In *Tropical Rain Forest: Ecology and Management.* ed. S.L. Sutton, T.C. Whitmore and A.C. Chadwick, pp. 93-105. Oxford: Blackwell.

Wolda, H. (1988). Insect seasonality: why? *Annual Review of Ecology and Systematics,* 19, 1-18.

Wolfe, J.A., and Upchurch, G.R. (1987). North American nonmarine climates and vegetation during the Late Cretaceous. *Palaeogeography, Palaeoclimatology, Palaeooecology,* 61, 33-77.

Woodhead, P.M.J. (1971). The growth and normal distribution of reef corals. In *Report of the Committee Appointed by the Commonwealth and Queensland Governments on the Problem of the Crown-of-thorns Starfish (* Acanthaster planci*).* ed. R.J. Walsh. pp. 1-31. Govt. of Australia

Woodley, J.D. (1980). Hurricane Allen destroys Jamaican coral reefs. *Nature,* 287, 5781.

Woodward, I. (1987). *Climate and Plant Distribution.* Cambridge: Cambridge University Press.

World Bank. (1989). *World Tables 1988-89 Edition. From the Data Files of the World Bank.* Baltimore, Maryland: Johns Hopkins University Press.

Wright, D.H. (1983). Species-energy theory: an extension of species-area theory. *Oikos,* 41, 496-506.

Wright, H.A., and Bailey, A.W. (1982). *Fire Ecology: United States and Southern Canada.* New York: Wiley.

Wright, H.E. (1974). Landscape development, forest fires, and wilderness management. *Science,* 186, 487-95.

Wright, H.E. Jr. (1971). Late Quaternary vegetation history of North America. In *The Late Cenozoic Glacial Ages.* ed. K. Turekian. pp. 425-64. Hartford: Yale University Press.

Wright, H.E. Jr. (1972). Interglacial and post-glacial climates: the pollen record. *Quaternary Research,* 2, 274-82.

Wu, H., Sharpe, P.J.H., Walker, J., and Penridge, L.K. (1985). Ecological field theory (ETF): A spatial analysis of resource interference among plants. *Ecological Modelling,* 29, 215-43.

Wyatt-Smith, J. (1954). Storm forest in Kelantan. *Malayan Forester* 17, 5-11.

Yorke, J.A., and Li, T. (1975). Period three implies chaos. *American Mathematical Monthly*, 82, 985-92.

Young, J.A., and Evans, R.A. (1974). Population dynamics of green rabbitbrush in disturbed big sagebrush communities. *Journal of Range Management*, 27, 127-32.

Young, J.A., and Tipton, F. (1990). Invasion of cheatgrass into arid environments of the Lahontan Basin. In *Proceedings - Symposium on Cheatgrass Invasion, Shrub Die-off, and Other Aspects of Shrub Biology and Management. Gen. Tech. Report INT-276*. pp. 4-10. Provo, Utah: U.S. Department of Agriculture Forest Service.

Yount, J.L. (1956). Factors that control species numbers in Silver Springs, Florida. *Limnology and Oceanography*, 1, 286-95.

Zahner, R. (1958). Soil water depletion by pine and hardwood stands during a dry season. *Forest Science*, 1, 258-64.

Zahner, R. (1970). Site quality and wood quality in upland hardwoods: theoretical considerations of wood density. In *Tree Growth and Forest Soils*, ed. C.T. Youngberg, and C.B. Davey, pp. 477-497. Oregon State University Press, Corvallis, Oregon.

Zangerl, A.R., Pickett, S.T.A., and Bazzaz F.A. (1977). Some hypotheses on variation in plant population and an experimental approach. *Biologist*, 59, 113-22.

Zedler, P.H. (1981). Vegetation change in chaparral and desert communities in San Diego County, California. In *Forest Succession Concepts and Applications*, ed. D.C. West, H.H. Shugart, and D.B. Botkin, pp. 406-24. New York: Springer-Verlag.

Zedler, P.H., Gautier, C.H., and McMaster, G.S. (1983). Vegetation change in response to extreme events: the effect of a short interval between fires in California chaparral and coastal scrub. *Ecology*, 64, 809-18.

Zeevalking, H.J., and Fresco, L.F.M. (1977). Rabbit grazing and diversity in a dune area. *Vegetatio*, 35, 193-6.

Zenkevitch, L.A., and Birstein, Y.A. (1960). On the problem of the antiquity of the deep-sea fauna. *Deep-Sea Research*, 7, 10-23.

Zhadin, V.I., and Gerd, S.V. (1961). *Fauna and Flora of the Rivers, Lakes, and Reservoirs of the U.S.S.R.* Tr. 1963. Jerusalem: Israel Program for Scientific Translations.

Ziegler, B.P. (1977). Persistence and patchiness of predator-prey systems induced by discrete event population exchange mechanisms. *Journal of Theoretical Biology*, 67, 687-713.

Ziegler, B.P. (1978). Multilevel multiformalism modeling: an ecosystem example. In *Theoretical Systems Ecology*, ed. E. Halfon, pp. 18-54. New York: Academic Press.

Zimmerman, M.H., and Brown, C.L. (1971). *Trees, Structure, and Function*. New York: Springer-Verlag.

Zimmermann, M.H., and Milburn, J.A. (1982). Transport and storage of water. In *Physiological Plant Ecology II. Water Relations and Carbon Assimilation*, ed. O.L. Lange, P.S. Nobel, C.B. Osmond, and H. Ziegler, pp. 135-51. Berlin: Springer-Verlag.

Zoebel, D.R. (1969). Factors affecting the distribution of *Pinus pungens*, an Appalachian endemic. *Ecological Monographs*, 39, 303-33.

Index

Note to index: reference to the various types of ecosystems will locate the families and genera where these are not listed separately